Industrial Applications of Biopolymers and their Environmental Impact

Editors

Abdullah Al-Mamun
Lead Engineer Plastic
Corporate R & D
Adler Pelzer Group
Germany

Jonathan Y. Chen
The University of Texas at Austin
Austin, Texas, USA

CRC Press
Taylor & Francis Group
Boca Raton London New York

CRC Press is an imprint of the
Taylor & Francis Group, an **informa** business
A SCIENCE PUBLISHERS BOOK

Cover credit: The editor of the book, Dr. Abdullah Al-Mamun, has provided the cover photo.

CRC Press
Taylor & Francis Group
6000 Broken Sound Parkway NW, Suite 300
Boca Raton, FL 33487-2742

Version Date: 20200116

International Standard Book Number-13: 978-1-4987-6965-5 (Hardback)

Library of Congress Cataloging-in-Publication Data

Names: Al-Mamun, Abdullah, 1975- editor. | Chen, Jonathan Y., editor.
Title: Industrial applications of biopolymers and their environmental impact / editors, Abdullah Al-Mamun, Jonathan Y. Chen.
Description: Boca Raton : CRC Press ; Taylor & Francis Group, [2020] | "A Science Publishers book." | Includes bibliographical references and index.
Identifiers: LCCN 2019043672 | ISBN 9781498769655 (hardcover)
Subjects: LCSH: Biopolymers--Industrial applications. | Biopolymers--Environmental aspects.
Classification: LCC TP248.65.P62 I5274 2020 | DDC 572/.33--dc23
LC record available at https://lccn.loc.gov/2019043672

Visit the Taylor & Francis Web site at
http://www.taylorandfrancis.com

and the CRC Press Web site at
http://www.routledge.com

Preface

A very important task of climate protection is to reduce CO_2 emissions. Existing and future technologies could reduce CO_2 emissions through new sources and efficient use of energy. Renewable resources are effective and interesting alternatives to fossil fuels and resulting products. Energy forms, processes and services enhance the value-added potential of agriculture and industry. The development of new plastics, especially biodegradable biopolymers, may be the only solution to environmental damage caused by non-degradable plastics. With bio-based polymers new functionalities are achieved. These include biodegradability, compostability and carbon footprint. The use of bioplastics can create a natural cycle in which plastics produced from renewable raw materials can later be recycled or returned to nature. This creates a natural balance. The degradability of bioplastics saves the large costs of waste disposal. The associated reduction in CO_2 emissions can also contribute to climate protection.

The development of biomaterials is an interactive process that involves the creation of increasingly safer, more reliable, comfortable expensive and more physiologically appropriate replacements for specific application in medical science or short & long term industrial applications. This book presents a number of chapters on common functional biomaterials for medical, packaging, commodity and industrial applications. Functional biomaterials are natural materials or synthetic or systems that perform functions biologically. The book concerns development, analysis and study of materials designed for existing applications and developing innovative applications in several industries. This book could be of interest to scientists, students and engineers. It should serve as a useful reference to engineers, materials scientists and medical personals with an interest in the mechanics and performance of biomaterials. The multidisciplinary nature of the book will facilitate the exchange of information between the wide ranges of disciplines required for the development of more robust functional biomaterials. The book provides an interrelationship between chemistry, morphology and properties and also provides a comprehensive review of synthesis process and modification process.

In summary, this book intends to widen and integrate the knowledge of chemistry of common biomaterials, influence of processing methods on key properties and find out relationship between structure and properties for the large scale production with consistent properties.

Abdullah Al-Mamun
Jonathan Y. Chen

Contents

Chapter 1

Modification of Polylactic Acid (PLA) and its Industrial Applications

A. Al-Mamun*[1], M. Nikusahle[2], M. Feldmann[2] and H.-P. Heim[2]

[1]Corporate Material Development, Adler Pelzer Group, Hagen, Germany
[2]Plastic Engineering, University of Kassel, Germany

1.1 INTRODUCTION

Since industrialization, the demand for fossil fuels, such as oil, natural gas, and coal have increased. This increased prices and greenhouse gases in the atmosphere. One of the most important tasks of climate protection is to reduce CO_2 emissions. Existing and future technologies could reduce CO_2 emissions through new sources and efficient use of energy. This could at least stabilize the climate and slow down the pace of climate change. Renewable resources are effective and interesting alternatives to fossil fuels and resulting products. Energy forms, processes, and services enhance the value-added potential of agriculture and industry. The development of new plastics, especially biodegradable biopolymers, may be the only solution to environmental damage caused by non-degradable plastics. With bio-based polymers new functionalities are achieved. These include biodegradability, compostability, and carbon footprint. The use of bioplastics can create a natural cycle (see Fig. 1.1), in which plastics produced from renewable raw materials can later be recycled or returned to nature.

This creates a natural balance. The degradability of bio-plastics saves the big costs of waste disposal. Figure 1.1 shows that the associated reduction in CO_2 emissions can also contribute to climate protection [1, 2].

From the beginning of the plastics age, additives have been used primarily to improve or extend polymer properties or to make the plastics more resistant to heat treatment during the forming process. The extension of the properties of the polymer by additives has played a significant role in the growth of plastics. Another reason for using the new additives is the increasing awareness of environmental protection to use more biopolymers [3].

**Corresponding author*: Email: a.mamun@pelzer.de

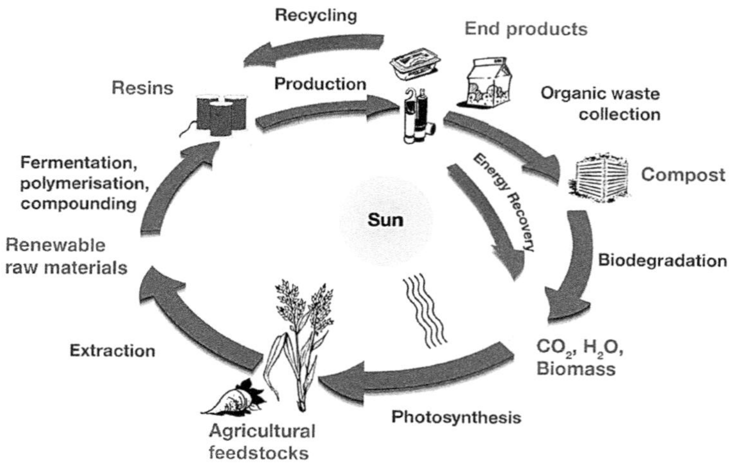

Figure 1.1 Life cycle of biodegradable plastics [1].

PLA can crystallize in various forms as α-, β-, γ-, and sc (stereo-complex). It has been found experimentally that the stereo-complex structure between PLLA and PDLA shows a significant improvement in thermal stability and mechanical properties [4]. For this purpose, various mixtures of different PLLAs and PDLA were prepared and investigated in order to produce a complete stereo-complex structure by optimizing the parameters in a twin-screw extruder. The different batches of stereo-complex mixers were then further processed in a hot press machine to produce square plates with a thickness of 4 mm. Test specimens were obtained from these plates. Subsequently, the specimens were examined in various thermal and morphological experiments. By morphological examination, the crystal structure and the crystallization were evaluated in the stereo-complex.

Furthermore, PLA properties were to improve the properties of PLA by using the impact modifier. Various mixtures of the PLA granules were mixed manually with different additives and then compounded in a twin-screw extruder. The test sample was produced by injection molding. Subsequently, the prepared specimens were examined for various mechanical and thermal properties.

1.1.1 Fundamental of Biopolymers

Today's conventional polymers are produced from petrochemicals. Due to limited oil resources, conventional polymers should gradually be replaced by biopolymers based on renewable sources. The term biopolymer often leads to misunderstandings. The best general definition of biopolymers today is: a polymer material consisting of bio-based (natural renewable) raw materials or/and biodegradable [5, 6]. Principally, there are three following biopolymer groups:

a. Degradable biopolymers based on petrochemicals
b. Degradable biopolymers based on renewable resources
c. Non-degradable biopolymers based on renewable resources

That means biodegradable plastics on both petrochemical raw materials as well as renewable resources, as shown in Fig. 1.2. However, not all polymers that are based on renewable raw materials are biodegradable [6], which means that parts of the biopolymers are not biodegradable. It should be noted that there are different degradation mechanisms for degradable biopolymers, such as the different terms of degradability and compostability, which are already more suggestive. Biodegradability has no dependence on the raw materials used, but depends solely on the chemical structure and molecular structures. It is independent of the origin of the monomers. Figure 1.3 shows a connection between biodegradable biopolymers from fossil and renewable raw materials [7].

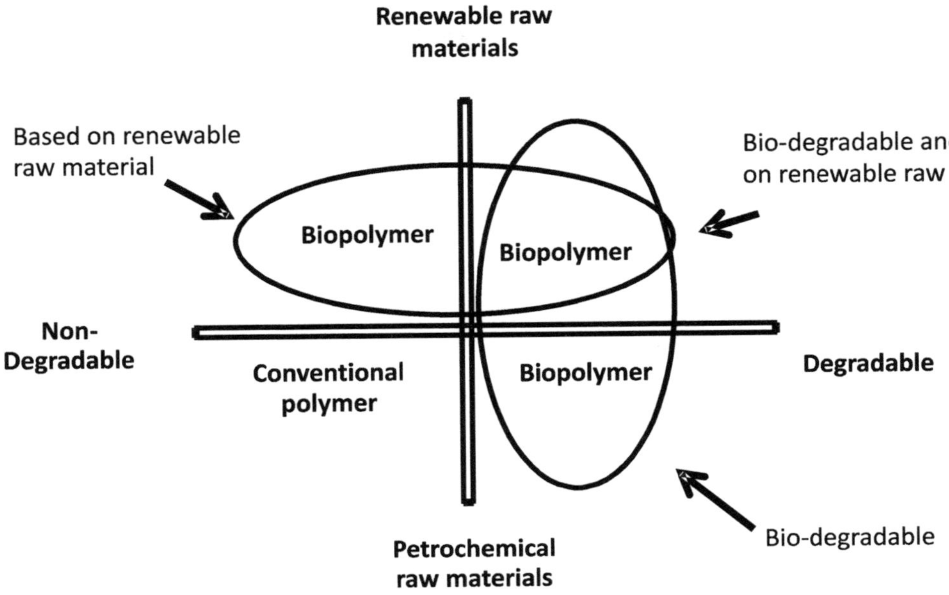

Figure 1.2 Three fundamentally different biopolymer groups [6].

Biopolymers based on different raw materials are shown in Fig. 1.2. Polymers that are based on renewable raw materials are not all biodegradable [5], which means all biopolymers are not biodegradable. It should be noted that there are different degradation mechanisms for degradable biopolymers, such as the different terms of degradability and compostability. Biodegradability has no dependence on the raw materials used, but depends solely on the chemical structure and molecular structures. It is independent of the origin of the monomers. Figure 1.3 shows a connection between biodegradable biopolymers from fossil and renewable raw materials [6].

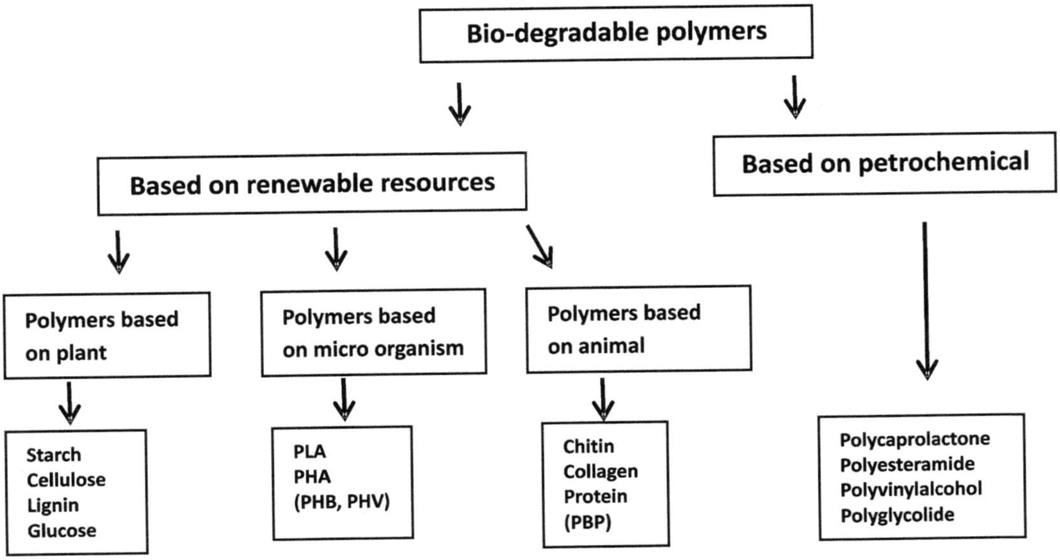

Figure 1.3 Classification of biodegradable polymers.

As renewable raw materials for biopolymers, cellulose, starch, sugar, vegetable oils and their derivatives, and in some cases lignins and proteins, can also be used as material components [6, 7].

Biodegradable plastics are most used in the packaging and catering sector. In addition, they are also used in agriculture, horticulture, medical sector, and in pharmaceuticals. The broadly introduced products are garbage bags, shopping bags, disposable tableware (cups, plates, cutlery), packaging films, bottles, fruit and vegetable trays, packaging aids, expandable foams, protection films, flowerpots, etc. [8].

The potential demand of the biopolymer is expanding day by day. The total biopolymer consumption in Western Europe is about 50 million tons per year [9]. Among them, about 42 million tons of this demand (85%) could be biopolymers. The production of bioplastics is expected to increase from approximately 890,000 tons in 2013 to over 2.22 million tons in 2020 [9].

1.2 POLYLACTIC ACID

Polylactide (PLA or polylactic acid) or polylactic acid is biodegradable, bioresorbable, and thermoplastic linear aliphatic polyester that can be produced from renewable resources. PLA is made from renewable and degradable resources, such as corn and rice, which can be good alternatives in the energy crisis and at the same time, it will reduce dependence on fossil fuels. Figure 1.4 discusses the cycle of PLA in nature.

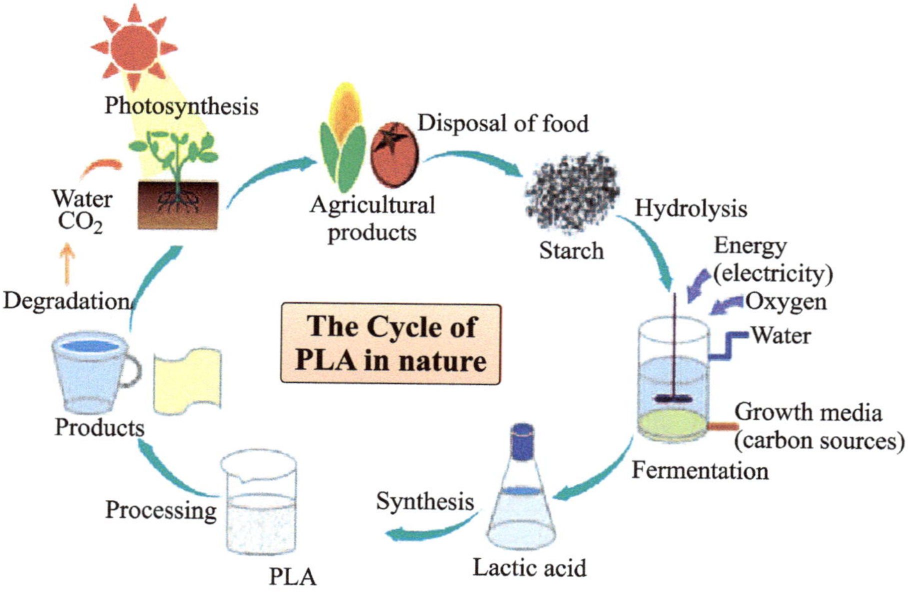

Figure 1.4 The cycle of PLA in nature.

PLA is neither toxic nor carcinogenic to the human body. Therefore, it is an excellent material for biomedical applications [10]. PLA is not only compostable and biocompatible, but can also be processed by most standard processes, such as film casting, extrusion, and blow molding or fiber manufacturing process. Due to the greater thermal processability compared with other biomaterials, the use of PLA as an environmentally friendly biomaterial with excellent properties in industrial areas, such as textiles and food packaging, might be beneficial [10, 5].

Lactic Acid (lactic acid) or 2-hydroxypropionic acid is an ubiquitary, natural, organic acid found in the two optically active forms of L (+) and D (–) lactic acid. Lactic acid has been produced by various synthetic routes since the 1960s. The world production volume of lactic acid of 70–90% is produced by fermentation processes, as these are inexpensive. In general, the fermentation refers to the conversion of biological materials by means of bacterial, fungal, or cell cultures or by the addition of enzymes [10, 6]. L-lactide acide and D-lactide acide, which are two isomers of lactic acid, are shown schematically in Fig. 1.5.

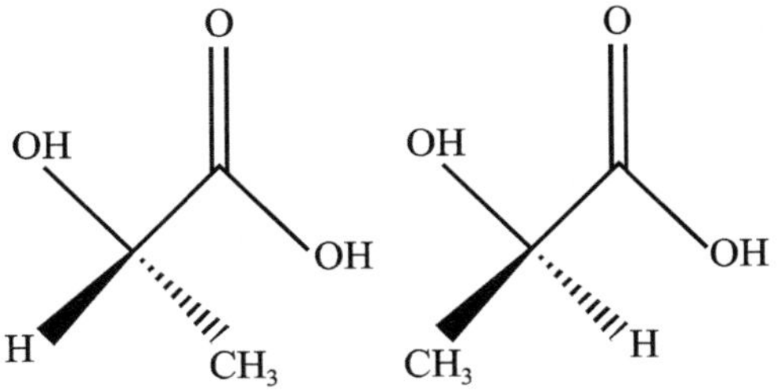

Figure 1.5 Isomers of lactic acid.

The synthesis of PLA requires pure L-lactic acid or D-lactic acid or a mixture of both components. The homopolymer of LA is a white powder. At room temperature, the glass transition temperature (T_g) value is about 55°C and the melt temperature (T_m) value is 175°C. Four different materials can be produced with the two isomers of LA. Poly (D-lactic acid) (PDLA) is an isomer, which has a regular chain structure. Poly (L-lactic acid) (PLLA) is an isomer, which is partially crystalline and also has a regular chain structure. Poly (D, L-lactic acid) (PDLLA) is a mesa isomer, which is amorphous, and meso-PLA, which is obtained by polymerization of meso-lactide [5].

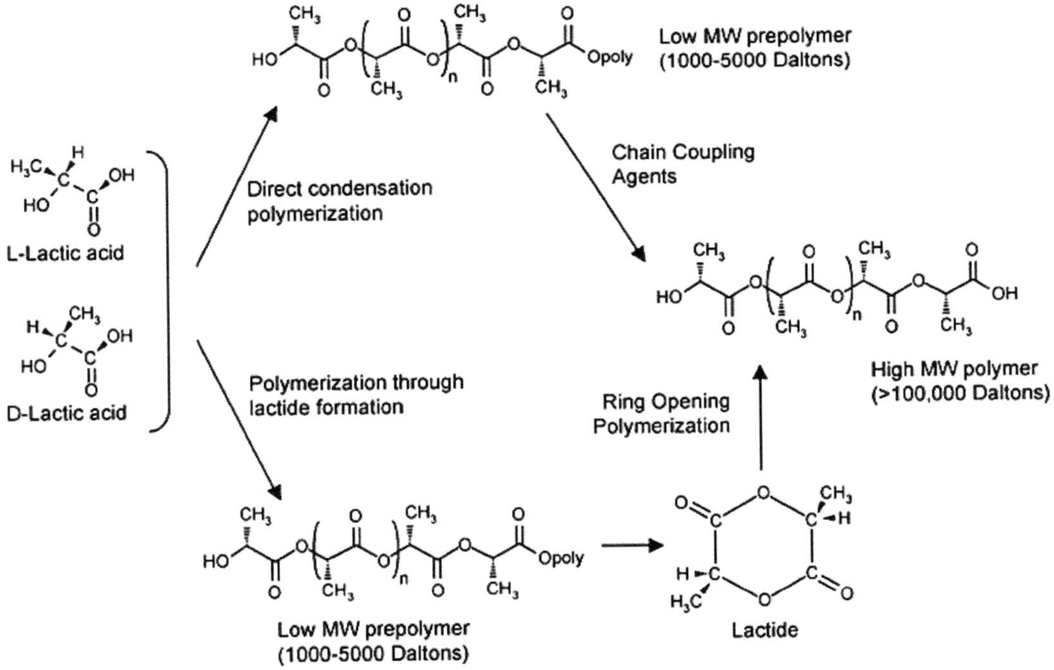

Figure 1.6 Polymerization of PLA.

Commercial PLA can be a homopolymer of poly (L-lactic acid) (PLLA) or a copolymer, namely poly (D, L-lactic acid) (PDLLA), which is made from L-lactic acid and D-lactic acid [3]. Two main synthetic methods are used to obtain PLA Fig. 1.6: direct polycondensation (including solution polycondensation and melt polycondensation) and ring opening polymerization (ROP). The most favorable way to produce PLA from lactic acid is direct polycondensation (oligocondensation).

However, since PLA produced through polycondensation is high in molecular weight, therefore, commercial production involves the use of ring-opening polymerization (ROP). The ROP of L-lactide is generally the most preferred route for producing the high molecular weight polylactide. This gives the possibility of accurate control of the chemical process (catalyst controlled), and thus varies the properties of the resulting polymers in a controlled manner [6, 10].

A stereo-complex type of the PLA (sc-PLA) consists of the two enantiomers of the PLA (PLLA and PDLA) and it is a high performance polymer because of its melting temperature (T_m = 230°C), which is 50°C higher than the PLLA or PDLA polymers. Recently, the increasing attention of production stereo-complexes PLA has influenced the demand of PLLA and PDLA production. As compared to the homo-polymer, the stereo complex not only gives a much higher melting point to corresponding PLLA and PDLA, but also exhibits better modulus of elasticity, tensile strength, and elongation at break in a particular molecular weight range. Thermal stability improved dramatically and the melt temperature of PLA increased to 230°C by stereo complexation [11]. This can be seen in Fig. 1.7.

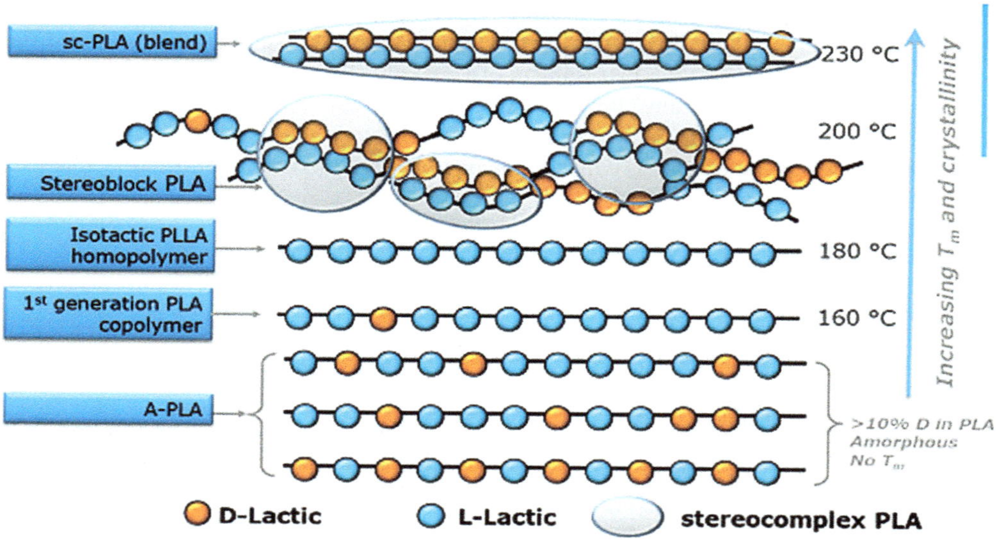

Figure 1.7 The increase of melt temperature of PLA due to stereo complexation.

The mechanical and thermal properties of the sc-PLA were compared with the other biopolymer in Table 1.1. The sc-PLA has the highest thermal and mechanical properties, although complete sc crystallinity could not yet reach the high molecular weight of enantiomeric polymers. The production of sc-PLA has just begun and therefore they cannot yet replace conventional oil-based polymers [4, 12].

Table 1.1 Mechanical and thermal properties of sc-PLA in comparison with other biopolymers [11].

Properties	*sc-PLA*	*PLLA*	*Polyhydroxybutyrate*	*Polyglycol*
T_m (°C)	220–240	170–190	188–197	225–230
T_g (°C)	65–72	50–65	2–5	40–45
ΔH_m (J/g)	142–155	93–203	146–160	180–207
Density (g/cm^3)	1.21–1.342	1.25–1.30	1.18–1.26	1.50–1.69
Tensile strength (MPa)	850–1200	120–1250	180–200	80–1000
Elasticity modulus (MPa)	800–1400	150–2000	180–200	50–900
Tensile strain (%)	20–40	10–26	50–70	30–40

1.3 ADDITIVE

Additives are essential ingredients in plastic formulations as they can affect the maintenance and/or modification of polymer properties. Therefore, they are used to guide the plastics with better performance and longer term applicability. Recent examples of the new technologies show that the additives can not only change the polymer itself and add new properties, but also, when incorporated into the plastic, provide high performance plastics with favorable shock, temperature, and flame properties [3, 4]. Various additives are – adhesion promoters, chain extenders, flame retardants, UV stabilizers, impact modifiers, etc. The following additives were selected and used for experiments with additive systems in Table 1.2. The additives were extruded with PLA 3051 using a twin-screw extruder and then test specimens produced by injection molding.

Table 1.2 List of used additives.

Name of additives	*Producer*
CESA-extend as chain extender	Clariant, DE
Effect of catalysitor on CESA	
Joncryl 3229 as compatibilizer	BASF, DE
Biomax strong 120 as toughness modifier	Dupont, DE
Biomax thermal 300 as heat deflection temperature	Dupont, DE
S547 as UV stabilisator	Sukano, CH
S550 as Impact modifier	Sukano, CH

1.3.1 CESA as Chain Extender

Chain extension is a method to obtain higher molecular weight polycondensates by a post polymerization reaction carried out either in the final polymerization processing step or during compounding, extrusion, or injection molding of the final product [13]. The chain extension is done by reacting end groups of polycondensates with a bi- or multifunctional reactive component. Depending on their chemistry, it can either couple two or more equal or unequal end groups. Low molecular weight bifunctional chain extender will lead to a strictly linear polymer because it will couple exactly two, typically equal end groups. A relatively high amount of linear chain extender is needed to get a significant increase in molecular weight of the resulting chain-extended polycondensate. Low molecular weight multifunctional extenders face the problem of a relatively small processing window, while overdosing of the additive leads to over crosslinking and gelation, which make the material impossible to process with regular extrusion or injection molding equipment [14]. The possible reaction between chain extender and PLA molecules are seen in Fig. 1.8.

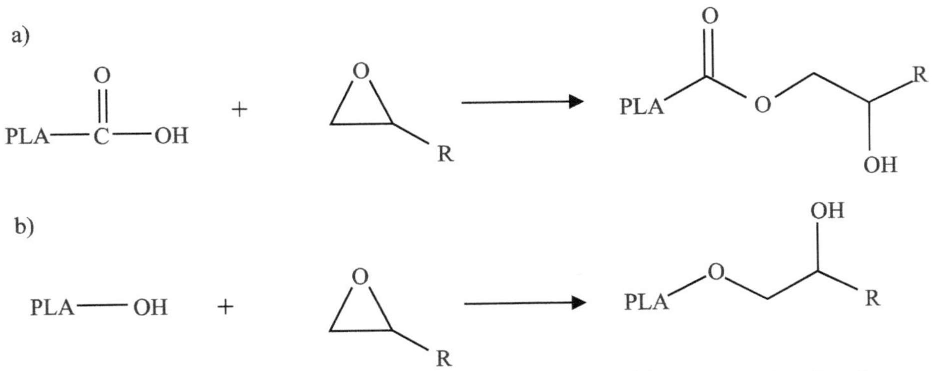

Figure 1.8 Possible reaction mechanisms between epoxide groups and carboxyl or hydroxyl end groups of PLA chain.

CESA-extend polymeric chain extenders comprise a portfolio of low and high functionalities that allow molecular weight modifications in a much larger processing window. This chain extension technology uses nonlinear chain extension. Epoxy-functional styrene-acrylic based or styrene-free acrylic based reactive polymers are used to modify the initial polymer with long chain branched structures. Even if it is not possible to obtain only linear structures, the relatively small amount of branched and hyper branched structures allow a significant improvement in mechanical, rheological, and physical properties while maintaining the optical properties of the polycondensates [15, 16].

During compounding, all polyesters are subjected to severe degradation of molecular weight due to hydrolysis, alcoholysis, and thermal degradation. The loss of molecular weight results in a decrease of physical and mechanical properties of the material.

The decrease of molecular weight appears during processing of the polyesters and polyamides after the initial polymerization process, as well as during the intended use of the polymer in an article and during recycling of the material. Specially, during processing or recycling at high temperatures, the loss of molecular weight is extremely fast. The use of both post-consumer materials and internal process regrind is limited to less demanding applications if the loss of molecular weight cannot be compensated. In general, it is difficult to achieve a very high molecular weight for biopolymers, such as PLA, PHA, and PHB, which as a consequence results in polymers with relatively low melt strength. This implies that the processing of these bio-plastics often show a very narrow processing window under which adequate parts can be made. Melt strength is a critical parameter for processing techniques, such as sheet extrusion or thermoforming, injection blow molding, film blowing, profile extrusion, extrusion blow molding, and sheet foam extrusion mostly combined with thermoforming.

The degradation is usually due to the thermo-mechanically induced processes in the cylinder and/or residual moisture, or impurities. As a result of the addition of a chain extender, there is a significant increase in molecular weight, improved process capability (such as melt strength), and ultimately, improvement in mechanical properties. An indication of this reaction may be, for example, the increase in viscosity and the resulting change in machine parameters, such as the torque on the screw or the injection pressure in injection molding. Quantification of this effect can be achieved by molecular weight determinations.

1.3.1.1 Influence of Chain Extender on Molecular Weight

The influence of chain extender can be seen in Fig. 1.9. It is clearly seen that even minor proportions of the chain extender lead to increase of the weight average molecular weight. One wt-% CESA extend causes an increase of Mw by a factor of 2. This behavior appears to be irregular and with additions of 10%, the increase of Mw would stabilize at about 500 kg/mol, and a further addition of CESA would result in no change, and in addition, a nearly identical trend of the injection pressure in function of the Mw.

Furthermore, it can be quantified that the ratio (relative injection pressure/CESA content) is linear (see in Fig. 1.9). It can be seen that for PLA 3051D, the addition of each wt% CESA increases the injection pressure by approximately 5 percent.

There are some other authors who used different types of chain extender with PLA [17]. It has been demonstrated that the higher molecular weight indeed brings a more robust processing window. For instance, the addition of 0.15% of CESA-extend to PLA provides the ability to double the line speed during film blowing and the addition of the chain extenders allow production of PLA foam sheets with a finer cell structure.

The influence of CESA on number average and weight average molecular weight of PLA was measured using gel permeation chromatography (GPC) with multi angle light scattering detectors. The GPC chromatograms of PLA are shown in Fig. 1.10. The variation of polydispersity of PLA (3052 D NatureWorks) with CESA content can be seen in Table 1.3.

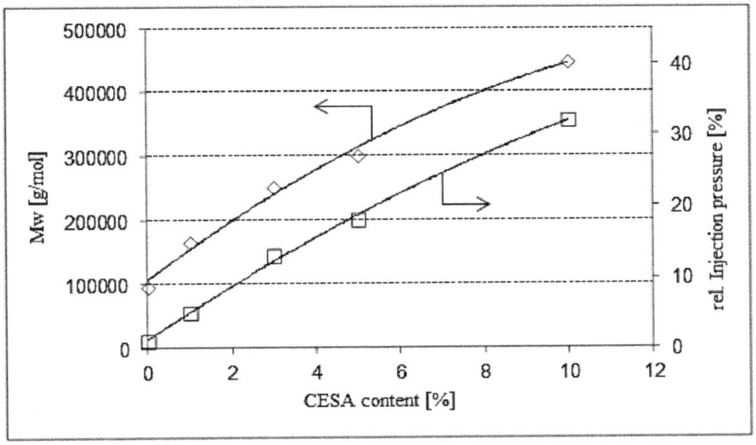

Figure 1.9 Influence of CESA on the change of molecular weight and injection pressure.

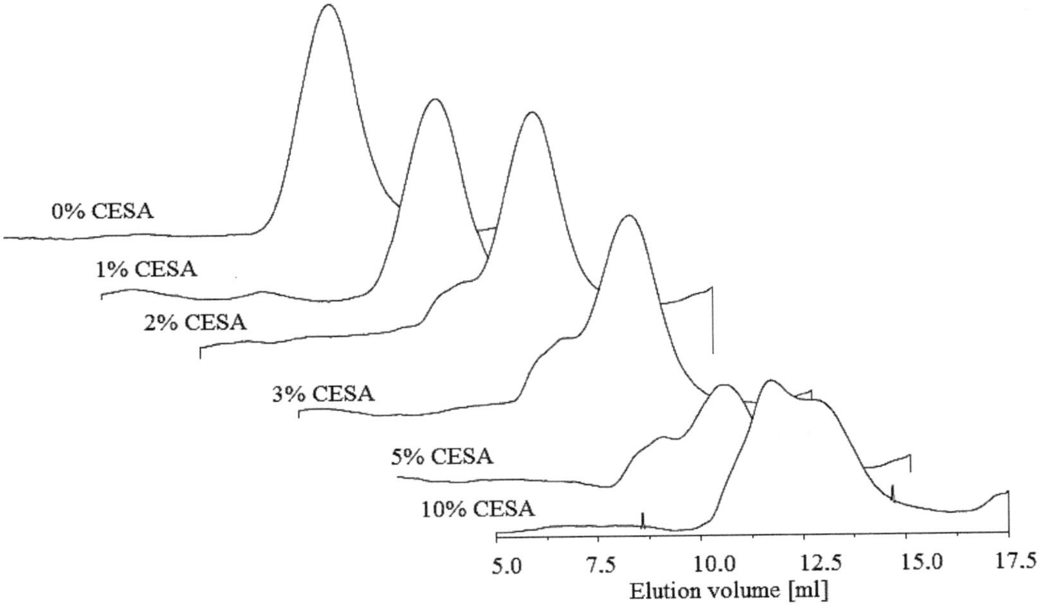

Figure 1.10 GPC chromatography of PLA with CESA content.

The term polydispersity (Mw/Mn) is used to describe the degree of non-uniformity of a molecular weight distribution. The distribution is usually described by the ratio of the average molecular weight by mass (of molecules in the different slices) and the other the average molecular weight by number (of molecules in the different slices of the distribution). For a perfectly uniform (monodisperse) sample consisting of exactly one and only one molecular weight, both the Mw and the Mn would be the same value. For real samples, the two numbers are, however, not the same, and the ratio of the two can be used to describe how far away the encountered distribution is from a uniform distribution. The ratio Mw/Mn is called the dispersity, which was formerly known as polydispersity index (PDI). It was observed in Table 1.3 that the molecular weight increased with increasing amount of CESA and dispersity also increased till addition of 5% CESA. Afterwards, the dispersity decreased slightly due to addition of 10% CESA. It means that the chain extension occurred ununiformly and irregularly due to addition of CESA.

Table 1.3 The number average and weight average molecular weight and polydispersity of PLA.

CESA [%]	*Mn* [g/mol]	*Mw* [g/mol]	*Dispersity* (Mw/Mn)
0	58000	77500	1.34
1	73000	105000	1.44
2	91000	238500	2.62
3	93500	297500	3.18
5	112500	352000	3.13
10	157500	461500	2.93

1.3.1.2 Influence of Chain Extender on Temperature and Molding Properties

The influence of CESA was observed in term of the melt viscosity and temperature change with processing time. The melt rheology of polymeric materials reflects the relationship between molecular structure and dynamic properties. The viscosity and elasticity of the molten PLA are of primary importance in plastic processing operations (extrusion, injection, and spinning). The reactivity chain extender and PLA (3052 D NatureWorks) were mixed using a laboratory kneader (Haake Rheomix 3010p, Thermo Fischer, Germany). The mixing unit is consisted of screw unit with a chamber volume of 310 cm^3. It was coupled with a driver (Rheocord, Germany). The rheomix chamber is connected with heating and cooling systems so that the processing temperature can be controlled precisely. Prior to mixing, PLA and chain extender were placed into the chamber. The torque or the momentum was measured with screw rotation 100 rpm at 210°C.

Figure 1.11 shows the effect of CESA on the torque over the measuring time. It can be seen that the chain extension is highly dependent on the concentration of the CESA extend. The reaction starts within a few minutes and is completed after about 10 minutes.

The torque of PLA reduced with processing duration and the addition of CESA-extend leads to the rebuilding of the molecular weight (Mw). As a result, the torque (viscosity) used to increase and the melt temperature (not shown) were found to increase. The melt temperature exceeds the target temperature by up to 25°C and 35°C due to addition of 5% and 10% CESA, accordingly.

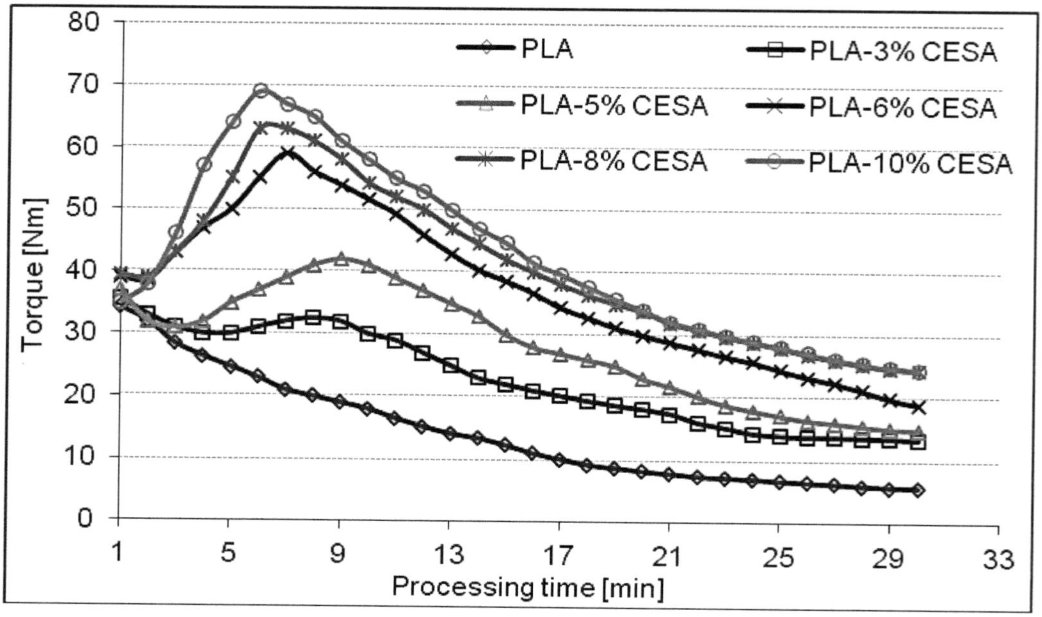

Figure 1.11 Effect of the chain extender (CESA) on the temperature and melt viscosity of PLA.

It means that PLA showed thermal degradation in correlation with the mixing duration. While either compounding, employing a polymer in a product, or recycling the material, all polyesters are subject to severe degradation of their molecular weight due to thermal, oxidative, and alcoholysis degradation. The loss of molecular weight results in a decrease of the physical and mechanical properties of the material [18].

1.3.1.3 Influence of Chain Extender on Mechanical Properties

PLA mixed with varying chain extender (CESA) contents was compounded using a twin screw extruder, and test specimens were prepared using injection molding. The effect of the chain extender on the Charpy impact strength is shown in Fig. 1.12. The notched Charpy and Charpy impact strength both increased continuously in addition of chain extender. The notched Charpy impact strength increased by 25% and 30% due to the addition of 6 wt% and 10 wt% of the chain extender, respectively. Furthermore, the Charpy impact strength increased by 35% and 55% owing to the addition of chain extender (6 wt% and 10 wt%), respectively. This is due to the increase of the molecular weight that promotes the enhancement of the entanglement of the macromolecules. Subsequently, the energy absorption of PLA increased.

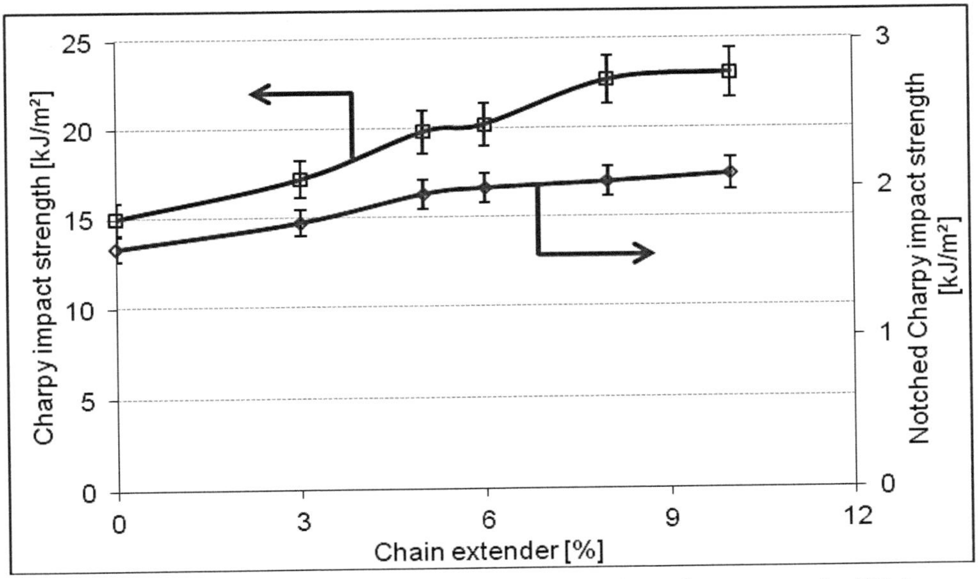

Figure 1.12 Effect of the chain extender (CESA) on Charpy impact strength of PLA.

The tensile properties of PLA were also measured using varying contents of the chain extender (Fig. 1.13). It was observed that the tensile strength slowly increased despite an increase of the chain extender (CESA) content. The tensile strengths increased by 6% and 11% owing to additions of 6 wt% and 10 wt% chain extender, respectively. The elongation at break were recorded at 10% and 15% after adding 6 wt% and 10 wt% chain extender, respectively.

Therefore, it can be summarized that the tensile properties did not improve as the molecular weight of PLA increased. The increase of polydispersity is a possible reason. It was explained previously that the chain extension process occurs in a random manner (linear, branching, and crosslinking). Thus, it formed a vigorous mixture of differing PLA chain lengths, chain architectures, and tacticity, owing to the addition of the chain extender. This mixture is thermodynamically immiscible and forms an unstable blend with internal micro- and meso-cracks.

However, it can be said that the amount of chain extender (CESA) has a significant effect on the melt viscosity and the melt strength, which provides more processing degree of freedom.

On the other hand, an excessive amount of chain extender caused a sudden increase of the melt viscosity. It is suggested based on the experimental findings that the most suitable amount of CESA is maximally 6 wt%. The amounts exceeding 6 wt% caused damage of the machine, because of a quick development of a high compounding pressure. The use of chain extenders with a functionality equal to, or larger than three has become attractive as a means of increasing efficiency of the extenders albeit at the "price" of going from linear to long chain branched structures [18]. The multi-functional extenders in this art have had very limited success. This is mostly due to their very narrow processing window, which begins with noticeable chain extension, moves onto long chain branching, and ends with catastrophic gelation [15].

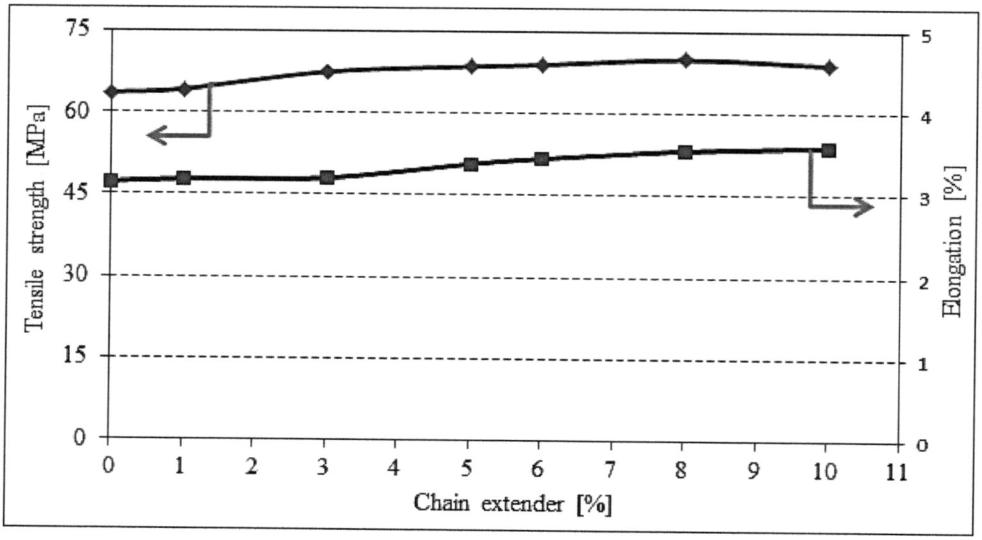

Figure 1.13 Effect of the chain extender (CESA) on tensile strength of PLA.

Moreover, the melt viscosity or the torque of control PLA reduced gradually with different slope at different temperatures. The torque of PLA reduced to 13.5 Nm at the end of processing when the processing temperature equalled 190°C; it further reduced to 5.7 Nm when the processing temperature equalled 210°C. When the processing temperature was at 210°C, the torque of 6 wt% chain extender mixed with PLA reached a peak point (59 Nm) at processing time 6 minutes (Fig. 1.13). In contrast, the torque of 6 wt% chain extender mixed with PLA reached a peak point (51 Nm) at processing time 9 minutes when the processing temperature was kept at 190°C (Fig. 1.14). Therefore, it is obvious that the degradation of PLA and reaction kinetics of chain extender depend on the processing temperature.

1.3.2 Effect of Catalysts on Reactivity of Chain Extender and their Optimization

PLA was mixed with the chain extender (6 wt%) and a catalyst using laboratory kneader (rheomix) with screw rotation 100 rpm at 190°C. The concentration of catalysts was considered according to suggestion of the manufacturer. The effect of the catalyst zinc stearate on the reactivity of chain extended (CESA) can be seen in Fig. 1.14. The concentration of zinc stearate used in this study ranged from 0.5% to 2% of CESA contents. No substantial changes of the torque were observed as a result of the addition of zinc stearate. In contrast, the torque of PLA displayed slightly improved kinetics due to the addition of 1% zinc stearate. However, after reaching the peak point, the torque of PLA (PLA-CESA-ZnSt) reduced in correlation, increasing processing time.

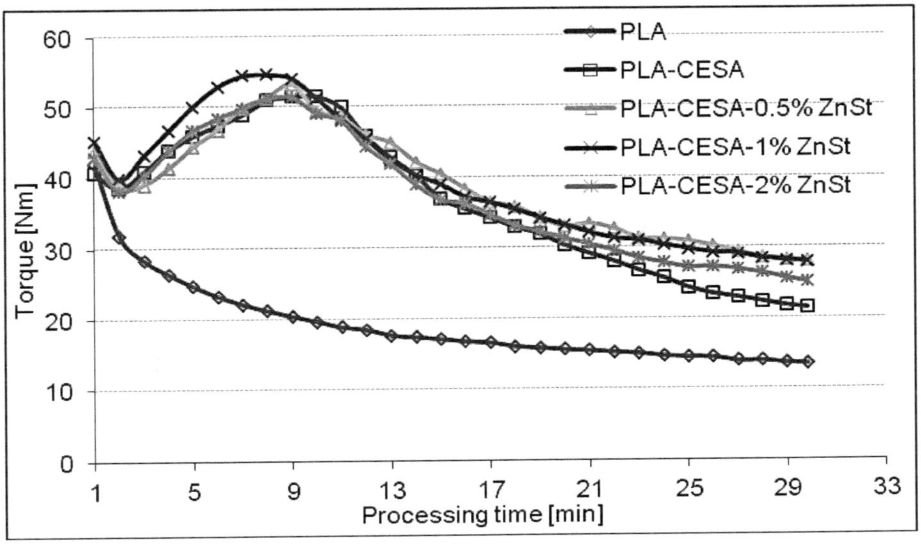

Figure 1.14 Effect of zinc stearate catalyst on reactivity of CESA and on the melt viscosity of PLA.

A similar tendency was observed for all zinc stearate concentrations. In nature, the epoxy-based chain extender is very reactive but the reaction kinetic depends on diffusion kinetics when it occurrs in melt or viscous phase. Zinc stearate is able to open the cyclic epoxy ring at a relatively low temperature. Moreover, the catalyst effect was not as effective as it was originally expected to be. A trace amount of moisture can cause hydrolysis of zinc stearate, causing zinc hydroxide and stearic acid to form. Zinc hydroxide readily reacts with carboxylic containing PLA intermediates which resulted from thermal degradation of PLA [19, 20]. Therefore, the catalytic activity of zinc stearate was found to hinder the chain extension.

Figure 1.15 shows the significant effect of N-methylimidazole (MIZ) on the reactivity of the chain extender (CESA). The torque of PLA increased strongly, and reached its peak after only three minutes of processing owing to the addition of 2% MIZ. The torque of PLA reached a maximum torque (67 Nm) after 2.30 minutes of processing as a result of adding 3% MIZ. The addition of 5% MIZ also led to an increase in the torque of PLA; A maximum torque (65 Nm) was reached after 2.25 minutes of processing. Therefore, the reaction kinetics evidently increased as a result of the addition of MIZ. The catalytic center of MIZ consists of (=NH) groups and the favorable results depend on the presence of an active hydrogen in the MIZ molecule, which could react with an epoxy group of the chain extender and thereby become incorporated into intermediates of PLA.

It is also very probable that the MIZ reacts with the epoxy group of the chain extender, and forms a mono adduct in the initial reaction. Subsequently, a reaction with another epoxy group takes place, and an alkoxide ion forms with an effective catalyst center, which can attack another epoxy group, and thus, can continue ring opening reaction [21, 22]. Therefore, the chain extension of PLA occurs very fast and increases the melt viscosity. After reaching the peak point, the melt viscosity of PLA reduced very rapidly as processing continued. Similar reactions were observed for all MIZ concentrations. The decrease of the melt viscosity could be attributed to the thermal decomposition of MIZ. This appears to be related to the trans-effect and hydrogen bonds in the imidazole complexes, which break down at a temperature around 220°C and subsequently produce free radicals [23]. The decomposition of MIZ can also occur at a lower temperature and a high pressure. Moreover, the set temperature of the system was 190°C. Due to the high shearing force the set temperature increased up to 10°C. The decomposition of PLA succeeded as a result of the development of free radicals.

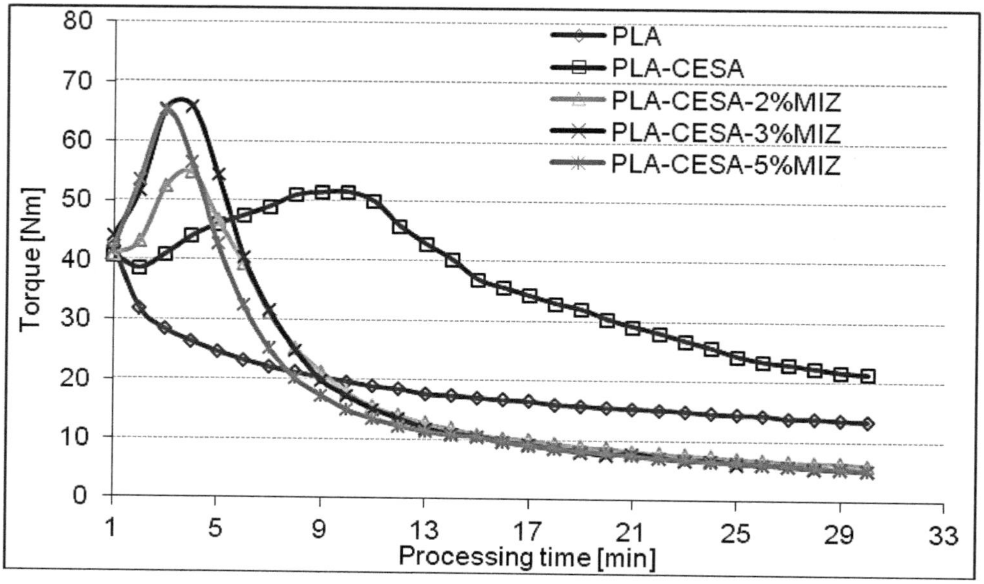

Figure 1.15 Effect of N-methylimidazole (MIZ) catalyst on reactivity of CESA and on the melt viscosity of PLA.

Additionally, these free radicals can also cause thermal decomposition of the epoxy-based chain extender, which, in turn, leads to the production of even more free radicals [24]. As a result, the vigorous decomposition of PLA and the melt viscosity of PLA reached a minimum after approximately 10 minutes of processing. The addition of 5% MIZ did not improve the melt viscosity of PLA and moreover, the maximum torque was found to even be a bit lower than that measured after the addition of 3% MIZ. This could be attributed to complex formation in between two or more MIZ molecules in the presence of carboxylic derivatives of PLA [25].

The influence of 1,4-Diazabicyclooctane (DAO) on the catalytic activity of chain extender (CESA) can be seen in Fig. 1.16. In this study, catalyst 1,4-Diazabicyclooctane (DAO) was considered 3%, 5%, and 6% of CESA contents. It was observed that DAO increased the reactivity of the chain extender and led to a completion of the chain extension reaction within 1.30 minutes. The torque or melt viscosity of PLA increased very sharply and reached a peak point for all concentrations of DAO.

DAO is a lewis base catalyst and very sensitive to accelerating the ring opening reaction of epoxy and consequently, speeding up chain extension reaction of PLA. After reaching the peak point, this property decreased very quickly. As previously described, alcoholic and carboxylic derivatives were produced as a result of the thermal degradation of PLA. The catalytic conversions of those alcohols to the corresponding aldehydes or ketones are produced by DAO. Moreover, the DAO catalyst takes part esterification reaction with high yield and produces anhydride type products. The successive chain secession of PLA occurred due to the resulting anhydride and ketones [26]. The catalytic effect of DAO increased the DAO concentration from 3% to 5%, and afterwards no significant catalytic effect was observed. Therefore, it could be noted that the optimal concentration of DAO is 5 percent.

The catalytic effects of all catalysts (optimal concentration) were compared and presented in Fig. 1.17. It was observed that zinc stearate helps to increase the kinetics of the chain extension reaction slightly. The chain extension reaction kinetics of PLA improved nearly indentically for both modes by N-methylimidazole and 1,4-Diazabicyclooctane, but the maximum torque of PLA found for N-Methylimidazole was higher than that of 1,4-Diazabicyclooctane. The torque of PLA fell instantly just after maximum value was reached. In comparison with the torque without a catalyst, it crosses down within 6 minutes in the case of N-Methylimidazole and within 3 minutes in the case of 1,4-Diazabicyclooctane. Thus, it is evident that the catalytic effects of N-Methylimidazole and 1,4-Diazabicyclooctane catalysts strongly depend on processing time.

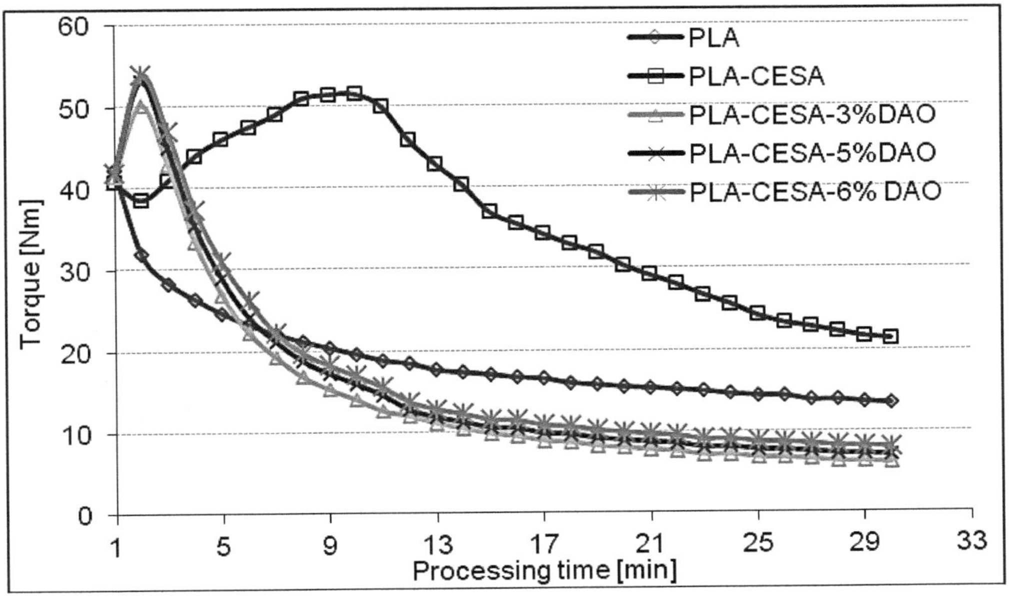

Figure 1.16 Effect of 1,4-Diazabicyclooctane (DAO) catalyst on reactivity of CESA and on the melt viscosity of PLA.

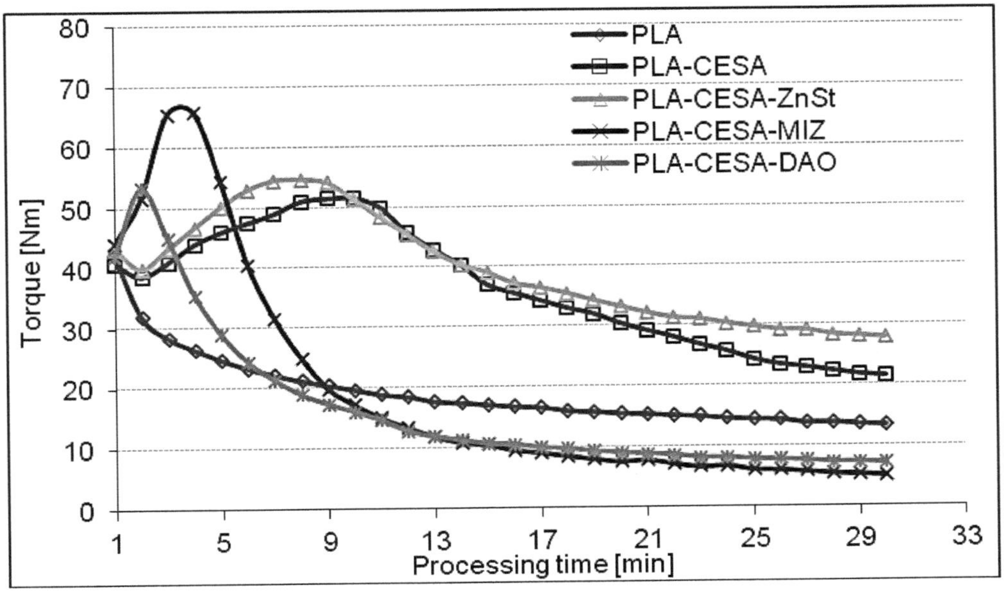

Figure 1.17 Effect of catalysts on reactivity of CESA and on the melt viscosity of PLA.

1.3.2.1 Effect of Catalysts on the Molecular Weight of PLA

The viscosity average molecular weights of PLA and chain extender modified PLA (with and without) catalyst are shown in Fig. 1.18.

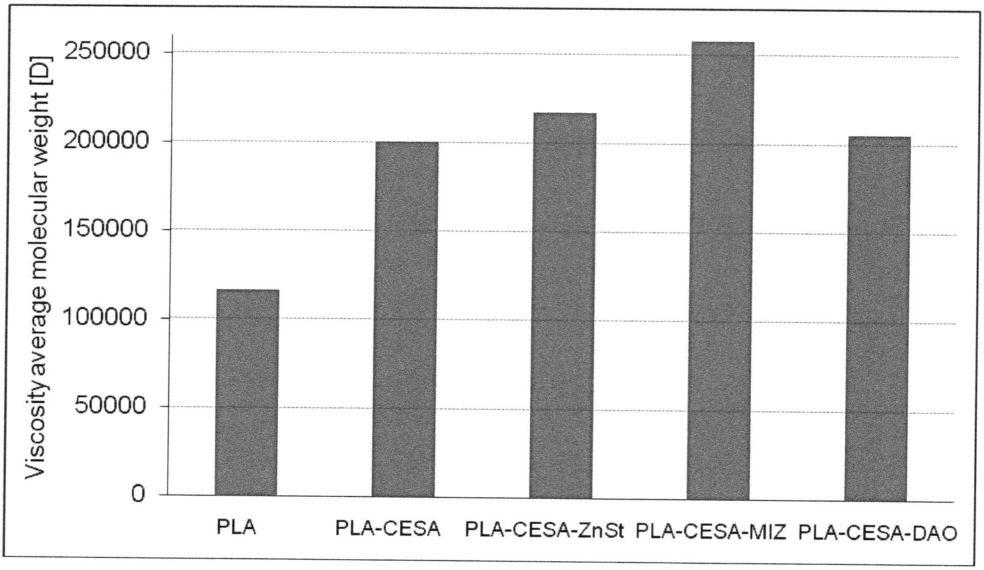

Figure 1.18 Effect of CESA and catalysts on the molecular weight of PLA.

The samples were considered at optimal concentration of catalysts. The molecular weight of PLA increased 75%, 87%, 120%, and 80% in comparison with control PLA due to the addition of chain extender, chain extender with zinc stearate, N-Methylimidazole, and 1,4-Diazabicyclooctane catalyst, respectively. The chain extender and catalysis chain extender showed a significant effect on the molecular weight of PLA. Therefore, it could be summarized that the chain extender extended short chain of PLA and resulted in the increase of molecular weight. The catalysts were found to further enhance the chain extension ability of the chain extender during the processing of PLA.

1.3.2.2 Effect of Catalysts on thermal Properties of PLA

Figure 1.19 illustrates DSC thermograms of PLA (3052D, Naturework) and PLA mixed with chain extender, with and without catalysts. Neat PLA showed two transitions while heating, which included the glass transition temperature at about 60.5°C, and the melt endothermic temperature at about 150.9°C. The crystallization peak temperature was at approximately 120°C.

The glass transition temperature shifted minimally (approximately 2 to 3°C) to the lower temperature due to the addition of the catalyst. Apart from that, the glass transition temperature of PLA remained almost the same despite the addition of zinc stearate catalyst. The melt enthalpy was found to decrease from 26 J/g (control PLA) to 20 J/g by means of addition of the chain extender.

This is due to the chain extension (linear, branch, or crosslinking) which occurred. Consequently, the chain extension led to increased disorder on the molecular level. The melt enthalpy increased from 20 J/g (PLA-CESA) to 23 J/g for the zinc stearate catalyst. This may be explained by the fact that the zinc stearate catalyst acts as a nucleating agent. In contrast, this property decreased from 20 J/g (PLA-CESA) to 13 J/g for the N-Methylimidazole catalyst and to16 J/g for 1,4-Diazabicyclooctane catalyst. A rapid increase of the melt viscosity was observed for the N-Methylimidazole and 1,4-Diazabicyclooctane catalysts and can be explained by the fast branching and crosslinking which occurred by means of both the catalysts. This is the reason the melt enthalpy decreased. Moreover, the melt enthalpy of PLA for N-Methylimidazole catalyst was minimal. This might be because the degree of crosslinking is higher for the N-Methylimidazole catalyst. This can be supported by a sudden increase of melt viscosity due to N-Methylimidazole catalyst. The nonlinear chain extension (branch and cross linking) contributes to a higher torque or melt viscosity. Two melting peak temperatures were also detected after the addition of the chain extender, and all three catalysts. This might be evidence of the mixture of various molecular weights PLA produced during the chain extension process.

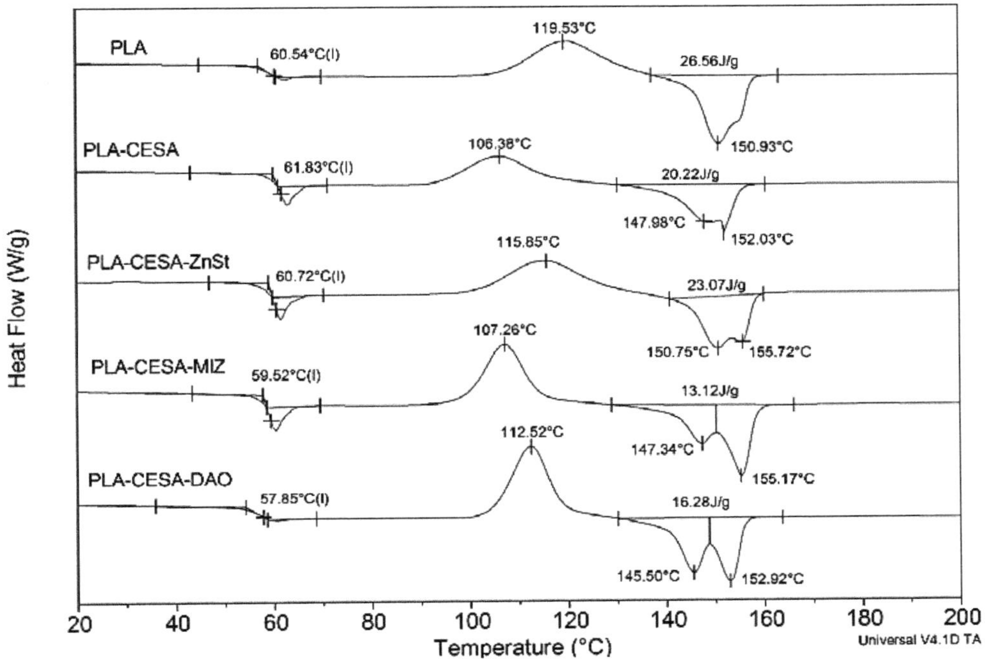

Figure 1.19 Effect of the chain extender and catalysts on the DSC of PLA.

1.3.2.3 Effect of Chain Extender and Catalysts on Mechanical Properties of PLA

The optimum amount of chain extender and catalysts were compounded with PLA using a twin screw extruder and the mechanical properties were measured subsequently. The tensile strengths of PLA, PLA mixed with chain extender, and with and without catalysts are presented in Fig. 1.20. The tensile strength increased by 12% in comparison with control PLA owing to the addition of the chain extender. The weight average molecular weight increased about three-fold and the polydispersity index increased approximately two-fold due to the addition of 6% chain extender [18]. On the other hand, a slight decrease of the crystallinity (melt enthalpy) due to the addition of 6 wt% chain extender was identified. The combination those effects could be reasons for an increase in tensile strength.

This property increased by 16% due to the addition of the zinc stearate catalyst, which may have caused a slight increase in crystallinity. The tensile strength of N-Methylimidazole catalysed PLA displayed a similar value as control PLA and decreased by 10% in comparison with chain extender modified PLA. This is caused by the decrease of the crystallinity. In the case of 1,4-Diazabicyclooctane catalysed PLA, this property scored the same value as chain extender modified PLA, and 13% better than that of control PLA.

The Charpy impact strengths of PLA and chain extender mixed PLA with and without catalysts are illustrated in Fig. 1.21. In comparison with control PLA, the Charpy impact strength increased by 32% as an effect of the chain extender. This is because of an increase in the molecular weight of PLA produced by chain extender. In comparison with the chain extender modified PLA, this property increased by 29% for zinc stearate catalyst, and by 23% for 1,4-Diazabicyclooctane catalyst system. Unlike PLA modified with the chain extender, the Charpy impact strength decreased by

12% for N-Methylimidazole catalysed system. Zn salts are able to catalyze transesterification between polyesters.

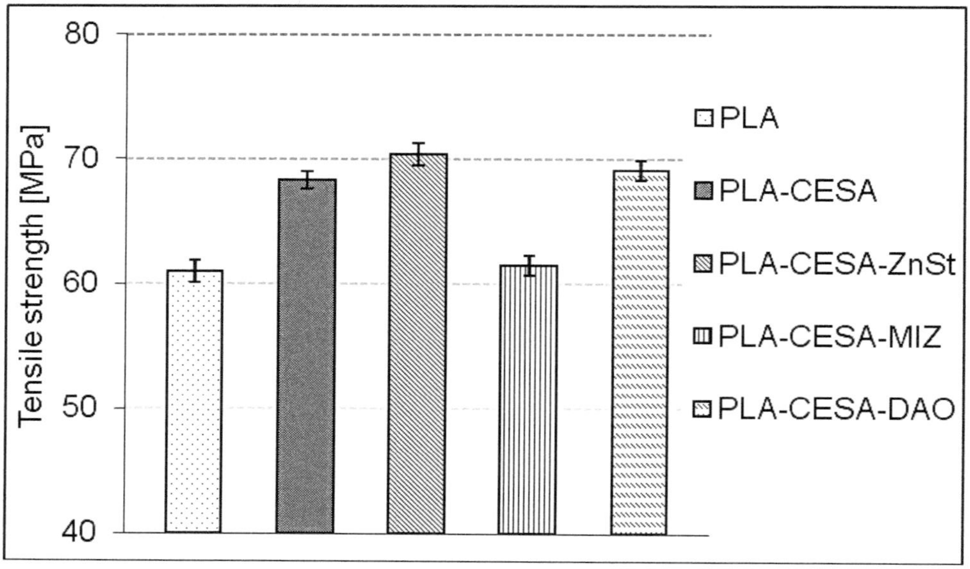

Figure 1.20 Effect of the chain extender and catalysts on tensile strength of PLA.

Therefore, the interfacial compatibilization may stem from possible transesterification between PLA and PLA intermediates or ester groups in chain extender catalysed by the Zn stearate. As a result, the secondary phase appears to change gradually from occluded subinclusions into the co-continuous phase [28]. Such changes in interface wetting were consistent with the variation of the impact toughness. This property reduced for N-Methylimidazole catalysed PLA. This could be because of random nonlinear chain extension caused by Methylimidazole. Accordingly, the transition of the molecular weight from a narrow to wide distribution occurs.

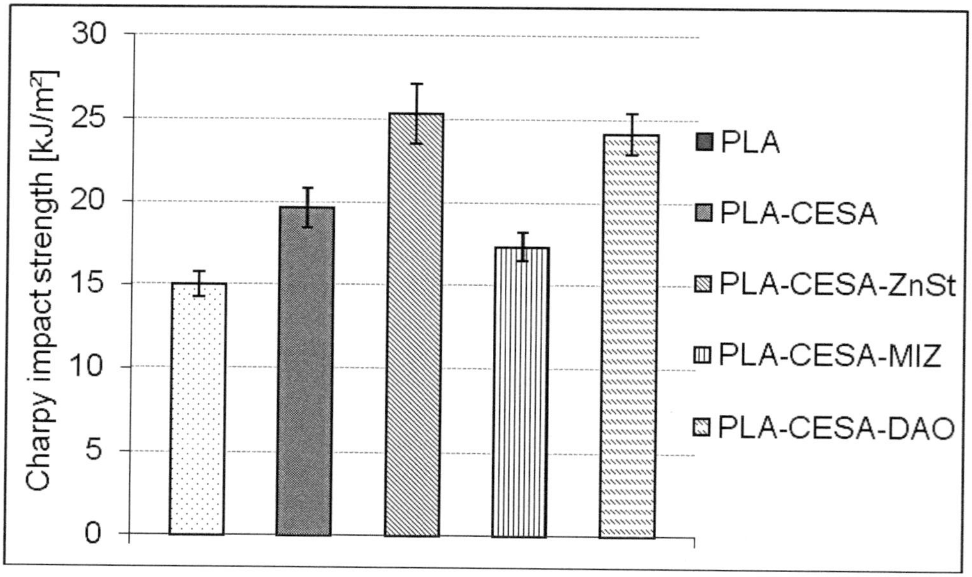

Figure 1.21 Effect of the chain extender and catalysts on Charpy impact strength of PLA.

1.3.2.4 Effect of Chain Extender and Catalysts on Environmental Stress Corrosion Resistance of PLA

Environmental stress cracking resistance (ESCR) tests were carried out using a specially designed plate with a 0.75% fixed stain chamber (Fig. 1.22). Tensile test specimens were placed into the chamber, and, subsequently, the chemical disinfectant (either Meliseptol or Melisept SF) was poured into the chamber. Then, the test specimens were stored in the chamber for 7 days at 23°C.

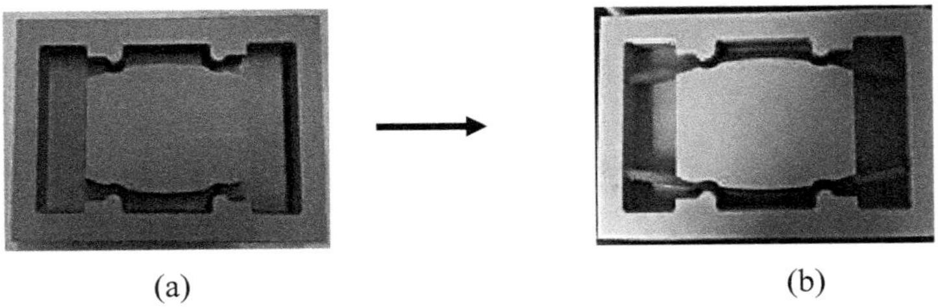

Figure 1.22 ESCR experiments; a) empty chamber, b) chamber with specimens.

After certain periods of time, the test specimens were taken out of the chamber and mechanical experiments were performed.

Environmental stress corrosion resistance (ESCR) is the premature initiation of cracking and embrittlement of a plastic due to the simultaneous action of stress and strain with specific aggressive fluids. ESCR is of utmost importance when testing medical applications. Therefore, the two most usable disinfectants (Meliseptol, Melisept SF) were applied in this study.

The effects of the disinfectants on the tensile strength can be found in Fig. 1.23. Control value means the value without applied stress and aggressive medium. Both disinfectants were found to have significant effects on tensile strength. The tensile strength reduced by about 30 to 35% with respect to their control value caused by melisept SF disinfectant. On the other hand, this property reduced about 35 to 40% by meliseptol in comparison with their control value. The increase of molecular weight of PLA due to addition of the chain extender showed limited (10%) resistance against disinfectants. However, Zinc stearate and 1,4-Diazabicyclooctane catalysed PLA showed better resistance (15% to 20%) than chain extended PLA or control PLA.

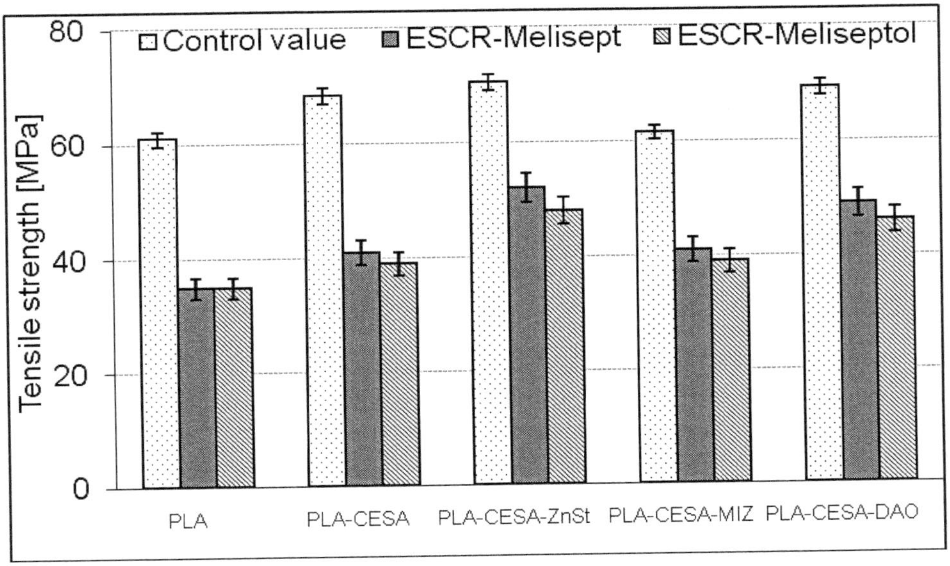

Figure 1.23 Effect of the chain extender (CESA) and catalysts on the ESCR of PLA.

The aggressive fluids lead to two failure mechanisms of cracking. The first and most obvious are those fluids that chemically attack and irreversibly degrade the polymer by means of oxidation, chlorination, and hydrolysis, and which lead to chemical modification, chain scission, and cross linking. PLA polymers are relatively stable against alcohol, salt, and disinfecting agents when no load is applied. They are known to be swelled minimally in selective solvents. However, when a load is applied, the first mechanism can be triggered. The second category does not modify or degrade the polymer, but promotes cracking via a physical process which is generally understood to be due to the selective absorption of the fluid within a micro-yielded or stress dilated zone and embrittle the polymer in the presence of stress. This reduces the yield strength, and consequently, reduces the tensile strength.

1.3.2.5 Effect of Chain Extender and Catalysts on Gamma Sterilization of PLA

Tensile test specimens (160*10*4 mm^3) were sterilized using a C-188 cobalt-60, manufactured by Nordion, Canada. The total dosing of gamma radiation was 17, 33 and 65 kGy. High-energy photons are emitted from an isotopic source (Cobalt 60), thus producing ionization (electron disruptions) throughout a product. In living cells, these disruptions result in damage to the DNA, and other cellular structures. These photon-induced changes at the molecular level cause the death of the organism or render the organism incapable of reproduction. The gamma process does not create residuals, or impart radioactivity in processed products. The gamma process can effectively sterilize a wide variety of products composed of different materials, with varying densities, configurations, and orientations. In general, the sterilization of medical devices is accomplished using gamma radiation in the range of 25 to 30 kGy.

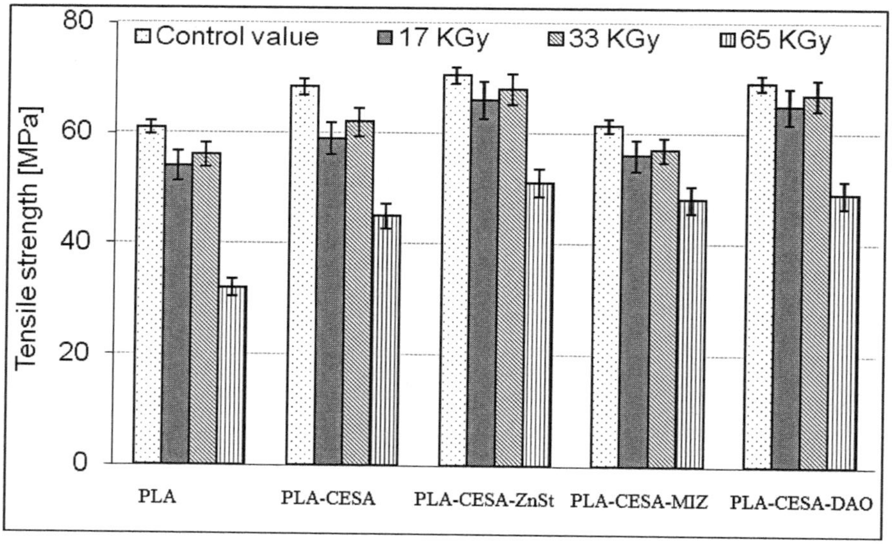

Figure 1.24 Effect of the chain extender (CESA) and catalysts on the sterilization of PLA.

The effect of gamma sterilization on the tensile strength of PLA and PLA modified with the chain extender with and without catalyst is shown in Fig. 1.24. In the case of unmodified PLA, the tensile strength was found to reduce 12% when sterilized using 17 kGy. The reduction of the tensile strength is due to the degradation of the main chain of PLA molecules, which are influenced by the production of PLA intermediates and free radicals. This property reduced only by 8% when sterilized using 33 kGy. This is because of crosslinking of the PLA intermediates. Moreover, the increase of crystallinity occurred as a result of 33 kGy gamma radiations. A similar observation was given in literature. Dadbin et al. clarified that gamma radiation at a dose of

30 kGy promoted formation of α crystalline phase [29]. This property reduced about 48% when 65 kGy gamma radiation was used. This is caused by random degradation of the PLA molecules. In this case, there may also be crosslinking of PLA intermediates, but the degradation reaction dominates the whole system. The mechanical properties decrease notably because random main chain scission of polymer occurred when a high amount of radiation was used [30]. The similar tendency was observed for chain extender and catalyst modified PLA. However, Zinc stearate and 1,4-Diazabicyclooctane catalysed PLA showed better resistance (15% to 20%) than chain extended PLA and 25% to 30% than unmodified PLA at all doses of gamma radiation.

1.3.3 Chain Extender as Compatibilizer

Many engineering thermoplastics and biopolymers are combined with other polymers to obtain the balance of properties which are required for specific applications or in reinforcing composite, polar fiber needs to wet with different polymers for optimizing surface energy. Polymer-polymer interactions in such systems are usually unfavorable and the interface between the two phases is thermodynamically not stable. The basic idea of compatibilization is to reduce interfacial energy between polymers in order to increase adhesion. The addition of compatibilizers also results in finer dispersions, as well as more regular and stable morphologies.

Thus reactive compatibilization is a powerful method for resolving that problem by increasing interfacial strength just as controlling and stabilizing resulting products morphology. The Joncryl® technology offers a new way for reactive compatibilization, since the functional group is able to react with the end groups of polymers. A fundamental requirement is that both polymer species have reactive end-groups, so that a graft copolymer between the two polymer chains is being made. As a result, the morphology can be manipulated over a very wide range [17, 31].

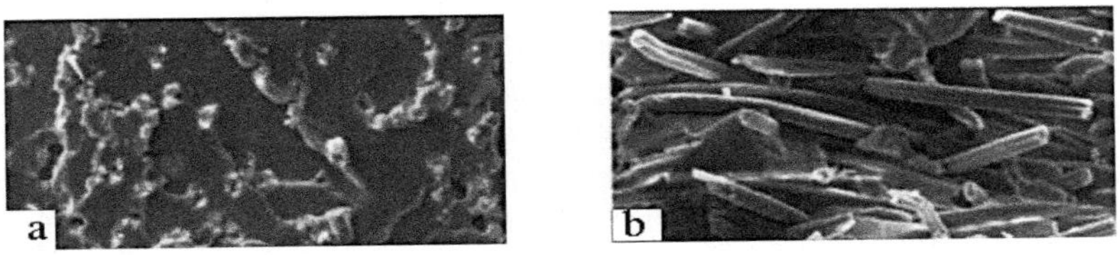

Figure 1.25 Interface of two phase system, a) without joncryl and b) with joncryl.

Joncryl 3229 is a maleinated chain extender based on styrene acrylic copolymer with moderate polarity which influences the interfacial region of two phase systems. For instance, the fiber pull outs became the leading failure mode in fiber reinforced PLA composites, which can be seen in SEM micrographs (Fig. 1.25).

The outcome of chain extender (Joncryl 3229) on Tensile strength and notched charpy impact strength of PLA can be seen in Fig. 1.26. The tensile strength reduced gradually due to addition of joncryl 3229, and this property reduced about 10% due to addition of 6% joncryl 3229. On the other hand, the notched charpy impact energy absorption increased ca. 30% due to addition of 6% joncryl 3229.

Villalobos et al. and Pilla et al. used joncryl chain extender after ring opening polymerization and have found increased molecular weight of PLA as well as improved rheological properties [31, 32].

Chain extender (CESA) could join together the thermally degraded matrix polymers and consequently increased the molecular weight of polymer to initial molecular weight and even more than initial. In general, mechanical properties are proportional to molecular weight of the matrix. It also increases the viscosity of the polymer which is directly proportional to molecular weight. In contrast, chain extender (joncryl) acts as compatibilizer which consequently improves the interface.

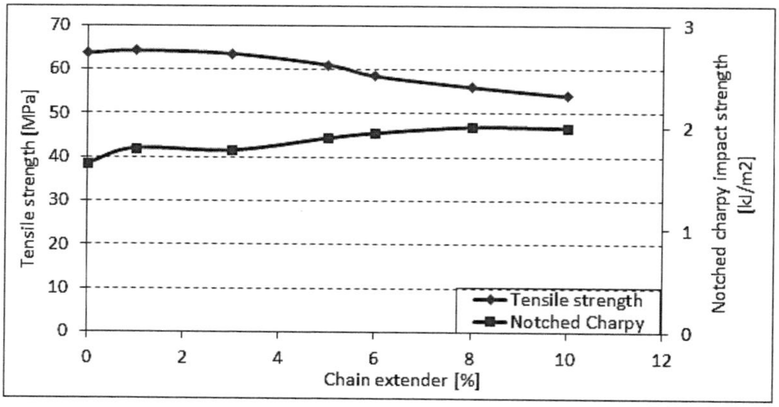

Figure 1.26 Effect of the chain extender (Joncryl) on tensile strength and notched charpy impact strength of PLA.

1.3.4 Impact Modification

PLA was compounded with impact modifiers (1 to 20 wt%) in a twin screw extruder. The temperature profile in the compounding process varied from 180°C to 200°C. Subsequently, the samples were processed by injection molding. In this study, sukano S550, Sukano AG, was used as impact modifier. The tests were carried out by means of an instrumented impact pendulum (DIN EN ISI 179-2/1eA). Figure 1.27 shows the change in impact strength (with/without notch) with respect to addition of impact modifier at room temperature and –30°C. It can be seen that the impact modifier (10 wt%) causes an increase of about 40% in impact resistance without a notch. The similar tendency of impact resistance without a notch was observed for both temperature conditions. The impact strength at room temperature is about 15% higher values in all modifier proportion in comparison with the impact strength at –30°C.

On the other hand, the notched impact strength remains nearly constant at both temperatures. It is known from the literature that the PLA is very sensitive to notches. This is also reflected in the corresponding measured values. So, there is no increase in the impact strength due to the modifier.

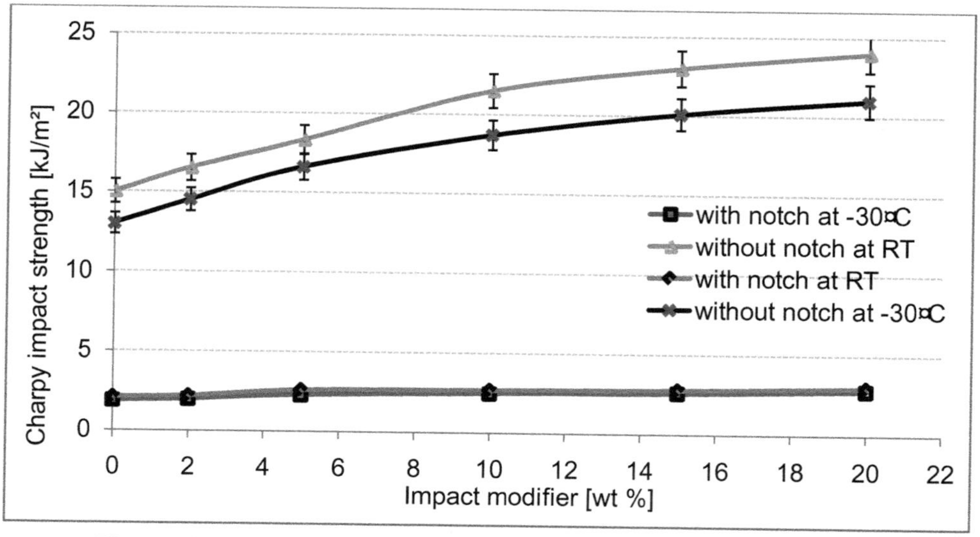

Figure 1.27 Effect of the impact modifier on Charpy impact strength of PLA.

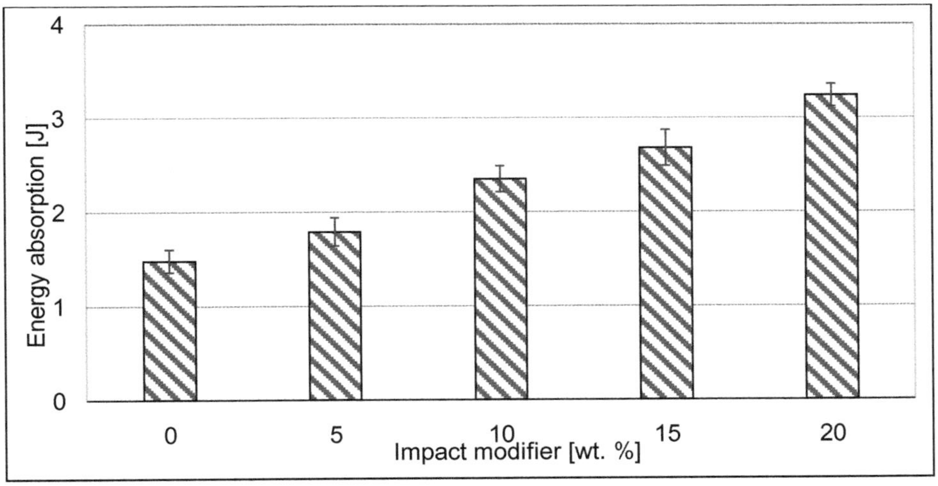

Figure 1.28 Effect of the impact modifier on energy absorption of PLA.

Figure 1.28 shows the energy absorption in the falling weight impact (penetration) test of PLA with different impact modifier fractions. The energy absorption increases significantly with the addition of impact modifier. The impact energy absorption increased by 60% due to addition of 10% impact modifier.

Further analyses of impact modified PLA were performed with a light microscope and scanning electron microscope. Figure 1.29 clearly shows that the impact modifiers are well distributed. On the right picture it can also be seen that there are many irregular pores. The pores can act as a crack starting point and accelerate crack growth. Apparently, the pores could be formed by the decomposition of the impact modifier (>200°C processing temperature).

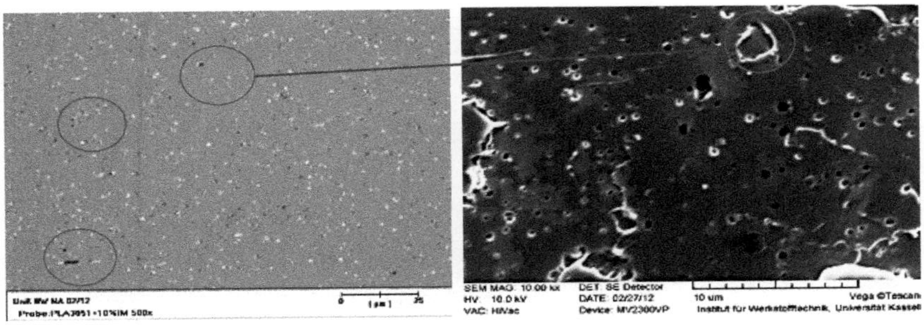

Figure 1.29 Effect of the impact modifier on micro structure of PLA.

1.3.5 Toughness Modification

Biomax® Strong 120 (BS) is a commercial modifier for PLA offered by the company DuPont. It is a copolymer of ethylene-acrylate and was developed to improve the toughness of PLA in packaging and industrial applications. Biomax® Strong 120 has the following advantages: increases toughness, reduces brittleness, increases flexibility, and improves impact resistance.

1.3.5.1 Impact Strength

The impact resistance of PLA (3001 D, Naturework) was carried out by instrumented impact pendulum test. The test was performed with 5 Joule pendulum according to DIN EN ISO 179-2/1eA, which is equipped with a piezo quartz computerized data acquisition system. In this study,

PLA 3052 D and PLA 3001 D were used. In Fig. 1.30, there is a significant increase in impact strength with the addition of the toughness modifier. The impact strength increases for PLA 3052 D by 115% with the addition of 10% toughness modifier. Whereas, this property increased ca. 90% for PLA 3001 D. Moreover, this property increased 147% for PLA 3052 D and 186% for PLA 3001 D due to addition of 20% toughness modifier.

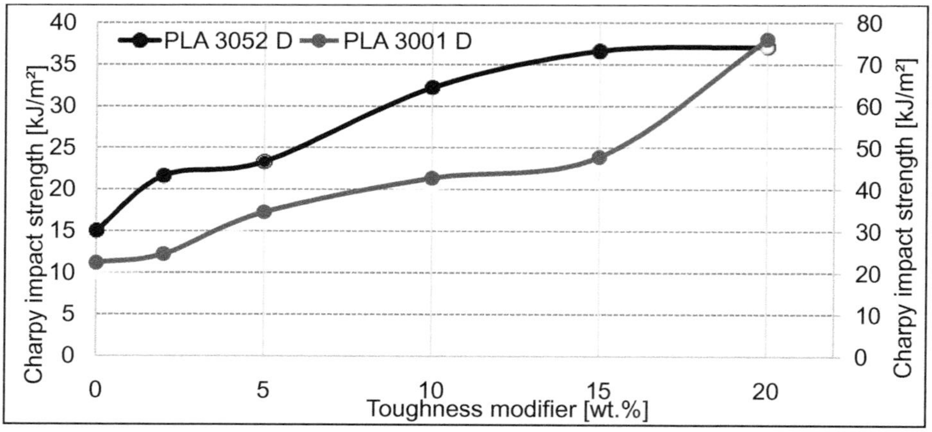

Figure 1.30 Effect of the toughness modifier on notched Charpy impact strength & energy absorption of PLA.

In the following image (Fig. 1.31), the red circles illustrate the part of the test specimens where breakage resistance was found to improve due to addition of toughness modifier. As a result, more energy is needed for a break, which means an increase in impact resistance. It is noteworthy that increasing the modifier decreases the strength (firmness) of PLA [34, 35]. By optimizing the quantity, however, it is possible to produce suitable materials, in particular for packaging, where high impact strength is a prerequisite.

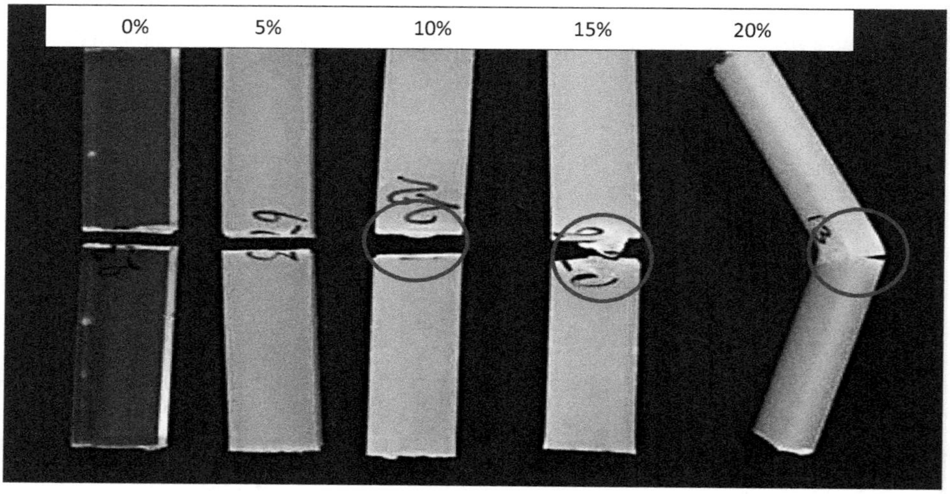

Figure 1.31 Test specimens of PLA 3001 D due to Charpy impact test.

1.3.5.2 Falling Weight Impact Strength

The falling weight impact test describes the force-displacement relation, which could help assess the multidimensional load-mechanical properties of a material. The method provides a stiffness parameter for the material. Furthermore, the response of a material to a given impact energy can be split into a dissipation and a storage energy contribution. The dissipation energy is defined as the energy absorbed by a material by means of damage initiation, deformation, development of deliminations, and fracture [33]. The storage energy describes the energy given back to impactor.

The deformability (puncture mode), impact test was carried out in accordance with DIN EN ISO 6603-2. In the experiment, the total mass of dart was 3.65 kg and the fall on the test specimens was at the rate of 4.43 m/s, and the height was adjusted 1.1 meter. Therefore, impact energy is high enough to puncture the material. So there was no storage energy left. The maximum force, displacement at maximum force, stiffness, and energy absorption of PLA are represented in Fig. 1.32. It was found that maximum force, displacement at maximum force, and energy absorption of PLA increased with increasing amount of toughness modifier (BS). As shown in this graph, the addition of 5% (by weight) modifier into the PLA matrix has no great effect, while above this concentration there is a significant increase in maximum force, displacement at maximum force, stiffness, and energy absorption of PLA.

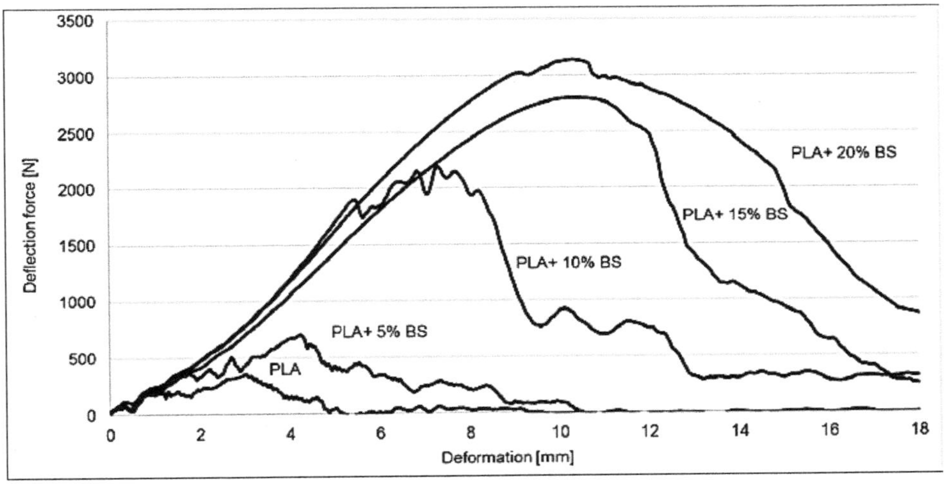

Figure 1.32 Effect of the toughness modifier on deflection force and energy absorption of PLA.

Therefore, it can be said that this modifier has a significant effect on impact properties. The modifier (copolymer of ethylene-acrylate) introduced elastic secondary phase, which incorporates an extension of those properties.

Figure 1.33 shows the specimens with 0%, 5%, 10%, 15%, and 20% modifier (BS). Pure PLA showed brittle fracture behavior. While the sample with 5% BS breaks even brittle, the samples with higher content of BS showed ductile fracture behavior. It was also seen that the sample with 20% BS content is not broken but cracked.

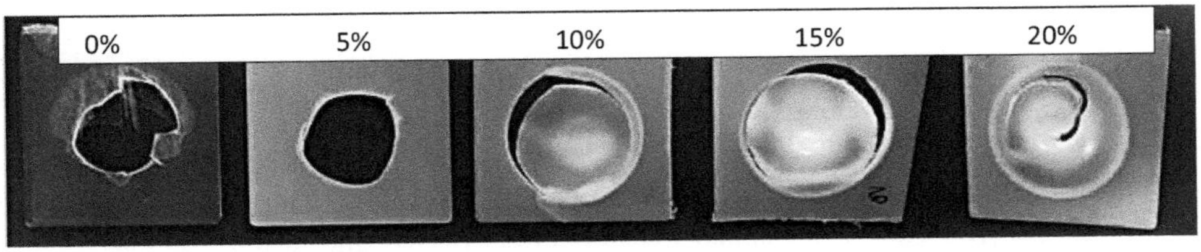

Figure 1.33 Test specimens of PLA due to impact (penetration mode) test.

Figure 1.34 shows that the energy absorption increases significantly with the addition of the toughness modifier for both the types of PLAs. The energy absorption increased 13-folds for

PLA 3001 D and 6-folds for PLA 3052 D due to the addition of 10% modifier, and energy absorption increased 26-folds for PLA 3001 D and 14-folds for PLA 3052 D due to the addition of 15% modifier.

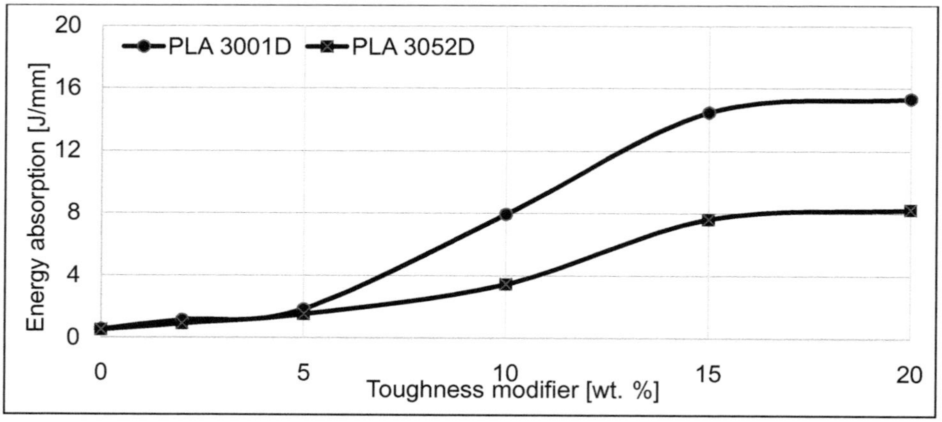

Figure 1.34 Effect of the toughness modifier on energy absorption of PLA.

1.3.5.3 Tensile and Flexural Strength

The addition of toughness modifier (Biomax Strong 120) showed (Fig. 1.35) a decrease in flexural strength and tensile strength of the PLA. Both strength properties reduced gradually and they reduced about 20% and 35% due to addition of 10% and 20% modifier, respectively. The reduction of strength property is due to increase of softness PLA incorporating with ethylene-acrylic soft polymer [34, 35].

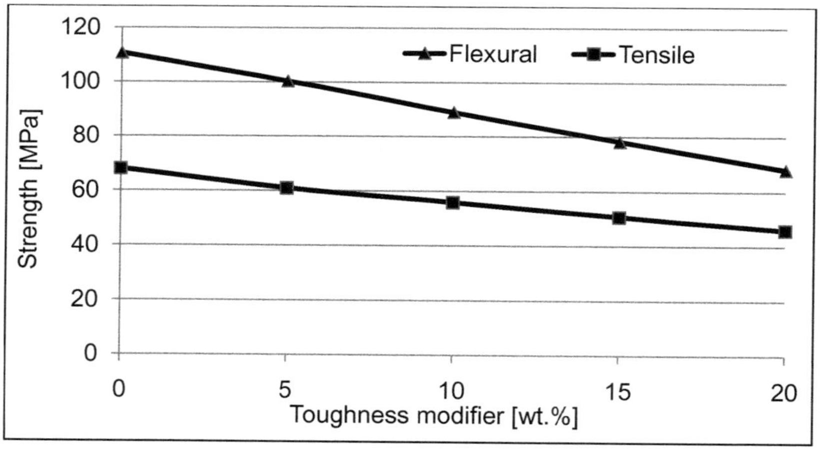

Figure 1.35 Effect of the toughness modifier on flexural & tensile strength of PLA.

1.3.5.4 Dynamic mechanical thermal analysis

The storage modulus of PLA can be seen in Fig. 1.36, and found to reduce with addition of modifier, from which it can be presumed that this effect occurs due to the low glass transition temperature modifier. The storage modulus reduced affectedly after 70°C, which is related to the glass transition temperature of PLA. Thereafter, the storage modulus showed an increase at temperatures above 100°C, which is different for each sample and thus appears to be dependent on the specific composition of test specimens.

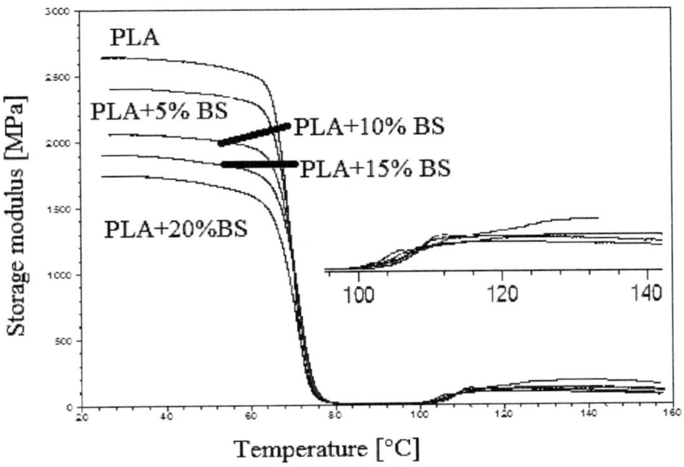

Figure 1.36 Effect of the toughness modifier on DMA of PLA.

1.3.5.5 Differential Scanning Calorimetry

The melting and crystallization behavior testing of the blends was performed using a DSC analyzer machine (DSC Q1000, TA Instrument, USA). The scan was carried out at the rate of 10°C/min from 0°C to 200°C with a flow rate of nitrogen gas of 30 mL/min.

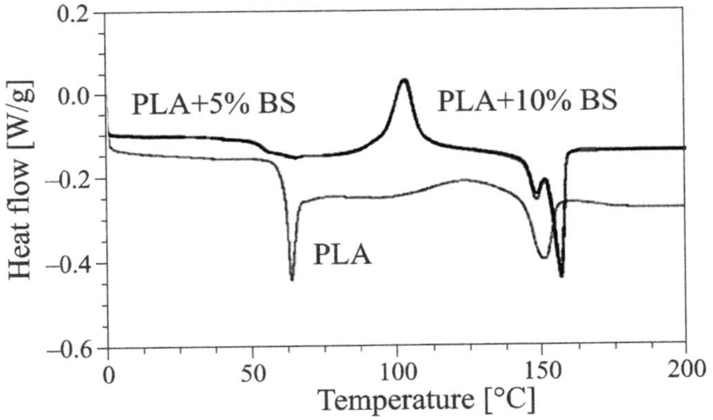

Figure 1.37 Effect of the toughness modifier on DSC of PLA.

The glass transition and melt temperature (Fig. 1.37) of PLA are detected at 60°C and 155°, respectively. The enthalpy value of PLA is 23 J/g, and respective crystallinity is about 24 percent. The glass transition temperature almost disappears due to addition of toughness modifier. There is no alteration notices between 5% and 10% toughness modifier modified PLA. The toughness modifier has an effect on recrystallization and a strong crystalline peak appears due to the addition of modifier. The crystallinity is almost always triggered at the glass transition temperature in most cases, proceeds slowly, and leads to higher degrees of crystallinity in the case of the native PLA. This result is therefore related to the formation of finer and more regular crystallites, and indicates the nucleating effect of the modifiers used [36, 37].

1.3.5.6 X-ray Diffraction

X-ray diffraction measurement was carried out by using a Shimadzu XRD 600 X-ray diffractometer with CuKα radiation (λ = 1.542 Å) operated at 30 kV and 30 mA. Data was collected within the range of scattering angles (2θ) of 10° to 40° at the rate of 2°/min. The basal spacing was derived from the peak position (d001 reflection) in the XRD diffractogram according to the Bragg equation ($\lambda = 2d \sin\theta$).

X-ray diffraction pattern of PLA and toughness modifier (Biostrong) modified PLA is shown in Fig. 1.38. The degree of crystallinity of PLA increased due to addition of toughness modified. It was found that the degree of crystallinity increased 5% and 11% due to addition of 5% and 10% modifier, respectively.

PLA showed two diffraction peaks (31, 14) detected at 2θ = 16.8° and 22.9°, respectively. Their corresponding d-spacing values were determined as 0.54 and 0.39 nm (not shown). The crystal peaks of PLA were less obvious with low absorption intensity, showing more amorphous scattering. This may be due to the different degrees of molecule deformation during the processing process. Toughness modified PLA showed two relatively sharp crystal peaks with strong absorption intensity, which led to a high degree of crystallinity. This result can be compared with DSC analysis. The crystallinity data obtained from DSC and XRD are confirming each other [38, 39].

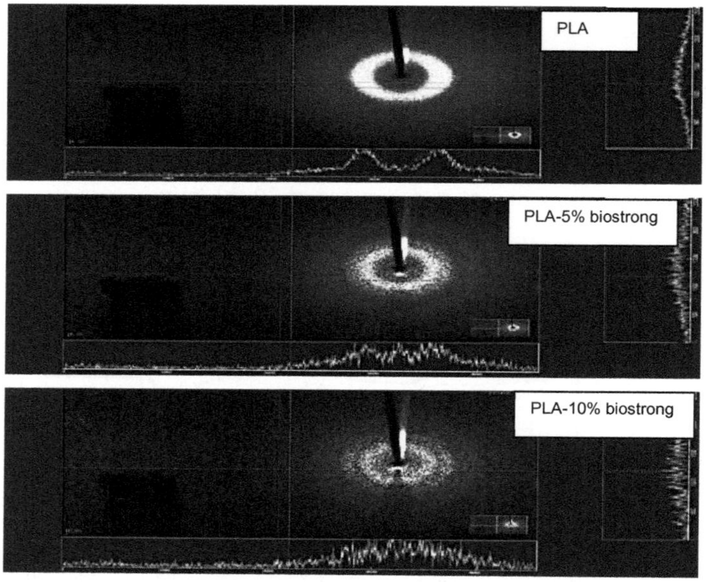

Figure 1.38 Effect of the toughness modifier on X-ray diffraction pattern of PLA.

1.3.6 Effect of Flame Retardant

While there is a diversity of PLA in electronic device housing applications, it is important to verify their burning behavior. Exolit AP 462 is a product based on ammonium polyphosphate. PLA with exolit AP 462 was compounded using twin-screw extruder and successively, test specimens were produced by injection molding. Exolit contents for PLA were optimised in terms of flame and heating wire behavior, limited oxygen index (LOI), and heat releasing rate. The phosphate based exolit improved flame behavior, LOI, and heat releasing properties of PLA.

1.3.6.1 Flammability Test

Determination of flame retardancy was performed using the UL 94 Vertical Test. This test is used extensively in the field of electrical engineering and electronic applications for the assessment of fire behavior and allows a classification of the tested materials under the influence of an external ignition source in the form of an open flame. The afterburning times, the afterglow behavior, and the dripping behavior of the test specimens were evaluated. For the classification of a flame-retardant plastic in class V-0, the following criteria must be met and can be seen Fig. 1.39.

For a set of five specimens, all specimens shall not burn after 10 seconds of flame exposure with an open flame of defined height for no longer than 10 seconds. The sum of the afterburn times of five test specimens may not exceed 50 seconds. In addition, no burning, dripping, or complete burning should take place and the sum of the afterburning and afterglow times of the respective test specimen must not exceed 30 seconds. Classification of V-1 requires that the individual afterburning times are not longer than 30 seconds and that the sum of the afterburning times of five samples is not greater than 250 seconds. In addition, no burning, dripping, or complete burning should take place and the sum of the afterburning and afterglow time of the respective test specimen must not exceed 60 seconds.

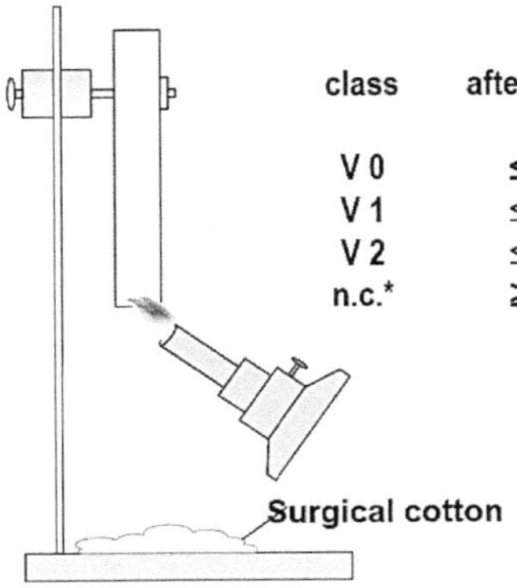

class	afterflame time	ignition of cotton
V 0	≤ 10 sec.	no
V 1	≤ 30 sec.	no
V 2	≤ 30 sec.	yes
n.c.*	≥ 30 sec.	yes

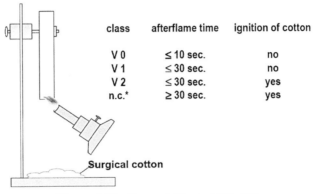

Figure 1.39 Flammability test (UL 94).

A classification in V-2 is made when it comes to a burning, dripping when fulfilling the other criteria, as they apply to the classification in the class V-1. If the above criteria are not met, the evaluation will be done n.c. = not classifiable as V-0, V-1 or V-2.

In Table 1.4, it can be seen that the pure PLA with 0.8 mm wall thickness is not classifiable. With an addition of 10% Exolite, the material was classified in the class V-2. Moreover, with an addition of 15% or more Exolite, the material met the criteria for the highest class V-0. Furthermore, PLA with a wall thickness of 1.6 mm was classified in class V-2. Likewise, with an addition of 10% Exolite or more, the material met the requirements for the highest class V-0.

Table 1.4 Flammability of PLA and modified PLA.

***Recipe* (%)**				
PLA	100	90	85	80
Exolit AP 462	–	10	15	20
Classification	↓	↓	↓	↓
Specimen thickness (0,8 mm)	n.c.	V-2	V-0	V-0
Specimen thickness (1,6 mm)	V-2	V-0	V-0	V-0

Therefore, it can be said that the flame retardant contents and sample wall thickness have an effect on flame properties of PLA. The intended field of application with the respective requirements determines whether such formulations can essentially find an application.

1.3.6.2 Glow Wire Test

The application finds solid materials, which are examined in the fire test to see if a flame forms and when it is extinguished. Furthermore, it is observed whether burning parts drip or splash during the test. The glow wire test is the flammability in the real environment. The investigation was carried out in accordance with IEC 60695-2-10-13. Before the first measurement and when installing a new filament loop, the equipment should be calibrated with silver foil. The silver foil is placed on the tip of the filament and the temperature is slowly increased with a low heating rate from 900°C. The silver foil should melt at 960 ± 15°C. Thereafter, the molten film is immediately removed by brushing to avoid alloying.

Before the start of the test, the room air system is to be put into operation. After starting the machine, the time of ignition, the flame height on the test specimen, and the time of possible extinction must be observed and recorded on the log (ti, flame height, te).

After 30 seconds, the cart is automatically removed from the filament. In addition, you can interrupt the test at any time with the stop button. The temperature at which a sample is found must be confirmed three times. If there is no flame, this must also be recorded in the protocol. Special observations (i.e., heavy smoke) are to be documented in the comments column. After each test, the filament must be cleaned with a steel brush.

Table 1.5 Glow wire flammability test of PLA and modified PLA.

Material	*Tem.* [°C]	*Thickness* [mm]	*Initial time* t_i [s]	*End time* t_e [s]	*Flame height* [cm]	t_e–t_i [s]	t_e–30 [s]	*GWFI status*
PLA	960	1.5	0	64	10	64	34	Fail
PLA+10% Exolit	960	1.5	0	35	7	35	5	Pass
PLA+15% Exolit	960	1.5	0	34	7	34	4	Pass
PLA+20% Exolit	960	1.5	0	32	3	32	2	Pass

In Table 1.5, it can be seen that the test specimen of PLA are non-existent. The modified PLA material passed the test with the addition of Exolit.

1.3.6.3 Limited Oxygen Index

The combustion processes take place with increased or reduced oxygen content in the ambient air with different intensities. Up to which oxygen concentration a flame retardant can prevent the burning of the specimen is a measure of the effectiveness of the flame retardant, and is called Limited Oxygen Index (LOI). The investigation was carried out according to DIN EN ISO 4589.

For this purpose, the entire top of the specimen is evenly coated with the flame for five seconds. If this does not produce a permanent flame, this process is repeated up to five times or until a permanent flame is created. Burning time and burned distance of the sample are measured. If the burn time is more than 180 seconds or the burned distance is more than 5 cm, the test is failed. According to the result, the oxygen concentration is increased or decreased. This is repeated until a limit between passed and failed for 1% oxygen is found. To determine the value, five more samples are measured at the boundary.

Table 1.6 Limited oxygen index of PLA and modified PLA.

Material	*Vol.%* O_2 (±2%)
PLA	22.3
PLA+10% Exolit	28.4
PLA+15% Exolit	28.6
PLA+20% Exolit	32.6

The Limited oxygen index of PLA and modified PLA can be seen in Table 1.6. The minimum oxygen concentration for PLA is about 22 vol.% (±2) to preserve the combustion of the specimen. It means that the test pieces of PLA are combustible in normal air. With an addition of Exolit, a higher concentration of oxygen is required. It means that the specimens of Exolit-modified PLA are not combustible in normal air.

1.3.7 Effect of UV Stabilizer

UV Stabilizers (S547-Sukano) were supplied by Sukano, Switzerland. One wt.% of UV stabilizers was compound with PLA and consequently, the test specimen was produced using injection molding machine. The influence of the UV stabilizer was carried out by means of a weathering test based on the VW-PV 2005 specification. The weathering test was examined at about 40°C, relative humidity 65%, and 8 hours under UV light per day. The tensile strength and flexural strength was found to decrease gradually with the exposure time and after 600 hours of exposure, both properties decreased about 25 percent.

The UV stabilizer improves UV resistance of modified PLA (Fig. 1.40). The PLA material is subject to degradation, resulting in noticeable decrease in strength properties, which however can be prevented by the addition of the UV stabilizer.

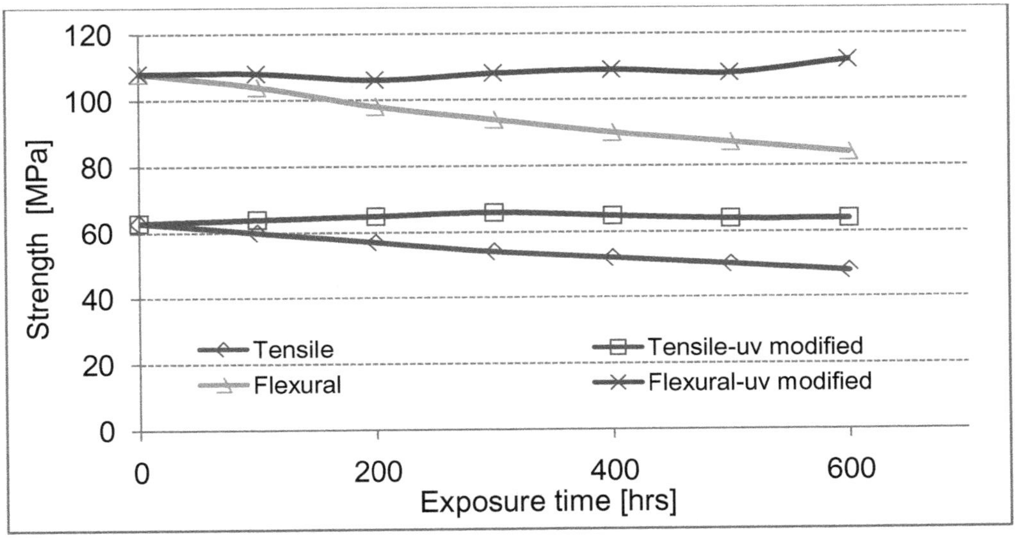

Figure 1.40 Weathering of PLA and UV modified PLA.

1.4 STEREO COMPLEX FORMATION OF PLA

A stereocomplex of the PLA (sc-PLA) was formed from two enantiomers PLLA and PDLA using lab scale knitter or twin screw extruder. The different blend ratio of the corresponding PLLA and PDLA are mixed together for the stereo complex formation. PLA 3001 D, NatureWorks, USA, was used as PLLA. Lapol® HDT, a product of Lapol ™, USA, was used as PLDA. Lapol® HDT is a thermoplastic renewable polymer that is miscible with PLA. The melt temperature of Lapol is in the range 170°C–190°C, density 1.25 g/cm^3, and molecular weight is 20,000 g/mole. It shows improved crystallization and increased heat deflection temperature while compounding with PLLA. It is able to dramatically increase the melting point of PLA to enable use in high temperature applications.

Prior to producing the test specimen, the stereo complex formation processing temperature was in the range 170°C–200°C. The test specimen was produced using mini injection molding machine or press technique. The stereocomplex PLA were characterized by thermal investigations (DSC), which provides initial information about the melt temperature, glass transition temperature, crystalline peak temperature, and melt enthalpy. Afterwards, the thermo-mechanical (DMA and HDT) properties, morphological investigations, such as SEM, and polarization microscopy and the macrostructures and crystal structures of the samples were determined.

1.4.1 Differential Scanning Calorimetry

In order to investigate the specific material properties and the thermal history, such as the glass transition temperature, melting temperature, enthalpy of fusion, and crystallization process of raw materials as well as product are measured by DSC methods. As shown in Fig. 1.41, the samples are usually heated twice at a defined heating rate (10°K/min) above t he melting point to 250°C and cooled once. As can be seen in Fig. 1.41 (left), the glass transition of PLA3001 D is found at about 64°C and the melt temperature is around 170°C. On the other hand, the glass transition of Lapol HDT was found at about 44°C in cycle 3 and the melt temperature is around 166°C.

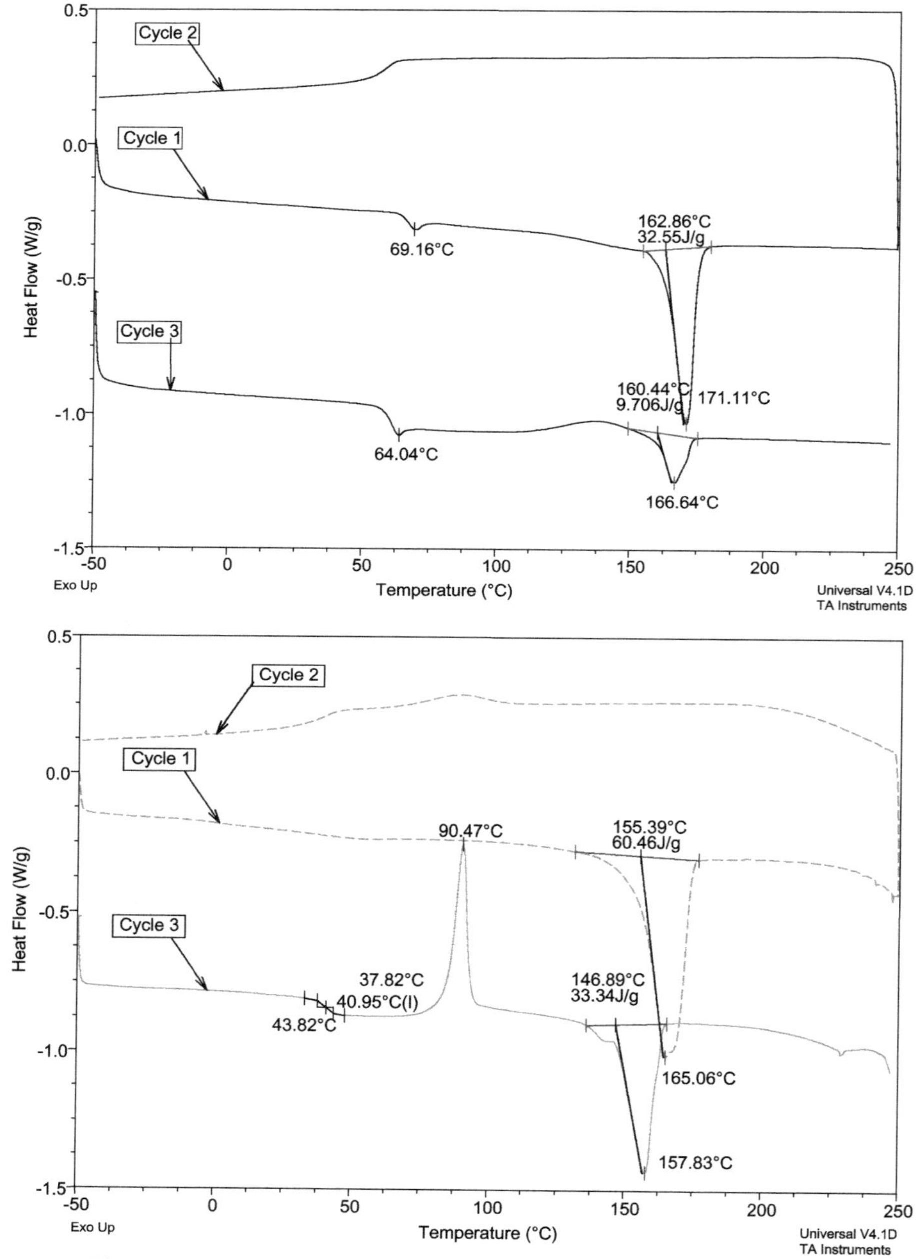

Figure 1.41 DSC diagram of PLA 3001 D (top) and Lapol HDT (bottom).

For Lapol HDT, no glass transition zone was found in the first cycle. Moreover, an exothermic cold crystallization peak was found at 90°C for Lapol HDT in the cycle 3. The temperatures above its glass transition temperature crystallize the molecular chains. For both PLA, the third cycle shows a decrease of the melt enthalpy and melting temperature. Both effects indicate chemical degradation processes.

In order to understand stereo-complex formation better, the first step, a corresponding mixture of 70/30 ratio PLLA and PDLA was mixed in lab extruder and the sample was investigated by DSC analysis. The heating and cooling DSC diagram of PLLA & PDLA mixture can be seen in Fig. 1.42. It can be seen that an exothermic crystallization peak appears for each heating cycle (cycle 1 & 3). This peak sharpness is fewer intensifiers than that of PDLA and it appears up at higher temperatures. The glass transition temperature became less significant and also flatter for cycle 3. In the cooling cycle (cycle 2), an exothermic peak is at about 152°C, which is not seen in any PLLA or PDLA DSC curves. It means that at this temperature some of the material crystallizes.

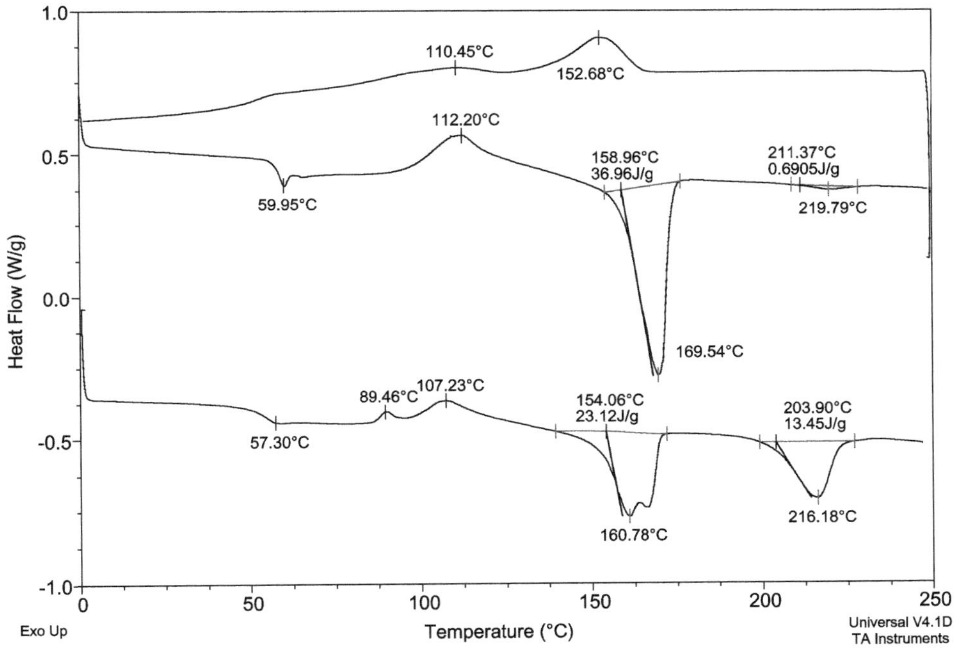

Figure 1.42 DSC diagram of mixture from 70% PLA 3001 D and 30% Lapol HDT.

In the cycle 3, two melting points appear at 160°C and 166°C. In comparison with the DSC diagrams of the pure material, the first peak belongs to PDLA and second peak belongs to PLLA. In the first heating cycle slightly and in the second heating cycle intensely appears an exothermic peak at about 216°C, which belongs to a mixture of sc-PLA [40].

Two different chains (PLLA and PDLA) are found to be combined and form the stereo crystals. This causes a large diffusion path and no regular chain folding that occurs in crystallization mechanisms [41].

With the previous knowledge, sc-PLA was produced using a twin-screw extruder as mixture ratio of 50/50 from PLLA and PDLA. The compounding temperature was in the range 190°C to 220°C, which were very close to the melting temperature of sc-PLA.

Figure 1.43 shows the DSC curve of the extruded samples. It is seen that the glass transition temperatures disappeared for heating and cooling cycles. Moreover, dual melting peaks were recorded at over 200°C, which is higher than that of pure semicrystalline PLA. Therefore, it can be said that it is a mixture of stereo complex PLA and no pure PLA is remaining. Different combinations of the PLA chains occurred and formed a mixture of stereoblocks and stereocomplexes PLA, which shows different melting temperatures. This behavior has been mentioned in several research works. Melt recrystallization process of sc-PLA is formed through two crystal combination, such as α and β or α and α [41, 42].

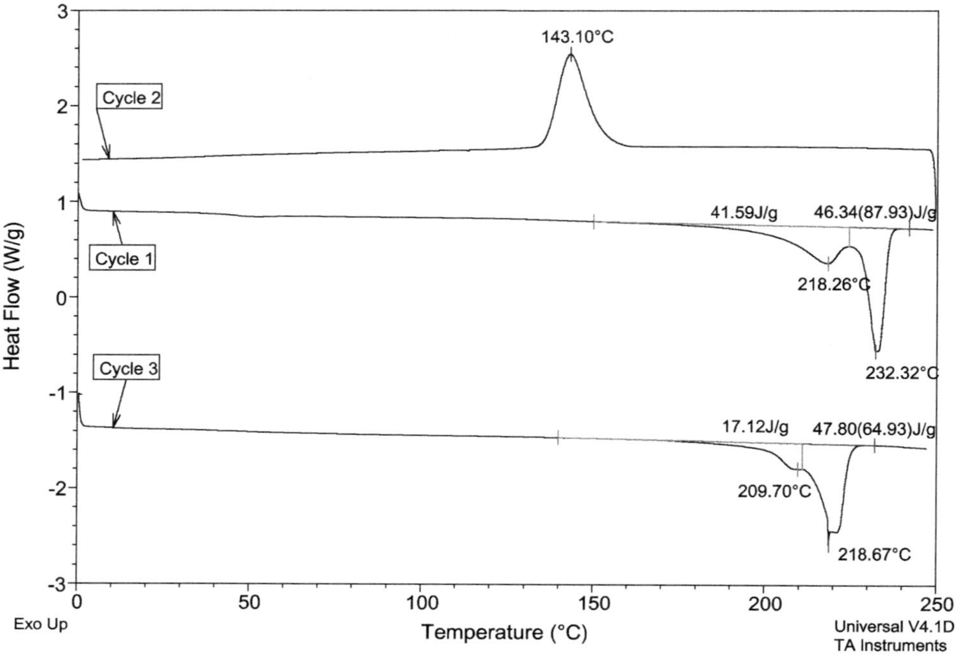

Figure 1.43 DSC diagram of mixture from 50% PLA 3001 D and 50% Lapol HDT.

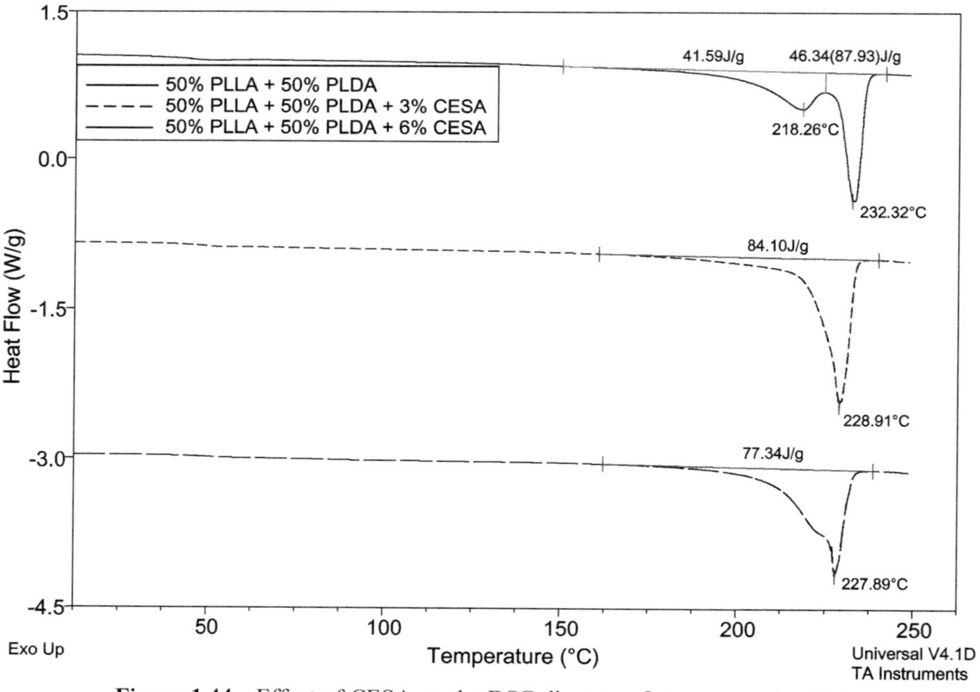

Figure 1.44 Effect of CESA on the DSC diagram of stereo complex PLA.

Taking into account the crystal structure of the stereo complex PLA, the next step is to produce homogeneous stereo complex PLA with improved properties using different additives. Figure 1.44 illustrates that by using CESA (a chain extender), not only does the glass transition temperature almost disappear, but there is also almost only one peak in the melting range. Furthermore, this diagram shows that melt enthalpy and melting temperature decreased by addition of 3% CESA. This change can be explained as follows by using CESA, a homogeneous mixture of stereo complex PLA is produced. This homogenization reduces the melting enthalpy and melting temperature of the entire material. However, uniform materials with only one melting peak can be seen in the middle curve of Fig. 1.44. As it is seen in the lower curves, increasing levels of CESA (6%) does not improve this effect. Therefore, the level of 3% CESA is considered for the next step.

The effect of Biomax Strong 120 (BS) as toughness modifier on stereo complex PLA was investigated. BS was added 10% to the stereo complex PLA compound. Figure 1.45 shows the DSC curves of stereo complex PLA, stereocomplex PLA with 10% BS (toughness modifier), and stereo complex PLA with 10% BS and 3% CESA.

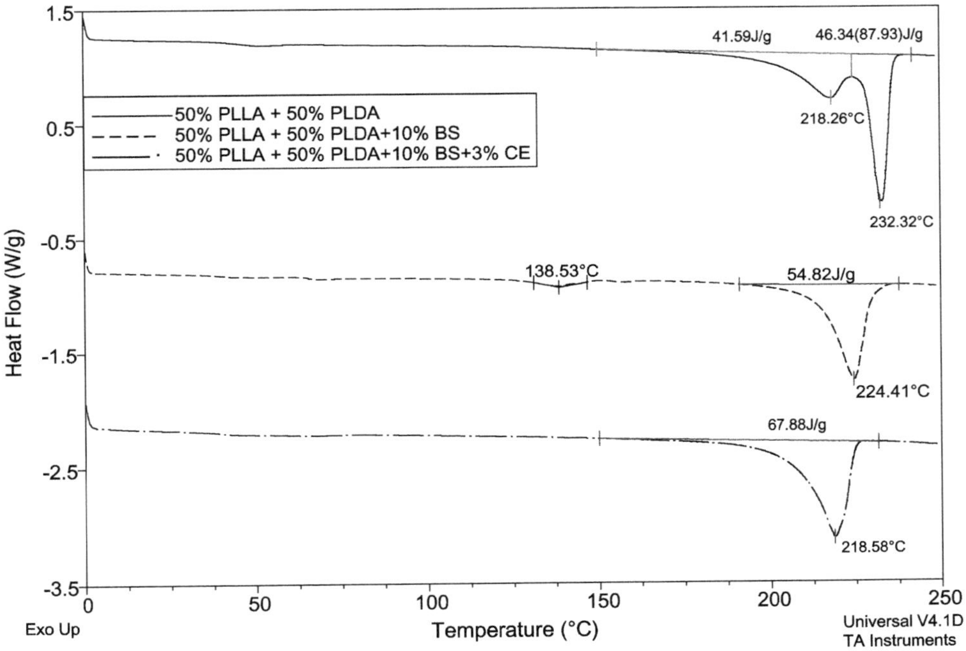

Figure 1.45 Effect of CESA and BS on the DSC diagram of stereo complex PLA.

Addition of the BS shows a peak at about 138°C in the DSC curve. This could be the melting temperature peak of toughness modifier (BS). When adding the toughness modifier, the melt enthalpy decreased by 30% and only one melting temperature appeared. It may be an effect of BS, which made materials more homogeneous. Further addition of 3% CESA can lead to a higher melt enthalpy and no glass transition temperature appeared. Due to the properties of the chain extender, a more homogeneous mixture of materials is produced. The melting peak temperature decreased 6°C.

Figure 1.46 compares the crystallization (cooling cycle of DSC diagram) process of stereo complex PLA with different PLLA. It can be seen that the crystallization peak temperatures of all batches are different. The highest crystallization peak temperature was observed for the stereo complex PLA made from PLA 3001 D, which is at 143°C. Due to the addition of chain extender (CESA), the crystallization peak temperature shifted about 3°C to the lower temperature range. It was further shifted about 18°C and 23°C to the lower temperature range for the stereo complex

PLA, made from PLA 3052 D and PLA 4042 D, respectively. Therefore, it means that different PLLA has significant effect on crystallization of stereo complex PLA.

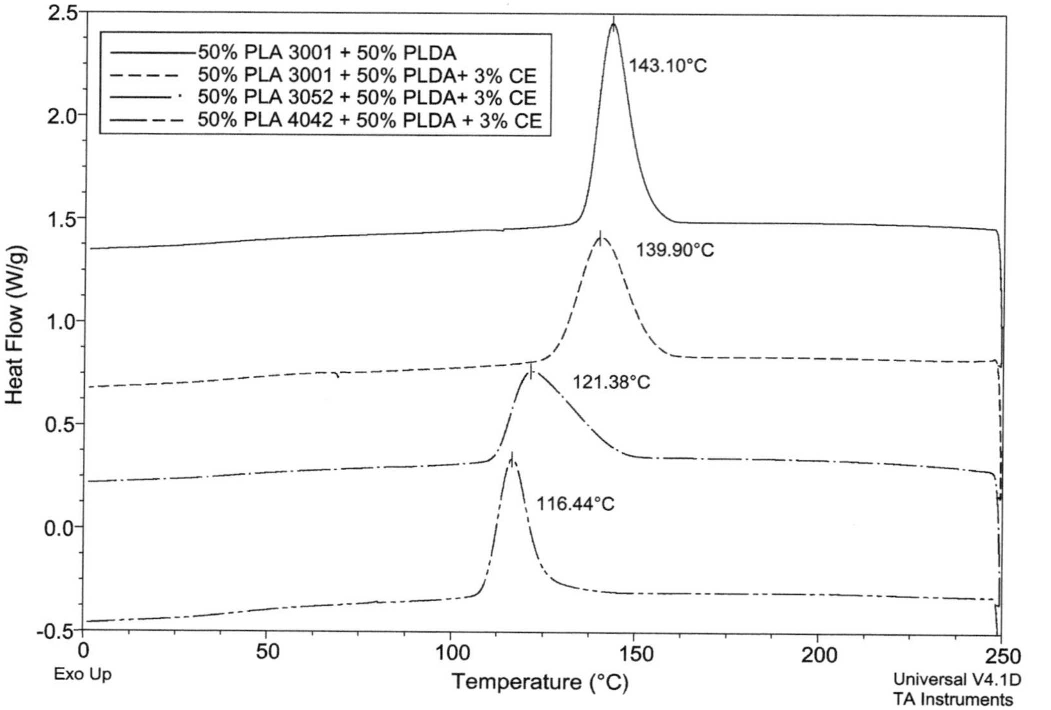

Figure 1.46 Effect of different PLLA on the crystallization of stereo complex PLAs.

The different PLLAs and PDLAs build different sc-PLAs, depending on their characteristics. A high-percentage stereocomplex consisting of PLLA and PDLA is dependent on the stereocomplex reaction between those components. It can be seen from the DSC diagrams that PLA3001 D has a better response with PDLA (Lapol-HDT) than other PLLAs.

1.4.2 Dynamic Mechanical Thermal Analysis

Here, the samples were tested for their heat distortion temperature (HDT) and their dynamic mechanical (DMA) properties.

PLA shows very low heat distortion temperature. It is very important to find ways to increase and improve the heat distortion temperature of the PLA. It has already been mentioned in various studies that there are several ways to improve the heat deflection temperature. One of the ways to improve the heat distortion resistance of a PLA is to increase the degree of crystallinity (by tempering or increasing the crystallization rate). The degree of crystallinity of PLA can be increased by adding PDLA, which acted as a nucleating agent, and thus suitable for the preparation of stereo blocks and stereo-complexes of PLA. According to DSC diagram, the sc-PLAs with or without additives (chain extender) were found to crystallize on cooling, which is barely visible in the DSC diagram of the pure PLA. This crystallization provides the materials very high heat deflection temperature. The heat deflection temperature (Method B) of stereo complex PLAs with different additives is seen in Fig. 1.47. The heat deflection temperature of stereo complex PLAs is 184°C, and due to addition of 3% CESA, the heat deflection temperature reduced to 166°C. It further reduced due to addition of 6% CESA. This can be explained by reduction of degree of crystallinity.

The heat deflection temperature dramatically reduced to 84°C due to addition of 10% BS, and by the further addition of 3% CESA with this mixture, this property improved. This can also be explained by the increase of degree of crystallinity.

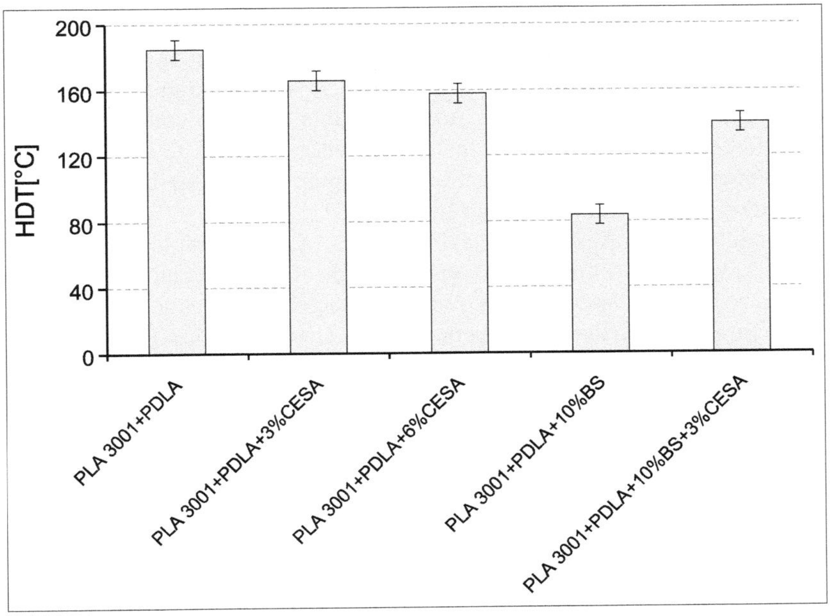

Figure 1.47 Heat deflection temperature of stereo complex PLAs.

The heat deflection temperature (Method B) of stereo complex PLAs from different PLLAs is seen in Fig. 1.48. The heat deflection temperature of stereo complex PLAs with 3% CESA from PLA 3001 is 166°C. This property for stereo complex PLAs with 3% CESA from PLA 3052 and from PLA 4042 are 104°C and 110°C, respectively. The prerequisite for achieving such a high proportion of stereo-complexes is the condition of having a high L content in PLA and also proportion of PDLA to the mixture. Therefore, it is obvious to achieve different results in the heat distortion temperature values by using different PLLAs.

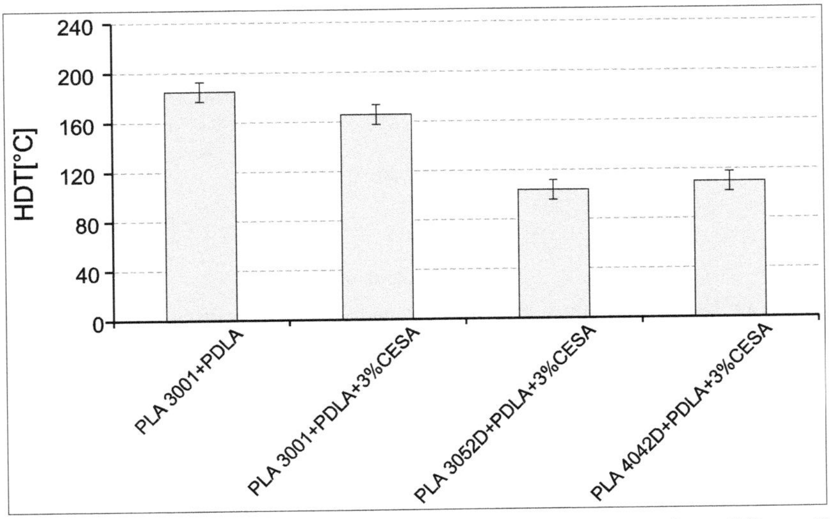

Figure 1.48 Heat heat deflection temperature of stereo complex PLAs from different PLLA.

The following diagrams compare the dynamic-mechanical properties of the different manufactured stereo complex PLAs and PLAs.

All batches of stereo-complex PLA show a very low loss modulus because of their high glass transition temperature compared to pure PLA.

Figure 1.49 shows a comparison among the storage modulus, loss modulus, and Tan δ of PLA 3001 and Stereo-complex PLA. The storage module of PLA is higher than that of Stereo-complex PLA. The storage module of PLA showed a dramatic drop at the glass transition temperature range. On the other hand, the storage module of stereo complex PLA showed slow downturns. In addition, the red lines show that it retained more than 30% of its maximum storage modulus at a temperature of 140°C. This means that pure PLA loses its mechanical properties after glass transition temperature faster than sc-PLA.

The loss modulus of PLA showed maximum loss of lose modulus in the glass transition region of the PLA, where the PLA loses its strength and softens. In contrast, in stereo complexes, a slight increase in the loss modulus occurred in the temperature range of 60–80°C. Therefore, DMA diagrams further show that the structure of the stereo-complex PLA is suitable for high service temperatures. This change in material behavior is based on a possible structural change.

Several literature sources have reported that the values of tensile strength, modulus of elasticity, and elongation of the stereo complex PLA are higher than the values of pure PLLA and PDLA [43, 44].

However, stereo-complex PLAs tend to show lower strength compared with the literature. It can be presumed that because of a higher degree of crystallinity and a tendency to nucleate the stereo-complex PLA, many micro cracks are formed in the specimen, thereby causing the low strength.

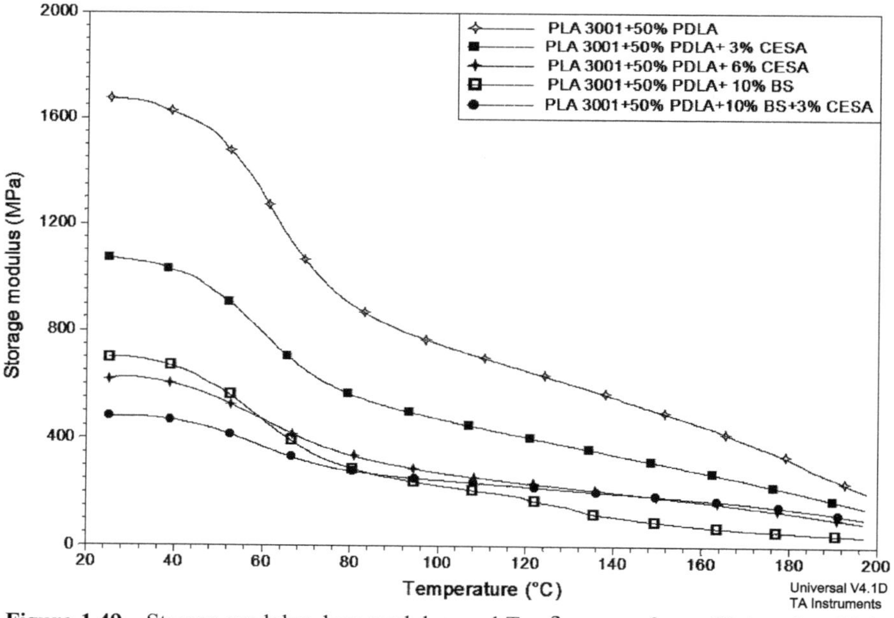

Figure 1.49 Storage modulus, loss modulus, and Tan δ curves of pure PLA and sc-PLA.

The storage modulus stereo-complex PLA from PLA 3001 with chain extender (CESA) and toughness modifier (BS) can be seen in Fig. 1.50. It was found that due to the addition of 3% CESA, the storage modulus reduced ca. 30% at the room temperature and this property reduced slowly with increasing temperature. This property reduced ca. 70% due to addition of 6% CESA. It can be explained that the degraded molecules cross linked each other and hindered stereo-complex formation.

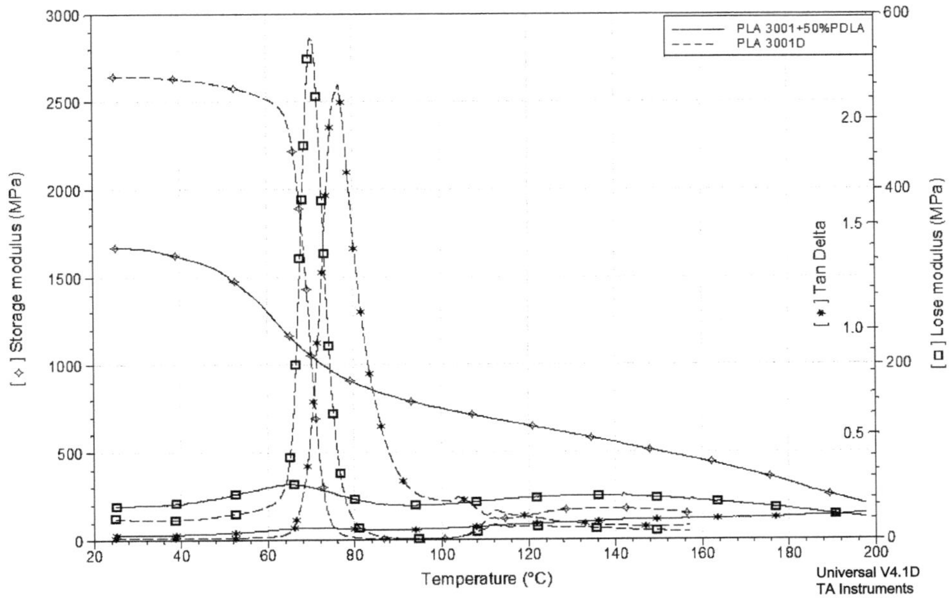

Figure 1.50 Storage modulus sc-PLA with chain extender and toughness modifier.

On the other hand, due to addition of 10% BS, the storage modulus reduced 65% in care with sc-PLA. It reduced sharply around glass transition temperature and then reduced slowly with the increasing temperature. Toughness modifier (BS) introduced a secondary phase within stereo-mixture, which can have a negative effect on crystallinity and homogeneity. This property reduced further due to addition of CESA to the above mixture at the room temperature but after glass transition temperature, this property found better than that of without CESA.

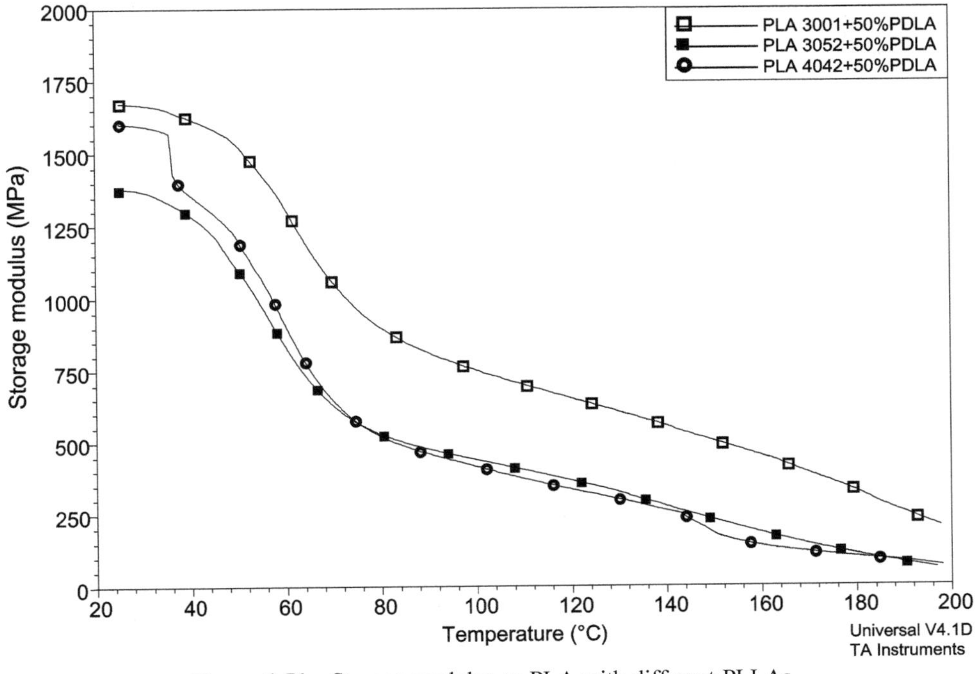

Figure 1.51 Storage modulus sc-PLA with different PLLAs.

To get the best mixture of stereo complex PLA, different PLLAs were used. Figure 1.51 compares the history of the storage modulus for differently prepared stereo-complex PLAs. PLA 3001 stereo-complexes showed highest storage modulus at all temperatures. The stereo complexes of PLA 3052 and PLA 4042 showed nearly similar storage modulus at all temperatures. As can be clearly seen in the diagram, all batches of the stereo-complex PLA have a better progression than commodity polymer. Therefore, it is still possible to heat these materials up to 200°C, which is unlikely to be higher for commodity polymer because of the low melting temperature.

1.4.3 Microscopy

In this step, morphological investigations were carried out with the aid of scanning electron microscopy to detect the stereo complex compounds at the fracture surface. Polar microscopy was also used to monitor crystallization in the stereo complex PLA while solidification occurred.

1.4.3.1 Scanning Electron Microscopy

The following pictures show the scanning electron micrograph of the fracture surface of the stereo-complexes PLA from PLA 3001 to determine the specific properties. Figure 1.52 shows fracture surfaces of stereo-complexes PLA in two different positions. The mixture of the stereo complex PLA is homogeneous, but in both pictures the cracks, holes, and cavities are clearly visible. The right picture shows spherical particles. The fracture surface indicates a brittle fracture.

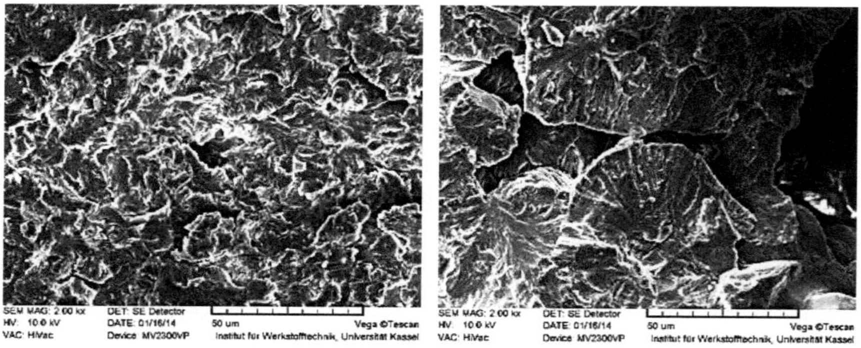

Figure 1.52 SEM of stereo-complexes PLA from PLA 3001.

As is mentioned previously, CSEA improved the melting properties of stereo-complexes PLA. It is shown in Fig. 1.53 that the homogeneity of stereo-complexes PLA was found to increase but many cracks, spherical particles, and phase separation remained on the surface. Those are significant reasons for the occurrence of brittle fracture.

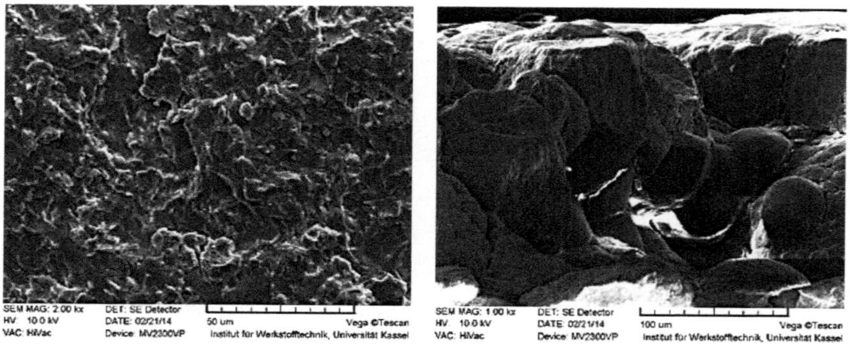

Figure 1.53 SEM of stereo-complexes PLA from PLA 3001 with chain extender (CESA).

Figure 1.54 presents the fracture surface of stereo-complex PLA with 10% toughness modifier (BS). The toughness modifier (BS) was found to be well distributed, but there was no homogeneous mixture. The immiscibility of BS components is a reason for micro cracking during fracture and consequently lowering mechanical properties and crystallinity. Part of the toughness modifier pulled out from material. This is due to the weak adhesion between toughness modifier and the stereo complex PLA.

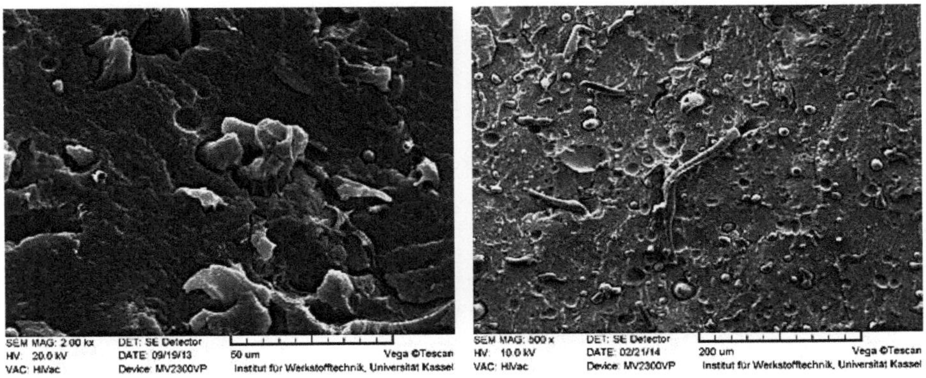

Figure 1.54 SEM of stereo-complexes PLA from PLA 3001 with toughness modifier (BS).

Figure 1.55 presents SEM of stereo-complex PLA with 10% toughness modifier (BS) and 3% CESA. By adding CESA, the component becomes more homogeneous and the adhesion between toughness modifier and the stereo complex PLA improves significantly. The cracks between toughness modifier and the stereo-complex PLA became fewer and few spherical particles were detected. That is why the improvement of mechanical properties and increase of crystallinity occurred.

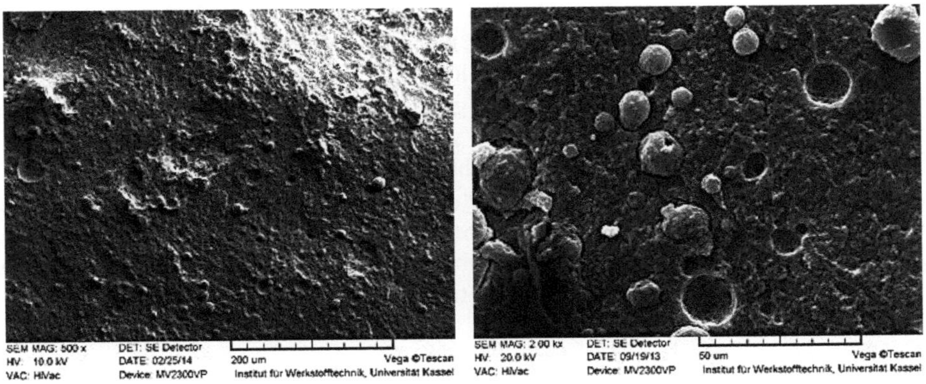

Figure 1.55 SEM of stereo-complexes PLA with toughness modifier (BS) and CESA.

Figure 1.56 shows SEM of stereo-complexes PLA from PLA 3052 with CESA. There are many visible spherical particles, holes, and micro cracks detected. The broken sphere showed a homogeneous mixture and good adhesion between spheres.

Figure 1.57 shows SEM of stereo-complexes PLA from PLA 4042 with CESA. There are many visible spherical particles, holes, and micro cracks detected. The fracture surface and broken sphere showed a homogeneous mixture and weak adhesion between spheres.

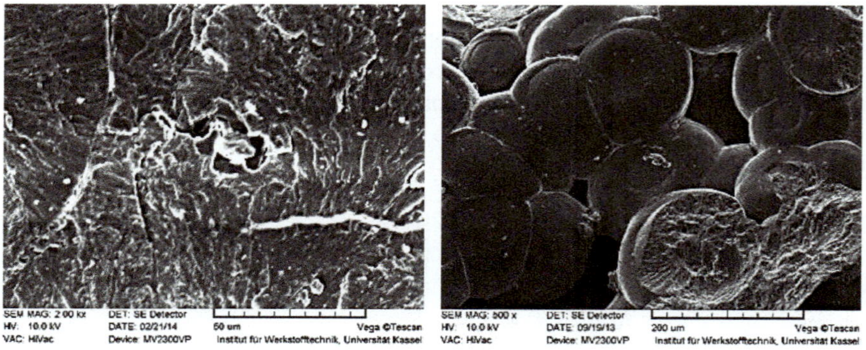

Figure 1.56 SEM of stereo-complexes PLA from PLA 3052 with CESA.

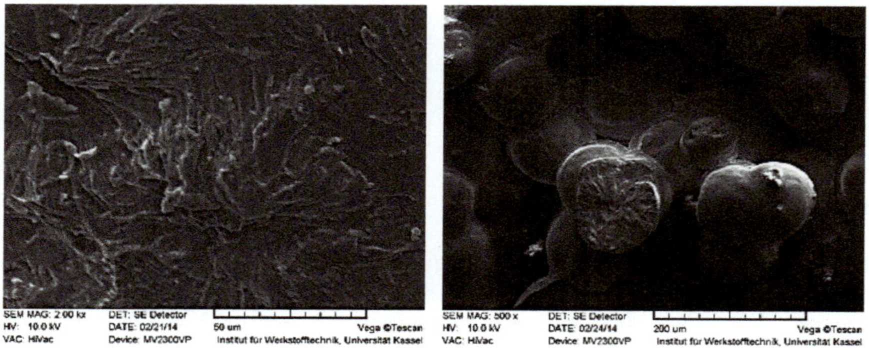

Figure 1.57 SEM of stereo-complexes PLA from PLA 3052 with CESA.

1.4.3.2 Polar Microscopy

By means of polar microscopy, the crystallization of the stereo-complex PLA was observed. Here, nucleation of the crystal upon solidification of the melt was recorded. Figure 1.58 shows the spherulites of the stereo-complex PLA, to which no additives have been added. The left image presents the Spherolite 2 minutes after the first germination. After a short time, the germination was completed and overlapping of spherolites was observed. There are also some defects of crystal found on the surface. As is shown in the DSC curves, this stereo-complex PLA has a crystallization peak (150°C) during the cooling phase. The size of the crystals is remarkably large, which indicates an increase in residual stresses and cracks between the crystals, which are consequentially responsible for low dynamical-mechanical properties. The crystal sphere size is about 100–150 μ. Different additives were used to overcome this situation.

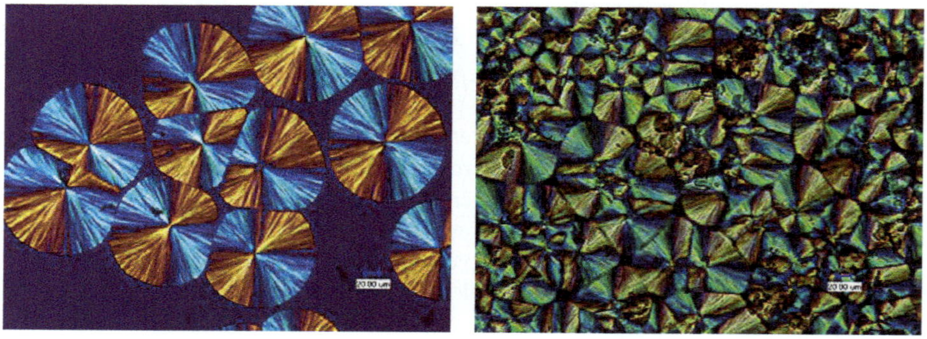

Figure 1.58 Polar micrograph of stereo-complex PLA from PLA 3001.

Figure 1.59 presents that there are huge numbers of nucleation in the melt and the growth kinetic of spherulites is also very fast by the addition of chain extender (CESA). However, the growth kinetic of spherulites also hindered each other. The crystal sphere size is about 20–30 μ. Crystals take the form of lamellae and many of the spherulites are cut off-center, in which the presence of static heterogeneities (impurities or molecular defects and mass polydispersity in polymeric materials) lead to a rejection of these components from the growth front to form channels similar to those found in eutectics. Therefore, the huge number of crystal spherulites could improve their properties and miscibility.

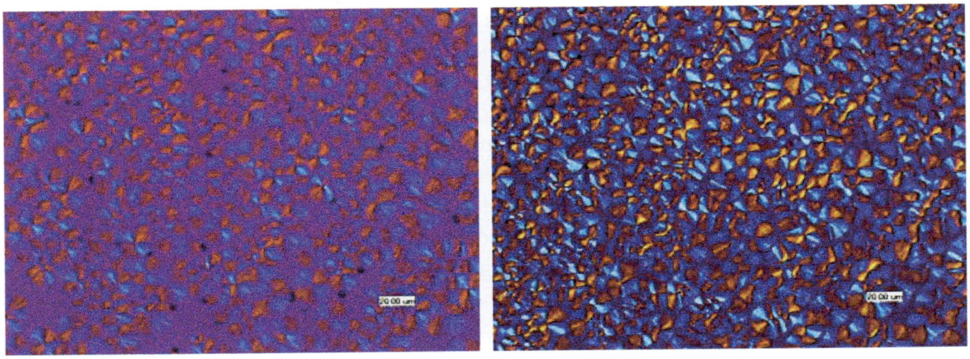

Figure 1.59 Polar micrograph of stereo-complex PLA with CESA.

The influence toughness modifier on spherulites formation and growth crystal can be seen in Fig. 1.60. The number and size of spherulites was also investigated. The toughness modifier had the same effect on the change in the structure of the spherolites. The spherulites remain small, but the growth rate was slower. The crystal sphere size is about 5–10 μ. In general, the high number and the relatively small size of spherulite provide improved mechanical and thermal properties. However, in this case, a huge number of crystals overlapped each other, and many defects were observed.

Figure 1.60 Polar micrograph of stereo-complex PLA with CESA and BS.

1.5 LIFE CYCLE ASSESSMENT OF PLA

The perceptions and preferences of consumers are changing the focus to sustainability. Therefore, environmental assessment tools are being used to evaluate the environmental footprint of systems and products. The environmental footprint is a quantitative measurement, unfolding how human activities can inflict different impacts on global sustainability, considering the environmental, social, and economic indicators [45]. Life cycle assessment can be charity to assess the environmental footprint of PLA. LCA is a method to evaluate the environmental performance of products and/or potential impacts of a system, including raw materials acquisition, production, use, and end life or disposal [46]. The cradle-to-grave approach will be used for LCA estimation [46].

LCA methodically evaluates each of the life stages of the product system, in which environmental inputs (resources) and environmental outputs (emission and waste) are produced and the impacts to human health and environment are calculated. LCA results are interpreted in relation to the objectives of the study [47]. The environmental footprint of PLA resins and/or PLA products can be evaluated using midpoint impact categories. Additionally, measuring key indicators, such as greenhouse gases emissions and non-renewable energy use, and comparing the data between PLA and commodity polymers (e.g., PET, PP, and PS) can give insights about PLA environmental performance. In 2003, NatureWorks LLC published the first cradle-to-grave lifecycle inventory data for its PLA (Ingeo™) based on the 140,000 tons/year plant design, in which they provided some information regarding the production technology [48]. In the year 2010, NatureWorks LLC published an updated eco profile based on the production technology improvements and also benchmarked the results for energy requirements and CO_2 emissions with data for a selection of petrochemical based polymers [49]. In the year 2015, the company published an updated PLA eco profile, providing a detailed description of the production of PLA resin (150,000 tons/year plant) and focused on the corn feedstock used to produce Ingeo™ and on the PLA intrinsic zero material carbon footprint [50].

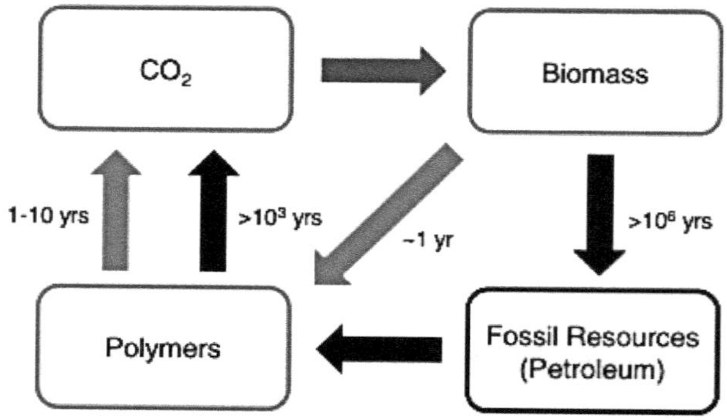

Figure 1.61 Carbon cycle of petroleum-based polymer and bio-based polymers [51].

Carbon cycle history of petroleum-based polymer and bio-based polymers can be seen in Fig. 1.61. Polymers based on biomass require 1 to 10 years to convert to CO_2 and require 1 year converting biomass to polymer. Therefore, it can be said there is zero carbon footprint in terms of raw materials. Another way is that the carbon footprint reduction occurs from the material itself and not from the process of converting the feedstock to products [51]. Biodegradable polymers made from bio-based materials, such as corn and corn starch, can be produced and converted into biomass in similar time frames. Fossil resources could be considered renewable, but it takes more than a million years for biomass to be converted into fossil fuels. Since the rate of consumption is much greater than the rate of replacement, a mass imbalance occurs in the carbon cycle.

Figure 1.62 shows the global warming potential (GWP), primary energy from non-renewable resources, such as oil, gas, coal, and uranium, and water uptake for 1 kg of Ingeo™ PLA resin [52]. One of the main value propositions for using PLA to replace other fossil-based polymers, is the lower GWP due to carbon sequestration during the corn-growing stage. Several authors have done LCAs regarding the performance of PLA in comparison with other materials, such as PET and PS for different applications, in which PLA could be a good substitute for clamshell containers, trays, and water bottles [53].

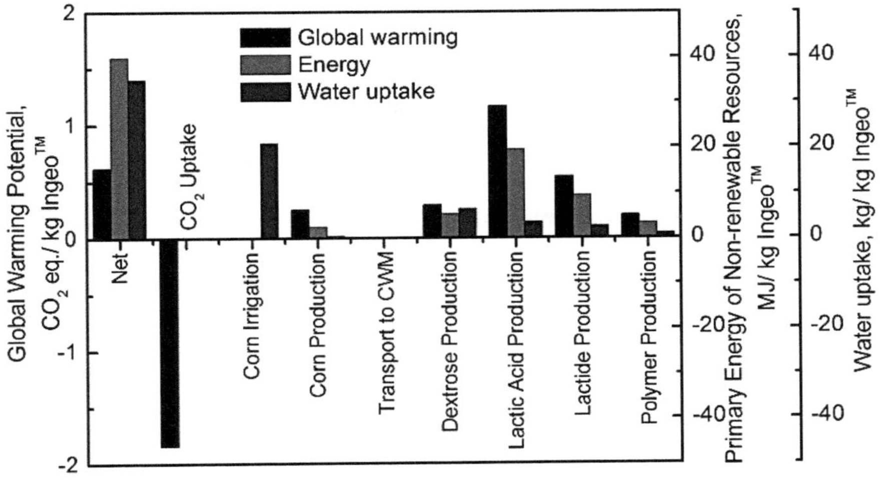

Figure 1.62 Global warming potential and primary energy and water requirements for Ingeo PLA resin [52].

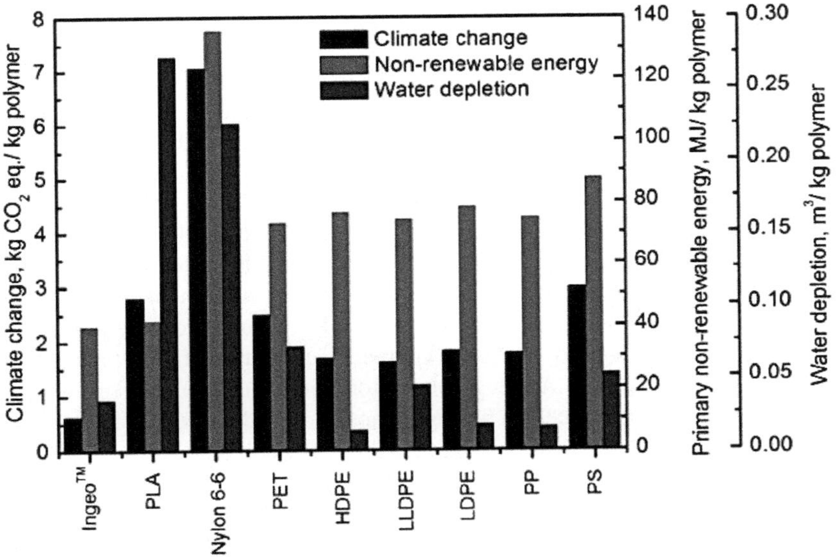

Figure 1.63 Greenhouse gas emission, nonrenewable energy, water depletion for 1 kg of PLA, and other commercial polymers.

Figure 1.63 shows general information about the EFP of PLA in comparison with other polymers. Such a comparison is effective only if: a) polymer weights in the studied applications are quite similar; b) contributions to impact categories are dominated by the polymer-pellet production; c) energy requirements for converting the polymer into product are relatively small or relatively similar; d) use phase is similar; e) the same recycling or end-of-life routes are employed; f) the same level of detail in the lifecycle inventory data-collection process was used; g) the same LCA methodology was used; h) the same database for upstream inventory data was used; and i) the same life cycle impact assessment methodology, indicators, and characterization factors were used [52].Given the above, the climate change of Ingeo™ and PLA have large differences since carbon sequestration has not been accounted for in PLA. If this factor is taken into consideration, a global warming potential of PLA would be 0.9 kg CO_2 eq per kg of polymer versus 0.62 kg

CO_2 eq per kg of resin for Ingeo™. Thus, a large benefit is obtained using the new reported data for Ingeo™ [52]. In the case of non-renewable energy, similar values are reported by Vink and Davies, and the current data are available in Ecoinvent 3.2. In the case of water depletion, the new values reported by Vink and Davies make sure to properly account for water uptake from river and ground for the Blair manufacturing plant, excluding the water for hydropower installations and rainwater. So, much lower water EFP is reported for Ingeo™, provided that water consumption is similar to that for polyolefins.

PLA can be derived from renewable resources, such as regular crops, plant-based materials, and biomass waste, and it could be treated in all levels of the waste management hierarchy. In this regard, however, there are still limitations due to the lack of suitable infrastructure for sorting, recycling, and/or composting PLA products at their end of life. So, efforts should be centered on working with industries, commodity groups, industry associations, and government groups to improve the recovery rate of PLA. Finally, life cycle assessment has been used to evaluate the environmental footprint of PLA, providing useful information about the environmental impacts that PLA may have during raw material acquisition, production, use, and disposal. Robust data exists about the PLA resin production from one producer, NatureWorks LLC. However, information is missing regarding the use and end-of-life scenarios of PLA parts.

1.6 APPLICATIONS OF PLA

Among bio-based totally biodegradable polymers, poly(lactic acid) (PLA) has been studied for use in different fields because of its compostability and renewability. The use of life cycle assessment principles helps to quantify the environmental benefits of PLA polymers. Most recent developments with PLA in the field of packaging show how this plastic material is moving from commodity to specialty applications, facing competition from polyolefins, particularly as barrier polymers for shelf-life enhancement. However, after successful completion of this project, modified PLA (chain extender and toughness modifier) is used for some industrial applications in medical equipment cover, medical accessory, shop and office equipment cover, drill bits holder, and office accessory. The applications of modified PLA are given below step-wise.

Figure 1.64 shows housing for perfusor compact automatic infusion system. This part is produced from impact modified PLA using conventional injection molding machine at B. Braun, Germany. Multi parts were produced in a single shot and production yield and cycle time was same as the standard part. The original part is made of ABS polymer.

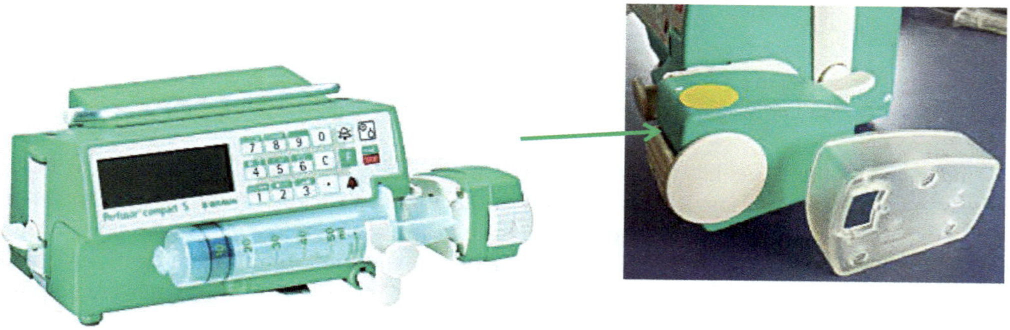

Figure 1.64 Medical equipment (Perfusor compact S) casing/housing.

Roller clamps and patient connectors (Fig. 1.65) were produced with a sample tool (4-fold) with chain extender and toughness modified PLA at the company B. Braun, Germany. The PLA showed a good demolding and same cycle time. The original part is made of HIPS polymer.

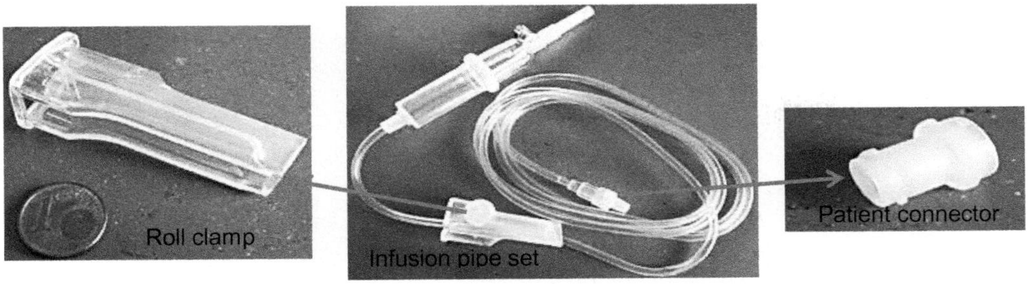

Figure 1.65 Accessory (roll clamp and connector) for infusion set.

Cover and housing (Fig. 1.66) of medical and cashier equipment were produced at the Wincor Nixdorf, Germany. These parts are made from fire retardant and toughness modified PLA. The cycle time of this part production was higher than the standard part.

Figure 1.66 Cover of medical and cashier equipment.

Figure 1.67 Office accessory (storage / utility box).

Boxes were produced using standard injection molding mahines at the Institute of Material Science (IfW), University of Kassel. This box was (Fig. 1.67) produced from PLA mixed with toughness modifier and bio-based colour maserbatch. The original box is made of PP.

Round tube plugs/stopper, cable guides (Fig. 1.68), and drill bits holders (Fig. 1.69) and covers were produced using standard injection molding machine with a serial tools (4-fold). Those parts were produced at OL Plastik, Germany from toughness modified PLA.

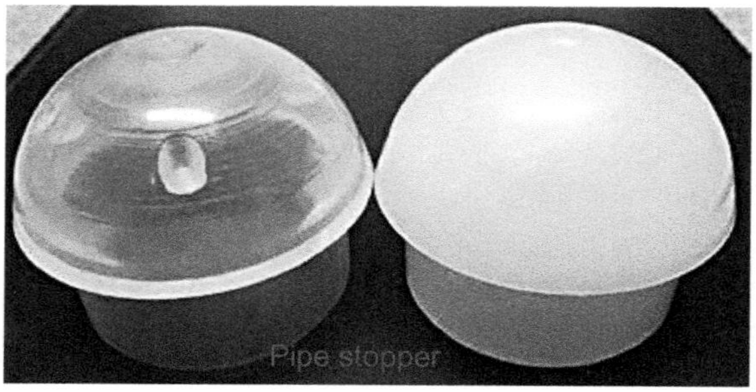

Figure 1.68 Fins or pipe stopper and cable guider.

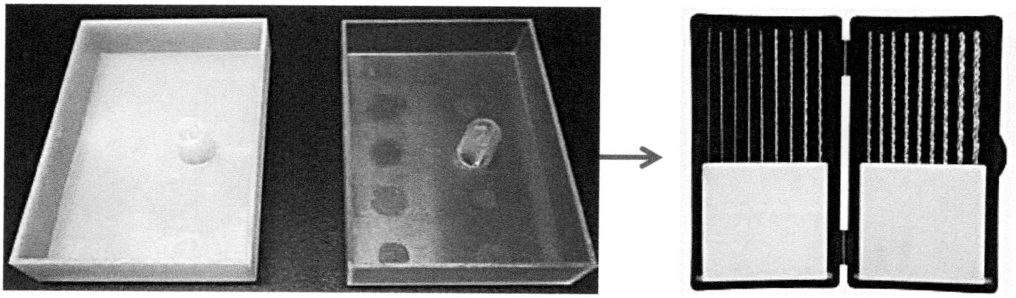

Figure 1.69 Storage box and holder for drill bits.

1.7 CONCLUSION

After characterization it can be said that the chain extender has a significant influence on the molecular weight, the melt flow behavior, and the injection pressure of molding. It can improve the thermal degradability of PLA. The catalysts have a significant influence on the reactivity of CESA. The catalysts improve the reaction kinetics and result in a higher molecular weight level in a short (2–3 minutes) time. The catalysts also have a significant influence on the impact strength properties.

The impact modifiers have significant effects on impact strength, energy absorption, and also on tensile strength and elongation properties. Toughness modifier (BS) shows a remarkable increase in impact strength, energy absorption, and tensile elongation at break. The increase in elongation at break and impact strength is linearly dependent on the amount of addition of BS.

The flame retardant was found to improve the flammability of PLA.

Toughness modified PLA shows a positive resistance to ESCR. The mechanical properties of modified PLA decreased by about 10 to 20% through 100 kGy gamma radiations.

The stereo-complex formation of PLA has a significant influence on the heat distortion resistance of PLA and showed very high heat resistance (HDT B). The properties of the stereo complex PLA are dependent on the components PLLA and PDLA. These can be used to produce different PLAs with their specific properties. The pictures of polarization microscopy showed a very high degree of crystallinity in all stereo complex PLAs. The additives may have influence on the size of the spherulites.

References

1. Ro, A.J., S.J. Huang and R.A. Weiss. Synthesis and properties of random poly(lactic acid)-based ionomers. Polymer 2009; 50(5): 1134-1143.

2. Hashima, K., S. Nishitsuji and T. Inoue. Structure-properties of super-tough PLA alloy with excellent heat resistance. Polymer 2010; 51: 3934-3939.
3. Ren, J. (ed.). Biodegradable Poly (Lactid Acid): Synthesis, Modification, Processing and Applications. Springer Verlag, Berlin, 2010; pp. 15-25.
4. Tingting, L. Structure Property Relationship of Crystalline Poly(lactic acid) S: DFT/DFPT Studies and Applications. Doktorarbeit, Sangapore, 2011.
5. Endres, H.J. Biopolymere als Energieträger. KunststoffeZeitung, U.A. 8/2010; S.103
6. Endres, H.J. and A. Siebert-Raths. Technische Biopolymere. Carl Hanser, Verlag, München, 2009, S.5-7 and S.103-108.
7. Wilhelm, H. and K. Reitinger. Nachwachsende Biopolymere als Substitution für Massenkunststoffe Wien. Bundesministerium für Verkehr, Innovation und Technologie, Juli 2003.
8. Wolfgang, B. Biologisch abbaubare Kunststoffe. Umweltbundesamt Pressestelle, 2009.
9. N.N., Global production capacities of bioplastics, Bioplastics market data, European Bioplastic Market, http://www.european-bioplastics.org/market.
10. Xiao, L., B. Wang, G. Yang and M. Gauthier. poly (lactid acid)-based-biopolymer: synthesis, modification and applications. *In*: D.N. Ghista (ed.). Biomedical Science, Engineering and Technology, Intech-Verlag, Canada, 2012, S.247-262.
11. Deligio, T. Cellulose derived nanocrstals PLA bioplastic nanocomposites. Materials, Plastics Today, 2011.
12. Auras, R., L.T. Lim, S.E.M. Selke and H. Tsuji (eds). Poly(Lactic Acid): Synthesis, Structures, Properties, Processing, and Applications. John Wiley & Sons, Verlag, New Jersey, 2010, S.59-65.
13. Blasius, J., W. George, G.A. Deeter and M.A. Villalobos. United State Patent nr. 6984694, January 10, 2006.
14. Nishii, M. and K. Hayashi. Solid state polymerization. Annuel Review of Materials Science 1975; 5: 135-149.
15. Scherzer, D., A. Eipper, C. Weiss, M. Yamamoto, G. Skupin and U. Witt. World Patent nr. WO 2006087346, 2006; 24.
16. Scholtens, B.J.R. and T. Loontjens. A new chain extender for high molecular weight polyester and polyamides. Polyester 2000: 5th World Congress, Zurich, Switzerland: Maack Business Service.
17. Frenz, V., D. Scherzer, M. Villalobos, A.A. Awojulu, M. Edison and R. Meer. Multifunctional polymers as chain extenders and compatibilizers for polycondensates and biopolymers. Society of Plastics Engineers Annual Technical Conference, Wisconsin USA, 2008, 1682-1687.
18. Mamun, A.A. and H-P. Heim, Journal of Biobased Materials and Bioenergy 2014; 8: 1
19. Lower, E.S. Zinc stearate: its properties and uses. Pigment and Resin Technology 1982; 11(6): 9-14. https://doi.org/10.1108/eb041810.
20. Heinz, W.E. and E.T. Smith, Stabilization of oxymethylene polymers. Nr-3488303, 1970, USA.
21. Blank, W.J., Z.A. He and M. Picci. Catalysis of the epoxy-carboxyl reaction. Proceedings International Waterborne, High Solids and Powder Coating Symposium. 2003, New Orleans, USA.
22. Song, X., X. Zhang, H. Wang, F. Liu, S. Yu and S. Liu. Methanolysis of poly lactic acid (PLA) catalyzed by ionic liquids. Polymer Degradation and Stability 2013; 98: 2760.
23. Ranninger, M.C.N., M.G. Andrade and M.A.A. Franco. Thermal decomposition of some imidazole and N-methyl substituted imidazole complexes of palladium(II). Journal of Thermal Analysis 2005; 14: 281.
24. Neiman, M.B., B.M. Kovarskaya, L.I. Golubenkova, A.S. Strezhkova, I.I. Levantovskaya and M.S. Akutin. The thermal degradation of some epoxy resins. Journal of Polymer Science 1962; 56: 383.
25. Heise, M.S., G.C. Martin and J.T. Gotro. Analysis of gel formation in imidazole-catalyzed epoxies. Polymer Engineering and Science 2004; 32: 529-534.
26. Baghernejad, B. 1,4-Diazabicyclo[2.2.2]octane (DABCO) as a useful catalyst in organic synthesis. European Journal of Chemistry 2010; 1: 54.
27. Liu, H., W. Song, F. Chen, L. Guo and J. Zhang. Interaction of microstructure and interfacial adhesion on impact performance of polylactide (PLA) ternary blends. Macromolecules 2011; 44: 1513.
28. Zhao, L., Y. Liu, Z. Wang , J. Li, W. Liu and Z. Chen. Synthesis and degradable property of novel sulfite-containing cycloaliphatic epoxy resins. Polymer Degradation and Stability 2013; 98: 2125.

29. Dadbin, S. and F. Naimian. Gamma radiation induced property modification of poly(lactic acid)/ hydroxyapatite bionanocomposites. Polymer International 2014; 63: 1063.
30. Zaman, H.U., M.A. Khan and R.A. Khan. Comparative experimental measurements of jute fiber/ polypropylene and coir fiber/polypropylene composites as ionizing radiation. Polymer Composites 2012, 33: 1077.
31. Villalobos, M., A. Awojulu, T. Greeley, G. Turco and G. Deeter. Oligomeric chain extenders for economic reprocessing and recycling of condensation plastics. Energy 2006; 31: 3227-3233.
32. Pilla, S., S.G. Kim, G.K. Auer, S. Gong and C.B. Park. Microcellular extrusion-foaming of polylactide with chain-extender. Polymer Engineering and Science 2009; 49: 1653-1660.
33. Bledzki, A.K., A. Reis, D. Passman and H-P. Heim. Influence of the compression molding process on impact bevavior. ANTEC 2010; 369-373.
34. Bledzki, A.K., A.A. Mamun and O. Faruk. Abaca fibre reinforced PP composites and comparison with jute and flax fibre PP composites. eEXPRESS Polymer Letters 2007; 1(11): 755-762
35. Bledzki, A.K., A.A. Mamun and J. Volk. Barley Husk and Coconut shell reinforced polypropylene composites: The effect of fibre physical, chemical and surface properties. Composite Science and Technology 2010; 70: 840-846.
36. Lu, Y., Y.C. Chen and P.H. Zhang. Preparation and characterosation of PLA/PCL composites micrpfibre membranes. Fibres & Textiles in Eastern Europe 2016; 3(117): 17-25.
37. Silverajah, V.S.G., N.A. Ibrahim, N. Zainuddin, Z.W. Yunus and H.A. Hassan. Mechanical, thermal and morphological properties of PLA/epoxidized palm olein blend. Molecules 2012; 17(10): 11729-11747.
38. Emad, A.J.A., H.S. Adeel and A.A. Saadon. New biopolymer nanoblends based on epoxidized soybean oil plasticized poly (lactic acid)/fatty nitrogen compounds modified clay: Preparation and characterization. Industrial Crops and Products 2011; 33(1): 23-29.
39. Ibrahim, N.A., W.M.Z. Wan Yunus, M. Othman and K. Abdan. Effect of chemical surface treatment on the mechanical properties of reinforced plasticized poly(lactic acid) biodegradable composites. Journal of Reinforced Plastics and Composites 2011; 30: 381-388.
40. Saeidlou, S., M.A. Huneault, H. Li and C.B. Park. Effect of nucleation and plasticization on the stereocomplex formation between enantiomeric poly(lactic acid)s. Polymer 2013; S0032-3861(13): 00814-8.
41. Saeidlou S., M.A. Huneault, H. Li, P. Sammut and C.B. Park. Evidence of a dual network/spherulitic crystalline morphology in PLA Stereocomplexes. Polymer 2012; 53: 5816e5824.
42. Vos, S.D. and A.S. Raths. Progress in processing of PDLA-nucleated PLA compounds. European Bioplastics Conference, Düsseldorf, December 2010.
43. Raths, A.S. H.J. Endres and M. Nelles. Modifizierung von Biopolymeren fürtechnische Anwendungen, speziell im Automobilbereich, Berlin, Juni 2010.
44. Yu, L. Biodegradable Polymer Blends and Composites from Renewable Resources. John Wiley & Sons-Verlag, New Jersey, 2009, S.165-190.
45. Čuček, L., J.J. Klemeš and Z. Kravanja. Areview of foot print analysis tools for monitoring impacts on sustainability. Journal of Cleaner Production 2012; 34: 9-20.
46. Potting, J. How to approach the assessment? *In*: M.A. Curran (ed.). Life Cycle Assessment Handbook: a Guide for Environmentally Sustainable Products. John Wiley & Sons, Inc., Cincinnati, OH, USA, 2012, pp. 391-412.
47. Weitz, K.A. Life cycle assessment and end of life materials management. *In*: M.A. Curran (ed.), Life Cycle Assessment Handbook: a Guide for Environmentally Sustainable Products, John Wiley & Sons, Inc., Cincinnati, OH, USA, 2012, pp. 249-266.
48. Vink, E.T., K.R. Rabago, D.A. Glassner and P.R. Gruber. Applications of life cycle assessment to NatureWorks™ polylactide (PLA) production. Polymer Degradation and Stability 2003; 80: 403-419.
49. Vink, E.T., S. Davies and J.J. Kolstad. The eco-profile for current Ingeo® polylactide production. Industrial Biotechnology 2010; 6: 212-224.
50. Narayan, R. Carbon footprint of bio plastics using bio carbon content analysis and life cycle assessment. Materials Research Society Bulletin 2011; 36: 716-721.

51. Castro-Aguirre, E., F. Iñiguez-Franco, H. Samsudin, X. Fang and R. Auras. Poly(lactic acid) mass production, processing, industrial applications, and end of life. Advanced Drug Delivery Reviews 2016; 107: 333-366.

52. Vink, E.T. and S. Davies. Life cycle inventory and impact assessment data for 2014 Ingeo™ polylactide production. Industrial Biotechnology 2015; 11: 167-180.

53. Müller, R.-J. Biodegradability of polymers: regulations and methods of testing. *In*: A. Steinbuchel (ed.). Biopolymer, General Aspects and Special Application, Wiley Publishers, 2008, pp. 366-388.

Chapter 2

Grain Waste Product as Potential Bio-fiber Resources

A. Al-Mamun

Corporate R&D, Adler Pelzer Group, Bochum, Germany

2.1 INTRODUCTION

The increased environmental awareness, growing global waste problem, geometrically increasing synthetic fiber raw material crude oil prices, and high processing costs trigger off to develop concepts of sustainability and also reconsider renewable resources. Natural fibers have already established a record of accomplishment as reinforcing material in manufacturing industries. New environmental legislation and consumer demands have forced manufacturing industries, e.g., automotive, packaging, construction, and others to substitute conventional reinforcing materials, which are non-renewable, such as glass fiber, inorganic fillers, etc. The main advantages of natural fibers are their economic viability, low density, carbon neutrality, sustainability, reduced tool wear, enhanced energy recovery, reduced dermal and respiratory irritation, and comparable specific strength and modulus properties [1–6]. However, their relatively low thermal degradation temperature, high level of moisture absorption, poor wettability, and insufficient adhesion between fibers and the polymer matrix lead to emitting odor and debonding with age.

These materials will have limited acceptance in modern industrial applications. The properties of natural fibers vary considerably depending on the fiber diameter, structure (e.g., proportion of crystalline fibrils and non-crystalline regions, spiral angle of fibrils), supramolecular structure (degree of crystallinity), degree of polymerization, crystal structure (type of cellulose, defects, orientation of the chains of non-crystalline and crystalline cellulose), void structure (pore volume, specific interface, size of pores), and finally, whether the fibers are taken from the plant stem, leaf or seed and on the growing conditions [7–15].

The most common classification depends on their botanical type or origin, natural fibers can be classified into bast, leaf, seed, core, grass, and wood fiber. Seed fibers are further classified into fiber, pod, fruit, and husk [5, 16, 17].

For Correspondence: Email: a.mamun@pelzer.de or mithu05@gmail.com

The sources and world production of the latest most usable natural fibers and the predictions of total demand of natural fiber in Germany, Europe, North America, and Asia in the year 2005 and the future predictions in the year 2010 and 2015 are presented in Table 2.1 (all arithmetical values are approximated). In comparison with 2010, the forecasting demand in the year 2015 will be increased, about 45% for Germany and 30% for North America, whereas the propagation will be increased 120% in Asia [18, 19]. The forecasting demands in the year 2020 will be increased, about 35% to 45percent.

Table 2.1 Production of natural fiber and estimated consumption [18, 19].

Region/Year	***Production* (tons)**				
	2000	**2005**	**2010**	**2015**	**2020**
Germany	22000	43500	75000	108000	120000
EU-30	28300	68000	98000	123000	135000
Asia	9000	14500	31000	69000	87000
North America	30000	45500	60000	78000	105000

2.1.1 Grains Waste Products

The utilization of feed stocks, such as lignocellulose and lignocellulosic waste materials as substitutes for existing fossil based raw materials will be a challenge in the coming century. There is propelling in the direction of reduced greenhouse gas emission. Fossil based raw materials contribute over 25% of greenhouse gas emission. The use of plant/crop-based feedstock for the raw materials of chemicals in the European Union could deliver greenhouse gas emission reductions of over 6 million tonnes per annum in the next decade [20]. The International Energy Agency has estimated and released that the total greenhouse emission was 494 million tonnes in European Union for 2008. So the estimated greenhouse gas emission reduction in EU was about 1.2% of the total greenhouse gas emission. As a consequence of these events, there has been coordinated R&D strategy across the globe for the utilisation of plant/crop-based products [21]. The functional waste material obtained from the renewable resources gained much more attention due to economics and limitation of wood fiber sources [22]. The effort has been based on the use of new waste sources, with the aim to obtain biologically active compounds which can be applied in different fields and applications.

These natural lignocellulosic crops and grain residues are compatible with the environment and could provide the sources for specialty chemicals [23]. The use of the crops and grain residues as a filler or reinforcement in the production of plastic composites alleviate the shortage of wood resources and can have the potential to start a natural fiber industry in countries where there are little wood resources left. The composite industries are looking into alternative low cost lignocellulosic sources, which can decrease overall manufacturing costs and increase properties of the materials [24–27].

Grain waste products, e.g., wheat husk, rye husk, rice husk, and coconut shell are waste products of food processing from grain and have sufficient fiber value (structural material e.g., cellulose). It contains about 25% to 50% structural materials, which is equivalent to common natural fiber and wood fiber [28, 29].

Wheat (*Triticum spelta;* German name is Dinkel) is a member of grass family. The grain color is dull yellow. It is a hexaploid species of wheat and is a hybrid of domesticated tetraploid emmer wheat (*Triticum dicoccum*) and the wild goat grass (*Aegilops tauschii*). On the other hand, wheat (*Triticum aestivum*; German name is Weizen) is known as common wheat. The grain color is light yellow and is allohexaploid. It means an allopolyploid with six sets of chromosomes, two sets from each of three different species [30, 31]. The hybridization and change of genetic code is because of increase of tolerance with different weather conditions and increased productivity [32].

Wheat has two distinct growing seasons. Winter wheat is sown in the fall and harvested in the spring or summer. Spring wheat is planted in the spring and harvested in late summer or early fall. Wheat classes are determined not only by the time of year they are planted and harvested, but also by their hardness, color, and the shape of their kernels. Each class of wheat has its own similar family characteristics, especially as related to milling and baking, or other food use [33, 34].

Wheat is the most common and important human food grain and ranks second in total production as a cereal crop. Wheat grain is a staple food used to make flour for leavened, flat and steamed breads; cookies, cakes, pasta, noodles, and couscous; and for fermentation to make beer alcohol or biofuel [35]. Wheat is planted to a limited extent as a forage crop for livestock and the straw can be used as fodder for livestock or as a construction material for roofing thatch. Wheat husk is a lignocellulosic waste product which is about 20–25% of wheat and some extents of wheat husk are used to feed the livestock, poultry occupants, and as fuel [36, 37].

Rye (*Secale cereale;* German name is Roggen) is a member of grass family. It is a diploid species and the tribe of Gramineae with two chromosomes. It is closely related to wheat tribe (Triticeae). Rye grows more rapidly than wheat, oat, barley, or various other annual grasses. Although it is usually regarded as a winter crop, several spring-sown varieties are available. It grows extensively as a grain and forage crop [38-40]. Rye grain is used for flour, rye bread, rye beer, whiskies, vodkas, and animal fodder. Non food part of rye is agro waste, which is about 20–25 wt% of rye [36, 37].

Rice (*Oryza sativa*; Italian name is Arborio) is the seed of a monocot plant of the grass family (Poaceae). It has the smallest cereal genome consisting of just 430 mega base pairs across 12 chromosomes. It contains two major subspecies: the sticky, short grained *japonica* or *sinica* variety, and the non-sticky, long-grained *indica* variety. Arborio is a cultivar of the japonica subspecies of *Oryza sativa* [41–43]. As a cereal grain, it is the most important staple food for a large part of the world's human population, especially in East Asia, South Asia, Southeast Asia, Latin America, Africa, and the West Indies. Rice is normally grown as a half-annual plant, although in tropical areas its life cycle is 3 to 4 months. Rice husk is a waste product of rice milling process, and a great resource as a raw biomass material for manufacturing added-value products. Rice husk is 15 to 25 wt% of rice [44].

Coconut (*Cocos nucifera* L.) is a member of the palm family. It is botanically classified into two major groups based on its stature- Tall palms and as Dwarf palms. The Tall can also be referred to as var. Typica (Nav) and the Dwarfs as var. Nana (Griff). Tall palms are widely planted and grow up to 30 meters in height. They are normally cross-pollinating and are therefore considered to be heterozygous and constitute the polymorphic population. They are slow to mature, flowering 6–10 years after planting and have economic life of 65–75 years. Dwarf palms grow up to a height of 10–15 meters, flowering 3–4 years after planting, and they are self-pollinating and considered to be homozygous and have economic life of 30–40 years [45, 46].

The coconut palm is used for decoration as well as for its many culinary and non-culinary uses; virtually every part of the coconut palm has some human use. The husk, or mesocarp, is composed of fibers, called coir. The inner stone or endocarp is the hardest part of the nut, called the shell. Adhering to the inside wall of the endocarp is the *testa*, with a thick albuminous endosperm, the white and fleshy edible part of the seed. Coconut shell is the non food part of coconut, which is hard lignocellulosic agro waste. Coconut shell is 90 to 150 gram (dry) per coconut, which is 15 to 20 wt% of coconut [47, 48].

Note: The quality of grain husk or grain waste product as a raw material for composite materials decisively depends on grain collection, grain storage, moisture content, and the way the food is processed. So, it is necessary to use modern technology for husk separation from grain. For this research work, grain husks (waste materials) were selected and these materials have no food value.

2.1.2 Sources and Availability

Grain waste products are semi and an annually renewable fiber and are available in abundant volume throughout the world. The world and European Union production of grains in the year 2010 was estimated and the data of this table was extracted from mainly FAO database and the quantity of grain waste products was determined associating a given production of crops. The sources and estimated production were given in Table 2.2. As of October 2010, the world production of wheat husk was 169 million tons, of which 35 million tons were produced in European Union and the world production of rye husk was 4.25 million tons, of which 2.4 million tons were produced in European Union [49]. In the year 2010, the approximate world production of wood fiber was 1750 million tons, bast fiber (jute, kenaf, flax, hemp) was 11 million tons, and leaf fiber (sisal, abaca, henequen) was 480 thousand tons [5].

Table 2.2 Grain waste products source and production [47, 49].

Name	*Main source*	*EU-27* (MT)	*World* (MT)
Wheat husk	EU-27, China, India, Russia, USA, Ukraine	34.5	169.5
Rye husk	EU-27, Russia, Belarus, Ukraine	2.4	4.25
Rice husk	China, India, Bangladesh, Indonesia, Vietnam, Thailand	0.35	89.6
Coconut shell	India, Indonesia, Philippines, Thailand, Sri Lanka, Africa	–	4.8

Note: So in comparison with the conventional natural fiber, the world production of the grain waste products and the straw are found to be the second highest after wood fiber.

2.1.3 Fiber Processing and Morphology

The morphology of grain waste products also depends on grain processing. Grain milling is the removal of husk or bran from corn. In general, dry milling process is used by grain processing industries. This process consists of several steps, e.g., break grain, separate husk or barn from endosperm, and grind to final product [50]. The main aim is to minimize loss of edible, mineral, and vitamin portion during milling [51]. To reach the aim, the milling process has needed to modify its name to wet milling process and the process cost became higher. In case of wet milling process, conditioning step was added to the dry milling process, where a certain percent of moisture (20%–35%) has to be maintained. Husk is the waste from milling process of grain. However, it is a value added green lignocellulosic material for composites.

The morphology and microstructure of husk are damaged during the milling process and lose their properties. Some industries developed a new enzyme technology to remove husk or barn from grain. It does not damage the edible portion with mineral and vitamin as well as husk [52]. Half shell husks from enzyme technology are followed with further milling process to get maximum fiber quality with desired fiber size and distribution, which consists of conditioning, grinding, and sorting.

The half-cell picture of grain is shown in Fig. 2.1. Moreover, quality of husks is strongly influenced by genetic factors, environmental factors, harvesting, transportation, drying of grain, storage, and moisture content during the storage [53]. Fiber sizes and shapes are one of the most important factors for composite materials. The effective surface area which may have influence on mechanical properties inversely depends on particle size and shape.

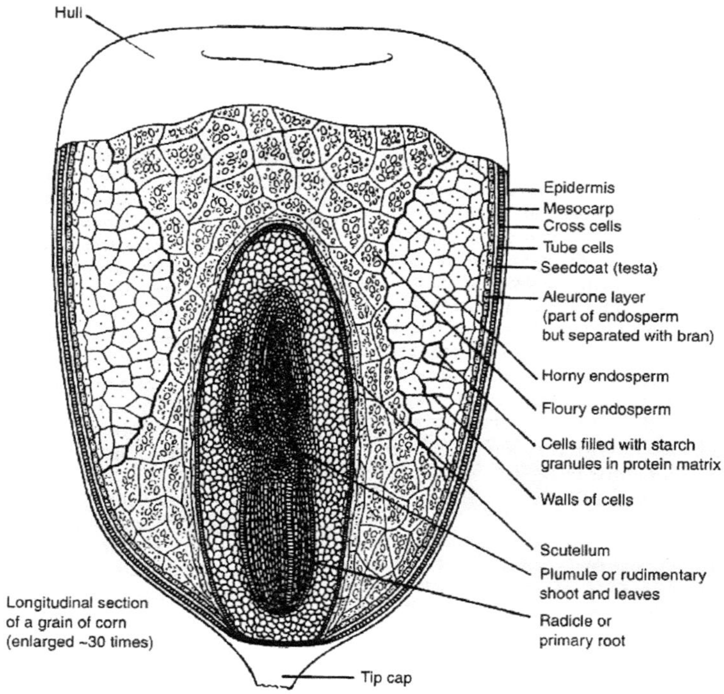

Figure 2.1 The internal structure of the grain [53].

2.1.4 Chemical Compositions

Grain waste products are three-dimensional, polymeric composites made up primarily of cellulose, hemicelluloses, lignin, and a small amount of protein, starch, fat, and ash. The chemistry of individual components is explained briefly.

Cellulose is a linear syndiotactic homopolymer composed of D-anhydrogluco pyranose units, which are linked together by β- (1$\rightarrow$ 4) glycosidic bonds. Taking the dimer cellobiose as the basic unit, it can be considered an isotactic polymer of cellobiose. Hydroxy groups at C–2, C–3, and C–6 positions are capable of undergoing the typical reactions known for primary and secondary alcohols. The vicinal secondary hydroxyl groups represent a typical glycol structure. The hydroxyl groups at both ends of the cellulose chain show different behaviors. The C–1 end has reducing properties, while the glucose end group with a free C–4 hydroxyl group is non-reducing. The hydroxyl groups including the bridging and the ring oxygen atom are predominantly involved in intra- and intermolecular interactions, mainly hydrogen bonds, and the bridging and the ring oxygen atom are involved in degradation reactions [54–56]. The molecular structure of cellulose is shown in Fig. 2.2.

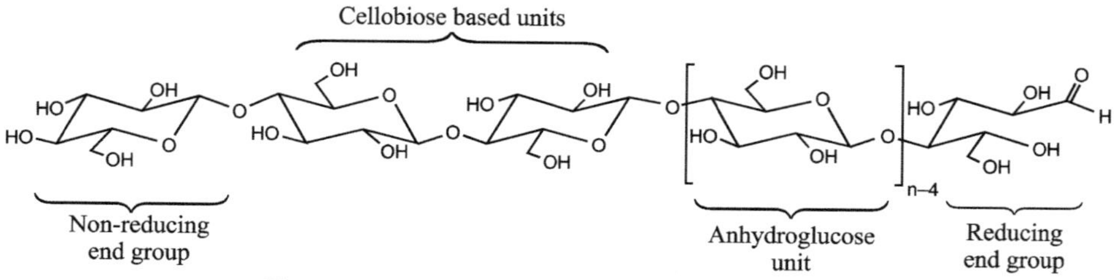

Figure 2.2 Molecular structure of cellulose [55].

The molecular weight or degree of polymerization of cellulose differ widely and depend on origin and fiber separation process. The degree of polymerization of cellulose from plants is ranged up to 12000 and is always polydispersed. The presence of intra and intermolecular hydrogen bond is of high relevance with regard to single chain conformation and crystal formation. The existence of hydrogen bonds between O–3–H and O–5′, and between O–2–H and O–6′ are responsible for the considerable stiffness of the cellulose chain, and stabilize the two-fold helix conformation of crystalline cellulose. Deviations from this two-fold helix are due to alternate hydrogen bond formation [57–58]. According to different isolation process and pre-treatment, cellulose lattices are classified as cellulose I, II, III, and IV depending on alternation of hydrogen bond and the chain polarity which leads to differing unit cell dimensions [59]. The hydrogen bond patterns of cellulose I and II are presented in Fig. 2.3.

Chemical processing or pre-treatment of cellulose proceeds during its whole course or at least in the initial phase usually as a heterogeneous reaction. The rate and final degree of conversion depend strongly on the availability of hydroxyl groups in the anhydroglucose units, so called accessibility. From the sight of cellulose structure, accessibility depends largely on the available inner surface and also on supramolecular order and fibrillar architecture [60].

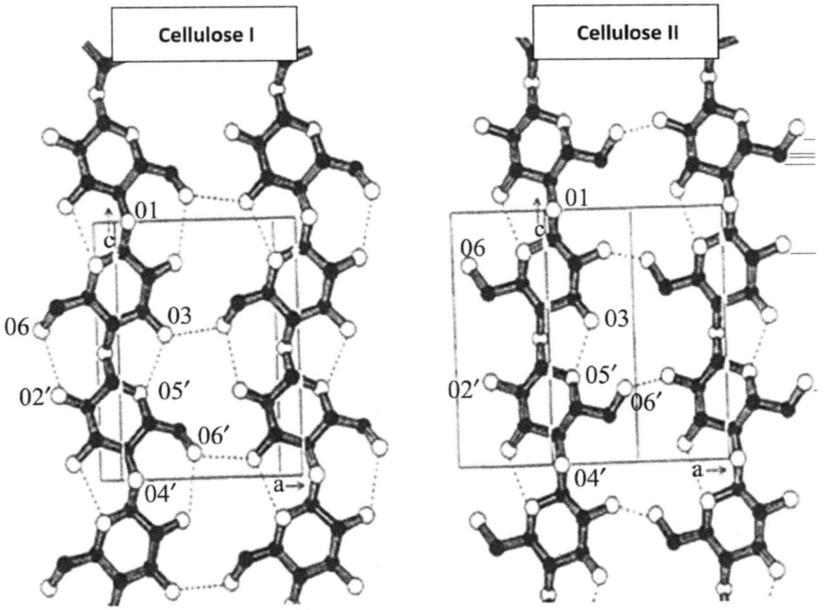

Figure 2.3 Hydrogen bond patterns of cellulose allomorphs [60].

Hemicellulose represents a class of heteropolymers with a selection of sugar molecules as monomeric units. The most common monomers are D-glucose, D-galactose, D-mannose, D-Xylose, and L-arabinose. Hemicelluloses contain most of the D-pentose sugars, and occasionally small amounts of L-sugars as well. Xylose is the sugar monomer present in the largest amount, but mannuronic acid and galacturonic acid also tend to be present [61, 62]. The structure of hemicellulose can be understood by first considering the conformation of the monomer units. There are three entries under each monomer, as shown in Fig. 2.4. In each entry, D and L refer to standard configurations for two optical isomers of glyceraldehyde and designate the conformation of hydroxyl group at C–4 for pentoses and C–5 for hexoses. The configuration of hydroxyl group on C–1 refers to α and β [63, 64].

Hemicellulosic polymers are branched and have a significantly lower molecular weight than cellulose. The degree of polymerization of hemicelluloses is ranged from several hundred to up to

3000. It is strongly bound to cellulose fibrils by hydrogen bonds. It is fully amorphous and has a lower frequency of interchain hydrogen bonding. Due to its open structure containing many –OH and –HOC–CH_3 groups, it is easily accessible [65].

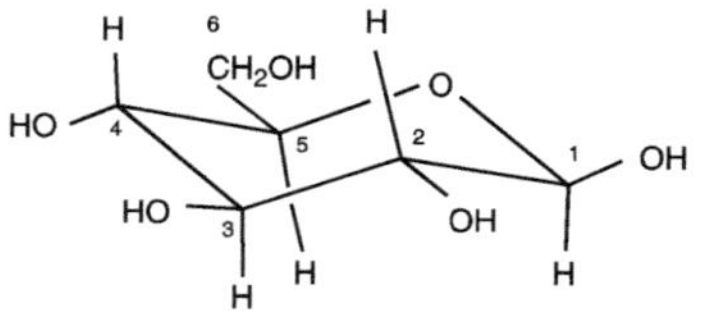

β-D-Glucose
β-D-Glucopyranose
β-D-Glup

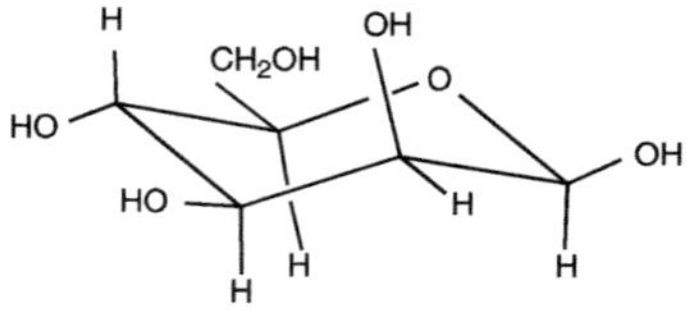

β-D-Mannose
β-D-Mannopyranose
β-D-Manp

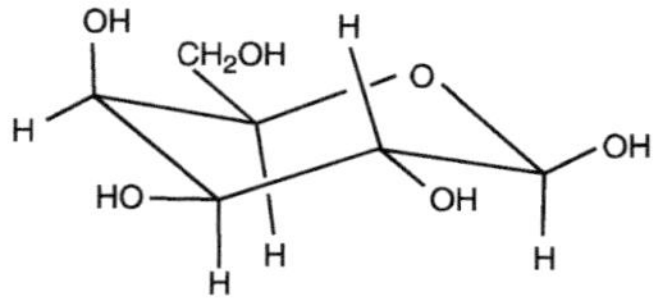

β-D-Galactose
β-D-Galactopyranose
β-D-Galp

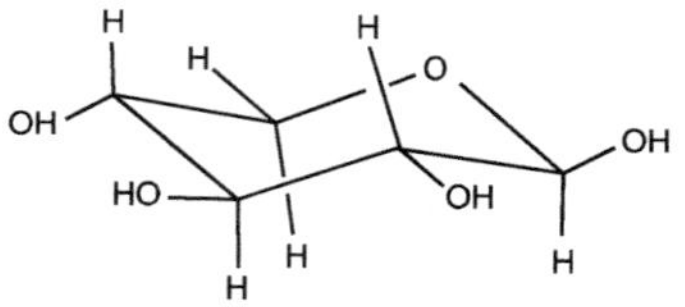

β-D-Xylose
β-D-Xylopyranose
β-D-Xylp

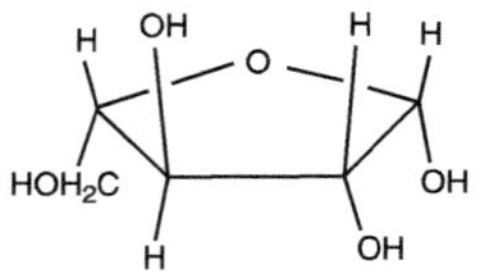

α-L-Arabinose
α-L-Arabinofuranose
α-L-Araf

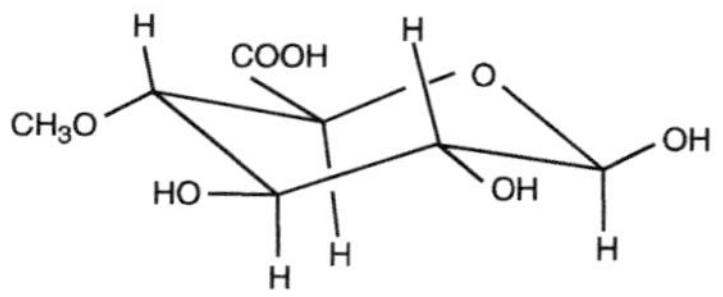

4-O-Methylgucuronic acid
4-O-Methylglucopyranosyluronic acid
4-O-Me-*α*-D-GlupA

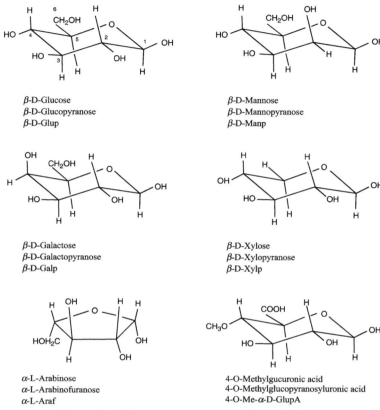

Figure 2.4 Monomer components of hemicelluloses.

Lignins are amorphous, highly complex, mainly aromatic polymers of phenyl propane units that are considered to be an encrusting substance. The three-dimensional polymer is made up of C–O–C and C–C linkages. The precursors of lignin biosynthesis are *p*-coumaryl alcohol (known as structure 1), coniferyl alcohol (structure 2), and sinapyl alcohol (structure 3) (Fig. 2.5). Those precursors of lignin depend on source of plant parts and plant age [66–67].

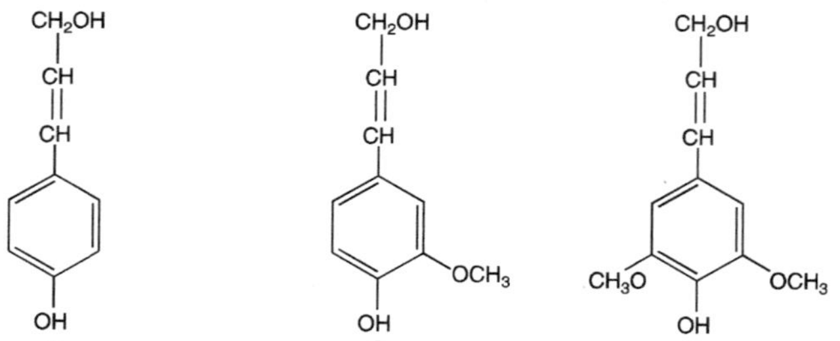

1 2 3

Figure 2.5 Structure of lignin precursors: 1) p-coumary alcohol, 2) coniferyl alcohol, and 3) sinapyl alcohol.

Lignins do not have a single repeating unit, but instead consist of a complex arrangement of substituted phenolic units. Lignins consist of mainly three basic building blocks of guaiacyl, syringyl, and *p*-hydroxyphenyl moieties, although other aromatic units also exist in many different types of plants. The molecular weight of lignin depends on the method of extraction and it is ranged from 260 to 50000000 [68]. Lignins are associated with the hemicelluloses forming lignin-carbohydrate complexes that are resistant to hydrolysis even under pulping conditions [69, 70].

Fats in plant parts are ester of high molecular weight long chain monohydroxy alcohol fatty acids. It is important for many forms of life, serving both structural and metabolic functions, and an important part of the diet of most heterotrophs. There are many different kinds of fats present in plant bark or in corn parts. All fats consist of fatty acids (chains of carbon and hydrogen atoms, with a carboxylic acid group at one end) bonded to a backbone structure, often glycerol (a backbone of carbon, hydrogen, and oxygen) [71, 72]. Chemically, this is a triester of glycerol, an ester being the molecule formed from the reaction of the carboxylic acid and an organic alcohol. Depending on source and genetic modification of corn seeds, waxes or fats associated with other active chemicals form different complex chemicals e.g., terpens, flavonoids, tannins, etc. [73]. Lipid is a category of fats, distinguished from fat by their chemical structure and physical properties. The properties of any specific fat molecule depend on the particular fatty acids that constitute it. Fatty acids with long chains are more susceptible to intermolecular forces of attraction (hydrogen bonding and van der Waals forces), raising its melting point. The molecular weight of fatty acid is ranged from several hundreds to several thousands. Lipids can easily undergo hydrolysis if they are not in form of complex structure [74, 75].

Proteins (also known as polypeptides) are organic compounds made of amino acids which are arranged in a linear chain and folded into a globular form. The amino acids in a polymer are joined together by the peptide bonds between the carboxyl and amino groups of adjacent amino acid residues. Most proteins fold into unique three-dimensional structures [76, 77]. The shape into which a protein naturally folds is known as its native conformation. It contains an unusual ring to the N-end amine group, which forces the CO–NH amide moiety into a fixed conformation. Proteins can also work together to achieve a particular function, and they often associate to form stable complexes [78]. Proteins content in cereal or plant products are different in forms which make metal complex with multivalent metal. In most cases protein and fat are bonded together by covalent or hydrogen bond (lipoprotein) and form a coat for plant cell [79]. Cereal protein affected by chemical depends on degree of cyclic structure of amino acid and chain length. The molecular weight of plant proteins is ranged from several hundreds to several thousands dalton [80, 81].

Starch or amylum is a carbohydrate that consists of a large number of glucose units joined together by glycosidic bonds. This polysaccharide is produced by all green plants as energy storage. Starch molecules arrange themselves in the plant in semi-crystalline granules. Pure starch is a white, tasteless, and odorless powder. It consists of two types of molecules: the linear and helical amylose and the branched amylopectin [82, 83]. Depending on grains, starch contains about 20% of amylose and 80% of amylopectin. Starch can be hydrolyzed into simpler carbohydrates by acids, various enzymes, or a combination of the two, and the resulting fragments are known as dextrins. It is insoluble in cold water or alcohol. The chain length or molecular weight depends on the source of the grain, and it is ranged from $2.5*10^8$ to $6*10^8$ dalton [84, 85].

Minerals include calcium (Ca), potassium (K), phosphorus (P), aluminium (Al), and magnesium (Mg), and inorganics include silicon (Si), which are minor parts but essential for plant cell [86]. Calcium in the cell wall exists mainly in pectin, forming calcium pectate to stabilize the structure of the cell wall. Calcium chelators can extract calcium ions from pectin in plant cell walls, resulting in the pectin becoming soluble, which leads to separation of fiber bundles. Other minerals are mostly associated with protein and fat molecules and form complex [87, 88].

Note: In a fiber, the chemical structure of individual components, surface chemistry, and bonds associated with them are important for fiber modification and for interfacial strength in composites.

2.1.5 Influence of Chemical Composition and Surface Morphology

The fiber surface morphology plays a vital role in case of composite materials. Every fiber has unique surface properties. External surface features of fibers such as contours, defects and damage, and surface layer are important. The fiber consists of mainly cellulose ($O/C=0.83$), hemicelluloses (polysaccharides with a range of monomers, where $O/C \approx 0.83$), lignin ($O/C=0.35$), pectin (O/C similar to hemicelluloses), and wax or lipids. It has long been postulated that the grain waste product surface contain lipid and proteinacious compound, and the lipid molecule is usually bonded to the protein molecule by ester or thioester bond. The amount of lipid on the fiber surface has an influence on hydrophobicity and surface tension. The more amount of lipid on the fiber surface means the more hydrophobic and more surface tension, as well as the smoother fiber surface forming a thin film [89–91].

Fibers change dimensions with changing moisture content because the cell wall polymers contain hydroxyl and other oxygen containing groups that attract moisture through hydrogen bonding. The hemicelluloses are mainly responsible for moisture sorption, but the accessible cellulose, non-crystalline cellulose, and surface of cellulose also play a major role. Water molecules absorbed by dry cellulose form a true cellulose hydrate and the reaction is exothermic, which provides the driving force for further absorption. The primary layer of water bound directly to the $-CH_2OH$ group forms a relatively strong hydrogen bond and attracts other water molecules by weaker hydrogen bonding. This weaker bound layer may be thickened by several molecular layers and shows some order [92–94].

Fibers are degraded biologically because organisms recognize the carbohydrate polymers mainly by hemicelluloses in the cell wall. Biodegradation of the high molecular weight cellulose and hemicelluloses weaken the fiber cell wall and fiber strength is lost as the fiber cellulosic polymer undergoes degradation through oxidation, hydrolysis, and dehydration reactions [95]. The photochemical degradation takes place primarily in the lignin component, which is responsible for the characteristic color changes. The lignocellulosic fibers burn with increasing temperature because the cell wall polymer undergoes pyrolysis reactions and give off volatile and flammable gas [96]. The fat, starch, and protein polymers are degraded by heat, much before the cellulose and hemicelluloses, and the cellulose and hemicelluloses polymers are degraded by heat much before lignin [97]. So, it is obvious that the chemical composition and amount play a vital role of fiber properties and composites properties if it is used as reinforcement. Fiber modifications and coupling agents are considered to optimize the interface.

Note: The fiber surface layer is a relatively small portion of fiber but it plays an important role in wettability and surface tension.

2.1.6 Surface Modification

Surface properties of natural fiber, including grain waste products, can be divided into two major groups: physical and chemical properties. Physical properties include morphology, roughness, smoothness, specific surface area, and permeability. Chemical properties include elemental, molecular, functional, and group composition. These two majors groups of properties determine the thermodynamics of the surface, such as free energy, acid base acceptor, and donor numbers [98]. Natural fiber surface treatment generally involves imparting an altered surface chemistry, changing thermodynamic properties, and creating beneficial microtopographical features without a deliberate coating of the surface. In most cases, effective surface treatments not only remove native surface material and leave behind more active function group to promote wetting, but also roughen the surface to some degree, therefore increasing surface area, and potentially enhancing mechanical interlocking. From a composite processing viewpoint, the use of a surface treatment is desirable to promote wettability and increase the interfacial bonding between matrix and fiber [99].

2.1.6.1 Biochemical Treatment

Biotechnology, one of the three defining technologies, has made advances in recent years and emerged as a frontline area of research and development, with an overwhelming impact on the society. The developments are very fast globally and new dimensions are being added every day. Virtually all chemical reactions in biological systems are catalysed by macromolecules called enzymes. The reactions rarely proceed at perceptible rates in the absence of enzymes, while reaction rates increase as much as a million times when enzymes are present. The ability of enzymes to function outside of a cell has greatly increased their use in a large variety of commercial products and reactions. The uses of enzyme technology in the processing of natural fiber and in the field of textile for modification are increasing substantially. A major reason for embracing this technology is the fact that application of enzyme is regarded as environmental friendly and the reactions catalysed are very specific with a focussed performance as a consequence. Other potential benefits of enzyme technology include cost reduction, energy and water saving, improved product quality, and potential process integration [100–102].

Three quarters of the market of enzymes is involved in the hydrolysis of natural polymers, of which two-third are proteolytic enzymes used in the detergent, dairy, and leather industries; and one third are carbohydrases used in the animal feed, baking, brewing, distilling, starch, and textile industries. Detergent manufacturers use 45%, food processing industries use 40%, paper industries use 4%, textile and fiber processing industries use 6%, leather industries use 2%, and others use 3% of all industrial enzymes [103–105].

The application of enzymes to modify the surface of chemical fiber and natural fiber, such as hemp, flax, wool, cotton has been widely researched by industry. Most of the industrial applications are aimed to improve surface properties by removing adsorbed components, such as lignin, fats, waxes, proteins, and non crystalline parts. In the textiles processing areas, such as desizing, scouring, and bleaching of cellulose and woollen fabrics are some examples of successful bio-treatments of textiles [106–118]. Not only unwanted adsorbed material may be removed, but modification of the fiber surface may also be accomplished by enzymes [103]. Besides defurring and antifelting treatments of textile, the so called biopolishing of cotton fabrics and garments is another good example [119–121].

Cellulase

Cellulase refers to a class of enzymes produced chiefly by fungi, bacteria, and protozoans that catalyze the cellulolysis, e.g., hydrolysis of cellulosic materials. However, there are also cellulases produced by other types of organisms, such as plants and animals. Several different kinds of cellulases are known, which differ in structures and reaction mechanisms. Cellulases cleave β-1, 4-glycosidic bonds in cellulose, which is apparently accessible as cellulose and amorphous cellulose portion in cellulosic materials. The enzyme class is divided into endo-cellulases (endo-glucases) and exo-cellulases (cellbiohydrolase). Endo-cellulases catalyze the endo-hydrolysis of 1, 4-β-D-glycosidic linkages in cellulose, lichenin (moss starch), and cereal β-D-glucans. Exo-cellulases hydrolyze the 1, 4-β-D-glycosidic link in cellulose and cellotetraose, releasing cellobiose from the nonreducing ends of the cellulose chain [103].

The cellulase enzyme molecule is composed of up to three types of functionally different domains: 1) the catalytically active core, which is a large spherical domain, 2) the linker domains, which is an elongated and flexible spacer, and 3) a spherical cellulose-binding domain. The presence of a cellulose-binding domain is of particular importance for binding of the enzyme on insoluble and crystalline cellulose, and for hydrolytic effects [122–124]. The catalytic reaction activity of cellulase is very selective to temperature and pH. As explained previously, the functionality and reaction mechanism of the enzyme depends on enzyme source. So, it is very important to find out the optimum parameter for every single system. From literature, it could be summarized that the cellulase enzymes are active in the temperature range of 25–70°C and in the pH range of 4–10. However, in the case of plant fiber, temperature 40–60°C and neutral pH or slightly basic medium are favorable [125]. Another important reaction parameter is dosage rate. Extremely high dosages of cellulases can cause fiber damage and it may appear as loss of fiber strength and excessive softening [125, 126]. Cellulases are widely used in textile industry, in food processing, and in laundry detergents [127, 128]. They are also used in the pulp and paper industry for various purposes, in the fermentation of biomass into biofuels, and in pharmaceutical applications [129, 130]. Cellulases are also used as dough conditioning agent or bioscouring. They can remove particular dirt, soil, and unwanted small molecules, and simultaneously enhance wettabity, machinability, and color brightness of textile fiber [131–133].

Pectinase

Microbial production of pectinolytic enzymes is mainly from filamentous fungi, yeasts, and filamentous and non filamentous bacteria. Generally, fungal enzymes are acidic in nature, while alkaline enzymes are produced by bacterial strains. Pectinases can be classified as esterases, eliminative depolymerises, and hydrolytic depolymerises with respect to their role in degradation of pectin [134]. Pectinesterases or pectin methyl hydrolases catalyze hydrolytic removal of the methyl ester group of pectin, forming pectic acid. Depolymerases degrade the α-1,4 glycosidic linkages unit adjacent galacturonic acid residues of pectin. Among the depolymerases, hydrolytic depolymerases act by hydrolysis, whereas eliminative depolymerises act on depolymerisation of pectin by transelimination, which results in galacturonide [135, 136].

Pectinases are effective to the degradation of pectin, which is a complex acidic polysaccharide present in the primary cell wall and middle lamella of higher plant tissues. The enzyme activity is favorable in alkali medium and in temperature range 40–60°C [137]. Alkaline pectinases are important for retting and degumming of plant fibers or natural fibers by removing intermolecular pectin, which acts as a cementing substance between the fibers [138–140]. They are also applied in paper and pulp industry, juice processing, vegetable oil extraction, and tea and coffee fermentation [141–143].

Lipase

Lipase was isolated from the fungus (*Thermomyces lanuginous)* with low level of enzyme expression. By using rDNA technique in the harmless host microorganism (aspergillus oryzae), the lipase expresses in acceptable yields in the current commercial production process [144]. While a diverse array of genetically distinct lipase enzymes are found in nature, and represent several types of protein folds and catalytic mechanisms, most are built on an alpha/beta hydrolase fold, and employ a chymotrypsin-like hydrolysis mechanism, involving a serine nucleophile, an acid residue (usually aspartic acid), and a histidine [145, 146]. Lipases hydrolyze triacrylglycerol (triglyceride) substrates, as are present in fats and oils. A triglyceride molecule is composed of three fatty acid moieties linked to a glycerol backbone by ester bonds. Lipases hydrolyze triglyceride to more hydrophilic mono and diglyceride, free fatty acid, and glycerol. These hydrolysis products are all insoluble in neutral and acidic medium, but all are soluble in alkali wash condition [147]. The best rate of enzyme activity was found to be above pH 8 because of the presence of free fatty acid in decomposed strains [148].

The enzyme activity will be more favored by a small amount of Ca^{++} ions. Most lipases act at a specific position on the glycerol backbone of lipid substrate [149, 150]. Lipases serve important roles in human practices as ancient as yogurt and cheese fermentation. However, lipases are also being exploited as cheap and versatile catalysts to degrade lipids in more modern applications. For instance, a biotechnology company has brought recombinant lipase enzymes to market for use in applications, such as baking, laundry detergents, and even as biocatalysts in alternative energy strategies to convert vegetable oil into fuel [151–153]. Pseudomonas and some fungal lipases also can be used for degradation of crude oil [154].

Laccase

Laccases are polyphenol oxidase enzymes that are found in many plants, fungi, and microorganisms. Laccases, such as ones produced by the fungus *Pleurotus ostreatus* can therefore be included in the broad category of ligninases. They belong to the family of blue multicopper oxidases. The laccase molecule is a dimeric or tetrameric glycoprotein, which usually contains four copper atoms per monomer distributed in three redox sites [155, 156]. Laccases act on phenols and similar molecules, performing one-electron oxidations. They play a role in the formation of lignin by promoting the oxidative coupling of lignols, a family of naturally occurring phenols. Laccases can be polymeric, and the enzymatically active form [157, 158].

Laccase catalyses oxidation of ortho and paradiphenols, aminophenols, polyphenols, polyamines, lignins, and aryl diamines, as well as some inorganic ions. Spectrophotometry can be used to detect laccases, using the substrates ABTS, syringaldazine, 2, 6-dimethoxyphenol, and dimethyl-p-phenylenediamine. Activity can also be monitored with an oxygen sensor, as the oxidation of the substrate is paired with the reduction of oxygen to water [159, 160]. The catalysis reaction of laccase happens in an acidic medium, and the most favorable pH range is 4–6. The reaction temperature is in the range of 30–55°C [161, 162].

The applications of laccase include the detoxification of industrial effluents, mostly from the paper and pulp, textile, and petrochemical industries [163–169]. They are used as a tool for medical diagnostics and as a bioremediation agent to clean up herbicides, pesticides, and certain explosives in soil. Laccases are also used as cleaning agents for certain water purification systems, as catalysts for the manufacture of anti-cancer drugs, and even as ingredients in cosmetics [170]. In addition, their capacity to remove xenobiotic substances and produce polymeric products makes them a useful tool for bioremediation purposes [171].

Protease

Proteases are the broad category of proteolytic enzymes that are found in many bacteria, fungi, and microorganisms. Protease produced by the organism *Staphylococcus avreus* or bacteria of genus *Bacillus* are excellent producer of protease, but several other strains also create significant amounts of enzymes [172]. Protease enzymes may have different properties from those of enzymes from other strains. As proteins are a highly heterogeneous population of molecules, due to 20 amino acids from which they are composed, the population of enzymes capable of hydrolyzing proteins is very diverse [173]. This population of enzymes is divided into two categories: the *peptidases* and the *polypeptidases* (synonymous with protease). The peptidases are exo-acting enzymes that remove one amino acid or a peptide from C- or N-terminus of protein, and the proteases are endo-acting that attack internal peptide bond of protein, yielding water soluble peptides [174].

Proteases catalyze the reaction of hydrolysis of protein with the participation of water molecule and produce amino acids or oligomer of amino acids. Protease reaction decisively depends on pH of medium. Proteases may be classified by the optimal pH in which they are active: acidic, basic, and neutral [175]. Proteases could be selected depending on the required product. The recently modified proteases are active in the pH range of 5 to 10 and in the temperature range of 30 to 60°C [176, 177]. Proteases are used in backing and bread industries. They reduce viscosity and improve the quality of cracker, biscuit, wafer, and pizza during the processing and also improve the flavor and color of bread. Proteases are also used in fish and meat processing, in detergents, and in beverage industries [178, 179].

Amylase

Amylases are the category of Amylolytic enzymes that are found in many plants, bacteria, fungal and microbial sources. Amylolytic enzymes are enzymes that break starch down into sugar. Naturally occuring starch consists of amylose (15–25%) and amylopectin (75–85%). Amylose composes of α-1, 4-glycosidic linkages and amylopectin composes of α-1, 4-glycosidic linkages and α-1, 6-glycosidic linked branch point occurs in every 17–26 glucose units. Amylase is present in human saliva, where it begins the chemical process of digestion. Foods that contain much starch but little sugar when they are chewed amylases turn some of their starch into sugar in the mouth. All amylases are glycoside hydrolases and act on α-1,4-glycosidic bonds. According to catalytic action, specific amylase proteins are classified as α-amylase, β-amylase, and γ-amylase [180].

Commercial α-amylases are from *Bacillus amyloliquefaciens* and the heat stable α-amylase from *Bacillus licheniformis* and from *Aspergillus* species. α-amylases catalyzes the endo-hydrolysis of 1, 4-α-glycosidic linkages in polysaccharides containing three or more 1, 4-α-linked glucose units [181]. The enzyme acts on starches, glycogen, and oligosccharides in a random manner. α-Amylases are characterized by attacking the starch polymer in endo fashion, randomly cleaving internal α-1,4 bond to yield shorter water soluble dextrins [182, 183]. They hydrolyse α-1,4 bonds but cannot cut α-1,6 linkages in amylopectin [184]. Pullulanase is an enzyme that is able to cleavage α-1,6 or debranch. The enzyme is efficient in the temperature range of 35 to 60°C and in mostly neutral pH medium or a bit acidic medium (pH 6.5) [185, 186]. α-Amylases are used in bread and cake making, fruit and juice industries, detergent, textile, alcohol production, medical diagnosis, protein and genetic techniques [187–189].

Xylanase

Xylanases are β-1, 4-xylan xylanohydrolase and their traditional name is pentosanases or hemicellulases. They are found in many plants, fungi, and microorganisms. They can catalyze the endo-hydrolysis of β-1, 4-xylosidic linkage in xylans in a random method, thus breaking down the hemicelluloses of plant cell wall. Most xylonases belong to the two structurally different glycosyl hydrolase groups with differences in molecular weight and structure [190, 191]. Xylanases have been reported to contain either a xylan-bonding domain or a cellulose-bonding domain. Some other bonding domains were found to increase the degree of hydrolysis of fiber bound-xylan. Neither a xylan- nor cellulose- binding domain was found to have a significant role in the action of xylanases in pulp fibers [192]. They are able to hydrolyze different types of xylan, showing only differences in the spectrum of end products [193]. The most characteristic of xylanases are their pH and temperature stability and activity, whereas those properties depend on source and host strains of enzyme. The most efficient temperature range is 20 to 70°C and pH range is 4 to 8, but in most cases a neutral medium is preferable [194, 195].

Commercial applications for xylanase include pulp and paper processing, the chlorine-free bleaching of wood pulp, crispbread, crackers; improve cereal products and juice, biomass conversation, the increased digestibility of silage and processing textile fiber [196–200].

Note: It was observed that the enzymes have been using in different fields from the ancient age but in the field of natural fiber modification they are very new. It is possible to remove the specific materials from the fiber surface without damaging the fiber structure. It means that the enzyme modification process could be able to enhance wettability without losing fiber strength.

2.2 INTERFACE & INTERPHASE

The interface is commonly defined as a perfect two-dimensional mathematical surface which divides two distinguished phases or components in a composite. The interface is a region of finite thickness (usually less than 0.1μm) [201]. The interface is characterized by an abrupt change in properties and in chemical composition. For the purposes of stress transfer from matrix to the reinforcement, one can assume that all stress transfer takes place at the interface, which is then characterized by a single property, e.g., interfacial shear strength. This approach is frequently used when there are chemical bonds which can control the mechanical response of a composite, while in natural fiber composites, phases are highly heterogeneous and there is assumed to be no chemical bond in between polar fiber and non polar thermoplastic [201, 202].

The mesophase concept was developed, which depends on a large number of parameters and possesses physical properties different from those of the bulk polymer. However, around an inorganic inclusion a complex situation exists due to imperfect bonding, surface topology of filler, stress gradients, voids, craze, and microcracks. Moreover, presence of a solid fiber in the polymer melt or monomer in the course of solidification or cure facilities causes physical changes in the morphology of the polymer phase in a region near solid surface. This results in a restriction of molecular mobility of molecules. In case of semi-crystalline polymer, there are crystalline and amorphous regions, which result in complex herterogenious interfaces [203].

The interphase is a three-dimensional layer in the immediate vicinity of a solid filler surface, possessing physical properties different from two main phases or components in a composite. In these layers, gradient of chemical composition, mechanical, morphological, and physical properties exist. The driving force has come from continuous change of composition and energy from bulk phase to the other. The reversible work required to create a unit interfacial area is the interfacial surface tension which is affected by the change of Gibbs free energy with respect to the interfacial area of total system and influence of temperature gradient, load gradient, and the total number moles of matter in the system [204].

In a natural fiber composites system, natural fibers are molecules rich in hydroxyl, carboxyl, and other functional groups. Consequently, the fibers are usually strong polar and exhibit significant hydrophilicity. In contrast, most polymer matrices are apolar and mostly hydrophobic. As a result there are significant problems of compatibility between the fiber and matrix, leading to poor dispersion, a weak interface, and ultimately inferior quality composites. Such problems could be tackled with the use of appropriate methods to improve adhesion [205]. To achieve this goal, first there is a possibility to alter the fiber surface and second to modify the matrix or use compatibilizer. In both cases, the main objective is to improve the wettability of matrix on the fiber surface and to promote adhesion by the introduction of chemical bond or the tuning of surface energy.

2.2.1 Wettability and Surface Energy

The wettability is related to the balance of surface energies or surface tension in solid-liquid-gas system. The balances of forces are applied to an ideal system in thermodynamic equilibrium. The condition required before the interfacial bonding can occur is that intimate contact between matrix and fiber can be obtained. In other words, good wetting of a surface is the prerequisite physical process which is required for a good adhesion. Therefore, wettability of the fiber by the matrix would be one of the most important factors when predicting the matrix-fiber adhesion and the wettability can be evaluated from the surface energy of fiber and matrix. Surface energy quantifies the disruption of intermolecular bonds that occurs when a surface is created. The surface energy may therefore be defined as the excess energy at the surface of a material compared to the bulk. In other words, surface energy is derived from the unsatisfied bonding potential of molecules at a surface. This is in contrast to molecules within a material which have less energy because they are subjected to interactions with molecules which are satisfied in all directions. Molecules at the surface will try to reduce this 'free energy' by interacting with molecules in an adjacent phase [206].

The manifestation of surface energy is a state of tension at the surface of a liquid, as a consequence of which work is required to increase the surface area of a liquid. One definition of surface energy is the work required to increase the surface area of a substance by unit area. However, when both phases are condensed, (e.g., solid-solid, solid-liquid, and immiscible liquid-liquid interfaces) the free energy per unit area of the interface is called the interfacial energy. The surface free energy of the solid-liquid interface is defined by contact angle between the solid-liquid interfaces. Liquids that form contact angles greater than 90° are called nonwetting, and liquids that form a contact angle less than 90° are termed wetting. When the contact angle is 0°, the liquid wets the solid and spreads over the surface spontaneously. Hence, the smaller the contact angle the better its wettability. Most polymers have low values of surface free energy (20 to 45 mJ/m^2), which decreases slightly with increased temperature [207, 208].

Fiber surfaces that have been exposed to the ambient environment act to minimize their surface free energy and adsorb material to lower their surface free energy. In some cases, this surface can have a surface free energy lower than that of the polymer matrix. In order to increase the fiber surface free energy, surface treatment is needed. Thus, a better wettability and a higher interfacial adhesion could be obtained, which result in higher mechanical, chemical, and physical properties [208].

2.2.2 Adhesion

Adhesion refers to the interaction of the adhesive surface with the substrate surface. It must not be confused with the bond strength. Certainly, if there is little interaction of the adhesive with the adherent, these surfaces will detach when force is applied. However, bond strength is more complicated because factors, such as stress concentration, energy dissipation, and weakness in surface layer often play a more important role than adhesion. Consequentially, the aspects of adhesion are a dominating factor in the bond formation process [209]. The theories of adhesion emphasize on both mechanical and chemical aspects. Chemical structure and interactions determine the mechanical properties and the mechanical properties determine the force that is concentrated on individual chemical bonds. Thus the mechanical and chemical aspects are linked and integrated parts of each other [209, 210]. Adhesion could be through mechanical interlocking and chemical bonding.

2.2.2.1 Mechanical Interlocking and Diffusion

In a mechanical interlock, the adhesion provides strength through reaching into the pores of the substrate. An example of mechanical interlock is Velcro; the intertwining of the hooked spurs into the open fabric holds the pieces together. This type of attachment provides great resistance to the pieces sliding past one another, although the resistance of peel forces is only marginal. The size of the mechanical interlock is not defined, although the ability to penetrate pores becomes more difficult and the strength become less when the pores are narrow. It should be noted that generally mechanical interlocks provide more resistance to shear forces than to the normal forces. Mechanical interlock strongly depends on the roughness of substrate surface [211]. If the concept of tentacles of adhesive penetrating into the substrate is transferred from the macro scale to the molecular level, the concept is referred to as the diffusion theory [212]. If there are also tentacles of substrate penetrating into the adhesive, the concept can be referred to as interdiffusion. This involves the intertwining of substrate and adhesion chains. This interface is strong since the forces are distributed over this intertwined polymer network. Interdiffusion can also work if only the adhesive forms tentacles into the substrate [213].

Note: The interface plays a pivotal role in determining the mechanical and physical properties of composites. A strong interface creates a material that displays exemplary strength and stiffness, but it provides very brittle nature with easy crack propagation through the matrix and fiber. A weaker interface reduces the efficiency of stress transfer from the matrix to the fiber and consequently, the strength and stiffness are not high but the toughness is considerably high.

2.3 ENVIRONMENTAL EFFECTS

As previously pointed out, natural fibers exhibit poor environmental and dimensional stability that prevent a wider use of natural fiber composites. The possibility for using these materials in outdoor applications makes it necessary to analyse their mechanical behavior under the influence of weathering action.

2.3.1 Hydrothermal Effect

Grain waste bio-fiber possess some disadvantages, such as higher moisture absorption which brings about dimensional changes, leading to microcracking, poor thermal stability, which may lead to thermal degradation during processing. Poor wettability and insufficient adhesion between untreated fibers and the polymeric matrix lead to debonding with age [214, 215].

The mechanical and physical properties of natural fiber reinforced composites can be reduced to great extent under moist conditions [216, 217]. This is a serious concern as there are potential outdoor applications, such as decking, fencing, and railing [218], where moisture absorption can have significant influence for these materials. The interfacial bonds between the natural fibers (which contain –OH, carboxyl –COOH, and other polar groups) and the relatively hydrophobic polymer (olefin) matrices would be weakened with higher water uptake. The weakened interface causes the reduction in the mechanical properties of the composites. The change in dimension of natural fibers with moisture gain is attributed to the interaction of the cell-wall polymers with water molecules via hydrogen bond formation [219].

The hemicelluloses and amorphous regions of the cellulose are generally responsible for the moisture absorption properties of fibers. Therefore, chemical and biochemical treatment of fiber is generally considered to remove hemicelluloses from natural fibers, because they are largely hydrophilic and they do not support load bearing mechanism. The study of the water absorption behavior of composites is necessary in order to estimate not only the consequences that the water absorption of fibers may have an effect on the composite properties, but also how this water uptake can be minimized. The way in which composite materials absorb water depends upon many factors, such as temperature, fiber volume fraction, orientation of the reinforcing fibers, fibers permeability, exposed area of the surface, diffusivity, and reaction between water and matrix [220].

Moisture absorption in composite materials can be conducted by several different mechanisms. The main process consists of diffusion water molecules inside the microgaps between polymer chains. The other common mechanisms are capillary transport into the gaps and flaws at the interfaces between fibers and polymer, due to the incomplete wettability and impregnation [221, 222]. In spite of the fact that all three mechanisms are active jointly during moisture exposure of the composite materials, the overall effect can be modelled conveniently using the diffusion mechanism only.

Apart from diffusion, two other minor mechanisms are active in moisture exposure of composite materials. The capillary mechanism involves the flow of water molecules into the interface between fibers and matrix. It is particularly important when the interfacial adhesion is weak, and the debonding of fibers and matrix has started. On the other hand, transport by microcracks includes the flow and storage of water in the cracks, pores, or small channels in the composite structure. These imperfections can be originated during the processing of the materials or due to environmental effects.

Scientific studies explain the hydrothermal effect of wood fiber reinforced virgin PP and recycled PP composites. They pointed out that due to hydrothermal ageing, tensile strength, Young's modulus, and hardness of both virgin and recycled composites were found to decrease, but impact strength and failure strain increased [223].

Effect of moisture absorption on the mechanical properties of PP composites with wood, sisal, coir, and luffa sponge was reported by Espert et al. [224]. They reported that the mechanical properties of those composites decrease after moisture absorption because it changes the structure and properties of fibers, matrix and the fiber-matrix interface.

The objective of this study is to observe the kinetic of moisture absorption and saturation moisture contents and the effect on mechanical properties of grain waste products reinforced composites.

2.3.2 Weathering Effect

UV exposure can cause changes in the surface chemistry of the composite known as photodegradation, which may lead to discoloration making the products aesthetically unappealing. Furthermore, prolonged UV exposure may ultimately lead to loss in mechanical integrity [225].

The composites are reinforced with fibers containing large amounts of lignin which are more susceptible to natural weathering than those with negligible amounts of lignin. This is because lignin and hemicelluloses existing in the middle lamellae of wood fiber are more susceptible to chemical degradation than cellulose. Hemicelluloses are responsible for moisture sorption and biological degradation in wood to a much greater extent than cellulose. Lignin is responsible for ultraviolet degradation [226, 227].

The photodegradation mechanisms of wood and plastic separately are well documented in the literature. However, the photodegradation mechanism of wood or other natural fiber composites is complicated, because each component, namely wood and plastic, may degrade via a different mechanism. The photodegradation is attributed to the degradation of its components, namely cellulose, hemicelluloses, lignin, and extractives [228]. Lignin contains chromophoric functional groups, such as carboxylic acids, quinines, and hydroperoxy radicals, which readily absorb UV and cause discoloration and yellowing in wood [229].

Table 2.3 The responsibility of lignocellulosics components on properties [230].

Strength	*Moisture absorption*	*Thermal degradation*	*Ultraviolet degradation*	*Biological Degradation*
Crystalline cellulose	Hemicellulose	Hemicellulose	Lignin	Hemicellulose
Non-crystalline cellulose	Non-crystalline cellulose	Cellulose	Hemicellulose	Non-crystalline cellulose
Hemicellulose	Crystalline cellulose	Lignin	Non-crystalline cellulose	Crystalline cellulose
			Crystalline cellulose	Lignin

Table 2.3 shows the effect of cellulosic components on the specific properties. The up to down direction indicates the order of responsibility.

The degradation of PP is due to photo-oxidation promoted by UV irradiation. Oxidative reaction initiated by UV radiation is represented as oxygen, and is used up before it can diffuse to the interior so that degradation is concentrated near the surface, even in polymers in which high UV levels are present in the interior. The photo oxidation process takes place mainly in the amorphous region because of the higher permeability to oxygen [231]. Ideally, composite materials are intended for long term use, should be tested in real time, and with realistic in-service environments. Often this is not viable because the time involved would significantly delay product development, and accelerated ageing techniques are required. During accelerated weathering, measured variables can include exposure time, exposure to UV irradiation over a specific wavelength range, and exposure to moisture as number of cycles or time. It is recommended that the performance of the materials after weathering be reported after a specific radiant exposure, the time integral of irradiance.

Ndiaye et al. studied the environmental degradation of wood PE/PP composites. The photooxidation initiated by forming free radicals and degradation mostly depends on the wood fiber contents [232].

Nicole et al. reported that exposing the wood fiber/HDPE composites to either UV radiation with water spray or UV radiation alone showed that the majority of the loss in mechanical properties after weathering was caused by moisture effects. The acceleration of oxidation reactions caused by water absorption results in swelling in the composites and weakening the interface between the wood and HDPE [233].

Seldén et al. studied UV aging of PP/wood fiber composites and found that the PP matrix and the PP/wood fiber composite both displayed good UV-resistance with regard to mechanical properties. The PP matrix displayed a 10% reduction in flexural strength and a 50% reduction in impact strength. The wood-fiber composites displayed a maximum 20% reduction in flexural strength and almost retained impact strength. The degradation of the composites was restricted to a thin surface layer, owing to the screening effect of the wood fibers. The degraded layer had a chalky appearance, due to degradation of the PP matrix, leading to chemicrystallization and extensive surface cracking. The rate of degradation of the PP matrix was approximately twice as high in samples with 50 wt% wood-fibers, compared with samples with 25 wt% fibers, owing to the higher number of chromophores in the former. DSC scans of degraded surface layers revealed a maximum 33°C decrease in PP melting temperature, due to molecular chain scission and the formation of extraneous groups, such as carbonyls and hydroperoxides [234].

2.4 MODIFICATION OF GRAIN WASTE PRODUCTS

2.4.1 Selection of Enzyme

Prior to enzyme modification, a mixture of enzymes were selected for specific grain waste products depending on their chemical contents. Amylase is an amylolytic enzyme which is able to remove amylose from starch. It does not remove amylopectin. Cellulase and xylanase (cellulolytic) enzymes are able to remove hemicellulose and part of accessible amorphous cellulose. Pectinase (pectinolytic) enzyme is able to remove pectin and hemicellulose and laccase (Ligniolytic) is able to remove lignin. Lipase enzyme is able to remove fat from fiber surface. To avoid complexity, name has given for those mixtures of enzymes as "mix". These given names are not similar to enzyme trade name.

2.4.2 Enzyme Treatment

2.4.2.1 Fungamix

Fungamix is a mixture of amylase-xylanase, laccase, and protease enzymes, and the contribution of enzymes ratio was 1:2:1 by weight, respectively. Fibers were submerged in water for half an hour prior to treatment. After that fibers were placed into autoclave with fiber to liquor ratio 1:7, under continuous agitation at 30 rpm for four hours. The enzyme treatments were incubated temperature starting at 40°C to 65°C with intervals of 5°C to optimise the temperature. The enzyme concentrations were 0.25, 0.50, 1.0, 1.25, 1.50, and 1.75 wt% per total volume of liquor. The liquor pH was adjusted 7.5 to using sodium hydroxide together with phosphate buffer. The liquor pH was increased adding sodium hydroxide and boiled for 30 minutes to deactivate the enzymes. Fibers were then filtered and dried in an oven.

2.4.2.2 Novamix

Novamix is a mixture of lipase, protease, and amylase-xylanase enzymes, and the contribution of enzymes ratio was 1:1:2 by weight, respectively. The treatment was incubated for four hours, adjusting liquor pH 8.0 with sodium hydroxide and citrate-phosphate buffer. The same treatment method was pursued for all enzymes mixtures.

2.4.2.3 Lipamix

Lipamix is a mixture of lipase, protease, and amylase-xylanase enzymes, and the contribution of enzymes ratio was 2:1:1 by weight, respectively. The treatment was incubated for five hours, adjusting liquor pH 8.0 with sodium hydroxide and citrate-phosphate buffer.

2.4.2.4 Laccamix

Laccamix is a mixture of laccase, protease, and cellulase enzymes, and the contribution of enzymes ratio was 2:1:1 by weight, respectively. The treatment was incubated for five hours, adjusting liquor pH 5.0 with acetate-phosphate buffer.

2.4.2.5 Palkomix

Palkomix is a mixture of pectinase, protease, and amylase-xylanase enzymes, and the contribution of enzymes ratio was 2:1:1 by weight, respectively. The treatment was incubated for six hours, adjusting liquor pH 6.0 with acetic acid and potassium dihydrogen phosphate buffer.

2.5 RESULTS AND DISCUSSION

2.5.1 Properties

Grain waste products are characterized and properties are explained below step by step. The performance of final product, which is from grain waste products, depends on the properties of individual components and their interfacial compatibility. So the contribution of fiber properties is very important for the properties of the final product.

2.5.1.1 Particle Size Distribution

Grain waste products are available in many different forms and produce different properties when added to thermoplastics. They may be used in the form of particles, fiber bundles, or single fibers, and may act as a filler or reinforcement for plastics. The specific particle size distribution in commercial grain waste varies among manufacturers, or the distribution range is broad from single manufacturer. They could be reprocessed to narrow the particle size distribution, raising the cost of the products. Therefore, typical commercial grades include a mixture of particle sizes. To get the intending properties of composites, it is important to measure the particle size distribution of reinforcing materials. The effective surface area of particle has an influence on properties of composites. The particle length distribution of grain waste products is represented in Fig. 2.6 . It was observed that 75–95% of all types of fibers were distributed in the range of 50–300 μm, but the distributions are not similar. Eighty three percent of wheat husk (D), 88% of rice husk, 77% of rye husk, 89% of coconut shell, and 92% of soft wood fiber lengths are lower than 300 μm, respectively. It could be explained that soft wood fibers are smaller than grain waste products. It was also observed that 23% of rye husk, 17% of wheat husk (D), and 8% of soft wood are bigger than 300 μm.

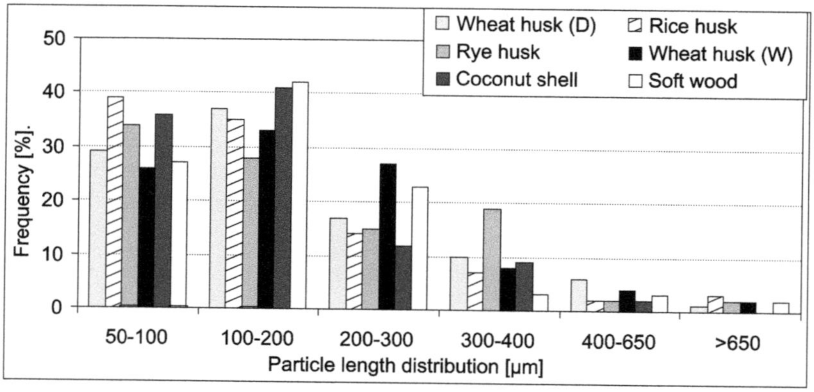

Figure 2.6 Particle length distribution of grain waste products.

The aspect ratios of grain waste products are given in Table 2.4. A broad range of aspect ratio was observed for grain waste products including soft wood fiber. The aspect ratio of wheat husk (D) is assigned in the range 1.4 to 4.6. A similar trend was also observed for wheat husk (W). On the other hand, coconut shell and rice husk were distributed in the lower range starting from about one. It means that some portions of particles are round shaped bimodal distribution. The aspect ratio of soft wood fiber is upper ranged starting from 2.2 to 5.8. In comparison with grain waste products, soft wood fibers are fibrous and longer.

Table 2.4 Aspect ratio of grain waste products and soft wood.

Fiber	*Wheat husk* (D)	*Rice husk*	*Rye husk*	*Wheat husk* (W)	*Coconut shell*	*Soft wood*
Aspect ratio	1.4–4.6	1.1–3.6	1.2–4.8	1.4–4.4	1.1–3.8	2.2–5.8

2.5.1.2 Bulk Density

The bulk density of grain waste products and soft wood fiber are shown in Table 2.5. The bulk density of coconut shell was observed at the highest (860 kg/m^3) and soft wood fiber was observed at the lowest (650 kg/m^3). The higher bulk density means a lower storage and transport space which could provide low cost handling. On the other hand, the bulk density of wheat husk (D), rice husk, rye husk, and wheat husk (W) was observed to be 750 kg/m^3, 840 kg/m^3, 700 kg/m^3, and 790 kg/m^3, respectively.

Table 2.5 Bulk density of grain waste products and soft wood.

Fiber	*Wheat husk* (D)	*Rice husk*	*Rye husk*	*Wheat husk* (W)	*Coconut shell*	*Soft wood*
Bulk density (kg/m^3)	750	840	700	790	860	650

2.5.1.3 Thermo Gravimetric Analysis

The thermogravimetric analysis relies on a high degree of precision in three measurements: weight, temperature, and weight change with temperature. The feature of this analysis is to determine degradation temperatures, absorbed moisture content of materials, the level of inorganic and organic components in materials, decomposition peaks temperature, and residues. A derivative weight loss curve can be used to tell the point at which weight loss is most apparent. This analysis of grain waste products and soft wood fibers is shown in Fig. 2.7. According to their chemical components and their crystallinity and molecular weight, every fiber showed different trends of thermal degradation. The DTG curves of all fibers showed initial peaks between 30°C and 130°C, which corresponds to the vaporization of water. After those peaks, the curves for rice husk, rye husk, and soft wood exhibit single decomposition step, and the decomposition peak temperatures are at 360°C, 306°C, and 383°C, respectively. On the other hand, the curves for wheat husk (D) and coconut shell exhibit two decomposition steps and the decomposition peaks temperature are at 309/367°C and 295/365°C consequently. The decomposition start temperature of coconut shell, wheat husk (D), rice husk, and soft wood was found to be 207°C, 205°C, 214°C, and 210°C, respectively.

The decomposition peak at around 300°C is due to the thermal decomposition of hemicelluloses and the glycosidic linkage of cellulose. The peak in the range 350–390°C is due to α- cellulose decomposition. The degradation of lignin initiates first and it continues with a slower rate than cellulose. The degradation of lignin depends on the molecular weight of lignin [235]. The peak corresponding to the lignin is slightly broader and it appears in the range 190–500°C, with maximum value at about 350°C. So there is a possibility to overlap the decomposition curves with each other [236, 237].

Since fibers are themselves composite materials, the decomposition temperature of fibers greatly depend on the amount of contents, number of hydrogen bonds associated with each other, molecular weight, and crystal/amorphous portion of specific content. The TGA experiment was associated with two steps. The first step is in inert medium (25°C to 600°C) and second step is in oxygen medium. The sudden step down at about 600°C is because of sudden oxidation of fiber contents.

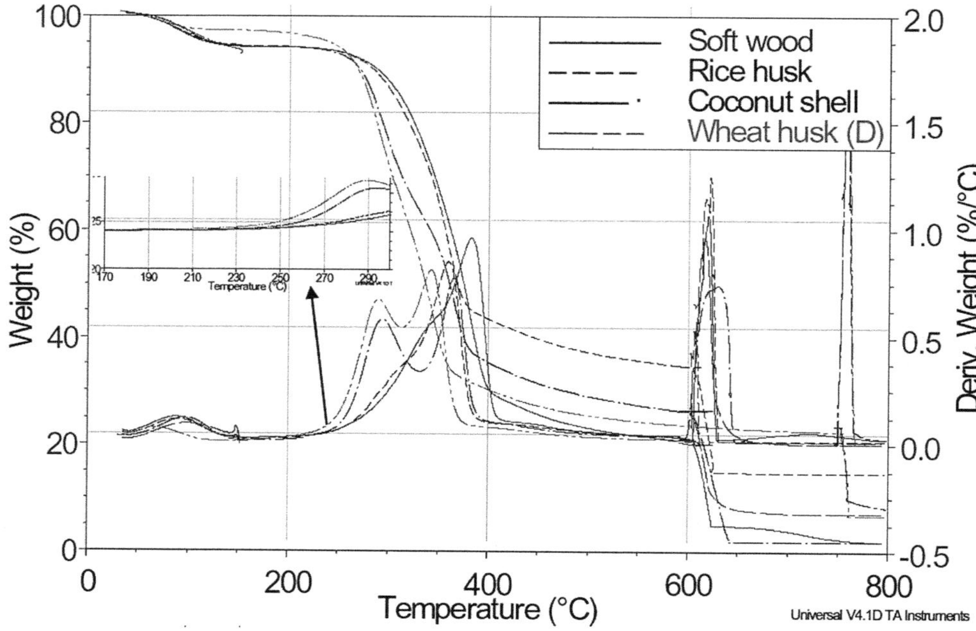

Figure 2.7 TGA of grain waste products and soft wood.

2.5.1.4 Moisture Absorption

Moisture uptakes in cell voids or lumens are known as free water and in the cell wall it is bond water. It depends on the specific gravity of fiber and specific gravity is vice-versa of lumen volume; lumen volume decreases as the specific gravity increases. As there is moisture uptake in fiber, there is no change in the volume (dimensions) of fiber until it reaches the fiber saturation point, and after that the dimensions change with respect to the relative humidity. That is why it is a dynamic property [238].

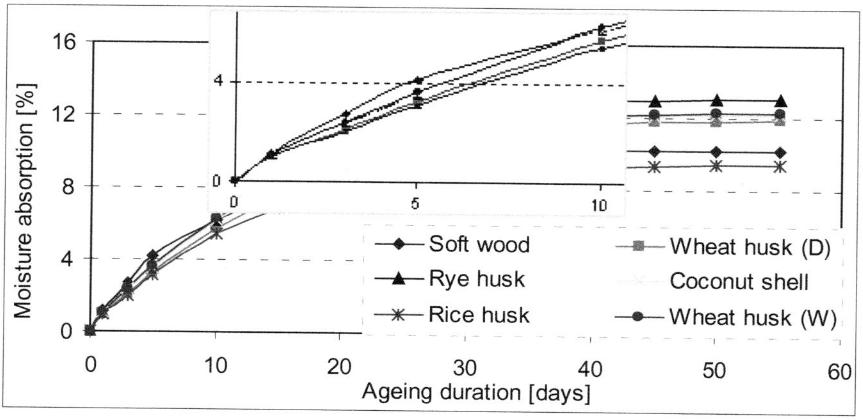

Figure 2.8 Moisture absorption of grain waste products and soft wood [23°C, RH 65%].

Moisture absorption of grain waste products and soft wood is illustrated in Fig. 2.8. During the early conditioning periods (till 10 days), the moisture uptake by soft wood fiber was found to be higher than moisture uptake by grain waste products. This may be due to the different surface morphology of fibers, which are composed of fat and protein: these, bonded together covalently, act as fiber coating. That is why moisture diffusion kinetic was hindered by surface contents and took a little more time to diffuse in the fiber. As it was shown in Table 2.5, all grain waste products contain certain amount of fat and protein. On the other hand, soft wood fiber contains very little amount of fat and protein on the surface. So, moisture uptake took place without initial hindering (nearly absent coating materials), depending on the overall chemical contents of soft wood fiber. Moisture uptake takes place mainly by hemicelluloses, noncrystalline cellulose, accessible cellulose, starch, and other polar surface molecules [239, 240].

The moisture uptake for all types of grain waste products was found to be at equilibrium after 45 days of conditioning periods. The equilibrium moisture content of rice husk was found to be 8% lower than moisture uptake by soft wood fiber. On the other hand, moisture uptake by all others grain waste products was found to be higher than soft wood fiber. Moisture absorptions were found to be higher 16% for wheat husk (D), 27% for rye husk, 19% for coconut shell, and 20% for wheat husk (W) than soft wood. A similar trend was observed at 95% RH, as shown in Fig. 2.9. The moisture uptake for all types of fiber was found to be at equilibrium after 55 days of conditioning periods and the equilibrium moisture content of wheat husk (D), rye husk, coconut shell, and wheat husk (W) were 10%, 22%, 18%, and 20% higher than moisture uptake by soft wood fiber, respectively. The equilibrium moisture content of rice husk was found to be lower than moisture uptake by soft wood.

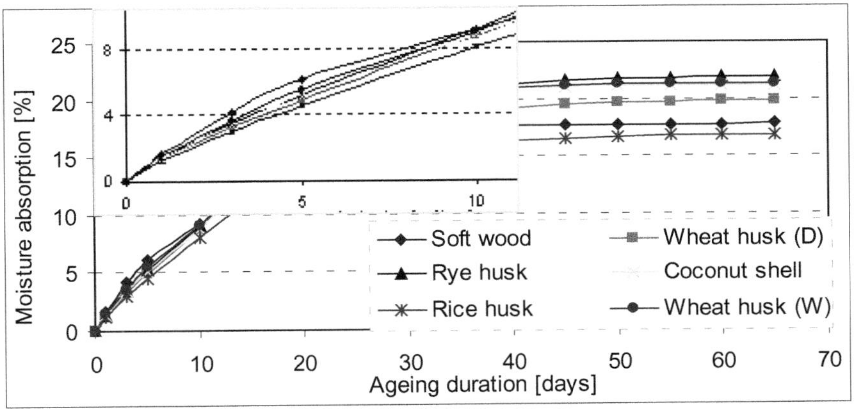

Figure 2.9 Moisture absorption of grain waste products and soft wood [23°C, RH 95%].

2.5.1.5 FT-IR

The FT-IR spectrums of wheat husk (D) and rye husk fibers are shown in Fig. 2.10. The spectrums of other grain waste products are not shown, but the summarized results are given in Table 2.6. The peaks in the range of 3460–3400 cm^{-1} are due to hydrogen bonded O–H stretching. The hydrophilic tendency of grain waste products was reflected in the broad absorption band (3100–3700 cm^{-1}), which is related to the –OH groups present in aliphatic or aromatic alcohol and present in their main components. The peaks in the range of 3000–2850 cm^{-1} are due to the C–H asymmetric and symmetric stretching from aliphatic saturated compounds. These stretching peaks are corresponding to the aliphatic moieties in cellulose and hemicelluloses [241].

In the triple bond region, the peaks in the range of 2400–2300 cm^{-1} are attributed to the P–H stretching, and P–OH stretching and the bands in the range of 2200–2100 cm^{-1} are reflected by Si–H stretching. It could be summarised that the bands in this region were representing the multivalent inorganic hydride or oxide [242]. In the double bond region, a shoulder peak in the range of 1738–1700 cm^{-1} in all spectra is associated with the C=O stretching of the acetyl and uronic ester groups of hemicelluloses or to the ester linkage of carboxylic group of the ferulic and p-coumaric acids of lignin [243].

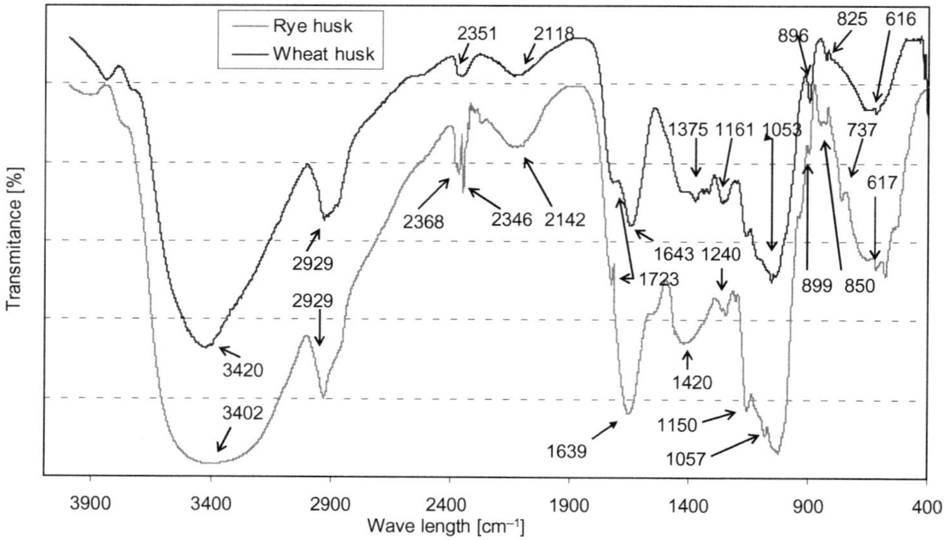

Figure 2.10 FT-IR of wheat husk (D) and rye husk.

Table 2.6 FT-IR of grain waste products.

Peak location range (cm^{-1})	*Assignment*	*Wheat husk* (D)	*Rice husk*	*Rye husk*	*Coconut shell*	*Wheat husk* (W)
3460–3400	O–H stretching	3420	3400	3402	3401	3400
3000–2850	C–H asymmetric and symmetric stretching in methyl and methylene groups	2929	2903	2929	2911	2921
2400–2300	P–H stretching and P–OH stretching	2351	–	2368/ 2346	2326	2364/ 2344
2200–2100	Si–H stretching	2118	2107	2142	2106	–
1738–1700	C=O stretching in acetyl and uronic ester groups or in carboxylic group of ferulic and coumaric acids	1723	1742	1723	1721	1724
1650–1580	N–H bending in primary amin or C=O stretching in hemicelluloses	1643	1639	1639	1621	1638
1500–1400	C–C stretching in aromatics ring	–	–	1420	–	–
1375–1350	C–H rocking in alkanes or C–H stretching in methyl and phenolic alcohol	1375	1363	–	1371	1373
1250–1150	Si–CH_2 stretching in alkane or C–C plus C–O plus C=O stretching in lignin	1161	1242/ 1170	1240/ 1150	1249	1239
1086–1030	C–O deformation in secondary alcohol and aliphatic ether or aromatic C–H in plan deformation plus C–O deformation in primary alcohol	1053	1041	1057	1032	1058
900–875	C–1 group frequency or ring frequency	896	895	899	893	897
850–550	Alkyl halides stretching	825	784	850	768	750
750–700	C–H rocking in alkanes	–	723	737	–	–
700–600	C–C triple bond : C–H bending in alkynes	616	604	617	603	605

The sharp peaks in the range of 1650–1580 cm^{-1} were reflected for amide I. The amide I band represents 80% of the C=O stretching of the amide group, coupled to the in-plane N–H bending and C–N stretching modes. The exact frequency of this vibration depends on the nature of hydrogen bond involving the C=O and N–H groups and the secondary structure of protein [241]. This peak also represents the C=O bonds in hemicellulose [243]. In the fingerprint region, the bands in the range of 1375 cm^{-1}–1350 cm^{-1} were assigned for the C–H rocking in alkanes or C–H stretching in methyl or phenolic alcohol. The peaks in the region of 1250–1150 cm^{-1} represents Si–CH2 stretching in alkanes or C–C plus C–O plus C=O stretching and the bands in the range of 1080–1030 cm^{-1} are attributed to the C–O deformation in secondary alcohol and aliphatic ether or primary alcohol in cellulose and lignin.

A small sharp peak in the range of 900–875 cm^{-1}, which is indicative of the C–1 group frequency or ring frequency, is the characteristic of β glycosidic linkages between the sugar units. The peaks, in the region of 850–550 cm^{-1}, stand for alkyl halide stretching, in the region of 750–700 cm^{-1}, for C–H rocking in alkanes, in the region of 700–600 cm^{-1}, for C–C triple bond: C–H bending in alkynes [241]. It could be noted that all grain waste products contain more or less similar chemical constituents but different peak areas, which are associated with different amounts. In many cases the peaks for same constituent were shifted, which is associated with the nature of hydrogen bonding and coupling effects.

2.5.1.6 EDX

The fiber consists of mainly cellulose (O/C=0.83), hemicelluloses (polysaccharides with a range of monomers, where O/C≈0.83), lignin (O/C=0.35), pectin (O/C similar to hemicelluloses), and wax [244]. Figure 2.11 shows the inspection spectra of surface elements acquired for rye husk. The elemental compositions and oxygen carbon atomic ratio of grain waste products and soft wood are given in Table 2.7.

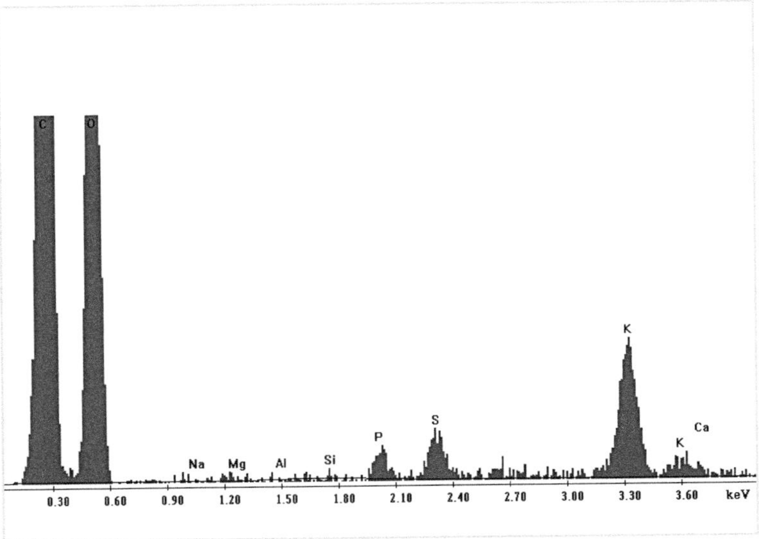

Figure 2.11 Elementary analysis of rye husk.

Grain waste products exhibit spectra containing mainly carbon, oxygen, and small amount of silicon, potassium, sulfur, phosphorus, sodium, magnesium, and aluminium. The relative atomic percent of the atoms were obtained from the peak area and corrected with an appropriate sensitivity factor.

Wheat husk (W) contains highest proportion of carbon atom as compared to other fibers, and rice husk contains lowest proportion of carbon atom. On the other hand wheat husk (W) contains lowest proportion of oxygen atom and soft wood fiber containns highest proportion of oxygen atom in comparison with other fiber. The higher proportion of carbon in fiber can be attributed to the presence of hydrocarbon rich waxy coating on the cuticle of fiber and lignin present on the surface.

Rice husk contains 10.8% silicon (as oxide form), which is much higher than other fibers. Wheat husk (D) contains double proportion of silicon as compared to soft wood, whereas rye husk, and coconut shell contain a little proportion of silicon compared with soft wood. The silicon content in fiber may have the influence on the properties of fiber reinforced composites. The highest oxygen-carbon ratio was observed for rice husk and the lowest was observed for wheat husk (W). On the other hand the oxygen-carbon ratio of rye husk was found to be lower than wheat husk (D) and coconut shell. The lower ratio of O/C in the wheat husk (W) and rye husk indicated a higher proportion of aliphatic and aromatics carbons near the surface than wheat husk (D) or coconut shell as well as soft wood. Hence, it is anticipated that wheat husk (W) and rye husk appear as a high compatibility with non-polar polymers compared with wheat husk (D) and soft wood fiber, whereas a large number of parameters are related to the compatibility.

Table 2.7 Elementary analysis of grain waste products and soft wood.

Fiber	*C (%)*	*O (%)*	*Si (%)*	*K (%)*	*S (%)*	*P (%)*	*O/C (%)*
Wheat husk (D)	74.9	22.4	2.1	0.6	–	–	0.30
Rice husk	64.9	24.1	10.8	0.2	–	–	0.37
Rye husk	75.6	18.9	0.2	3.3	1.3	0.7	0.25
Coconut shell	74.3	21.9	0.2	1.4	0.5	1.7	0.29
Wheat husk (W)	76.4	18.3	–	1.9	0.7	2.7	0.24
Soft wood	73.9	25.0	1.1	–	–	–	0.34

Note: Elementary analysis could provide information about surface energy, polarity, and consequently wettability.

2.5.1.7 Chemical Composition

Grain waste products have similar structure and chemical components as natural fibers. So, they behave like natural fibers and the properties, e.g., moisture absorption, dimensional stability, thermal degradation, biodegradation, photochemical degradation, and pyrolysis reactions to give off volatile and flammable gas at increasing temperature. All those properties depend on chemical compositions of fiber. So, it is obvious that the chemical composition and amount plays a vital role in fiber properties as well as composites properties if it is used as a reinforcing agent [245, 246].

The chemical compositions of grain waste products and soft wood are presented in Table 2.8. The structural material (cellulose) contained by wheat husk (D) is 36%, by rice husk is 47%, by rye husk 26%, by wheat husk (W) is 23%, by coconut shell is 34%, and by soft wood is 42 percent.

Table 2.8 Chemical contents of grain waste products and soft wood.

Compositions (%)	*Wheat husk (D)*	*Rice husk*	*Rye husk*	*Wheat husk (W)*	*Coconut shell*	*Soft wood*
Cellulose	36	47	26	23	34	42
Hemicellulose	18	14	16	21	21	22
Lignin	16	22	13	14	27	31
Starch	9	3	17	19	0	0.5
Protein	6	2	10	7	2	0.5
Fat	5	2	7	9	5	0.5

Whereas, another material starch contained by wheat husk (D), rice husk, rye husk, and wheat husk (W) are 9%, 3%, 17%, and 19%, respectively. Rice husk contains no starch, but soft wood contains less than 1 percent.

It is also observed that wheat husk (D), rice husk, rye husk, wheat husk (W), and coconut shell contain 6%, 2%, 10%, 7%, and 2% of protein and 5%, 2%, 7%, 9%, and 5% of fat, respectively. Protein and fat contents of soft wood are less than 1 percent. On the other hand, hemicelluloses and lignin content of soft wood are 22% and 31%, that are higher in contentthan grain waste products. The rest of the compositions of fibers are inorganic and ash contents.

Note: Chemical contents are accountable for fiber physical, chemical, and thermal properties.

2.5.2 Enzyme Modification of Fiber

According to chemical contents of grain waste products, wheat husk (D) was modified by novamix, rice husk was modified by fungamix, rye husk was modified by lipamix, coconut shell was modified by laccamix, and wheat husk (W) was modified by palkomix. The enzyme modification process parameters, e.g., temperature, dosing rate, and duration were optimized in terms of weight loss of fiber. The main aim of the modification was to remove fat, protein, extractives, lignin, hemicellulose, and amylose from the fiber surface without damaging the fiber structure. In this section, rice husk modification data are shown in the graphical form and other data are shown in the tabular form.

2.5.2.1 Optimization of Temperature

The temperature optimization of modification process depends on weight loss of the fiber. A certain percent of enzyme (0.5 wt%) and three different treatment durations were considered for temperature optimisation. The temperature range was considered according to the product data sheet (lower and upper limits). In case of rice husk, the temperature range from 40°C to 60°C were taken and weight loss was measured at every 5°C interval, which is shown in Fig. 2.12. It was observed that for all three treatment durations, the weight loss was higher at about 50°C. It may be concluded that 50°C is the optimum temperature for this system. The optimum temperatures of other grain waste products are given in Table 2.9.

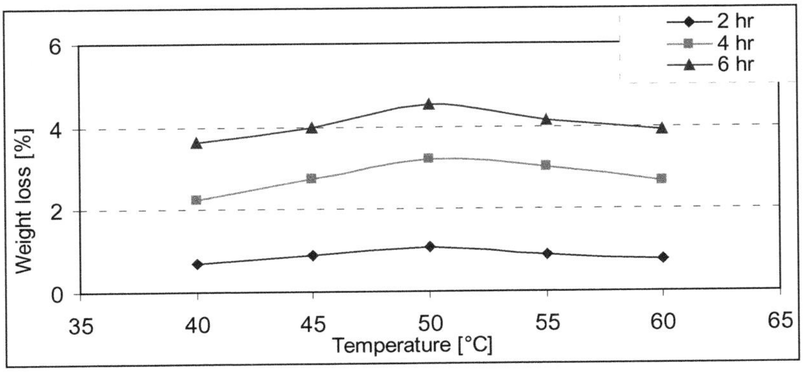

Figure 2.12 Optimal temperature of rice husk modification.

Table 2.9 Optimal temperature, dosing rate, and duration of grain waste products.

Fiber	*Enzyme*	*Temperature* (°C)	*Dosing rate* (wt%)	*Duration* (hr)
Wheat husk (D)	Novamix	60	1.25	4
Rice husk	Fungamix	50	1.0	5
Rye husk	Lipamix	40	1.50	4
Coconut shell	Laccamix	40	1.0	4
Wheat husk (W)	Palkomix	60	1.25	6

2.5.2.2 Optimization of Dosing Rate and Modification Duration

Figure 2.13 shows the enzyme activity on treatment duration and enzyme dosing rate. It can be seen that the activity was found to increase approximately linearly with treatment duration at the lower concentration of enzyme (till 1 wt%). After that the enzyme activity increased a little bit with respect to the treatment duration. On the other hand, the enzyme activity increased slightly after treatment duration of five hours. The slope of the curve decreases with treatment time. This is because of production of a wide range of water soluble products, those molecules have an influence on enzyme activity. It may be summarised that the optimum enzyme dosing rate is 1 wt% per litre of liquor and the optimum treatment duration is 5 hours.

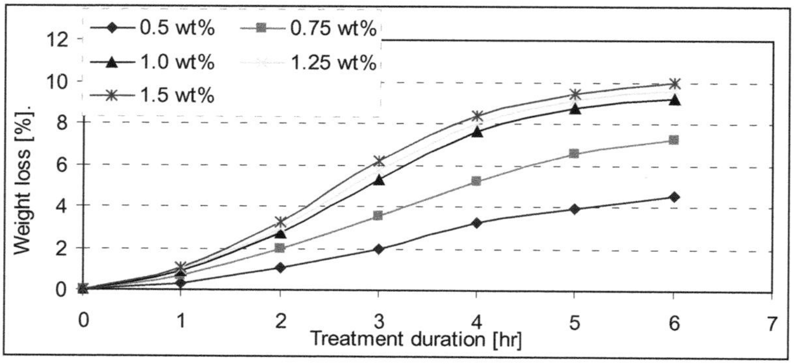

Figure 2.13 Optimal dosing rate and duration of rice husk modification [50°C].

The optimized dosing rates and durations for other grain waste products were determined at optimum temperature and are given in Table 2.9.

2.5.3 Fiber Characterization after Modification

2.5.3.1 FT-IR

Rice husk was analysed using FT-IR to know the changes chemical constituents due to fungamix treatment. The FT-IR spectrums of this fiber are shown in Fig. 2.14.

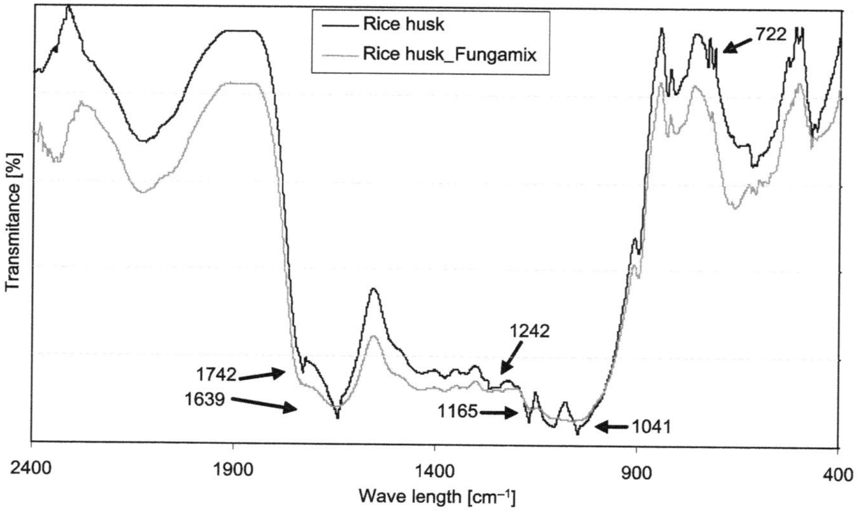

Figure 2.14 FT-IR spectrum of fungamix treated and untreated rice husk.

A shoulder peak at 1742 cm^{-1} in the untreated rice husk spectrum is assigned to the C=O stretching of the acetyl and uronic ester groups of hemicelluloses or to the ester linkage of carboxylic group of the ferulic and *p*-coumaric acids of lignin. On the other hand, for the treated rice husk, the shoulder peak was nearly absent, which indicates the reduction of hemicelluloses or lignin. The sharp peaks at 1639 cm^{-1} for untreated rice husk were reflected for amide I.

The broadening of peaks at 1639 cm^{-1} for enzyme treated rice husk indicated reduction of protein content. Natural corn fat contains three functional groups: carboxylic functional groups and two different hydroxylic groups. The combination of sharp peaks at 1165 cm^{-1}, 1041 cm^{-1}, and 722 cm^{-1} for untreated rice husk were associated CO–O–C asymmetric stretching, C–O–P stretching, and CH_2 rocking, respectively, which indicates the fat and lipids content. On the other hand, for fungamix treated fiber those peaks were absent. So, due to modification, the protein, fat, hemicellulose, and lignin were removed from the fiber surface. The summarized results of other grain waste products are given in Table 2.10.

To avoid difficulties the fiber names were abbreviated as follows: wheat husk (D) as "WH (D)", rich husk as "RH", rye husk as "Rye H", coconut shell as "CS", wheat husk (W) as "WH (W)", and soft wood as "SW".

Table 2.10 Selected peaks of treated and untreated grain waste products.

Peaks (cm^{-1})	*WH (D)*	*WH (D)-Novamix*	*Rye H*	*Rye H-Lipamix*	*CS*	*CS-Laccamix*	*WH (W)*	*WH (W)-Palkomix*
1738-1700	Sharp	Absent	Sharp	Absent	Sharp	Broden	Sharp	Absent
1650-1580	Sharp	Broden	Sharp	Broden	Broden	Absent	Sharp	Absent
1250-1150	Sharp	Broden	Broden	Absent	Sharp	Absent	Sharp	Broden
1086-1030	Broden	Absent	Sharp	Broden	Broden	Absent	Broden	Absent
750-700	-	-	Sharp	Broden	-	-	-	-

It was observed that due to modification, the selected peaks were either absent or broadened for all grain waste products. It means that the extractives, protein, fat, lignin, and hemicelluloses were removed from fibers.

2.5.3.2 Particle Size Distribution

Particle length distribution was measured after modification for wheat husk (W), and is shown in Fig. 2.15. The summarized results of changes of particle size for other grain waste products are given in Table 2.11. It can be scrutinised that the enzyme modified fiber length distributions were shifted to the smaller particle size. This is because of the removal of unwanted components from the surface of the fibers. A few percent of measurement error could be considered for this measurement.

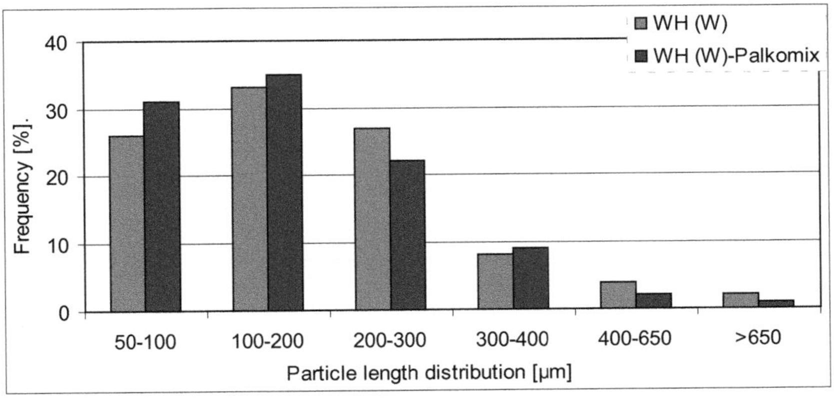

Figure 2.15 Particle size distribution of wheat husk (W) before and after treatment.

Table 2.11 Changes of particle size distribution due to modification.

Fiber size (μm)/%	50–100	100–200	200–300	300–400	400–500	>650
WH (D) → WH (D)-novamix	29→ 32	37→ 39	17→ 18	10→ 5	6→ 2	1→ 1
RH→ RH-fungamix	39→ 41	35→ 38	14→ 13	7→ 4	2→ 2	3→ 2
Rye H→ Rye H –lipamix	34→ 36	28→ 32	15→ 16	19→ 14	2→ 1	2→ 1
CS→ CS-laccamix	36→ 38	41→ 42	12→ 15	9→ 5	2→ 1	0→ 0

2.5.3.3 Moisture Absorption

The moisture absorption of treated and untreated wheat husk (W) and coconut shell is shown in Fig. 2.16 and the summarized results of other grain waste products before and after modification are shown in Table 2.12. The moisture uptake of all modified and unmodified grain waste products was found to be at equilibrium after 55 days. Due to modification moisture uptake was reduced by wheat husk (D) for 49%, rice husk for 54%, rye husk for 39%, coconut shell for 41%, and wheat husk (W) for 35%. The reduction of moisture uptake is due to removal of hemicelluloses, amorphous cellulose, and others polar molecules from the fiber surface. Additionally, –OH groups of cellulose and hemicelluloses are supposed to be modified by sodium hydroxide during the enzyme deactivation process.

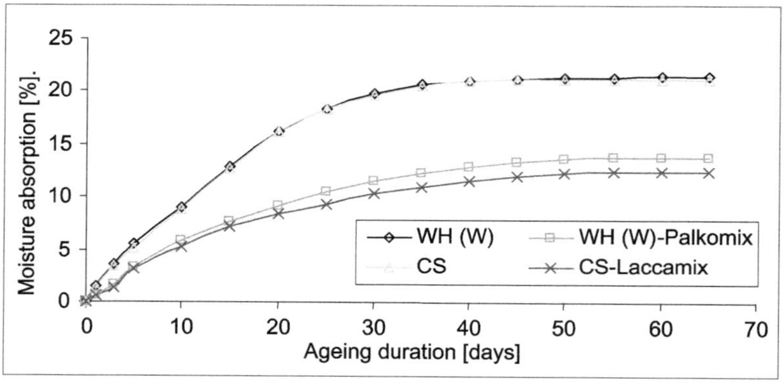

Figure 2.16 Moisture absorption of modified and unmodified wheat husk (W) and coconut shell [23°C, RH 95%].

Table 2.12 Moisture absorption of grain waste products before and after modification.

Fiber	*Equilibrium moisture content* (%)
WH (D)	19±3
WH (D)-novamix	10±2
RH	16±2
RH-fungamix	7±1
Rye H	21±3
Rye H-lipamix	15±2

2.5.3.4 Surface Morphology

The fiber surface layer is a relatively minor portion of the fiber, but it plays an important role in wettability and surface tension. It has long been postulated that the cereal fiber surface contains extractive, lipid, and proteinaceous compound, and the lipid molecule is usually bonded to the

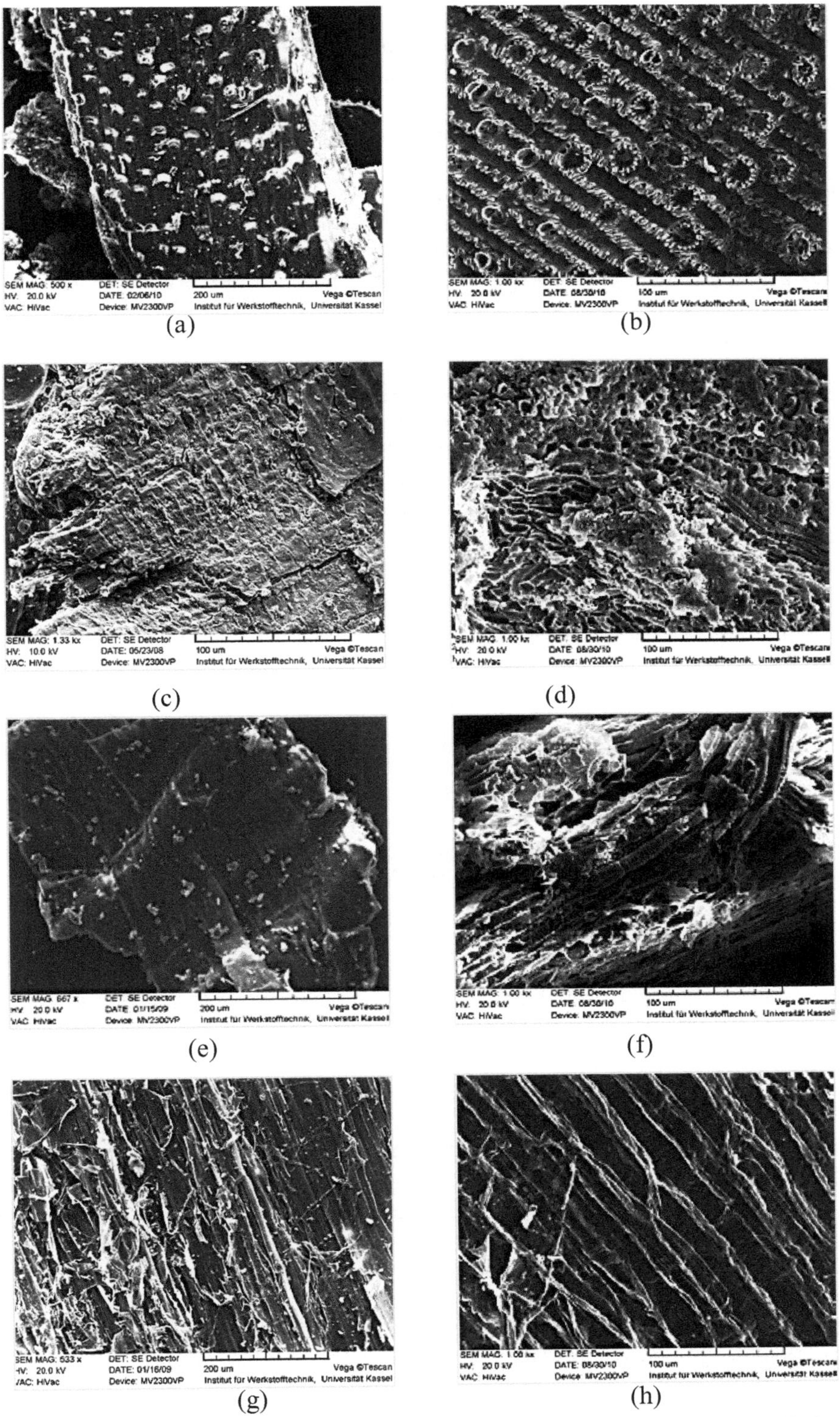

Figure 2.17 Surface morphology of (a) wheat husk (D), (b) modified wheat husk (D), (c) rice husk, (d) modified rice husk, (e) rye husk, (f) modified rye husk, (g) wheat husk (W), and (h) modified wheat husk (W).

protein molecule by ester or thioester bonds [244]. The amount of lipid on the fiber surface has an influence on hydrophobicity as well as on surface tension. The more amount of lipid on the fiber surface means the more hydrophobic and more surface tension, as well as smoother the fiber surface forming a thin film.

The surface morphologies of unmodified and enzyme modified grain waste products are shown in Fig. 2.17. A number of cracks, defects, and some damages were observed on the fiber surface for all cases of unmodified fiber. Wheat husk (W) fiber surface was relatively rough and a number of fiber damages were observed. Rye fiber surface was the smoothest of all fibers. Fat and proteinaceous molecules produced thick coat on fiber surface. On the other hand, rice husk surface was partially smooth but a lot of cracks present on the surface. There are many knobs and small holes on wheat husk (D) surface.

Fat, protein, lignin, and hemicelluloses are removed from wheat husk (D) surface due to novamix modification, which can be seen in Fig. 2.17(b). The spring-like things were exhibited on the surface of fiber. These are supposed to be micro fibrils. These micro fibrils are short and arranged orderly. Some defects and holes were observed, but no fiber damage occurred by the enzyme modification.

In Fig. 2.17(d), fungamix modified rice husk was observed. It was observed that the fiber surface howed to be full of pores, which is because of removal of packing materials. Coating materials are still on fiber surface in some places.

A number of holes, defects, and some damages were observed on the lipamix modofied rye husk surface due to removal of coating materials (fat and protein), as well as packing and binding materials, which can be seen in Fig. 2.17(f). On some places on the rye husk surface, existence of coating materials was observed.

There are many damages present on wheat husk (W) surface, which can be seen in Fig. 2.17(h). Some spring-like micro fibrils were also observed and some places on the fiber surface were also covered by coating materials

Note: It was clearly observed that for all cases the unwanted materials were found to be removed from the fiber surface. However, in some cases pores and holes were found on the fiber surface, which provides improvement of wettability and mechanical interlinking or anchoring with matrix material.

2.6 CONCLUSIONS

This study inspected the feasibility of utilizing grain waste products as alternative fillers for soft wood fiber as reinforcement for composites material. The following conclusions could be drawn:

- Grain waste products are cheaper than soft wood fiber, and an enormous amount is available.
- Grain waste products' chemical compositions and surface properties have an effect on composite properties.
- Enzyme modification was found to be an efficient modification process for grain waste products. Due to enzyme modification, almost all properties (physical, mechanical, chemical, and electrical) were found to improve significantly and the odor and emission property was reduced significantly for all composites.
- Enzyme modification could be an alternative for traditional coupling agents.
- Wheat husk (D) composites showed almost best properties in all grain waste products composites and all those properties are more or less comparable with soft wood fiber composite.

So grain waste products could be valuable raw materials in the field of composite materials, where the source of soft wood fiber is becoming limited.

References

1. Suddel, B.C. and W.J. Evans. Natural fibre composites in Automotive Applications. *In*: A.K. Mohanty, M. Misra and L.T. Drzal (eds). Natural Fibers, Biopolymers and Biocomposites. Taylor & Francis, New York, 2005, pp. 231-259.
2. Frank, R.R. Bast and other plant fibres, Overview. Woodhead Publishing Limited, Cambrige, England, 2005, pp. 1-24.
3. Mohanty, A.K., M. Misra, and L.T. Drzal. Sustainable bio-composites from renewable resources: Opportunity and challenges in the green materials world. Journal of Polymer Environment 2002; 10: 19-26.
4. Bledzki, A.K., V.E. Sperber and O. Faruk. Natural and wood fibre reinforcement in polymers. Rapra Technology Ltd., Shropshire, 2002, pp. 28-30.
5. Rowell, R.M. Natural-fibre: Types and properties. *In*: K.L. Pickering (ed.). Properties and Performance of Natural-fibre Composites. Cambrige, England: Woodhead Publishing Limited, 2008, pp. 3-64.
6. Bledzki, A.K., A.A. Mamun, A. Jaszkiewicz and K. Erdmann. Polypropylene composites with enzyme modified abaca fibre. Composite Science and Technology 2010; 70: 854-860.
7. Bledzki, A.K. and J. Gassan. Composites reinforced with cellulosic fibre. Progress of Polymer Science 1999; 24: 221-274.
8. Paiva, M.C., I. Ammar, A.R. Campos, R.B. Cheikh and A.M. Cunha. Mechanical, morphological and interfacial characterization. Composites Science & Technology 2007; 67(6): 1132-1138.
9. Yao, F, Q. Wu, Y. Lei, W. Guo and Y. Xu. Thermanl decomposition kinetics of natural fibres: activation energy with dynamic thermogravimetric analysis. Polymer Degradation and Stability 2008; 93(1): 90-98.
10. Arifuzzaman Khan, G.M., M. Shaheruzzaman, M.H. Rahman, S.M. Abdur Razzaque, S. Islam and S. Alam. Surface modification of okra fibre and its physico-chemical characteristics. Fibers and Polymers 2009; 10(1): 65-70.
11. Bogoeva-Gaceva, G., M. Avella, M. Malinconico, A. Buzarovska, A. Grozdanov and G. Gentile. Natural fiber eco-composites. Polymer Composites 2007; 28(1): 98-107.
12. Pott, G.T. Natural fibre with low moisture sensitivity. *In*: F.T. Wallenberger (ed.). Natural fibers, Plastrics and Composites. Kluwer Academic Publishers, London, England, 2004, pp. 105-122.
13. Kumar, P., R. Chandra and S.P. Singh. Interphase effect on fibre-reinforced polymer composites. Composite Interfaces 2010; 17: 15-35.
14. Sgriccia, N., M.C. Hawley and M. Misra. Characterization of natural fiber surfaces and natural fiber composites. Composites: Part A 2008; 39: 1632-1637.
15. Baltazar-y-Jimenez, A. and A. Bismarck. Wetting behaviour, moisture up take and electrokinetic properties of lignocellulosic fibres. Cellulose 2007; 14: 115-127.
16. Batra, S.K. Other long vegetable fibre: Abaca, banana, sisal, henequen, flax, ramie, hemp, sunn and coir. *In*: M. Lewin (ed.). Handbook of Fiber Chemistry. Taylor and Francis, Boca Raton, FL, 2007, pp. 453-520.
17. Bismarck, A., S. Mishra and T. Lampke. Plant fibre as Reinforcement for green composites. *In*: A.K. Mohanty, M. Misra and L.T. Drzal (eds). Natural Fibers, Biopolymers and Biocomposites. Taylor & Francis, New York, 2005, p. 45.
18. Huda, M.S., L.T. Drazel, D. Ray, A.K. Mohanthy and M. Misra. Natural fibre composites in automobile section. *In*: K.L. Pickering (ed.). Properties and Performance of Natural Fibre Composites. Woodhead PublishingCambridge, England, 2008, pp. 221-268.
19. Piotrowski, S. and M. Carus. Natural fibre in technical application. *In*: J. Müssig (ed.). Industrial Applications of Natural Fibre. John Wiley & Sons, Hoffenheim, Germany, 2010; 63-71.
20. Mousavioun, P. and W.O.S. Doherty. Chemical and thermal properties of fractional bagasse soda lignin. Industrial Crops and Products 2010; 31: 52-58.
21. Chen, H., Y. Yang and J. Zhang. Biotechnological potential of cereal straw and bran residues. *In*: P.S. Nigam and A. Pandey (eds). Biotechnology for Agro-Industrial Residues Utilisation. Springer Science+Business Media, Germany, 2009, pp. 327-340.

22. Belgacem, M. and A. Gandini (eds). Report on Monomers Polymers and Composites from Renewable Resources. Elsevier, Amsterdam, 2008.
23. Vandamme, E.J. Agro-Industrial residue utilization for industrial biotechnology products. *In*: P.S. Nigam and A. Pandey (eds). Biotechnology for Agro-Industrial Residues Utilisation. Springer Science+Business Media, Germany, 2009, pp. 3-13.
24. Flaspohler, D.J., C.R. Webster and R.E. Froese. Bioenergy, biomass and biodiversity. *In*: B.D. Solomon and V.A. Luzadis (eds). Renewable Energy from Forest Resources in the USA. Routledge, Canada, 2009, pp. 133-162.
25. Ince, P.J. Fiber Resources. *In*: J. Burley, J. Evans and J.A. Yiongquist (eds). Encyclopedia of Forest Resources. Elsevier, New York, 2004, pp. 102-143.
26. Thamae, T., R. Marien, L. Chong, C. Wu and C. Baillie. Developing and characterizing new materials based on waste plastic and agro-fibre. Journal of Material Science 2008; 43: 4057-4068.
27. Panthapulakkal, S. and M. Sain. Agro-residue reinforced high-density polyethylene composites: Fiber characterization and analysis of composite properties. Composites: Part A 2007; 38: 1445-1454.
28. Wrigley, C., H. Corke and C.E. Walker. Encyclopedia of Grain Science. Elsevier, New York, 2004, p. 3.
29. Lehrack, U. and J. Volk. The innovative potential of cereal products for technical application. Symposium Fibre conference proceedings, Poznan, Poland, 2002.
30. Akeret, O. Plant remains from bell beaker site in Switzerland and the begainings of triticum spelta cultivation in Europe. Veget Hist Archaeobot 2005; 14: 279-286.
31. Santos, M.M., F. Gananca, J.J. Slaski and M.A.A. Pinheiro de Carvalho. Morphological characterization of wheat genetic resources. Genetic Resources and Crop Evolution 2009; 56: 363-375.
32. Hancock, J.F. Cereal grains. *In*: Plant Evolution and the Origin of Crop Species. Oxan, UK: CABI Publishing, 2004, pp. 174-194.
33. Serna-Saldivar, S.O. Cereal grains: Properties, processing and nutritional attributes. CRC Press, Boca Raton, FL, 2010, pp. 1-25.
34. Hanus, H., K.U. Heyland and E.R. Keller. Handbuchd des Pflanzenbaues, Band:2 Getreide und Futtergräser. Eugen Ulmer KG, Stuttgart, Germany, 2008.
35. Cauvain, S.P. and P. Cauvain. Bread Making. CRC Press, Boca Raton, FL, 2003, p. 40.
36. Smith, A.E. Handbook of Weed Management Systems. Marcel Dekker, New York, 1995, p. 411.
37. Daniel, Z. and H. Maria. Domestication of Plants in the Old World, 3rd Ed., Oxford: University Press 2000, p. 75.
38. Börner, A., V. Korzun, A.V. Voylokov, A.J. Worland and W.E. Weber. Genetic mapping of quantitative trait loci in rye (*Secale cereale* L.). Euphytica 2000; 116: 203-209.
39. Stojlowski, S., B. Myskow, P. Milczarski and P. Masojc. A consensus map of chromosome 6R in Rye (*Secale cereale* L.). Cellular and Molecular Biology Letters 2009; 14: 190-198.
40. Altpeter, F., V. Korzun. Rye. *In*: E.C. Pau and M.R. Davey (eds). Biotechnology in Agriculture and Forestry, vol. 59. Transgenic crops IV. Springer, Berlin, Germany, 2007, pp. 107-117.
41. Tripathy, J.N., J. Zhang, S. Robin, T.T. Nguyen and H.T. Nguyen. QTLs for cell-membrane stability mapped in rice (*Oryza sativa* L.). Theoretical and Applied Genetic 2000; 100: 1197-1202.
42. Vinod, M.S., N. Sharma, K. Manjunatha, A. Kanbar, N.B. Prakash and H.E. Shashidhar. Candidate genes for drought tolerance and improved productivity in rice (*Oryza sativa* L.). Journal of Bioscience 2006; 31: 69-74.
43. Numa, H., T. Tanaka and T. Itoh. Bioinformatics and database of the rice genome. *In*: H.Y. Hirano, A. Hirai, Y. Sano and T. Sasaki (eds). Biotechnology in Agriculture and Forestry, vol. 62. Rice Biology in the Genomics Era. Springer, Berlin, Germany, 2008, pp. 13-20.
44. Owen. S. and R. Owen. The Rice Book, 2nd Ed. Frances Lincoln Limited, London, UK, 2003.
45. Baudouin, L. and P. Lebrun. Coconut (*Cocos nucifera* L.) DNA studies support the hypothesis of an ancient Austronesian migration from Southeast Asia to America. Genetic Resources and Crop Evolution 2009; 56: 257-262.
46. Manimekalai, R. and P. Nagarajan. Interrelationships among coconut (*Cocos nucifera* L.) accessions using RAPD technique. Genetic Resources and Crop Evolution 2006; 53: 1137-1144.

47. Chavan, J.K. and S.J. Jadhav. Coconut. *In*: D.K. Salunkhe (ed.). Handbook of Fruit Science and Technology: Production, Composition, Storage, and Processing. Marcel Dekker, New York, 1995, pp. 485-505.
48. Duke, J.A. *Cocos nusifera* L. (*Arecaceae*) coconut. *In*: Handbook of Nuts. CRC Press, Boca Raton, FL, 2000, pp. 100-106.
49. World production data sheet, Foreign Agricultural Service, United States department of Agriculture, USA, www.fas.usda.gov
50. Izydorczyk, M.S. and J.E. Dexter. Milling and processing. *In*: C. Wrigley, H. Corke and C.E. Walker (eds). Encyclopedia of Grain Science. Elsevier, New York, 2004, pp. 57-76.
51. Watson, S.A. and P.E. Ramstad. Corn Chemistry and Technology. American Association of Cereal Chemists, St. Paul, MN, 1987.
52. Mares, D.J., K. Mrva and G.B. Fincher. Enzyme activities. *In*: Wrigley, C., H. Corke and C.E. Walker (eds). Encyclopedia of Grain Science. New York: Elsevier, 2004; pp. 357-365.
53. Henry, R.J. Harvesting, storage and transport. *In*: C. Wrigley, H. Corke and C.E. Walker (eds). Encyclopedia of Grain Science. Elsevier, New York, 2004, pp. 46-57.
54. Nishiyama, Y., P. Langan and H. Chanzy. Crystal structure and hydrogen bonding system in cellulose I from synchrotron X-ray and neutron fibre diffraction. Journal of American Chemical Society 2002; 124(31): 9074-9082.
55. Klemm, D., B. Philipp, T. Heinze, U. Heinze and W. Wagenknecht. General considerations on structure and reactivity of cellulose. *In*: Comprehensive Cellulose Chemistry, vol. 1. Wiley-VCH, Weinheim, 1998, pp. 9-34.
56. Fink, H.-P. and B. Philipp. Models of cellulose physical structure from the viewpoint of the cellulose I→II transition. Journal of Applied Polymer Science 1985; 30: 3779-3790.
57. Fink, H.-P., D. Hofmann and H.J. Purz. Zur Fibrillarstruktur nativer Cellulose. Acta Polymerica 1990; 41: 131-137.
58. Atalia, R.H., B.E. Dimick and S.C. Nagel. Studies on polymorphy in cellulose. *In*: C. Jett and J. Arthur (eds). Cellulose Chemistry and Technology, vol. 48. American Chemical Society, Washington, 1977, pp. 30-41.
59. Nehls, I., W. Wagenknecht, B. Phillipp and D. Stscherbina. Characterization of cellulose and cellulose derivatives in solution by high resolution 13C-NMR spectroscopy. Progress of Polymer Science 1994; 19: 29-78.
60. Klemm, D., B. Philipp, T. Heinze, U. Heinze and W. Wagenknecht. Principles of cellulose reactions. *In*: Comprehensive Cellulose Chemistry, vol. 1. Wiley-VCH, Weinheim, 1998, pp. 130-164.
61. Sjostrom, E. Introduction to carbohydrate chemistry. *In*: Wood Chemistry; Fundamental and Applications, 2nd Ed. Academic Press, California, 1993, pp. 21-50.
62. Rowell, R.M., R. Pettersen, J.S. Han, J.S. Rowell and M.A. Tshabalala. Cell wall chemistry. *In*: R.M. Rowell (ed.). Handbook of Wood Chemistry and Wood Composites. CRC Press, Florida, 2005, pp. 35-76.
63. Sjostrom, E. Wood polysaccharides. *In*: Wood Chemistry; Fundamental and Applications, 2nd Ed. Academic Press, California, 1993, pp. 51-71.
64. Timell, T.E. Recent progress in the chemistry and topochemistry of compression wood. Wood Science and Technology 1982; 16: 83-122.
65. Saka, S. Chemical composition and distribution. *In*: D.N.-S. Hon and N. Shiraishi (eds). Wood and Cellulosic Chemistry. Marcel Dekker, New York, 1991, pp. 59-88.
66. Boerjan, W., J. Ralph and M. Baucher. Lignin biosynthesis. Annual Review of Plant Biotechnology 2003; 54: 519-546.
67. Lewis, N.G and E. Yamamoto. Lignin: occurrence, biogenesis and biodegradation. Bioresource Technology 2005; 95(6): 673-686.
68. Klemm, D., B. Philipp, T. Heinze, U. Heinze and W. Wagenknecht. Cell wall chemistry. *In*: Comprehensive Cellulose Chemistry, vol. 1. Wiley-VCH, Weinheim, 1998, pp. 35-74.
69. Reale, S., F. Attanasio, N. Spreti and F.D. Angleis. Lignin chemistry: Biosynthetic study and structural characterization of coniferyl alcohol oligomers formed in vitro in a micellar environment. Chemistry—A European Journal 2010; 16: 6077-6087.

70. Guerra, A., R. Mendonca, A. Ferraz, F. Lu and J. Ralph. Structural characterization of lignin during pinus taeda wood treatment with ceriporiopsis subvermispora. Applied and Environmental Microbiology 2004; 70(7): 4073-4078.
71. Fisher, E.A., C.B. Blum, V.I. Zannis and J.L. Breslow. Independent effects of dietary saturated fat and cholesterol on plasma lipids, lipoproteins, and apolipoprotein. Journal of Lipid Research 1983; 24: 1039-1048.
72. Mozaffarian, D., M.B. Katan, A. Ascherio, M.J. Stampfer and W.C. Willett. Trans fatty acids and cardiovascular disease. New England Journal of Medicine 2006; 354(15): 1601-1613.
73. Kays, S.E., W.R. Windham and F.E. Barton. Prediction of total dietary fibre by near-infrared reflectance spectroscopy in high fat and high sugar containing cereal products, Journal of Agricultural and Food Chemistry 1998; 46(3): 854-861.
74. Cuatrecasas, P. Interaction of wheat germ agglutinin and concanavalin A with isolated fat cells. Biochemistry 1993; 12(7): 1312-1323.
75. Czech, M.P. and W.S. Lynn. Topography of the fat cell plasma membrane. Biochemistry 1993; 12(19): 3597-3601.
76. Fakirov, S. Gelatin and gelatin based biodegradable composites: manufacturing, properties and biodegradation behaviour. *In*: S. Fakirov and D. Bhattacharyya (eds). Handbook of Engineering Biopolymers. Carl Hanser Verlag, Munich, 2007, pp. 419-464.
77. Milewski, S. Protein structure and physicochemical properties. *In*: Z.E. Sikorski (ed.). Chemical and Functional Properties of Food Proteins. CRC Press, Florida, 2000, pp. 31-56.
78. Baker, P.J., J.S. Haghpanah and J.K. Montclare. Elastin based protein polymers. *In*: H. Cheng (ed.). Polymer Biocatalysis and Biomaterials II. American Chemical Society, Washington, 2008, pp. 37-61.
79. Ebringerova, A., Z. Hromádková and G. Berth. Structural and molecular properties of a water-soluble arabinoxylan-protein complex isolated from rye bran. Carbohydrate Research 1994; 264: 97-109.
80. Horax, R., P. Chen and M. Jalaluddin. Preparation and characterization of protein isolation from cowpea. Journal of Food Science 2004; 69: 114-121.
81. Salcedo, C.B., C.J.A. Osuna, L.F. Guevara, D.J. Dominguez and L.O. Paredes. Optimization of the isoelectric precipitation method to obtain protein isolate from amaranth seeds. Journal of Food Chemistry 2002; 46: 71-81.
82. Yoo, S.H. and J. Jane. Structural and physical characteristics of waxy and other wheat starches. Carbohydrate polymer 2002; 49: 297-305.
83. Yuryev, V.P., A.V. Krivandin, V.I. Kiseleva, L.A. Wasserman, N.K. Genkina, J. Fornal, W. Blaszczak and S. Alberto. Structural parameters of amylopectin clusters and semi-crystalline growth rings in wheat starches with different amylose content. Carbohydrate Research 2004; 339(16): 2683-2691.
84. Mukerjea, R. and J.F. Robyt. Starch biosynthesis: sucrose as a substrate for the synthesis of a highly branched component found in 12 varieties of starches. Carbohydrate Research 2005; 340(13): 2206-2211.
85. Anglelier-Coussy, H., J.L. Putaux, S. Molina-Boisseau, A. Dufresne, E. Bertoft and S. Perez. The molecular structure of waxy maize starch nanocrystals. Carbohydrate Research 2009; 344(12): 1558-1566.
86. Stuart, T., Q. Liu, M. Hughes, R.D. McCall, H.S.S. Sharma and A. Norton. Structural biocomposites from flax: Effect of bio-technical fibre modification on composite properties. Composites: Part A 2006; 37: 393-404.
87. Dörnenburg, H. Evaluation of immobilisation effects on metabolic activities and productivity in plant cell. Process Biochemistry 2004; 39(11): 1369-1375.
88. Chabanon, G., I. Chevalot, X. Frambiosier, S. Chenu and I. Marc. Hydrolysis of rapeseed protein. Process Biochemistry 2007; 42(10): 1419-1428.
89. Bledzki, A.K., A.A. Mamun and J. Volk. Physical, chemical and surface properties of wheat husk, rye husk and soft wood and their polypropylene composites. Composites Part A: Applied Science and Manufacturing 2010; 41: 480-488.
90. Bledzki, A.K., A.A. Mamun, M. Lucka and V.S. Gutowsk. The effects of acetylation on properties of flax fibre and its polypropylene composites. eXPRESS Polymer Letters 2008; 2(6): 413-422.
91. Stevens, M.P. Polymer Chemistry; An Introduction. Oxford University Press, Oxford, 1990.

92. Pott, G.T. Natural fibres with low moisture sensitivity. *In*: F.T. Wallenberger and N.E. Weston (eds). Natural Fibres, Plastics and Composites. Kluwer Academic Publishers, New York, 2004, pp. 105-120.
93. Young, R.A. Wood and wood products. *In*: J.A. Kent (ed.). Handbook of Industrial Chemistry and Biotechnology, vol. 2. Springer, New York, 2007, pp. 1234-1293.
94. Satyanarayana, K.G. and F. Wypych. Characterization of natural fibres. *In*: S. Fakirov and D. Bhattacharyya (eds). Handbook of Engineering Biopolymers. Carl Hanser Verlag, Munich, 2007, pp. 3-47.
95. Huang, S.J. and P.G. Edelman. An overview of biodegradable polymers and biodegradation of polymers. *In*: G. Scott and D. Gilead (eds). Degradable Polymers: Principles and Applications. Chapman & Hall, London, 1995, pp. 18-24.
96. Scott, G. Photo-biodegradable polymers. *In*: G. Scott and D. Gilead (eds). Degradable Polymers: Principles and Applications. Chapman & Hall, London, 1995, pp. 169-183.
97. Avella, M., G. Rota, E. Martuscelli, M. Raimo, P. Sadocco and G. Elegir. Poly(3-hydroxybutyrate-co-3-hydroxyvalearate) and wheat straw fiber composites: Thermal, mechanical properties and biodegradation behaviour. Journal of Material Science 2000; (35): 829-836.
98. Tshabalala, M.A. Surface characterization. *In*: R.M. Rowell (ed.). Handbook of Wood Chemistry and Wood Composites. CRC Press, Florida, 2005, pp. 187-211.
99. Isogai, A. Chemical modification of cellulose. *In*: D.N.S. Hon and N. Shiraishi (eds). Wood and Cellulosic Chemistry, 2nd Ed. Marcel Dekker, New York, 2001, pp. 599.
100. Filer, K. Industrial production of enzymes for the feed industry. *In*: S. Roussos, C.R. Soccol, A. Pandey and C. Augur (eds). New Horizons in Biotechnology. Kluwer Academic Publishers, Dordrecht, 2003, pp. 1-15.
101. Bron, S., R. Meima, J.M. Van Dijl, A. Wipat and C.R. Harwood. Manual of industrial microbiology and biotechnology. *In*: A.L. Demain and J.E. Davices (eds). ASM Press, Washington, 1999, pp. 392-416.
102. Young, R.A. Wood and wood products. *In*: J.A. Kent (ed.). Handbook of Industrial Chemistry and Biotechnology, vol. 2. Springer, New York, 2007, pp. 1234-1293.
103. Aehle, W. Industrial enzymes. *In*: Enzymes in Industry: Production and Application, 2nd Ed. Wiley-VCH, Weinheim, 2004, pp. 101-257.
104. Hansen, C.A. Industrial uses of biotechnology. *In*: The Application of Biotechnology to Industrial Sustainability: Sustainable Development. OECD Publishing, Paris, 2001, pp. 17-24.
105. Bailey, J.E. and D.F. Ollis. Enzyme kinetics. *In*: Biochemical Engineering Fundamentals, 2nd Ed. McGraw Hill Publishing, New York, 1986, pp. 984-997.
106. Pickering, K.L., Y. Li and R.L. Farrell. Fungal and alkali interfacial modification of hemp fibre reinforced composites. Key Engineering Materials 2007; 335: 493-496.
107. Saleem, Z., H. Rennebaum, F. Pudel and E. Grimm. Treating bast fibres with pectinase improves mechanical characteristics of reinforced thermoplastic composites. Composites Science and Technology 2008; 68: 471-476.
108. Rajan, A., J.D. Sudha and T.E. Abraham. Enzymatic modification of cassava starch by fungal lipase. Industrial Crops and Products 2008; 27: 50-59.
109. Fu, G.Z., A.W. Chan and D.E. Minns. Preliminary assessment of the environmental benefits of enzyme bleaching for pulp and paper making. The International Journal of Life Cycle Assessment 2005; 10(2): 136-142.
110. Sena-Martins, G., E. Almeida-Vara and J.C. Duarte. Eco-friendly new products from enzymatically modified industrial lignins. Industrial Crops and Products 2008; 27: 189-195.
111. Ouajai, S. and R.A. Shanks. Solvent and enzyme induced recrystallization of mechanically degraded hemp Cellulose. Cellulose 2006; 13: 31-44.
112. Li, Y. and K.L. Pickering. The effect of chelator and white rot fungi treatments on long hemp fibre reinforced composites. Composites Science and Technology 2009; 69: 1265-1270.
113. Li, Y., K.L. Pickering and R.L. Farrell. Analysis of green hemp fibre reinforced composites using bag retting and white rot fungal treatments. Industrial Crops and Products 2009; 29: 420-426.
114. Li, Y. and K.L. Pickering. Hemp fibre reinforced composites using chelator and enzyme treatments. Composites Science and Technology 2008; 68: 3293-3298.

115. Janardhnan, S. and M. Sain. Isolation of cellulosic microfiber-an enzymatic approach. BioResources 2006; 1(2): 176-188.

116. Ouajai, S. and R.A. Shanks. Morphology and structure of hemp fibre after bioscouring. Macromolecular Bioscience 2005; 5: 124-134.

117. Liu, H., J.Y. Zhu and S.Y. Fu. Effects of lignin-metal complexation on enzymatic hydrolysis of cellulose. Journal of Agriculture and Food Chemistry 2010; 58: 7233-7238.

118. Valchev, I., S. Nenkova, P. Tsekova and V. Lasheva. Use of enzymes in hydrolysis of maize stalks. BioResources 2009; 4(1): 285-291.

119. Stuart, T., Q. Liu, M. Hughes, R.D. McCall, H.S.S. Sharma and A. Norton. Structural biocomposites from flax-Part I: Effect of bio-technical fibre modification on composite properties. Composites: Part A 2006; 37: 393-404.

120. Bajpai, P.K. Solving the problems of recycling fibre processing with enzyme. BioResources 2010; 5(2): 1311-1325.

121. Csiszar, E., A. Losonczi, G. Szakacs, I. Rusznak, L. Bezur and J. Reicher. Enzymes and chelating agent in cotton pretreatment. Journal of Biotechnology 2001; 89: 271-279.

122. Semedo, L.T.A.S., R.C. Gomes, E.P.S. Bon, R.M.A. Soares, L.F. Linhares and R.R. Coelho. Endocellulase and exocellulase activities of two Streptomyces strains isolated from a forest soil. Applied Biochemistry and Biotechnology 2000; 84-86: 267-276.

123. Hoshino, E., M. Chiwaki, A. Suzuki and M. Murata. Improvement of cotton cloth soil removal by inclusion of alkaline cellulase from *Bacillus* sp. KSM-635 in detergents. Journal of Surfactants and Detergents 2000; 3: 317-326.

124. Oliveira da Silva, L.A. and E.C. Carmona. Production and characterization of cellulase from Trichoderma inhamatum. Applied Biochemistry and Biotechnology 2008; 150: 117-125.

125. Andreaus, J., H. Azevedo and A. Cavaco-Paulo. Effects of temperature on the cellulose binding ability of cellulase enzymes. Journal of Molecular Catalysis B: Enzymatic 1999; 7: 233-239.

126. Comfort, D.A., S.R. Chhabra, S.B. Conners, C.J. Chou, K.L. Epting, M.R. Johnson, K.L. Jones, A.C. Sehgal and R.M. Kelly. Strategic biocatalysis with hyperthermophilic enzymes. Green Chemistry 2004; 9: 459-465.

127. Pazarlioğlu, N.K., M. Sariişik and A. Telefoncu. Treating denim fabrics with immobilized commercial cellulases. Process Biochemistry 2005; 40(2): 767-771.

128. Anish, R., M.S. Rahman and M. Rao. Application of cellulases from an alkalothermophilic *Thermomonospora* sp. in biopolishing of denims. Biotechnology and Bioengineering 2007; 96: 48-56

129. Lecourt, M., J.C. Sigoillot and M. Petit-Conil. Cellulase-assisted refining of chemical pulps: Impact of enzymatic charge and refining intensity on energy consumption and pulp quality. Process Biochemistry 2010; 45(8): 1274-1278.

130. Pèlach, M.A., F.J. Pastor, J. Puig, F. Vilaseca and P. Mutjé. Enzymic deinking of old newspapers with cellulase. Process Biochemistry 2003; 38(7): 1063-1067.

131. Jorgensen, H., J.B. Kristensen and C. Felby. Enzymatic conversion of lignocellulose into fermentable sugars: challenges and opportunities. Biofuels, Bioproducts and Biorefining 2007; 1(2): 119-134.

132. Yang, S., W. Ding and H. Chen. Enzymatic hydrolysis of rice straw in a tubular reactor coupled with UF membrane. Process Biochemistry 2006; 41(3): 721-725.

133. Csiszár, E., K. Urbánszki and G. Szakács. Biotreatment of desized cotton fabric by commercial cellulase and xylanase enzymes. Journal of Molecular Catalysis B: Enzymatic 2001; 11(4): 1065-1072.

134. Ricard, M. and I.D. Reid. Purified pectinase lowers cationic demand in peroxide bleached mechanical pulp. Enzyme and Microbial Technology 2004; 26: 499-504.

135. Maijala, P., M. Kleen, C. Westin, K. Poppius-Levlin, K. Herranen, J.H. Lehto, P. Reponen, O. Mäentausta, A. Mettälä and A. Hatakka. Biomechanical pulping of softwood with enzymes and white-rot fungus Physisporinus rivulosus. Enzyme and Microbial Technology 2008; 43(2): 169-177.

136. Cao, J., L. Zheng and S. Chen. Screening of pectinase producer from alkalophilic bacteria and study on its potential application in degumming of ramie. Enzyme and Microbial Technology 1992; 14(12): 1013-1016.

137. Riou, C., G. Freyssinet and M. Fevre. Purification and Characterization of Extracellular Pectinolytic Enzymes Produced by Sclerotinia sclerotiorum. Applied and Environmental Microbiology 1992; 58(2): 578-583.
138. Paulo, A.C. and G. Gübitz. Catalysis and processing. *In*: A.C. Paulo and G. Gübitz (eds). Textile Processing with Enzymes. CRC Press, Boca Raton, 2003, pp. 86-119.
139. Sharma, H.S.S., L. Whiteside and K. Kernaghan. Enzymatic treatment of flax fibre at the roving stage for production of wet-spun yarn. Enzyme and Microbial Technology 2005; 37: 386-394.
140. Ossola, M. and Y.M. Galante. Scouring of flax rove with the aid of enzymes. Enzyme and Microbial Technology 2004; 34: 177-186.
141. Rai, P., G.C. Majumdar and S. Gupta S. Optimizing pectinase usage in pretreatment of mosambi juice for clarification by response surface methodology. Journal of Food Engineering 2004; 64: 397-403.
142. Sun, Y., Z. Wang, J. Wu, F. Chen, X. Liao and X. Hu. Optimizing enzymatic maceration in pretreatment of carrot juice concentrate by response surface methodology. International Journal of Food Science & Technology 2006; 41: 1082-1089.
143. Basu, G., S.S. De and A.K. Samanta. Effect of bio-friendly conditioning agents on jute fibre spinning. Industrial Crops and Products 2009; 29(2): 281-288.
144. Boel, E. Lipases: structure, mechanism, and genetic engineering. GBF Monographs 1991; 16: 207-219.
145. Winkler, F.K., A. D'Arcy and W. Hunziker. Structure of human pancreatic lipase. Nature 1990; 343: 771-774.
146. Egmond, M.R. and C.J. Bemmel. Impact of structural information on understanding of lipolytic function. Methods in Enzymology 1997; 284: 119-129.
147. Gutiérrez-Ayesta, C., A.A. Carelli and M.L. Ferreira. Relation between lipase structures and their catalytic ability to hydrolyse triglycerides and phospholipids. Enzyme and Microbial Technology 2007; 41(1): 35-43.
148. Chakraborty, K. and R. Paulraj. Purification and biochemical characterization of an extracellular lipase from pseudomonas fluorescens. Journal of Agricultural and Food Chemistry 2009; 57(9): 3859-3866.
149. Edwinoliver, N.G., K. Thirunavukarasu, S. Purushothaman, C. Rose, M.K. Gowthaman and N.R. Kamini. Corn steep liquor as a nutrition adjunct for the production of aspergillus niger lipase and hydrolysis of oils thereof. Journal of Agricultural and Food Chemistry 2009; 57(22): 10658-10663.
150. Kim, H.R., I.H. Kim, C.T. Hou, K. Kwon and B. Shin. Production of a novel cold active lipase from pichia lynferdii. Journal of Agricultural and Food Chemistry 2010; 58(2): 1322-1326.
151. Hasan, F., A.A. Shah and A. Hameed. Industrial applications of microbial lipases. Enzyme and Microbial Technology 2006; 39(2): 235-251.
152. Dandik, L. and H.A. Aksoy. Applications of Nigella sativa seed lipase in oleochemical reactions. Enzyme and Microbial Technology 1996; 19(4): 277-281.
153. Wu, H.S. and M.J. Tsai. Kinetics of tributyrin hydrolysis by lipase. Enzyme and Microbial Technology 2004; 35(6): 488-493.
154. Gupta, R., N. Gupta and P. Rathi. Bacterial lipases: an overview of production, purification and biochemical properties. Applied Microbiology and Biotechnology 2004; 64(6): 763-781.
155. Ransbarger, D. and F. Xu. Activation of laccase by penicillin and derivatives. Process Biochemistry 2006; 41(9): 2082-2086.
156. Litthaue, D., M.J. Vuuren, A. Tonder and F.W. Wolfaardt. Purification and kinetics of a thermostable laccase from Pycnoporus sanguineus (SCC 108). Enzyme and Microbial Technology 2007; 40(4): 563-568.
157. Weihua, Q. and C. Hongzhang. An alkali-stable enzyme with laccase activity from entophytic fungus and the enzymatic modification of alkali lignin. Bioresource Technology 2008; 99(13): 5480-5484.
158. Chun-Han, Ko. and C. Shiao-Shing. Enhanced removal of three phenols by laccase polymerization with MF/UF membranes. Bioresource Technology 2008; 99(7): 2293-2298.
159. Kurisawa, M., J.E. Chung, H. Uyama and S. Kobayashi S. Laccase-catalyzed synthesis and antioxidant property of poly (catechin). Macromolecular Bioscience 2003; 3(12): 758-764.
160. Kudanga, T., E.N. Prasetyo, P. Widsten, A. Kandelbauer, S. Jury, C. Heathcote, J. Sipilä, H. Weber, G.S. Nyanhongo and G.M. Guebitz. Laccase catalyzed covalent coupling of fluorophenols increases lignocellulose surface hydrophobicity. Bioresource Technology 2010; 101(8): 2793-2799.

161. Zhou, G., J. Li, Y. Chen, B. Zhao, Y. Cao, X. Duan and Y. Cao. Determination of reactive oxygen species generated in laccase catalyzed oxidation of wood fibres from Chinese fir (Cunninghamia lanceolata) by electron spin resonance spectrometry. Bioresource Technology 2009; 100(1): 505-508.
162. Vikineswary, S., N. Abdullah, M. Renuvathani, M. Sekaran, A. Pandey and E.B.G. Jones. Productivity of laccase in solid substrate fermentation of selected agro-residues by Pycnoporus sanguineus. Bioresource Technology 2006; 97(1): 171-177.
163. Couto, S.R., J. Luis and T. Herrera. Industrial and biotechnological applications of laccases: A review. Biotechnology Advances 2006; 24: 500-513.
164. Cristóvão, R.O., A.P.M. Tavares, A.S. Ribeiro, J.M. Loureiro, R.A.R. Boaventura and E.A. Macedo. Kinetic modelling and simulation of laccase catalyzed degradation of reactive textile dyes. Bioresource Technology 2008; 99(11): 4768-4774.
165. Pazarlıoğlu, N.K., M. Sariişik and A. Telefoncu. Laccase: production by trametes versicolor and application to denim washing. Process Biochemistry 2005; 40(5): 1673-1678.
166. Aracri, E., J.F. Colom and T. Vidal. Application of laccase-natural mediator systems to sisal pulp: An effective approach to biobleaching or functionalizing pulp fibres. Bioresource Technology 2009; 100(23): 5911-5916.
167. Jurado, M., A. Prieto, Á. Martínez-Alcalá, Á.T. Martínez and M.J. Martínez. Laccase detoxification of steam-exploded wheat straw for second generation bioethanol. Bioresource Technology 2009; 100(24): 6378-6384.
168. Liu, N., S. Shi, Y. Gao and M. Qin. Fiber modification of kraft pulp with laccase in presence of methyl syringate. Enzyme and Microbial Technology 2009; 44(2): 89-95.
169. Valls, C. and M.B. Roncero. Using both xylanase and laccase enzymes for pulp bleaching. Bioresource Technology 2009; 100(6): 2032-2039.
170. Dias, A.A., R.M. Bezerra and A.N. Pereira. Activity and elution profile of laccase during biological decolorization and dephenolization of olive mill wastewater. Bioresource Technology 2004; 92(1): 7-13.
171. Morozova, O.V., G.P. Shumakovich, S.V. Shleev and Ya.I. Yaropolov. Laccase–Mediator systems and their applications: A review. Applied Biochemistry and Microbiology 2007; 43(5): 523-535.
172. Han, X.Q. and S. Damodaran. purification and characterization of protease q: a detergent- and urea-stable serine endopeptidase from bacilluspumilus. Journal of Agricultural and Food Chemistry 1998; 46(9): 3596-3603.
173. Genckal, H. and C. Tari. Alkaline protease production from alkalophilic *Bacillus* sp. isolated from natural habitats. Enzyme and Microbial Technology 2006; 39(4): 703-710.
174. Numata, K., P. Cebe and D.L. Kaplan. Mechanism of enzymatic degradation of beta-sheet crystals. Biomaterials 2010; 31(10): 2926-2933.
175. Çalık, P., E. Bilir, G. Çalık and T.H. Özdamar. Influence of pH conditions on metabolic regulations in serine alkaline protease production by *Bacillus licheniformis*. Enzyme and Microbial Technology 2002; 31(5): 685-697.
176. Ghorbel, B., A. Sellami-Kamoun and M. Nasri. Stability studies of protease from Bacillus cereus BG1. Enzyme and Microbial Technology 2003; 32(5):513-518.
177. Zhu, W., D. Cha, G. Cheng, Q. Peng and P. Shen. Purification and characterization of a thermostable protease from a newly isolated *Geobacillus* sp. YMTC 1049. Enzyme and Microbial Technology 2007; 40(6): 1592-1597.
178. Sugimoto, M., K. Ishihara and N. Nakajima. Structure and function of an isozyme of earthworm proteases as a new biocatalyst. Journal of Molecular Catalysis B: Enzymatic 2003; 23(2-6): 405-409.
179. Li, M., M. Ogiso and N. Minoura. Enzymatic degradation behavior of porous silk fibroin sheets. Biomaterials 2003; 24(2): 357-365.
180. Zareian, S., K. Khajeh, B. Ranjbar, B. Dabirmanesh, M. Ghollasi and N. Mollania. Purification and characterization of a novel amylopullulanase that converts pullulan to glucose, maltose, and maltotriose and starch to glucose and maltose. Enzyme and Microbial Technology 2010; 46(2): 57-63.
181. Balkan, B. and F. Ertan. Production of α-amylase from Penicillium chrysogenum under solid state fermentation by using some agricultural waste products. Food Technol Biotechnol 2007; 45(4): 439-442.
182. Wong, D.W.S., S.B. Batt and G.H. Robertson. Microassay for rapid screening of α-amylase activity. Journal of Agricultural and Food Chemistry 2000; 48(10): 4540-4543.

183. Chakraborty, S., A. Khopade, C. Kokare, K. Mahadik and B. Chopade. Isolation and characterization of novel α-amylase from marine *Streptomyces* sp. D1. Journal of Molecular Catalysis B: Enzymatic 2009; 58(1-4): 17-23.

184. Selvakumar, P., L. Ashakumary and A. Pandey. Biosysnthesis of glucoamylase from aspergillus niger by solid state fermentation using tea waste as the basis of a solid substrate. Bioresource Technology 1998; 65: 83-85.

185. Apar, D.K. and B. Özbek. α-Amylase inactivation by temperature during starch hydrolysis. Process Biochemistry 2004; 39(9): 1137-1144.

186. Prakash, B., M. Vidyasagar, M.S. Madhukumar, G. Muralikrishna and K. Sreeramulu. Production, purification, and characterization of two extremely halotolerant, thermostable, and alkali-stable α-amylases from Chromohalobacter sp. TVSP 101. Process Biochemistry 2009; 44(2): 210-215.

187. Goesaert, H., P. Leman, A. Bijttebier and J.A. Delcour. Antifirming effects of starch degrading enzymes in bread crumb. Journal of Agricultural and Food Chemistry 2009; 57(6): 2346-2355.

188. Yamada, R., Y. Bito, T. Adachi, T. Tanaka, C. Ogino, H. Fukuda and A. Kondo. Efficient production of ethanol from raw starch by a mated diploid Saccharomyces cerevisiae with integrated α-amylase and glucoamylase genes. Enzyme and Microbial Technology 2009; 44(5): 344-349.

189. Murthy, G.S., E.D. Sall, S.G. Metz, G. Foster and V. Singh. Evaluation of a dry corn fractionation process for ethanol production with different hybrids. Industrial Crops and Products 2009; 29(1): 67-72.

190. Lv, Z., J. Yang and H. Yuan. Production, purification and characterization of an alkaliphilic endo-β-1,4-xylanase from a microbial community EMSD5. Enzyme and Microbial Technology 2008; 43(4-5): 343-348.

191. Beaugrand, J., D. Crônier, P. Thiebeau, L. Schreiber, P. Debeire and B. Chabbert. Structure, chemical composition, and xylanase degradation of external layers isolated from developing wheat grain. Journal of Agricultural and Food Chemistry 2004; 52(23): 7108-7117.

192. Yin, L.J., H.H. Lin, Y.I. Chiang and S.T. Jiang. Bioproperties and purification of xylanase from *Bacillus* sp. YJ6. Journal of Agricultural and Food Chemistry 2010; 58(1): 557-562.

193. Yan, Q., S. Hao, Z. Jiang, Q. Zhai and W. Chen. Properties of a xylanase from Streptomyces matensis being suitable for xylooligosaccharides production. Journal of Molecular Catalysis B: Enzymatic 2009; 58(1-4): 72-77.

194. Qiu, Z., P. Shi, H. Luo, Y. Bai, T. Yuan, P. Yang, S. Liu and B. Yao. A xylanase with broad pH and temperature adaptability from Streptomyces megasporus DSM 41476, and its potential application in brewing industry. Enzyme and Microbial Technology 2010; 46(6): 506-512.

195. Taneja, K., S. Gupta and R.C. Kuhad. Properties and application of a partially purified alkaline xylanase from an alkalophilic fungus *Aspergillus nidulans* KK-99. Bioresource Technology 2002; 85(1): 39-42.

196. Liavoga, A.B., Y. Bian and P.A. Seib. Release of D-xylose from wheat straw by acid and xylanase hydrolysis and purification of xylitol. Journal of Agricultural and Food Chemistry 2007; 55(19): 7758-7766.

197. Ramos, J., M. González, F. Ramírez, R. Young and V. Zúñiga. Biomechanical and biochemical pulping of sugarcane bagasse with ceriporiopsis subvermispora fungal and xylanase pretreatments. Journal of Agricultural and Food Chemistry 2001; 49(3):1180-1186.

198. Figueroa-Espinoza, M.C., C. Poulsen, J.B. Soe, M.R. Zargahi and X. Rouau. Enzymatic solubilization of arabinoxylans from isolated rye pentosans and rye flour by different endo-xylanases and other hydrolyzing enzymes. Journal of Agricultural and Food Chemistry 2002; 50(22): 6473-6484.

199. Gonçalves, A.R., R.Y. Moriya, L.R.M. Oliveira and M.B.W. Saad. Xylanase recycling for the economical biobleaching of sugarcane bagasse and straw pulps. Enzyme and Microbial Technology 2008; 43(2): 157-163.

200. Wang, R., Y. Xue, X. Wu, X. Song and J. Peng. Enhancement of engineered trifunctional enzyme by optimizing linker peptides for degradation of agricultural waste products. Enzyme and Microbial Technology 2010; 47(5): 194-199.

201. Wu, S. Polymer Interface and Adhesion. Marcel Dekker, New York, 1982, pp. 1-28.

202. Jancar, J. Engineered interphases in polypropylene composites. *In*: H.G. Karian (ed.). Handbook of Polypropylene and Polypropylene Composites. Marcel Dekker, New York, 1999, pp. 367-419.

203. Theocaris, P.S.. The concept and properties of mesophase in composites. *In*: H. Ishida and J.L. Koenig (eds). Composites Interfaces. Elsevier, New York, 1986, pp. 329-349.
204. Piggott, M.R. Interface properties and their influence of fibre reinforced polymers. *In*: T.L. Vigo and B.J. Kinzig (eds). Composite Applications: The Role of Matrix, Fibre and Interface. VCH Publishers, New York, 1992, pp. 221-276.
205. Zafeiropoulos, N.E. Engineering the fibre-matrix interface in natural fibre composites. *In*: K.L. Pickering (ed.). Properties and Performance of Natural Fibre Composites. CRC Press, New York, 2008, pp. 127-162.
206. Comyn, J. Surface analysis and adhesion bonding. Analytical Proceedings 1996; 30: 27-28.
207. Belgacem, M.N. Characterization of polysaccharides, lignin and other woody components. Cellulose Chemistry and Technology 2000; 34: 357-383.
208. Dillingham, R.G. and B.R. Oakley. Surface energy and adhesion in composite-composite adhesive bonds. The Journal of Adhesion 2006; 82: 407-426.
209. Schulz, J. and M. Nadine. Theories and mechanisms of adhesion. *In*: A. Pizzi and K.L. Mittal (eds). Handbook of Adhesive Technology, 2nd Ed. Marcel Dekker, New York, 2003, pp. 53-68.
210. Pocius, A.V. Adhesion and Adhesives Technology: An Introduction. Hanser, Munich, 2002.
211. Packham, D.E. The mechanical theory of adhesion. *In*: A. Pizzi and K.L. Mittal (eds). Handbook of Adhesive Technology, 2nd Ed. Marcel Dekker, New York, 2003, pp. 69-94.
212. Wool, R.P. Diffusion and Autoadhesion. *In*: M. Chaudhury and A.V. Pocious (eds). Adhesive Science and Engineering-2: Surfaces, Chemistry and Applications. Elsevier, Amsterdam, 2002, pp. 351-401.
213. Berg, J.C. Semi-imperical strategies for predicting adhesion. *In*: M. Chaudhury and A.V. Pocious (eds). Adhesive Science and Engineering-2: Surfaces, Chemistry and Applications. Elsevier, Amsterdam, 2002, pp. 1-73
214. Gassan, J. and A.K. Bledzki, Effect of moisture content on the properties of silanized jute-epoxy composites. Polymer Composites 1997; 18(2): 179-184.
215. Chow, C.P.L., X.S. Xing and R.K.Y. Li. Moisture absorption studies of sisal fibre reinforced polypropylene composites. Composites Science and Technology 2007; 67: 306-313.
216. Thwe, M.M. and K. Liao. Durability of bamboo-glass fiber reinforced polymer matrix hybrid composites. Composite Science and Technology 2003; 63: 375-387.
217. Lu, X., M.Q. Zhang, M.Z. Rong, D.L. Yue and G.C. Yang. Environmental degradability of self-reinforced composites made from sisal. Composite Science and Technology 2004; 64: 1301-1310.
218. Clemons, C. Wood-plastic composites in the United States: The interfacing of two industries. Forest Products Journal 2002; 52(6): 10-18.
219. Rowell, R.M., S.E. Lange and R.E. Jacobson. Weathering performance of plant-fiber thermoplastic composites. Molecular Crystals and Liquid Crystals 2000; 353: 85-94.
220. Mishra, S. and M. Sain. Long-term performance of natural fibre composites. *In*: K.L. Pickering (ed.). Properties and Performance of Natural Fibre Composites. CRC Press, New York, 2008, pp. 461-502.
221. Guo, G., G.M. Rizvi, C.B. Park and W.S. Lin. Critical processing temperature in the manufacture of fine-celled plastic/wood-fiber composite foams. Journal of Applied Polymer Science 2004; 91: 621-629.
222. Smith, P.M. and M.P. Walcott. Opportunities for wood/natural fiber-plastic composites in residential and industrial applications. Forest Products Journal 2006; 56(3): 4-11.
223. Beg, M.D.H. and K.L. Pickering. Reprocessing of wood fibre reinforced polypropylene composites. Part II: Hydrothermal ageing and its effects. Composites Part A 2008; 39: 1565-1571.
224. Espert, A., F. Vilaplana and S. Karlsson. Comperison of water absorption in natural cellulosic fibres from wood and one-year crops in polypropylene composites and its influence on their mechanical properties. Composites Part A 2004; 35: 1267-1276.
225. Stark, N.M. and L.M. Matuana. Ultraviolet weathering of photostabilized woodflour filled high-density polyethylene. Journal of Applied Polymer Science 2003; 90(10): 2609-2617.
226. Hon, D.N.-S. Weathering and photochemistry of wood. *In*: D.N.-S., Hon and N. Shiraishi (eds). Wood and Cellulosic Chemistry, 2nd Ed. Marcel Dekker, New York, 2000, pp. 512-546.
227. Matuana, L.M., D.P. Kamdem and J. Zhang. Photoaging and stabilization of rigid PVC/wood-fiber composites. Journal of Applied Polymer Science 2001; 80: 1943-1950.

228. Matuana, L.M. and D.P. Kamdem. Accelerated ultraviolet weathering of PVC/wood fiber composites. Polymer Engineering Science 2002; 42(8): 1657-1666.

229. Muasher, M. and M. Sain. The efficiency of photostabilizers on the color change of wood filler plastic composites. Polymer Degradation and Stability 2006; 91: 1156-1165.

230. Beg, M.D.H. The improvement of interfacial bonding, weathering and recycling of wood fibre reinforced polypropylene composites. PhD thesis 2007

231. La Mantia, F.P. Photooxidation and stabilization of photooxidized polyethylene and its homopolymer blends. Journal of Applied Polymer Science 2004; 91: 2244-2255.

232. Ndiaye, D., E. Fanton, S. Morlat-Therias, L. Vidal, A. Tidjani and J.L. Gardette. Durability of wood polymer composites: Part 1. Influence of wood on the photochemical properties. Composites Science and Technology 2008; 68: 2779-2784.

233. Nicole, M.S. and L.M. Matuana. Influence of photostabilizers on wood flour– HDPE composites exposed to xenon-arc radiation with and without water spray. Polymer Degradation and Stability 2006; 91(12): 3048-3056.

234. Seldén, R., B. Nyström and R. Lngström. UV aging of poly(propylene)/woodfibre composites. Polymer Composites 2004; 25(5): 543-553.

235. Marcovich, N.E., M.M. Reboredo and M.I. Aranguren. Modified woodflour as thermoset fillers II. Thermal degradation of woodflours and composites. Thermochimica Acta 2001; 372: 45-57.

236. Cyras, V.P., C. Vallo, J.M. Kenny and A. Vazquez. Effect of chemical treatment on the mechanical properties of starch based blends reinforced with sisal fibre. Journal of Composites Materials 2004; 38: 1387-1399.

237. Negri, A.P., H.J. Cornell and D.E. Rivett. A model for the surface properties of keratin fibers. J Text Res 1992; 63: 109-116.

238. Feist, W.C. and H. Tarkow. A new procedure for measuring fibre saturation points. Forest Products Journal 1967; 17(10): 65-68.

239. Bledzki, A.K., A.A. Mamun and J. Volk. Barley Husk and Coconut shell reinforced polypropylene composites: The effect of fibre physical, chemical and surface properties. Composite Science and Technology 2010; 70: 840-846.

240. Rowell, R.M. and H.P. Stout. Jute and kenaf. *In*: M. Lewin and E.M. Pearce (eds). Handbook of Fibre Chemistry. Marcel Dekker, New York, 1998, pp. 466-502.

241. Stuart, B. Biological Applications of Infrared Spectroscopy. John Wiley & Sons, UK, 1997.

242. Sun, X.F., F. Xu, R.C. Sun, P. Fowler and M.S. Baird. Characteristics of degraded cellulose obtained from stream exploded wheat straw. Carbohydrate Research 2005; 340: 97-106.

243. Mwaikambo, L.Y. and M.P. Ansell. Chemical modification of hemp, sisal, jute, and kapok fibers by alkalization. Journal of Applied Polymer Science 2002; 84: 2222-2234.

244. Stevens, M.P. Polymer Chemistry; An Introduction. Oxford University Press, UK, 1990; 1-39.

245. Bledzki, A.K., A.A. Mamun, K. Erdmann and J. Volk. Characterization of grain waste products and properties of their biodegrade composites. Progress and development: 13th European conference on composite materials proceedings, Stockholm, Sweden, 2008.

246. Cherian, B.M., L.A. Pothan, T. Nguyen-Chung, G. Menning, M. Kottaisamy and S. Thomas. A novel method for the synthesis of cellulose nanofibril whiskers from banan fibres and characterisation. Agricultural and Food chemistry 2008; 56: 5617-5627.

Chapter 3

Bio-based Polyamides

M. Feldmann

Institute of Material Engineering, Polymer Engineering, University of Kassel
Moenchebergstrasse 3, 34125 Kassel, Germany

3.1 INTRODUCTION

Polyamides have been established in the field of technical applications for many years, and offer a large range of application options, owing to their various structural forms and correlating properties. This chapter exclusively focuses on semi-crystalline, aliphatic polyamides, which are especially employed in injection molding, extrusion, and thermoforming. An in-depth look at the processing of semi-finished textile products will not be provided here, but remains of great significance for polyamides as a whole.

The most important polyamides in the field of compact components are PA 6 and PA 6.6, which are often reinforced with fibers. In the case of polyamides reinforced with short fibers, processing is often completed using injection molding or extrusion. Organic sheets (flat thermoplastic composite materials reinforced with continuous fibers) are gaining popularity. In such materials, the polyamide usually plays an important role as the matrix material. For instance, these composite materials made of continuous glass or carbon fibers can be used for local reinforcement (i.e., in injection molded components), or can be made into flat (structural) components using a compression molding process.

The raw materials used for polyamides can be petro-based or bio-based. However, there are very few bio-based polyamides available in the market. According to the statistics provided in [1], a capacity of only 93 ktons of bio-based polyamides is available from manufacturers worldwide. If you look at the global annual capacity of polyamide producers in 2014, which equaled 4.847 ktons of PA 6 and 2.187 ktons of PA 6.6 [2], this is only a ratio of approximately 1.3%.

In the group of polyamides, bio-based polyamides are by no means a novelty. They have been around since the middle of the 20th century [3, 4, 5], and were employed in several fields. The term "bioplastic" has only been used for these types for a few years, since this aspect recently started gaining more and more significance. These materials can wholly or partially consist of renewable resources, and are not biodegradable. Recently, new types were introduced on the market in

For Correspondence: Email: Feldmann@uni-kassel.de, Phone: +49 561/804-2867

addition to the already known bio-polyamides, and they complement the property spectrum of the existing polyamides. In the field of co-polyamides, a biogenic monomer is frequently selected. Monomers made of castor oil taken from the castor oil plant are very common.

The most well-known and commercially available bio-polyamides are PA 6.10, PA 10.10, PA 11, and PA 4.10. PA 6.10 and PA 4.10 are partially bio-based, and the others are completely made of renewable resources. Furthermore, completely or partially bio-based types exist, which are either not at all available or available only in a limited manner commercially. At present, almost every large polyamide manufacturer offers one or several bio-based polyamides in their product portfolio under corresponding trade names (Table 3.1). The bio content is listed in wt.-%, and refers to the carbon atoms in the polymer. The bio content has direct influence on the reduction of excess carbon dioxide emission, which shows a lower GWP (global warming potential). For example, EVONIK lists a GWP of 2.8 (kg CO_2) / (kg material) for Vestamid Terra DS (PA 10.10), and a GWP of 4.1 kg CO_2) / (kg material) for Vestamid Terra HS (PA 6.10). DSM even advertises the CO_2 neutrality of its product EcoPaXX in the information sheet when referring to an entry from cradle to gate life cycle assessment, and when taking into account that the plant absorbed CO_2 while growing.

Almost all bio-based polyamides are commercially available as compounds with corresponding filling materials (i.e., glass fibers and inorganic fillers).

Table 3.1 Bio-based polyamides–Producers, trade names, and characteristics [4–14].

Type	*Producer*	*Trade name*	*Bio content**	*Reinforcement options*	*Other options*
PA 4.10	DSM	EcoPaXX	70%	glass fiber	
PA 6.10	DuPont	Zytel RS	63%	glass fiber	Blend PA 6.10/ PA 10.10
	BASF	Ultramid® Balance	> 60%	glass fiber	Blend PA 6.10/ PA 6.6
	EMS-Chemie	Grilamid 2S	62%	-	
	AKRO-Plastic	Akromid S	70%	glass fiber	
	EVONIK	VESTAMID® Terra HS	62%	-	
PA 10.10	EMS-Chemie	Grilamid 1S	99%	-	
	DuPont	Zytel RS	100%	-	Blend PA 6.10/ PA 10.10
	EVONIK	VESTAMID® Terra DS	100%	-	
PA 10.12	EVONIK	VESTAMID® Terra DD	45%	-	
PA 11	Arkema	Rilsan®	100%	glass fiber reinforced	
Amorphous PA	EMS-Chemie	Grilamid BTR	54%	-	
	Arkema	MIDO Rilsan®	54%	-	
PPA**	EMS-Chemie	Grivory HT3	48%	Neat	
	EVONIK	VESTAMID® HTplus TGP 3537	50%	Neat	
	Arkema	Rilsan® HT	70%	Neat	

*Manufacturer information in [wt.-%]
**Polyphthalamide

Various studies exist that estimate the worldwide annual production capacity of bioplastics. According to European Bioplastics [1], a production capacity of 7.848 million tons is expected for 2019. In 2014, the production capacity equaled 1.7 million tons. This would equal a four-fold to five-fold increase in the capacities. According to Greiner [16], annual sales and comprehensive employment of these materials are currently curbed by the fact that the costs for a bio-based polyamide are significantly higher than those for a conventional one. Generally, they fulfil the requirements, and would offer better material properties in most cases. Greiner states that bio-based polyamides have good chances to increase applications due to a limited amount of crude oil storage and price developments.

According to calculations made by [1] in 2014, the worldwide production of bio-based polyamides equaled 93 kt. Thus, the bio-based polyamides represented 5.5% of all bioplastic production (1.697 million tons) see Fig. 3.1.

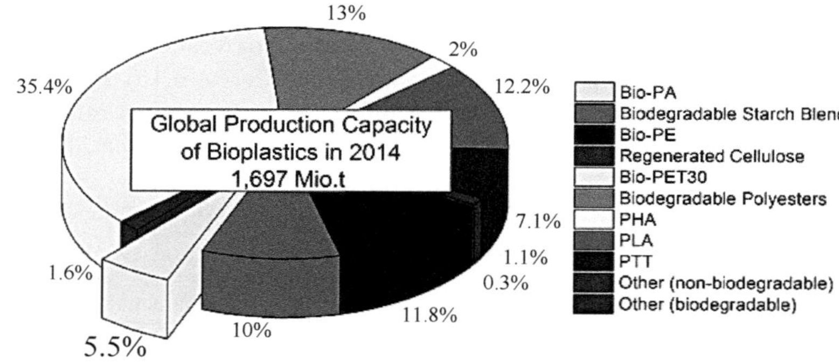

Figure 3.1 Global production capacity of bioplastics in 2014, according to [1].

3.2 RAW MATERIALS AND THE POLYMERIZATION PROCESS

Developments that took place in the crude oil market in the middle of the 20th century and the inexpensive manufacturing of monomers based on crude oil resulted in a decline of the significance of biogenic raw materials [17]. Renewable raw materials have been flooding the market owing to the increase in the population's environmental awareness in the last 20 years, and since the discussions about CO_2 emissions and sustainability arose [18, 20]. Thus far, the castor oil plant is the most important raw material for bio-based polyamides.

The fruits of the castor oil plant contain approximately 40 to 50% castor oil, from which approximately 90% fatty acid ($C_{18}H_{34}O_3$) can be obtained. The obtained fatty acid is split into a C8 and a C10 molecule by means of alkaline hydrolysis (see Fig. 3.3). The valuable sebacic acid (C10) can be used as the basic monomer for type X.10 polyamides. The C8 molecule is, for example, employed in softening agents [21, 22]. Type X.20 polyamides are known from literature, but have only been synthesized on a laboratory scale so far [23]. Also, Jasinska et al. [24] developed additional, new polyamides that are based on castor oil. The castor oil plant can mainly be found in India, China, and Brazil, and produces approximately 555,000 tons of the CO_2 neutral oil every year [4]. The basic materials are used in raw material form for the synthesis of plastics, and are literally experiencing a "Renaissance", such as PA 6.10, which has been explicitly available from different polymer manufacturers as a biopolyamide since approximately 2007 [4, 25, 26]. Succinic acid and sebacic acid in particular are considered biogenic raw materials of the future [23, 24, 27].

The procedure for the manufacture of sebacic acid monomer for bio-based polyamides is based upon alkaline oxidation of the castor oil or castor oil acid [15, 17, 20]. PA 6.10 is the product of polycondensation polymerization made from petro-based 1,6 hexamethylenediamine and bio-based 1,10 decanoic acid (sebacic acid). In this combination, the ratio of incorporated renewable raw materials equals approximately 63% [22]. The property profile of PA 6.10 ranges between that of the high-performance polyamide 6.12 and the standard polyamides PA 6 and

PA 6.6. Owing to its higher melting temperature, PA 6.10 possesses a higher heat deformation resistance than its relative PA 10.10 [25].

In addition to sebacic acid, the diamine can also be bio-based and can be manufactured from castor oil. For example, completely bio-based PA 10.10 can be produced [4]. PA 10.10 is the product of polycondensation of bio-based 1,10 decamethylenediamine and bio-based 1,10 decanoic acid (sebacic acid). The properties of PA 10.10 range between those of the high performance, long-chained polyamides, such as PA 12 and PA 12.12, and the standard polyamides PA 6 and PA 6.6, which have shorter chain lengths. Like PA 6.10, PA 10.10 is semi-crystalline. For this reason, it possesses a high mechanical strength and outstanding chemical resistance. It only absorbs a small amount of moisture. As a result, it displays little variation in its mechanical properties and a high level of dimensional stability when subject to alternating levels of surrounding moisture [28]. Figure 3.2 depicts the structural formula of the bio-based polyamides PA 6.10 and PA 10.10.

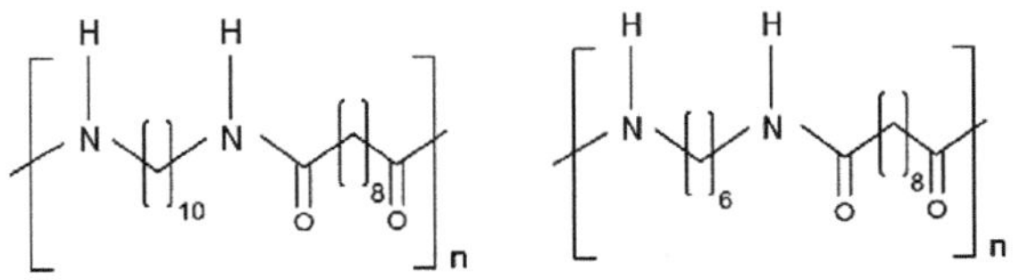

Figure 3.2 Chemical structures of the bio-based polyamides 10.10 (left) and 6.10 (right).

One benefit of using the castor oil plant as a raw material source is the fact that it is not competing for the environment which is used to grow foods. Thus, there is no "ethical conflict", because the areas in which they are cultivated are not suitable for food crops due to the low levels of precipitation. The fruits of the castor oil plant cannot be used as a food source [4, 26].

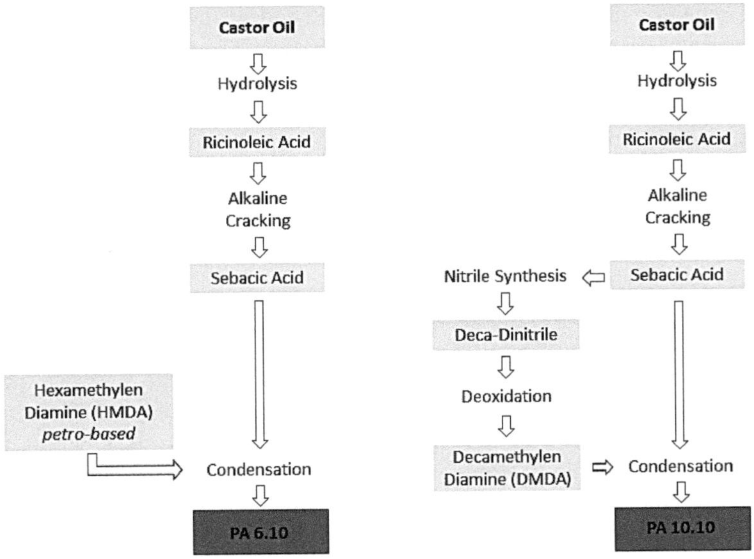

Figure 3.3 Schematic flow chart of the synthesis of partly bio-based polyamide 6.10 (left) and completely bio-based 10.10 (right), according to [19].

Figure 3.3 illustrates the steps of manufacturing castor oil-based polyamides 6.10 and 10.10. Starting with the castor oil, which is derived from the seeds of the castor oil plant, alkaline hydrolysis is carried out to produce sebacic acid (C10) from the C18 molecule. Subsequently, petro-based hexamethylenediamine is added for polycondensation to produce PA 6.10. In order to produce completely bio-based PA 10.10, the sebacic acid is processed in additional steps to create decamethylenediamine, which reacts with the sebacic acid in the final polycondensation step, producing PA 10.10.

In addition to the raw material "castor oil", which was already used 6,000 years ago [20], new biotechnical approaches have been developed in recent years. For example, the so-called drop-in solutions include bio-based PA 6 (see Chapter 4.6.3) and PA 6.6, which were made with fermentative adipic acid and caprolactam. These polymers are chemically identical to the available products based on fossil resources. For example, the raw material used for the bio-based polyamides is corn. The bacteria synthesize the corresponding monomer needed for polycondensation for the corresponding polyamide from the corn starch (similar to as is done in the case of PLA). However, these products are still in the research phase, and are not available in large series or commercially (see Chapter 4.6.6) [15, 29, 30].

Additional special polyamides are known from literature. However, they have only been created on a laboratory scale thus far. In the future, they could be produced based on bio-based succinic acid. Succinic acid is considered to have great potential as a bio-based building block. One example for this is PA 4.4, based on fermentative succinic acid from biomass [31, 32].

Habot et al. [33] also developed a new, semi-crystalline co-polyamide based on rapeseed and aliphatic diamines. Furthermore, lignocellulose is considered another promising, future raw material source for bio-based polyamides and other plastics, because it is available in large amounts, and also does not compete with foods, such as corn or rapeseed. By means of pre-treatment steps and enzymatic treatment, the previously mentioned fermentation for the required building blocks can be carried out subsequently [34].

3.3 STRUCTURE AND PROPERTIES

Polyamides are generally characterized by high strength and stiffness properties, and excellent heat deformation resistance. The amide group (CO–NH, Fig. 3.4) is characteristic of all polyamides, and is decisive for their behavior. There are four monomer groups that are used to manufacture polyamides: dicarbonic acids, diamines, lactams, and amino acids. Furthermore, as was also the case for petro-based polyamides, the macroscopic properties of bio-based polyamides are dependent on the monomers and their number of C-atoms.

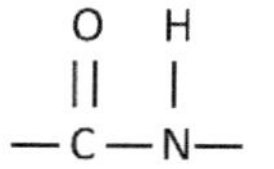

Figure 3.4 Carbonamide group according to [5].

In the case of aliphatic polyamides, a differentiation is made between polyamides that consist of one monomer, i.e., PA 6 with six C-atoms in the monomer (here ε-caprolactam), and polyamides that consist of two monomers, i.e., PA 6.10 with six C-atoms in the diamine (hexamethylenediamine) and ten C-atoms in the dicarbonic acid. When dealing with polyamides, it is especially important to take the decisive factor of moisture absorption into account, since it is crucial for the mechanical values and the geometrical dimensions. The stiffness and the strength of the material decline as the moisture content rises, while the elongation and toughness increase. As a consequence, many properties are a result of the moisture content.

How much moisture a polyamide absorbs is essentially dependent upon the CH_2/CONH relation (Table 3.2). Principally, it can be said that as the CH_2/CONH relation increases as a result of the larger distance between the polar groups, the absorption of moisture, strength, and stiffness reduce, while the impact strength increases. An even number of C-atoms in polyamides (i.e., PA 6, PA 6.10, etc.) results in higher intermolecular forces, because the CO- and NH-groups are located across from one another. A higher melting point also results from this. In contrast, an uneven number of C-atoms in polyamides (i.e., PA 11) causes a higher impact strength, but a comparably low melting point.

The degree of crystallinity is also of great significance, because, unlike in other polymers, it can be strongly influenced by processing, and can result in moisture absorption and other properties [5, 29]. In the group of aliphatic polyamides, the bio-based polyamides based on castor oil close the gap between the petro-based polyamides that are already established on the market (see Table 3.2). Apart from the mechanical properties (see Table 3.3), this also refers to the melting temperature and the moisture absorption behavior.

Table 3.2 Moisture absorption, melting temperature, and density depending on the CH_2-CONH-relation of selected polyamides (gray highlighted areas refer to bio-PA) (Source: Product Datasheets).

Type	*CH_2/NHCO-relation*	*Melting temperature*	*Density*	*Moisture uptake*	*Comments*
	-	[°C]	[g/cm³]	[%] @23°C 50% RH.	
PA 4.6	4	295	1.21	3.8	
PA 6.6	5	260	1.15	2.8	
PA 6	5	222	1.14	3	
PA 4.10	6	250	1.09	2	EcoPaXX Q170E
PA 6.10	7	220	1.07	3.3*	Vestamid Terra HS16
PA 6.12	8	215	1.06	1.3	
PA 10.10	9	200	1.04	1.8*	Vestamid Terra DS16
PA 10.12	10	190	1.03	1.6*	Vestamid Terra DD16
PA 11	10	189	1.03	1.6*	Rilsan BMNO
PA 12	11	178	1.01	0.7	

*Saturated according to ISO 62

Originally, PA 6.10 was regarded as the bio-based counterpart to conventional PA 6, since it possesses many material properties that resemble those of crude oil-based PA 6. However, in addition to the ratio of renewable raw materials it contains (>60%), PA 6.10 also absorbs less moisture and is resistant to other media. Since it absorbs less moisture, it displays stronger dimensional stability, and other material properties that are less dependent on the absorption of moisture. Accordingly, PA 6.10 can be ranked between the standard polyamides PA 6 or PA 6.6 and the higher ranked PA 6.12 concerning its technical characteristics. Regarding its properties, fully bio-based PA 10.10 ranks between the standard types PA 6 and PA 6.6 and the long-chained PA 12 [4].

Table 3.3 Mechanical properties of selected bio-based polyamides–dry state (Source: Product Datasheet).

Type	*Tensile strength @ yield point*	*Tensile modulus*	*Elongation@ yield point*	*Notched charpy strength*	*Comments*
	[MPa]	[GPa]	[%]	[kJ/m²]	
PA 4.10	80	3.0	4.5	5.4	EcoPaXX Q170E
PA 6.10	61	2.1	5	6	Vestamid Terra HS16
PA 10.10	54	1.7	5	7	Vestamid Terra DS16
PA 10.12	40	1.3	5	12	Vestamid Terra DD16
PA 11	40	1.09	21	8	Rilsan BMNO

In addition to the aforementioned bio-polyamides, a blend of bio-based PA 6.10 and crude-oil-based PA 6.6 is offered by BASF that has a property profile that is situated between those of the two polymers. Owing to their strong resistance to hot oil and moisture, as well as aggressive media, it is suitable for oil tubs and radiator caps [30].

Thus, the bio-based polyamides currently available cover a broad spectrum of mechanical properties (see Table 3.3), and represent a supplement to already existing types. The drop-in solutions made of renewable raw materials, i.e., PA 6, offer the same material properties as fossil polyamides. Due to the commercial availability, explicit values of bio-PA 6 have not been listed.

The partially pronounced polarity of the polyamides results in good adhesion to reinforcement fibers when dealing with composite materials. For example, when combined with glass or cellulose fibers, additional coupling agents are often not needed. The adhesion strongly depends upon the CO/NH relation, and declines as the CH_2 ratio increases [31, 32] (also see Fig. 3.10).

3.4 COMPOSITES, BLENDS, AND ADDITIVES

Bio-based polyamides are already offered by numerous manufacturers in the form of glass fiber reinforced variants (Table 3.1). They display a property spectrum that is quite similar to that of conventional polyamides (Table 3.4). With regard to the tensile modules, the values range from 7.3 GPa to 9.5 GPa. In the case of the tensile strength, the range begins at 137 MPa and extends to 170 MPa. Frequently, such compounds can be directly employed as a bio-based alternative for existing applications. Owing to their chemical structure, additional special fiber sizings are usually not needed.

Unlike the conventional polyamides 6 and 6.6, some of the bio-polyamides absorb less moisture, and accordingly, their mechanical values reduce less (especially their strength and tensile moduli) when they are in a moist/conditioned state. Therefore, stronger dimensional stability is on hand, and several bio-polyamides actually surpass the values of comparable, petro-based PAs (i.e., PA 6.10 and PA 6) when wet rather than when they are dry.

Besides glass fibers for reinforcement, additional reinforcement fibers are conceivable for the enhancement of the mechanical properties. This includes carbon fibers, but also cellulose and natural fibers (see Chapter 4.6).

Table 3.4 Mechanical properties of selected glass fiber reinforced bio-based polyamides (Source: Product Datasheet).

Type	*Tensile strength* [MPa]	*Young's modulus* [GPa]	*Elongation@ break* [%]	*Notched charpy strength* [kJ/m²]	*Comments*
PA 4.10 GF30	170	9.5	4	11	EcoPaXX Q-KG6
PA 6.10 GF30	146	8.3	3	9	VESTAMID TerraHS16-GF30
PA 10.10 GF 30	130	7.5	5	19	VESTAMID TerraDS16-GF30
PA 11 GF30	137	7.3	6	24	Rilsan BZM 30 O TL

Additivation is also of great significance for bio-based polyamides, especially with regard to the hydrolysis resistance, UV, and temperature resistances. Otherwise, a strong decrease in the mechanical properties can occur due to aging effects. Additives used for petrochemically manufactured polyamides can be used for stabilization.

3.5 INDUSTRIAL APPLICATION

Among other things, bio-based polyamides (especially PA 11) are used for the manufacture of cables and hoses, since they possess an excellent resistance to chemicals. As of late, PA 10.10 and 6.10 can be found in commercially available components. For instance, this includes injection molded components as well as extruded, technical profiles. The company Technoform Bautec Kunststoffprodukte GmbH creates such profiles based on bio-based polyamides for window construction (Fig. 3.5). TI Automotive GmbH uses bio-based PA 11 to manufacture fuel pipelines and Alfred Pracht Lichttechnik GmbH produces parts for luminaires based on polyamides from renewable resources (Fig. 3.5). Yet, there are also consumer products that are commercially available, such as toothbrushes.

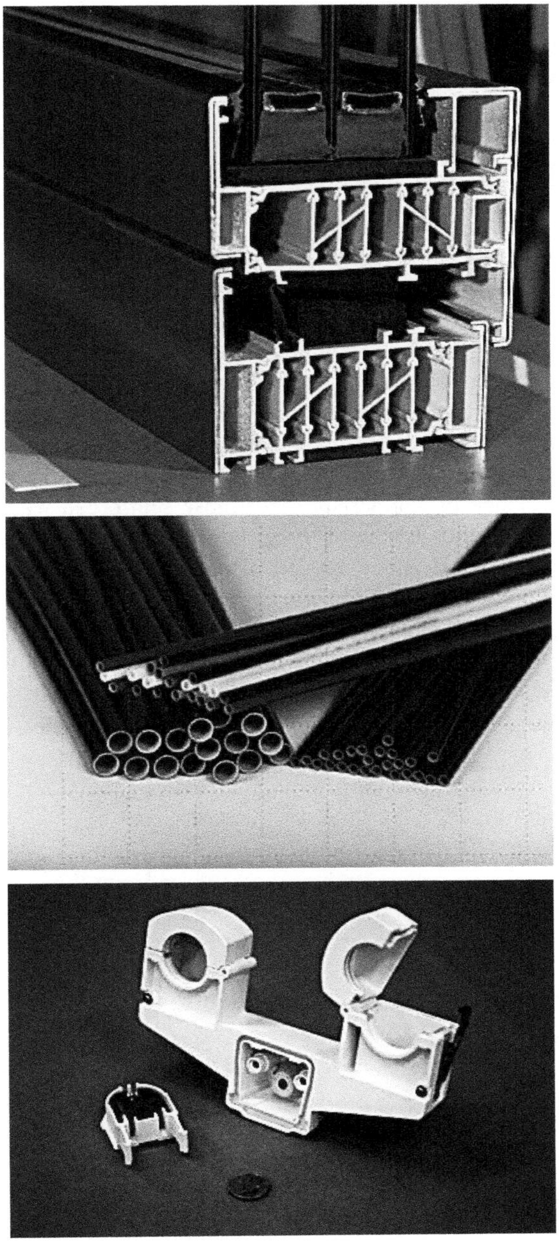

Figure 3.5 Industrial components based on polyamides from renewable resources: Technical profile made by the company Technoform Bautec Kunststoffprodukte GmbH based in a window construction (top), media lines with a PA 11 layer made by TI Automotive GmbH (center), and part of a luminaire (flame retardant) made by Alfred Pracht Lichttechnik GmbH (bottom).

According to various press reports that concern glass fibers, they are also employed in the automotive industry. For instance, the company A.Raymond uses glass-reinforced Ultramid® Balance (~60% bio-based), which is made by BASF SE, to make a quick connector by means of injection molding. In this case, the media resistance against fuel and zinc chloride, and the hydrolysis resistance are specifically named as benefits.

One significant branch of industry that polyamides are widely distributed throughout is the textile industry. Here, bio-based polyamides, such as PA 6.10, are used for diverse products.

3.6 DEVELOPMENT AND FUTURE TRENDS

3.6.1 Composites Based on Bio-Based Polyamides and Cellulose- or Natural Fibers

Polyamides are often employed when they have been reinforced with fibers. Thus far, glass fibers are the most significant reinforcement fibers for polyamides. This applies to conventional polyamides as well as bio-based polyamides. However, the option to use bio-based reinforcement fibers is available for the latter, since it results in an increase of the biogenic ratio, and contributes to the creation of a completely bio-based composite material. Reinforcement can be accomplished using short or continuous biogenic fibers. Furthermore, it is also common for an enhancement of the properties (especially the E-modulus) to be achieved by incorporating bio-based filling materials. The transition between the reinforcement variants can be fluent.

Regardless of the fiber length, the moisture absorption behavior in composites with cellulosic fibers is of special importance in comparison with those reinforced with glass fibers (Fig. 3.6). The moisture absorption property increases with the ratio of cellulosic reinforcement fibers in comparison with the unreinforced material [39]. This effect is the opposite when moisture is absorbed by glass fiber composites. In glass fiber composites, this leads to mechanical properties that are less moisture-dependent. Composites that contain regenerated cellulose fibers display an increased impact strength, and have a weakened strength and E-modulus when wet due to fibers swelling with moisture, and due to a higher level of elongation-at-break [32].

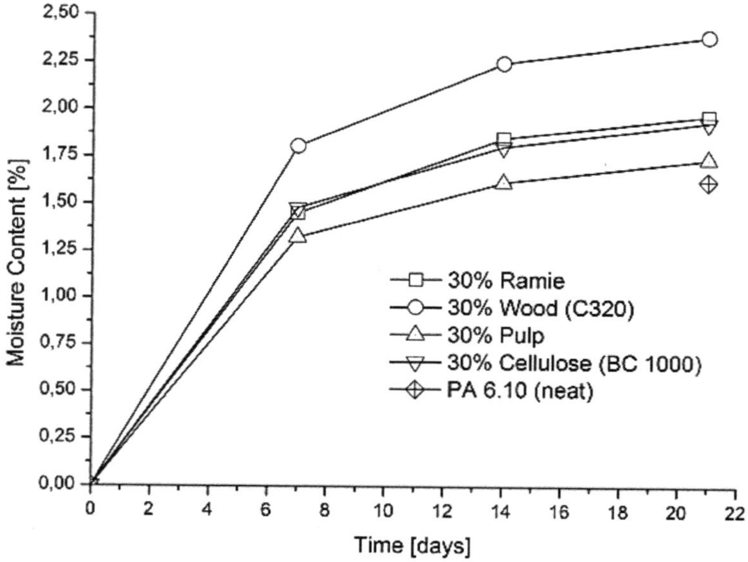

Figure 3.6 Moisture absorption of selected PA 6.10 composites during the first 21 days according to ISO 1110@ 62°C and 70% RH and [45].

3.6.2 Short Fiber Reinforced Composites

As is already known for polyethylene and polypropylene, various cellulosic fibers can be used as reinforcement fibers. They have property spectrums that are similar to those of glass fiber reinforced systems, or even better with regard to several specific properties. If the matrix material is switched for a technical plastic, such as polyamide, the correlating higher melting temperature must be considered. In many cases, it can result in the degeneration of the cellulose fibers, which, in turn, reduces their reinforcement effect [35, 41]. Accordingly, special attention must be paid to the processes, the process settings, and the raw materials in order to achieve an effective reinforcement of polyamides with cellulosic fibers. In particular, pure cellulose fibers combined with polyamides display very good composite properties.

In addition to the co-rotating twin screw extruder process [38], one- or two-step pultrusion processes have proven to be especially suitable for the preparation of granulate in laboratory scale [31, 32]. It leads to an enhancement of the mechanical properties in the course of reducing the thermal and mechanical loads on the cellulosic fibers (Fig. 3.7). This can be attributed to the reduced degradation of the fibers, which, in turn, results in the stronger properties. Figure 3.6 provides an exemplary illustration of the influence of the pultrusion process on the tensile strength and notched impact strength. It is evident that both properties were found to be increased significantly using a one-step (gentler) compounding method instead of a two-step method. In the case of PA 6.10 with 30 wt.-% regenerated cellulose fibers, a 107% higher tensile strength and 27% higher notched impact strength were achieved.

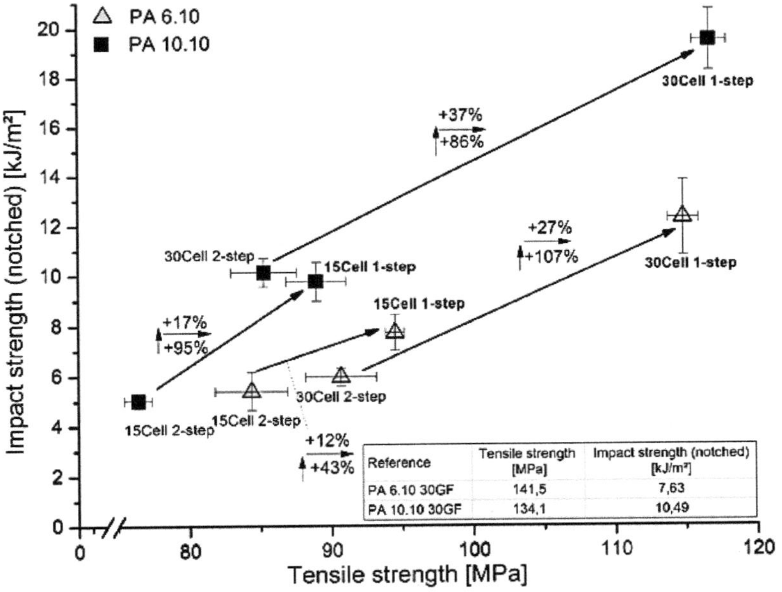

Figure 3.7 Notched impact strength and tensile strength of bio-based polyamides with cellulose fibers: the influence of different compounding methods.

Besides the development or improvement of preparation methods, approaches exist that reduce the processing temperature of polyamides by means of the incorporation of additives. In doing so, temperature-sensitive fibers were able to be incorporated using lower temperatures. In his dissertation, Amintowlieh describes the reinforcement of PA 6 with wheat straw (15 wt.-%) [42]. In order to solve the issue of processing the material at melting temperatures which is higher than the degradation temperature of the reinforcement fibers, additives (LiCl as well as N-Butylbenzenesulfonamide) were used. For instance, the melting point was brought down to

191°C by adding 4% LiCl. The stiffness of the materials increased in comparison with those of native PA 6. However, the strengths of the composites were lower. Furthermore, Amintowlieh also deduced that the thermal degradation of the fibers is decisive for the composite strength in contrast to the crystallinity.

Technical plastics are often employed in areas in which higher usage temperatures are required. Employing short fibers enables the area of application to be expanded considerably. This applies to the widely used glass fibers, but also to the cellulosic fibers (here, regenerated cellulose fibers were incorporated by using a pultrusion process), as is shown in Fig. 3.8. The HDT-A values show for biocomposites based on PA 6.10 and PA 10.10 with a 30 wt.-% fiber ratio comparable values to those of materials with 30 wt.-% glass fiber reinforcement. A fiber content of 15 wt.-% already leads to a considerable enhancement (>160°C) in comparison with an unreinforced plastic. The HDT-A values of PA 6.10 with 30 wt.-% exceed 200°C, and are close to the melting temperature (~220°C). In the case of polyamides that contain 30 wt.-% abaca fibers, the HDT-A value equaled 132°C (not shown here).

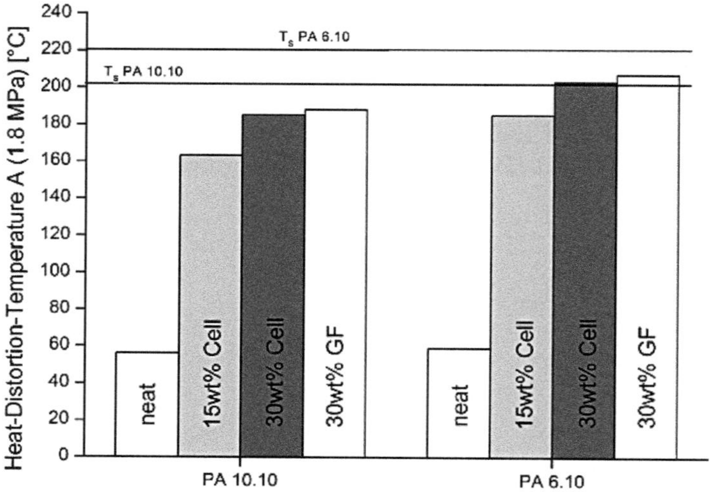

Figure 3.8 Heat distortion Temperature (HDT-A) of reinforced bio-based Polyamide 6.10 and 10.10 [37, 38].

In addition to regenerated cellulose fibers, lignocellulose fibers are also used as reinforcement. However, significant increases in the strength and impact strength have not been successfully achieved thus far. For one, this is due to the low thermal resistance of the fibers, which leads to a reduction of the mechanical properties. However, it is also due to the structure of the fibers themselves. Here, the fiber structure is meant in particular, which differs strongly from that of regenerated cellulose fibers. Cellulose fibers display a diameter of approximately 12 μm, while that of other natural fibers is significantly higher.

At the same time, regenerated cellulose fibers display average fiber lengths of >1 mm, even in injection molded samples. Thus, they display fiber lengths that clearly exceed those of natural fibers (see Fig. 3.9). One reason for this is the fact that they, even if they have comparable mechanical properties, possess a higher flexural strength solely based on their larger diameter, and shorten more during processes, i.e., on an extruder, because they lack flexibility and elongation-at-break. Moreover, natural plant fibers themselves are an erroneous composite material, and consist of cellulose, hemicellulose, and lignin. This often leads to premature material failure. This is especially when the orientation of the fiber/particle is opposite to the load.

In contrast to regenerated cellulose fibers, bio-based polyamides PA 6.10 and 10.10 that were reinforced with a content of 30 wt.-% abaca fibers display no significant increase in their tensile strengths (53 MPa @PA 6.10 and 69 MPa @PA 10.10) or their notched impact strengths (2.5 kJ/m² @PA 6.10 and 4 kJ/m² @PA 10.10). This can be explained by the fact that the two different composites display different failure mechanisms and the two reinforcement fibers have different fiber geometries (especially aspect ratio). The composite with long and thin regenerated cellulose fibers, and consequently, a significant higher aspect ratio, displays fiber pull-outs on the fracture surface. This mechanism leads to work-intensive failure behavior. The surfaces of the abaca fibers and many other natural fibers are rough, easily allowing forces to be transferred into the fibers very well without a coupling agent. Therefore, no work-intensive pull-outs occur. Additionally, the aspect ratio of abaca fibers is much lower and long pull-outs are not possible.

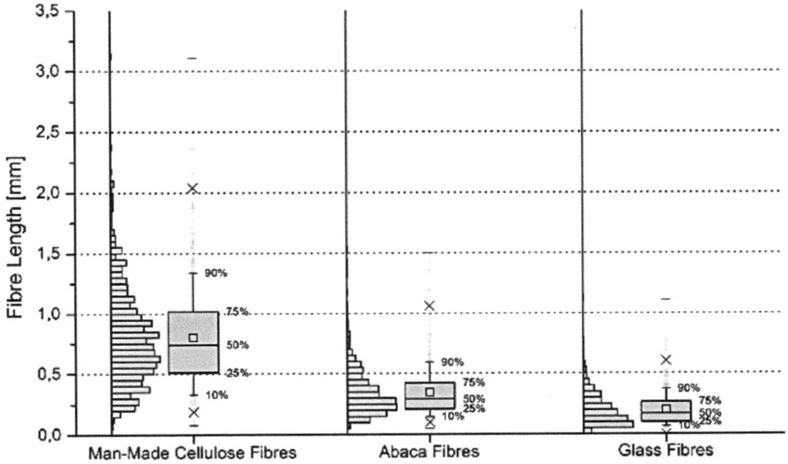

Figure 3.9 Fiber length distribution in injection molded specimens made of PA 6.10 composites with different reinforcement fibers (30 wt.-%), according to [37, 38].

3.6.2.1 Fiber Matrix Adhesion

In contrast to polyolefin matrix materials, the fiber-matrix adhesion is of special significance in bio-based polyamides due to their polarity. Basically, moderate adhesion is noted without introducing an additional coupling agent. The varying polarity of the matrix must be taken into account. For example, the polarity of bio-based PA 10.10 is lower than that of PA 6.10 or PA 4.10 due to its lower number of amide groups. This is visible in the REM images (Fig. 3.10). In the same conditions, polypropylene has a considerably larger number of fiber pull-outs than polyamide 6, which has very few pull-outs. Bio-based polyamide 6.10 ranks in between conventional matrix materials. The result can be nearly attributed to the polarity (CH_2/NHCO relation) of the various matrix materials (see Table 3.2 as well).

Figure 3.10 SEM pictures of PA 6, PA 6.10, and PP composites with 30 wt.-% man-made cellulose fibers [37].

3.6.2.2 Samples Made by Injection Molding and Extrusion

Figure 3.11 shows a selection of different components made of bio-based polyamides that contain regenerated cellulose fibers. The components were manufactured using standard tools and on standard machines in the context of the research project of FNR e.V/BMEL. They were made using close-to-standard injection molding and extrusion procedures. Modifications were only made regarding the processing temperature and additional process parameters. The component (b1) evidently demonstrates that the addition of 5 wt.-% of PLA to the compound (PA 10.10 20Cell) results in a lightening of the color. However, this addition does not lead to any noteworthy alteration of the thermomechanical properties. By means of (b3), it was also able to be shown that conventional color master batches can also be used for the biocomposites to obtain the desired color. As a result, the series components can hardly be told apart from their bio-based counterparts. However, in the case of component b, the component weight was able to be reduced by 35% in comparison with the standard material (PA 6.6 GF30), as is shown in Fig. 3.11. Hot runner systems were used for all injection molded components (a, b, c). Several components were examined after manufacture to assess their performance characteristics, and it was shown that the required specifications were fulfilled.

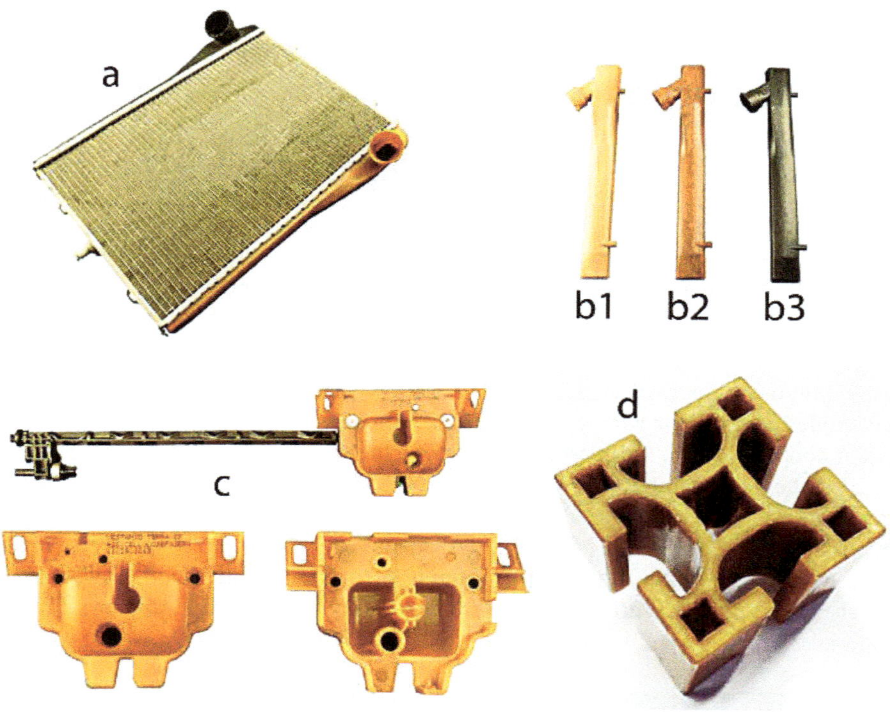

Figure 3.11 Completely bio-based moisture distributor in a cooling application made of PA 10.10 with 20 wt.-% cellulose fiber (a), with native color (b2), lightened with PLA (b1) and colored black (b3). Glove compartment lock made of PA 10.10 and 20 wt.-% cellulose fibers. Extruded construction profile made of PA 6.10 with 30 wt.-% cellulose fiber (d).

3.6.3 Continuous Cellulose Fiber Reinforced Composites

In the field of thermoplastics reinforced with continuous fibers, so-called organic sheets have been an interesting alternative to classic composite materials with a duroplastic matrix. They are already employed in structural components in the automotive industry, and have potential for application in large scale production because they require short cycle times during processing.

Polyamides have especially established themselves as matrix materials. They can be replaced by bio-based variants with no future complications. Similarly, cellulosic reinforcement fibers can be used instead of the classic reinforcement fibers (glass and carbon fibers). This can have many advantages, especially regarding lightweight construction (lower density of fibers), but also good impact loading and low risk crash behavior. However, the strength and elastic modulus are lower compared to conventional reinforcement fibers.

Laminates made of bio-based polyamides and regenerated cellulose fibers, which were made using a film stacking process, display comparable penetration energies as those that contain the same weight ratio of glass fibers (see Fig. 3.12) [39, 40]. In contrast to composites reinforced with cellulose fibers, the strength of the composites with glass fibers was found to increase more quickly, which can be attributed to the higher elastic modulus of the fibers, and which can be seen in Fig. 3.13.

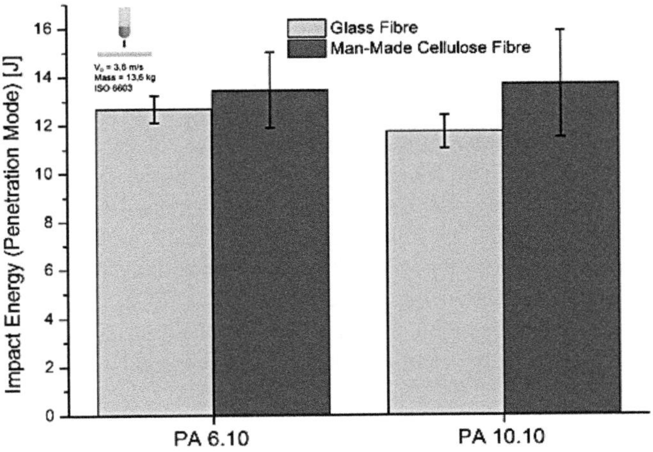

Figure 3.12 Impact energy (dart drop test - penetration mode) of continuous fiber reinforced (0/90 fabric 280 g/m²) bio-based polyamides 6.10 and 10.10, according to ISO 6603 and [46, 47].

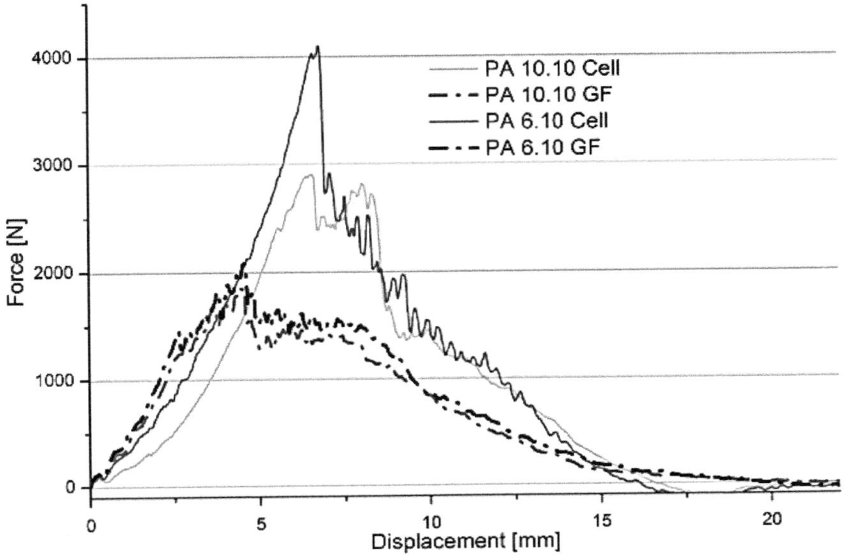

Figure 3.13 Force displacement graph of continuous fiber reinforced (0/90 fabric 280 g/m²) bio-based polyamides 6.10 and 10.10 (according [46, 47]).

Contrary to the expectations based on knowledge of quasi-statistical experiments, cellulose fiber composites display higher maximal forces and comparable elongation-at-break values. Despite the higher strength of glass fibers, these composites displayed evident and premature declines in force during the dart drop tests. In addition to the results shown here, the fracture surfaces of the cellulose fiber reinforced polyamides is decidedly softer than that of materials reinforced with glass fibers.

3.6.4 Bio-Based Polyamides with Different Bio-Based Fillers

Like the so-called WPCs, which are usually based on polyolefin plastics, wood fibers/flour can also be incorporated in polyamides. The main focus is on filling the material without any special requirement for increasing the strength. The processing conditions must of course be taken into account, because a degradation of the fibers can occur, which could have a negative effect on odor and emission properties.

The study carried out by Zierdt et al. [36] explained the mechanical properties of composites produced from beech wood flour in bio-based PA 11. An increase of the elastic modulus (~ 5GPa @ 50 wt.-% fiber content) were recorded for the composite that had a matrix, with a melting temperature below 200°C and modified beech wood. Also, a tensile strength of 65 MPa @ 50 wt.-% fiber content was able to be achieved. As is already known for polyolefin WPC materials, the elongation-at-break and Charpy impact strength reduce as the fiber ratio increases.

Similar results were obtained by the author for PA 6.10 and 10.10 with fiber ratios of 30 wt.-% [38]. The bio-based polyamides 10.10 and 6.10 especially displayed an increase in their elastic moduli when reinforced with wood flour (ArboCel C320). In contrast, the strength and impact strengths declined, or remained at the same level as the matrix. Generally, it is important to add that the positive influence of the reinforcement component reduces as the melting temperature increases.

The use of filling materials other than bio-based fillers is also being studied. Battegazzore et al. [42] used nanoclay and rice husk ash as reinforcements, and reported that the highest mechanical values were obtained (especially the HDT, elastic modulus, and stress at yield point) when nanoclay (10 wt%) and risk husk ash (5 wt.-%) were incorporated.

3.6.5 Overview of Selected Properties of Biocomposites with Different Fibers and Fillers

Table 3.5 provides an overview of selected mechanical properties of various biocomposites that are based on bio-polyamides. The selected composite materials display a broad spectrum of mechanical properties. The spectrum can be broadened and the application field can be enlarged when taking further fiber ratios into account.

3.6.6 Bio-Based Caprolactams from Fermentation

One petro-based polyamide that is widely used in plastics and textile technology is PA 6. A plastic can only be referred to as a real drop-in solution, and can only be used if it is chemically identical and not just comparable, such as PA 6.10. The main building block for the manufacture of PA 6 is caprolactam, which is polymerized to PA 6 by means of ring opening polymerization. In classic PA synthesis, it comes from a fossil raw material source. Bio-based caprolactam is obtained by means of fermentation of sugar. Various raw material sources are options for manufacturing bio-based caprolactam based on renewable raw materials. These include corn, sugar cane, and sugar beet.

Table 3.5 Selected mechanical properties (dry) of injection molded biocomposite materials based on bio-polyamides – an overview.

Polymer	*Fiber content and fiber*	*Tensile strength*	*Tensile modulus*	*Notched charpy strength*	*Processing*	*Reference*
	wt.-%	[MPa]	[GPa]	[kJ/m²]	-	-
PA 10.10	30 Man-made Cellulose Crdenka	114.8	5.0	19.5	1-step Pultrusion	[37]
PA 10.10	30 Man-made Cellulose	104.6	4.6	12.5	Twin-Screw Extruder	[44]
PA 10.10	10Clay+ 5RHA	44.0	2.7	–	Twin-Screw Extruder	[48]
PA 10.10	30 Abaca	68.7	5.3	3.9	2-step Pultrusion	[38]
PA 6.10	30 Man-made Cellulose	116.6	5.5	12.3	1-step Pultrusion	[37]
PA 6.10	10Clay+ 5RHA	52.1	3.4	–	Twin-Screw Extruder	[48]
PA 6.10	30 Abaca	50.6	5.8	2.2	2-step Pultrusion	[38]
PA 11	50 modified Beech	65	5.0	2.9	Internal Mixer	[43]

Lignocellulose is also imaginable as a future raw material source [33], since it competes less with foods than the other sources do. Figure 3.14 depicts the process steps and interim products of the manufacture of bio-based PA 6. Another benefit of bio-based PA 6, apart from the chemically identical structure to classic PA 6, is the worldwide production capacity. All that needs to be done is a switching of the petro-based monomer for the bio-based one. This manufacture path can also be used for the widely used PA 6.6, and is generally comparable with drop-in activities in the field of polyolefins [29, 30, 48].

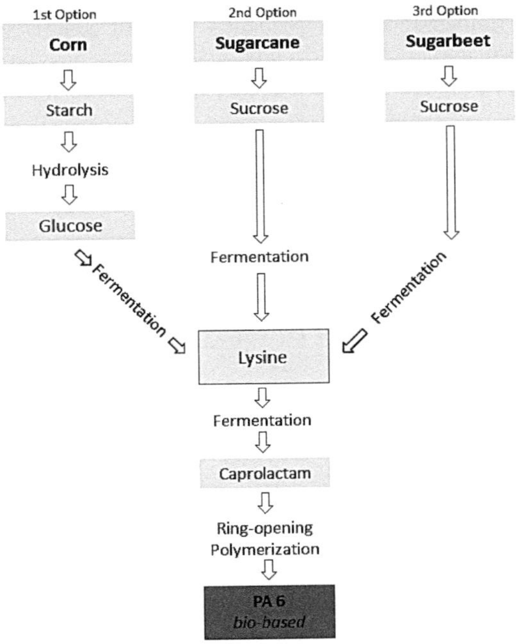

Figure 3.14 Schematic flow chart of the synthesis of bio-based polyamide 6, according to [19, 49].

3.7 CONCLUSIONS

Bio-based polyamides represent a large group of bio-based plastics, and have already partially become established in the market owing to their special properties. These especially include those polyamides that are manufactured from the raw material castor oil, and which represent a supplement for the property profile of conventional polyamides based on fossil raw materials. At present, these bio-polyamides are commercially available from many polymer manufacturers with varying bio ratios that can range up to 100%. According to current numbers, they only represent about 1% of the entire polyamide market. This can be attributed to their comparably high price. At present, the capacities for bio-polyamides are being expanded. For example, bio-polyamides are employed in injection molded and extruded components, but also in the textile industry. Real drop-in solutions, such as bio PA 6 and bio PA 6.6, which have identical chemical structures and property profiles, are not available on the market in larger quantities yet. They are made from fermented sugar, and, long-term, can also be made out of lignocellulose.

Cellulosic fibers as reinforcement options are increasingly being focused upon in science in the field of fiber reinforced bio-based polyamides, because they are a logical conclusion when taking the entire situation into account, and when considering that they have excellent potential to be used in lightweight applications owing to their low-density structures. The bandwidth in this case ranges from real fibers to particle-like filling materials. However, the thermal resistance of the fibers always poses a challenge. Thus, the reinforcement effect is often restricted to individual mechanical properties, but can also exceed the values of common composite materials in isolated cases. For this reason, special attention must be paid to the compounding process and the selection of fibers.

In comparison with many other biocomposites, those based on polyamides can display a higher usage temperature and better heat deformation resistance, which makes them interesting for many fields of application. As is common knowledge for polyamides, these properties have to be assessed while taking the moisture content into account. The first bio-based polyamides reinforced with cellulose fibers are now commercially available.

In principle, it is difficult to ascertain the ecological balance of bio-based polyamides, since varying balance limits or fundamentals are frequently used for comparison. On the whole, several manufacturers list a significantly improved ecological balance in comparison with petro-based polyamides.

3.8 ACKNOWLEDGMENTS

The authors would like to thank the BMEL, the FNR e. V., and the Hessen State Ministry of Higher Education, Research and the Arts - Initiative for the Development of Scientific and Economic Excellence (LOEWE) - special research project “Safer Materials” for the financial support, and express their gratitude to the companies for providing the materials.

References

1. Bioplastics facts and figures - European Bioplastics, Institute for Bioplastics and Biocomposites, Nova-Institute. 2015.
2. PCI Nylon Yellow Book. 2014.
3. Polyamide from renewable raw materials. Sustainable Development Newsletter, January 2008.
4. Häger, H. and J. Limper. Hightech-Kunststoffe vom Acker. EVONIK Science Newsletter 2009; elements 28: 24-30.
5. Dominingshaus, H. Die Kunststoffe und ihre Eigenschaften. Springer Verlag, Berlin, 2005.

6. Craig, M. and K. Weining. Verantwortungsvolle Innovation Jetzt. Engineering Design 2009; 2: 2-3.
7. Product Information. Rilsan, Arkema, 2005.
8. N.N., New high performance synthetic fiber of 100% vegetable origin produced from Rilsan PA11. Press Release, Arkema, 2009.
9. DSM Engineering Plastics: Biobasierte Polyamide in 5 neuen Ausführungen. Plastiker-News vom 2.12.2010.
10. Datenblatt Grilamid TR; EMS Chemie. Domat/EMS, 201.
11. Datenblatt Grivory GV; EMS Chemie. Domat/EMS, 2011.
12. N.N., Datenblatt Grivory und Grilon LFT; EMS Chemie. Domat/EMS, 2011.
13. N.N., AKRO-PLASTC: PA aus nachwachsenden Rohstoffen schließt Lücke zwischen PA 6 und PA 12. Plastiker-News vom 04.08.2009.
14. Product Information. Ultramid Balance, BASF SE, 2010.
15. Shen, L., J. Haufe and M.K. Patel. Product overview and market projection of emerging bio-based plastics. PRO-BIB 2009 - Final Report, Utrecht, June 2009.
16. Greiner, R. Polymere, Biopolymere und ihre Bedeutung für den Fahrzeugbau; Statement zum Pressegespräch anlässlich des VDI-Kongresses. Kunststoffe im Automobilbau, Mannheim, 17.03.2011.
17. Eierdanz, H. Perspektiven nachwachsender Rohstoffe in der Chemie. VHC Verlagsgesellschaft mbH, Weinheim, 1996.
18. Beier, W. Biologisch abbaubare Kunststoffe. Pressestelle Umweltbundesamt, Dessau, 2009.
19. Website: http://ifbb.wp.hs-hannover.de/; 28.04.2016.
20. Endress, H.-J. Marktchancen, Flächenbedarf und künftige Entwicklungen. Kunststoffe 2011; 101(9): 105-110.
21. Mutlu, H. and M. Meier. Castor oil as a renewable resource for the chemical industry. European Journal of Lipid Science and Technology 2010; 112(1): 10-30.
22. Ogunniyi, D.S. Castor oil: a vital industrial raw material. Bioresource Technology 2006; 97(9): 1086-1091.
23. Mutlu, H. and M.A. Meier. Unsaturated PA X, 20 from renewable resources via metathesis and catalytic amidation. Macromolecular Chemistry and Physics 2009; 210(12): 1019-1025.
24. Jasinska, L., M. Villani, J. Wu, D. van Es, E. Klop, S. Rastogi and C.E. Koning. Novel, fully biobased semicrystalline polyamides. Macromolecules 2011; 44(9): 3458-3466.
25. Yamamoto, M. and A. Schneller. Industrielle Perspektiven von Biopolymeren. Chemie Ingenieur Technik 2010; 82(8): S.1245-1249.
26. Werner, T. Technische Kunststoffe aus nachwachsenden Rohstoffen. Kunststoffe 2009; 99(10): 54-61.
27. Bechthold, I., K. Bretz, S. Kabasci, R. Kopitzky, A. Springer. Succinic acid: a platform chemical for biobased polymers from renewable resources.
28. Product Information VESTAMID Terra – Polyamide aus nachwachsenden Rohstoffen. Evonik Degussa GmbH, Juni 2010.
29. Dierks, S. Natürliche Ressourcen Nutzen. Kunststoffe 2011; 101(8): 26-29.
30. Bluhm, R. Polyamide (PA) – Dynamische Entwicklung. Kunststoffe 2011; 101(10): 46-54.
31. Frost, J.W. Redefining chemical manufacture. Industrial Biotechnology 1, 2005.
32. Bechthold, I., K. Bretz, S. Kabasci, R. Kopitzky and A. Springer. Succinic acid: a new platform chemical for biobased polymers from renewable resources. Chemical Engineering & Technology 2008; 31(5): 647-654.
33. Hablot, E., B. Donnio, M. Bouquey and L. Avérous. Dimer acid-based thermoplastic bio-polyamides: reaction kinetics, properties and structure. Polymer 2010; 51(25): 5895-5902.
34. Kawaguchi, H., T. Hasunuma, C. Ogino and A. Kondo. Bioprocessing of bio-based chemicals produced from lignocellulosic feedstocks. Current Opinion in Biotechnology 2016; 42: 30-39.
35. Hellerich, W. G. Harsch and S. Haenle. Werkstoff-Führer Kunststoffe. Hanser Verlag, München, 2004
36. Product Information. Ultramid Balance, BASF SE, 2010.
37. Feldmann, M. and A.K. Bledzki. Bio-based polyamides reinforced with cellulosic fibres – Processing and properties. Composites Science and Technology 2014; 100: 113-120.

38. Feldmann, M. Biobasierte Polyamide mit Cellulosefasern Verfahren – Struktur – Eigenschaften. Dissertation. Institut für Werkstofftechnik der Universität Kassel, Kassel, 2012.
39. Keil, U. Vestamid Terra bio-polyamide protects high-performance cables. Press Release EVONIK, 3. März 2010.
40. N.N., polyvanced: Schläuche aus Maiszucker, Rizinusöl und Zuckerrohr; Plastiker-News vom 08.03.2010
41. Feldmann, M. The effects of the injection moulding temperature on the mechanical properties and morphology of polypropylene man-made cellulose fibre composites. Composites Part A 2016. doi:10.1016/j.compositesa.2016.04.022. In Press.
42. Amintowlieh, Y. Nylon-6/agricultural filler composites. Dissertation, University of Moistureloo, Moistureloo, 2010.
43. Zierdt, P., T. Theumer, G. Kulkarni, V. Däumlich, J. Klehm, U. Hirsch and A. Weber. Sustainable wood-plastic composites from bio-based polyamide 11 and chemically modified beech fibers. Sustainable Materials and Technologies 2015; 6: 6-14.
44. Feldmann, M., H.P. Heim and J.C. Zarges. Influence of the process parameters on the mechanical properties of engineering biocomposites using a twin-screw extruder. Composites Part A: Applied Science and Manufacturing 2016; 83: 113-119.
45. Feldmann, M., A.A. Mamun, A.K. Bledzki and H.-P. Heim. Bio-based polyamides with innovative fibers for engineering parts materials – processing – characterization – applications. Antec 2012, 2.-4. April 2012, Orlando/FL, 2012.
46. Feldmann, M. and F. Verheyen. Impact behaviour of continuous biaxial reinforced composites based on bio-polyamides and man-made cellulose fibres. International Polymer Processing 2016; 31(2): 198-206.
47. Feldmann, M., F. Verheyen and H.-P. Heim. Continuous-fiber-reinforced engineering thermoplastics from renewable raw materials - processing and properties. Conference: 2nd Bioplasic Materials - SPE, At Schaumburg/Chicago/Ill, October 2014.
48. Battegazzore, D., O. Salvetti, A. Frache, N. Peduto, A. De Sio and F. Marino. Thermo-mechanical properties enhancement of bio-polyamides (PA10. 10 and PA6. 10) by using rice husk ash and nanoclay. Composites Part A: Applied Science and Manufacturing 2016; 81: 193-201.
49. Frost, J.W. U.S. Patent No. 8,367,819. Washington, DC: U.S. Patent and Trademark Office, 2013.

Chapter 4

PHB Production, Properties, and Applications

M.A.K.M. Zahari, M.D.H. Beg*, N. Abdullah and N.D. Al-Jbour

Faculty of Chemical and Natural Resources Engineering, Universiti Malaysia Pahang, Gambang 26300, Kuantan, Malaysia

4.1 INTRODUCTION

Synthetic plastics are often designed to mimic the properties of natural materials, i.e., they are highly resistant to the microbial degradation in landfills. This has become a problem for municipalities worldwide because municipal landfills lose capacity due to the accumulation of synthetic plastics [1]. On the other hand, since most of the available plastics are petroleum based, then the finite supply of fossil fuel and the increasing petroleum price, leads to make the bioplastic an appropriate approach and developing area of significant concern.

Bioplastic is a biopolymer that can be produced from naturally occurring materials. A number of researchers have identified the natural materials which could be processed further to produce the biodegradable plastics. Various types of microorganisms that have shown a capability of producing such materials produce products of different characteristics and at different efficiencies. Lactic acid, which is further polymerized into poly lactic acid (PLA), is one of the important products in the field of biodegradable plastics. Another important polymer produced naturally is polyhydroxyalkanoate (PHA).

Polyhydroxyalkanoates (PHAs) are bacterial polymers that are formed as naturally occurring storage polyesters by a wide range of microorganisms, usually under unbalanced growth conditions [2]. The mechanical properties of PHAs make them suitable replacements for different petrochemically produced plastics, such as polyethylene and polypropylene, but advantageous to these commodity plastics PHAs are completely degradable to carbon dioxide and water through natural microbiological mineralization. PHAs can be produced by biotechnological processes under controlled conditions. A series of PHAs with different monomeric compositions in addition to different physical and chemical properties could be produced by different types of microorganisms [3]. PHAs are considered of high interest because they have some properties similar

*Corresponding author: Email: mdhbeg@ump.edu.my.

to synthetic plastics [1]. In addition, the degradation product of PHA is a common intermediate compound in all higher organisms. PHA is nontoxic and biocompatible with animal tissues, and then it is suitable to be used in different surgical applications [1]. The material properties of PHAs are highly dependent on the constituting monomer units and their molecular weights. To date, more than 150 monomer units with different (*R*)-pending groups have been reported [4, 5]. One of the well known and most studied biodegradable polymer in the PHA family is P(3HB).

Poly-3-hydroxybutyrate P(3HB) and its copolymers with 3-hydroxyvalerate (3HV), poly (3-hydroxybutyrate-co-3-hydroxyvalerate) (PHBV), are the best known representatives of the PHA family. These polyesters have attracted widespread attention, as environmentally friendly polymers which can be used in a wide range of agricultural, industrial, and medical applications [6]. P(3HB) belongs to the polyesters class that was first isolated and characterized in 1925 by French microbiologist Maurice Lemoigne [7]. It is produced by different types of microorganisms, such as *Cupriavidus necator* or *Bacillus megaterium*. The polymer is primarily a product of carbon assimilation (from glucose or starch) and is employed by microorganisms as a form of energy storage molecule to be metabolized when other common energy sources are not available. Figure 4.1 shows the generic formula for PHAs, where x is 1 (for all commercially – relevant polymers) and R can be either hydrogen or hydrocarbon chains of up to around C16 in length [8]. A wide range of PHA homopolymers, copolymers, and terpolymers have been produced, in most cases at the laboratory scale. A few of them have attracted industrial interest and have been commercialized in the past decade.

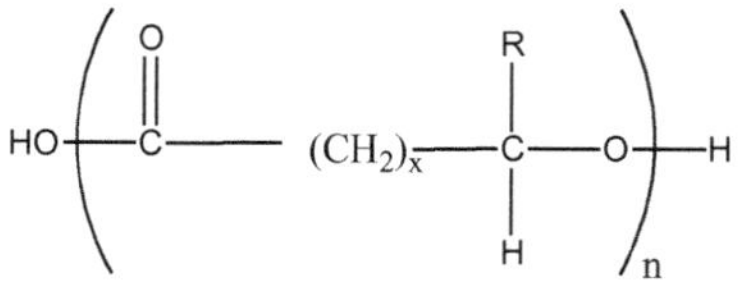

Figure 4.1 Molecule structure of PHA.

The poly-3-hydroxybutyrate, P(3HB) form of PHB is probably the most common type of polyhydroxyalkanoates, but many other polymers of this class are produced by a variety of organisms: these include poly-4-hydroxybutyrate, P(4HB), poly-3-hydroxyvalerate, P(3HV), poly-3-hydroxyhexanoate, P(3HHx), poly-3-hydroxyoctanoate, P(3HO), and their copolymers. Some exemplary molecular structures are shown in Table 4.1 [8]. Copolymers of PHAs vary in the type and proportion of monomers, and are typically random in sequence. Poly(3-hydroxybutyrate–*co*-3-hydroxyvalerate), P(3HB-*co*-3HV) is made up of a random arrangement of the monomers R=1 (methyl) and R=2 (ethyl). Poly(3-hydroxybutyrate-*co*-3-hydroxyhexanoate), P(3HB-*co*-3HHx), consists of the monomers R=1 (methyl) and R=3 (propyl). Poly(3-hydroxybutyrate-*co*-3-hydroxyalkanoate)s have co-polymer content varying from 3–15 mole % and chain length from C7 up to C19 [8].

Table 4.1 Structures of polyhydroxyalkanoates (PHAs) [8].

PHA short name	*PHA full name*	*x*	*R*
PHB	P(3HB)	1	$-CH_3$
PHV	P(3HV)	1	$-CH_2CH_3$
PHBV	P(3HB-*co*-3HV)	1	$-CH_3$ and $-CH_2CH_3$
PHBHx	P(3HB-*co*-3HHx)	1	$-CH_3$ and $-CH_2CH_2CH_3$
PHBO	P(3HB-*co*-3HO)	1	$-CH_3$ and $-(CH_2)_4CH_3$
PHBD	P(3HB-*co*-3HD)	1	$-CH_3$ and $-C_6H_8CH_3$
PHBOd	P(3HB-*co*-3HOd)	1	$-CH_3$ and $-(CH_2)_{14}CH_3$
	P(3HB-*co*-4HB)	2	$-CH_3$ and $-H$
	P(3HB-*co*-4HV)	2	$-CH_3$ and $-CH_3$

4.2 P(3HB) BIOSYNTHESIS PATHWAYS BY *Cupriavidus necator* (Formerly Classified as *Ralstonia eutropha*)

Among the different types of PHA studied, P(3HB) homopolymer has been extensively examined as a model PHA for process development [7]. The biosynthetic pathway of P(3HB) consists of three enzymatic reactions catalyzed by three different enzymes, as shown in Fig. 4.2. The first reaction consists of the condensation of two acetyl coenzyme A (acetyl-CoA) molecules into acetoacetyl-CoA by β-ketoacylCoA thiolase (encoded by phbA). The second reaction is the reduction of acetoacetyl-CoA to (R)-3-hydroxybutyryl-CoA by an NADPH-dependent acetoacetyl-CoA dehydrogenase (encoded by phbB). Lastly, the (R)-3-hydroxybutyryl-CoA monomers are polymerized into P(3HB) by P(3HB) polymerase, encoded by phbC [2].

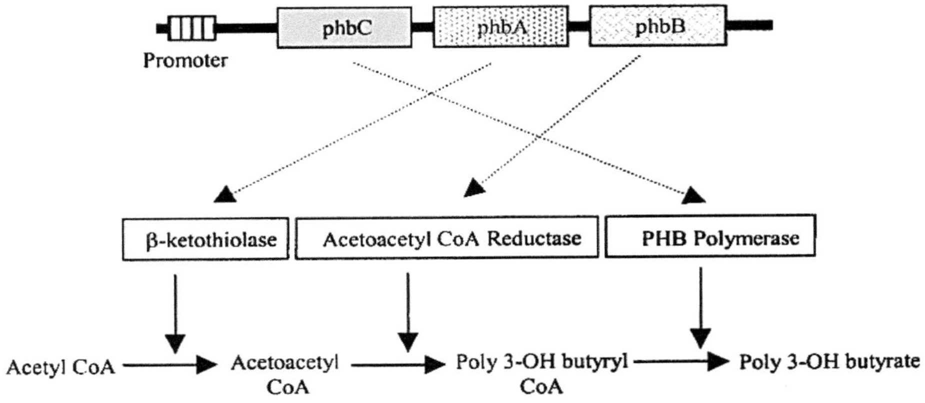

Figure 4.2 Biosynthetic pathway of poly(3-hydroxybutyrate). P(3HB) is synthesized by the successive action of b-ketoacyl-CoA thiolase (phbA), acetoacetyl-CoA reductase (phbB), and PHB polymerase (phbC) in a three-step pathway. The genes of the phbCAB operon encode the three enzymes. The promoter (P) upstream of phbC transcribes the complete operon (phbCAB) [2].

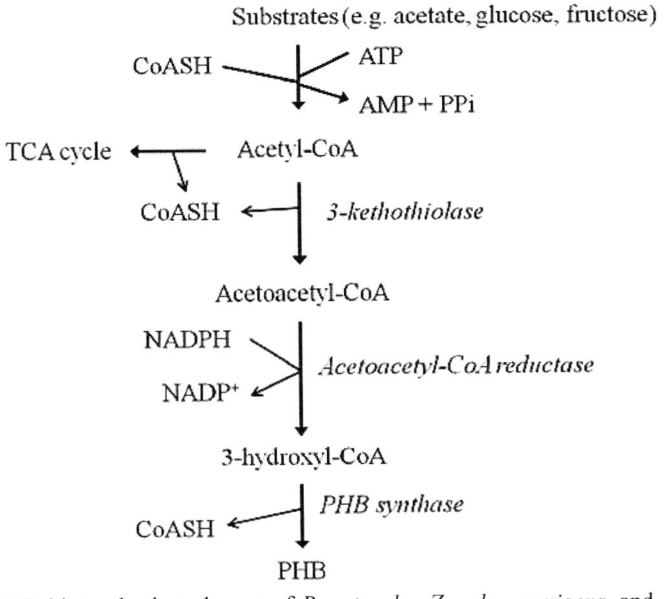

Figure 4.3 PHB biosynthesis pathways of *R. eutropha*, *Zooglea ramigera*, and *Az. beijerinckii* Source [7, 9].

The biosynthesis pathways of *R. eutropha*, *Zoogloea ramigera*, and *Azotobacter beijerinckii* are well established [7]. Most of the organisms synthesize P(3HB) using this pathway. Firstly, a substrate is condensed to acetyl-coenzyme A (acetyl-CoA). Two moles of acetyl-CoA are then used to synthesize one mole of P(3HB). Acetyl-CoA is subjected to a sequence of three enzymatic reactions as mentioned above, 3-ketothiolase, NADPH dependent acetoacetyl-CoA reductase, and P(3HB) polymerase for P(3HB) synthesis. Figure 4.3 shows a schematic of P(3HB) biosynthesis by these organisms [7, 9]. The P(3HB) biosynthetic pathway was dependent on the types of bacterium and precursor used to form the polymeric materials. As described by Braunegg et al., there were eight possible pathways in the formation of PHAs at various range of hydroxyalkanoic incorporated with P(3HB) homopolymer [10].

4.3 FACTORS AFFECTING THE PRODUCTION OF P(3HB)

Besides key enzymes and P(3HB) biosynthetic pathways naturally present in the microorganism, there are several other factors related to the biosynthesis of P(3HB) and affecting the production of P(3HB), such as the type of microorganism, cultivation conditions (carbon sources, pH, nitrogen and other nutrient sources, and oxygen requirements), in addition to that the genetically modified organism will influence the growth and biosynthesis of P(3HB) production.

4.3.1 Microorganisms and Carbon Sources

After the discovery of P(3HB) from the bacterium *Bacillus megaterium* in 1926, a large variety of different polyhydroxyalkanoates (PHAs) have been reported possessing different numbers of main chain carbon atoms and different types of pendent groups. PHAs are divided into two groups based on the number of constituent carbon atoms in their monomer units – short-chain length (SCL) PHAs and medium-chain-length (MCL) PHAs. The former consists of monomers with 3-5 carbon atoms and the latter contains monomers with 6–14 carbon atoms. Recently, there have been reports of several bacteria that are able to synthesize PHAs containing both SCL- and MCL-monomer units. PHA SCL is stiff and brittle with a high degree of crystallinity, whereas PHA MCL is flexible, with low crystallinity, tensile strength, and melting point [11].

More than 300 different microorganisms are known to synthesize and intracellularly accumulate PHAs [12]. Approximately, more than 100 genera of bacteria were identified for potentially produced PHA with various monomer substitutes. Types of the potential PHA accumulating genera are shown in Table 4.2. Out of 100 genera, several genera were identified having the potential of producing PHAs at industrial scales, such as *Ralstonia* sp., *Pseudomonads*, *Bacillus* sp., *Alcaligenes* sp., and *Burkholderia* sp. Among them, *Ralstonia euthropha*, H16 was a bacterium of choice for PHA production because of their rapid growth, in addition to having a wide range of PHA synthase substrates specificity [13].

Among the numerous bacteria known to produce P(3HB), *Cupriavidus necator*, *Methylobacterium organophilum*, *Alcaligenes latus*, and a recombinant *E. coli* have proven to be suitable for industrial applications [4]. *C. necator* and *M. organophilum* require nutrient limitation to accumulate P(3HB), whereas *A. latus* and recombinant *E. coli* do not. *M. organophilum* can utilize an inexpensive carbon source, methanol, for the biosynthesis of P(3HB), while *C. necator* (in fact a mutant strain of *C. necator* H16) and recombinant *E. coli* use glucose as a carbon source, and *A. latus* uses sucrose.

Among the bacterial species producing large concentrations of P(3HB), *C. necator* has been used in semi-commercial production, first by Zeneca BioProducts, and later by Monsanto [4]. Actually it grows well in a relatively inexpensive minimal medium, and accumulates a large amount of P(3HB) under unbalanced growth conditions, and using several carbon sources (except glucose). As glucose is often the most favorable carbon substrate among a number of sugars, a glucose-utilizing mutant strain of *C. necator* was developed and used for the production of P(3HB) and P(3HB-*co*-3HV) [4]. Various *C. necator* strains used in PHA research are detailed in Table 4.3.

Table 4.2 Types of PHA accumulating bacteria [14].

Acinetobacter	Escherichia	Paracoccus
Actinomycetes	Ferrobacillus	Pedomicrobium
Alcaligenes	Gamphospheria	Photobacterium
Aphanothese	Haemophilus	Pseudomonas
Aquaspirillum	Halobacterium	Rhizobium
Asticcaulus	Hyphomicrobium	Rhodobacter
Azomonas	Lamprocystis	Rhodococcus
Azospirillum	Lampropedia	Rhodopseudomonas
Azotobacter	Leptothrix	Rhodospirillum
Bacillus	Methanomonas	Sphaerotilus
Beggiatoa	Methylobacterium	Spirillum
Beijerinckia	Methylocystis	Spirulina
Beneckea	Methylomonas	Stella
Caryophanon	Methylovibrio	Streptomyces
Caulobacter	Micrococcus	Synthrophomonas
Chlorofrexeus	Microcoleus	Thiobacillus
Chlorogloea	Microcystis	Thiocystis
Chromatium	Moraxella	Thiodictyon
Chromobacterium	Mycoplana	Thiopedia
Clostridium	Nitrobacter	Thiosphaera
Corynebacterium	Nitrococcus	Vibrio
Derxia	Nocardia	Xantobacter
Ectothiorhodospira	Oceanospirillum	Zoogloea

Table 4.3 Various *C. necator* strains used in PHA research [4].

C. necator strain	*Characteristics*	*Sources and references*
H16	Wild-type, prototrophic	ATCC17699 [77]; DSM428; NCIMB 10442
NCIMB 11599	Glucose-utilizing mutant of H16	ICI [83]; [37]; [20]
PHB^{-4}	PHA-negative mutant of H16	DSM541 [84]
F11-1-116-EtOH	Alcohol-utilizing mutant of NCIMB 11599	ICI [85]
DSM 545	Glucose-utilizing mutant of H1(DSM 529)	DSM545 [86]
DSM 11348	Mutant from DSM 531, Increased growth on succinate and glucose	DSM 11348 [87]

Among the type of glucose-utilizing mutant as shown in Table 4.3, *C. necator* strain NCIMB 11599 has been widely used in the production of P(3HB) using glucose-based synthetic media and complex renewable feedstocks, such as tapioca hydrolysates [15], saccharified waste potato starch [16], wheat hydrolysate [17], and pulp fiber sludge [18]. It should be stressed that the P(3HB) concentration and content obtained when complex renewable feed stocks were used were lower than those obtained when synthetic media with purified glucose was used [17]. For instance, 94 g/L of P(3HB) was obtained from saccharified waste potato starch [16], and 162.8 g/L of P(3HB) was obtained from wheat hydrolysate [19]. In comparison, 232 g/L of P(3HB) was produced using *C. necator* NCIMB 11599 from pure glucose [20]. However, it is worth mentioning that by using a cheaper carbon source, the final production costs could be significantly lowered. For example, when cheap carbon sources, such as whey, cane molasses, and hemicelluloses hydrolysate are used to produce P(3HB), the final production cost may be reduced by 20%, compared to glucose [4].

Despite its basic attractiveness as a substitute for petroleum-derived polymers, the major hurdle commercial production and application of PHA is facing in consumer products is the

Table 4.4 List of microorganisms synthesized P(3HB) from glucose and renewable sources.

Polymer	*Microorganism*	*Type of waste/ Nitrogen source*	*Fermentation condition*	*Max. P(3HB) production (g/L)*	*P(3HB) content (%)*	*Productivity* ($gPHBl^{-1}h^{-1}$)	*P(3HB) yield $Y_{P/S}$* (gg^{-1} substrate)	*References*
P(3HB)	*Ralstonia eutropha* NCIMB 11599	Wheat-based biorefinery	Batch, fed-batch	162.8	93	0.89	0.47	[19]
P(3HB)	*Bacillus cereus* strain	Pea shell slurry/casein hydrolysate	Batch expt, shaken flask, 72 h	3.37	41	NG	NG	[23]
P(3HB)	*C. necator* DSM 545	Waste glycerol/ $(NH_4)_2SO_4$	2L STR, fed-batch	38.1	50	1.1	NG	[16, 24]
P(3HB)	*Ralstonia eutropha* NCIMB 11599	saccharified waste potato starch/ $(NH_4)_2SO_4$	Phosphate limitation, fed-batch	94	46	1.47	0.22	
P(3HB)	*Bacillus sp.*	Molasses/peach pulp	Shake flask expt	NG	7.92/ 7.78	NG	NG	[25]
P(3HB)	*Alcaligenes latus* ATCC 29714	Maple sap/ $(NH_4)_2SO_4$	Batch expt., 27h shake flask 20 L STR (10 L working volume)	3.41 3.26	77.6 77	NG	0.34 0.32	[26]
P(3HB)	*Bacillus megaterium*	Date syrup/beet	48 h	1.76	52	NG	NG	[20, 27]
P(3HB)	*Ralstonia eutropha* NCIMB 11599	Glucose	DO-stat, Fed-batch, 74 h 60 L fermenter	232	82	3.14	0.38	

NG = Not given

high cost of bacterial fermentation, making bacterial PHA 5–10 times more expensive than the petroleum-derived polymers, such as polypropylene and polyethylene, which costs approximately US$ 0.25–0.5 kg^{-1} [21, 22]. The most significant factor for increasing the production cost of PHA is the cost of substrate (mainly carbon source). Instead of pure substrate, which was being used as a carbon source, many researchers are now converting their interest to the utilization of renewable sources as a raw material for the production of P(3HB). This was due to the fact that the utilization of waste materials upgraded to the role of starting material for PHA biosynthesis constitutes a viable strategy for cost-efficient biopolymer production and helps the industry to overcome disposal problems. A list of microorganisms that utilized renewable sources for P(3HB) production are summarized in Table 4.4.

4.3.2 pH

pH is among various culture conditions that can easily be manipulated. In a shake flask experiment, initial medium pH is normally being adjusted to the desired level before starting fermentation. In a larger scale of 2 L fermenter, the pH of the medium can be regulated automatically by addition of base and acid. During growth and PHA accumulation stage, pH of the medium changing over time that reflects substrates consumption and intermediate by the formation of products depends on the bacterial strains and carbon sources used. The importance of initial medium pH was described elsewhere [28–30], in which the optimal pH shall be determined at desired level to obtain optimal growth and desired by-products.

Several studies have reported the influence of the growth, PHA content, and compositions by just regulating the initial medium pH. [28] investigated the PHA production from *R. eutropha* H16 (ATCC 17699 [28]) fed with butyric and valeric acids. Optimum conditions for P(3HB) production using butyric acid by this organism were at the concentration of 3 g/L butyric acid and pH of 8.0. PHV or other PHAs was not reported in this study. P(3HB) content of 75% was obtained under those conditions, while lower P(3HB) contents were achieved when pH was kept at 8 and butyric acid concentrations were 0.03, 0.3, 10 g/L, i.e., P(3HB) contents were 44%, 55%, and 63%, respectively. When butyric acid was kept at 3 g/L, P(3HB) content was lower as pH changed from the optimum value (pH 8). P(3HB) contents of 53% and 58% were obtained at pH values of 6.9 and 7.5, respectively, while no production of P(3HB) was obtained when pH was greater than 8.4. When propionic and butyric were used as substrates, 3HB-3HV copolymers were produced, however, the production of copolymers was lower than that of homopolymer produced by butyric alone. PHA content of 48% and 35% were obtained as the fraction of valeric acid was increased to 40% and 100%, respectively. Loo and Sudesh, (2007) have reported that 3HV incorporated in the P(3HB-*co*-3HV) copolymer could be improved from sodium valerate by adjusting the initial medium pH at acidic conditions [30]. In another study, compositions of the 4-hydroxybutyric (4HB) incorporation in the poly(3-hydroxybutyrate-*co*-4-hydroxybutyrate) [P(3HB-*co*-4HB) copolymer also could be regulated by the selection of the suitable initial medium pH using 1,4 butanediol as the carbon source [29].

4.3.3 Nitrogen and Other Nutrients Sources

Bacteria that are used for the production of PHAs can be divided into two groups based on the culture conditions required for PHA synthesis. The first group of bacteria requires the limitation of an essential nutrient, such as nitrogen, phosphorus, magnesium, or sulfur for the synthesis of PHA in the presence of excess carbon source. *Alcaligenes eutrophus*, *Protomonas extorquens*, and *Protomonas oleovorans* are some of the bacteria included in this group. The second group of bacteria, which include *Alcaligenes latus*, a mutant strain of *Azotobacter vinelandii*, and recombinant *E. coli*, do not require nutrient limitation for PHA synthesis and can accumulate polymer during growth [22, 31]. During the growth phase, essential nutrient supplies are needed to support growth and cells maintenance. Nitrogen sources are as important as other nutrients in developing new cells.

Bourque et al. (1995) investigated the production of P(3HB) by *Methylobacterium extorquens* ATCC 55366 using methanol as a sole carbon and energy source in a fed-batch fermentation system [12]. Biomass production and the growth rate of *M. extorquens* were affected by the mineral composition supplied. The absence of $(NH_4)_2SO_4$ or $MnSO_4$ and the absence of a combination of $CaCl_2$, $FeSO_4$, $MnSO_4$, and $ZnSO_4$ had a negative impact on biomass production and the specific growth rate of this bacterial strain. High concentrations of $(NH_4)_2SO_4$ were toxic to bacterial growth, while increases in concentrations of $MgSO_4$, $FeSO_4$, and a trace element mixture improved the bacterial growth rate. Highest P(3HB) production from this study was obtained under nitrogen starvation conditions [32]. Suzuki et al. (1986) reported the maximum PHB production of 66% of dry weight by *Pseudomonas sp.* using methanol as a sole carbon and energy source [33]. In order to obtain the high content of P(3HB), a proper medium composition was utilized. In this study, concentrations of phosphate and ammonium were maintained at low levels. Nitrogen deficiency was found to be the most effective way to stimulate the accumulation of P(3HB) [33].

4.3.4 Oxygen Requirement

An appropriate level of dissolved oxygen in the medium could be achieved by controlling the rate of air and impeller speed in the bioreactor system. Sufficient dissolved oxygen level also could be achieved by a certain agitation speed in shake flask fermentation study. Oxygen is required by the microorganism for assimilation of carbon sources for energy and cells duplication. Besides, it is required in the growth stage of bacterium. The limitation of dissolved oxygen concentration was found to increase the rate of P(3HB) production.

According to what was reported by Lafferty et al. [79] P(3HB) production was stimulated under oxygen limitation by *R. eutropha* and *Azotobacter beijerinckii* In a chemostat culture of *Az. beijerinckii,* P(3HB) synthesis is stimulated by oxygen deficiency. The mitigation of oxygen deficiency in the presence of excess glucose leads to a partial degradation of PHB. In this case, the rates of O_2 consumption and CO_2 evolution increase rapidly, and the cells accumulate some intermediate products of glucose metabolism. These data were interpreted in a way indicating that the high NADH/NAD+ ratio in cells grown under unbalanced conditions with respect to oxygen inhibits the enzymes of glucose catabolism and the Krebs cycle, thus suppressing the oxidation of acetyl-CoA in the Krebs cycle and stimulating its conversion into P(3HB). The mitigation of oxygen deficiency leads to the cessation of P(3HB) synthesis and oxidation of acetyl-CoA in the Krebs cycle [34, 35].

4.4 FED-BATCH FERMENTATION STRATEGY FOR THE BIOSYNTHESIS OF P(3HB)

It is important to produce P(3HB) with high productivity and high yield to reduce the overall cost. Fed-batch cultivations have been carried out to improve P(3HB) productivity. Standbury and Whitaker [36] introduced the term fed-batch cultures which are fed continuously, or sequentially, with medium, without the removal of culture fluid. Thus, the volume of the culture increases with time. Standbury and Whitaker [36] has reviewed the very wide range of fermentation process utilizing fed-batch systems. The use of fed-batch by the fermentation industry takes advantages of the fact that residual substrate concentration could be maintained at a very low level in such a system. The low residual level of substrates could be advantageous in removing the repressive effects of rapidly utilized carbon sources and maintaining the conditions in the culture within the aeration capacity of the fermenter, and finally avoiding the toxic effects of a medium component [36].

The fed-batch culture strategy, which is the most popular method to achieve high cell density and consequently the high productivity of the desired product, has been applied for the efficient production of P(3HB). Examination of conditions for the cultivation of *Cupriavidus necator* has suggested that it is important to maintain the concentration of carbon source at an optimal value, and to apply the nutrient limitation triggering of P(3HB) accumulation at an optimal point [20, 37]. A high cell density culture of *C. necator* NCIMB11599 has been studied extensively. In order to maintain the glucose concentration within the optimal range, several feeding strategies for fed-batch cultures were developed [20, 37]. As the increase of pH in response to carbon depletion was found to be slow, a dissolved oxygen (DO)-stat feeding with nitrogen limitation strategy was applied in order to improve P(3HB) production. However, as the pH was controlled by adding NaOH under nitrogen limitation conditions, the toxicity of highly concentrated hydroxide had a negative effect on cell growth at high cell density. Therefore, a fed-batch culture strategy of *C. necator* NCIMB11599 with DO-stat feeding strategy under phosphorus limitation was examined [20]. The pH was maintained by adding NH_4OH instead of NaOH, whereby was found that maintaining phosphate and magnesium concentrations above 0.35 g/l and 10 mg/l, respectively was important to obtain a high P(3HB) concentration. Under these conditions, a final cell concentration, P(3HB) concentration, and P(3HB) content of 281 g/l, 232 g/l, and 80 wt.%, respectively, were obtained, and this resulted in a high productivity of 3.14 g P(3HB)/l/h [20].

4.5 PRODUCTION COST OF P(3HB)

Commercial applications and wide use of PHA is hampered due to its price. The cost of PHA using the natural producer *A. eutrophus* is USD 16 (RM 48 at USD 1 = RM 3) per kg, which is 18 times more expensive than polypropylene [38]. At present, the raw material costs account for as much as 40 to 50% of the total production cost for PHA [4, 24, 39]. Use of lower cost carbon sources, recombinant *E. coli*, or genetically engineered plants should all lead to reductions in production cost [40]. Akiyama et al. (2003) have estimated the production cost for the fermentative production of two types of PHAs using a detailed process simulation model [41]. According to their calculations the annual production of 5,000 tons per year of poly(3-hydroxybutyrate-*co*-3-hydroxyhexanoate), P(3HB-*co*-3HHx) (also referred to as PHBHx) from soybean oil as the sole carbon source is estimated to cost from USD 3.50 (RM 10.50) to USD 4.50 (RM 13.50) per kg, depending on the presumed process performance. Microbial production of P(3HB) from glucose at a similar scale of production has been estimated to cost USD 3.80 (RM 11.40) – USD 4.20 (RM 12.60) per kg [41].

Another renewable carbon source that is gaining higher interest as an alternative to pure glucose for the production of P(3HB) is starch. The advantage of starch as a carbon source is that its price is lower than that of glucose. Choi and Lee (1999) [80] estimated that on a production scale of 100,000 tons of P(3HB) per year, production costs would decrease from USD 4.91 (RM 14.73) to 3.72 (RM 11.16) kg^{-1} if hydrolyzed corn starch (USD 0.22 (RM 0.66) kg^{-1}) was used instead of glucose (USD 0.49 (RM 1.47) kg^{-1}) [12]. For a plant with an annual production of 10,000 ton P(3HB), the costs of equipment for the fermentation, extraction, and purification plant including utilities is estimated at approximately USD 40 (RM 120) million. However, it is worth mentioning that by using a cheaper carbon source, the final production costs could be significantly lowered. For example, when cheap carbon sources, such as whey, cane molasses, and hemicelluloses hydrolysate are used to produce P(3HB), the final production cost may be reduced by 20% compared to glucose [4].

The price of PHAs in general is presently much higher than starch plastics and other bio-based polyesters due to high raw material costs, high processing costs (particularly the purification of the fermentation broth), and small production volumes. Today the price has considerably decreased compared to five years ago. Tianan currently offers its PHBV at USD 4.40 (RM 13.20)/kg (for orders over 50 Mt, FOB Ningbo harbor). For comparison, five years ago Biomer offered its P(3HB) at EURO 20 (RM 79 at EURO 1 = RM 3.95) per kg [8] and Metabolix's PHBV was estimated at EURO 10 (RM 39.50) – 12 (RM 47.40) per kg [42]. Tianan expects that the price will drop to USD 4.00 (RM 12.00)/kg in 2010 and USD 3.52 (RM 10.56)/kg in 2020 along with their capacity expansions [81]. Kaneka expects the price of its PHBHx will drop to EURO 3.40 (RM 13.43)/kg in 2020 [8].

4.6 PRODUCTION OF P(3HB) FROM LOW COST (WASTE-BASED) SUBSTRATES

The application of biotechnological processes for industrial production can be regarded as promising for sustainable development, although for a range of products biotechnological production strategies have not yet passed the test of economic viability. This is often caused by the cost of the raw materials. Here, a viable solution strategy can be identified in the utilization of a broad range of waste and surplus materials that can be upgraded to the role of feedstock for the biomediated production of desired end products. Such materials are mainly produced in agriculture and industrial branches that are closely related to agriculture [10, 22, 39]. In addition to that, it was estimated that the economics of PHA production is determined to a great extent (up to 50% of the entire production costs) by the cost of the raw materials. This is caused by the fact that PHA accumulation occurs under aerobic conditions, resulting in high losses of the carbon substrate by intracellular respiration. Hence, only a maximum amount of less than 50% of the carbon source is directed towards biomass and PHA formation [39].

In order to reduce the production cost of P(3HB), the product development of the biodegradable plastics could be obtained from renewable resources. Selection of a suitable substrate is an important factor for optimizing P(3HB) production and it will affect the content of P(3HB), in addition to its composition and polymeric properties. Many waste streams from agricultural and agro-industry (e.g., starch, maple sap, and wheat-based biorefinery) are potentially useful substrates and possibly may contribute to an economic P(3HB) production.

4.6.1 Production of P(3HB) from Starch

Starch is a renewable carbon source available in large quantities throughout the globe. Prior to fermentation, starch is hydrolyzed to glucose by a two-step process, liquefaction and saccharification. Commercial enzymes are used and represent a significant expense in the glucose production process. Many studies, some of which include genetic approaches, have been carried out to produce ethanol or lactic acid by direct fermentation of starch [43].

Kim and Chang (1998) have studied the direct production of P(3HB) from starch by using *A. chroococcum* [44]. The effects of culture volume on cell growth and P(3HB) production were investigated in flask cultures. Batch and fed-batch fermentations were carried out to obtain high concentrations of cells and P(3HB). From their study in shake flask culture, P(3HB) content increased up to 74% of dry cell weight with increasing culture volume, whereas in batch culture, P(3HB) content increased to 44% with O_2 limitation. In fed-batch culture, cell concentration of 71 g/l with 20% P(3HB) was obtained without O_2 limitation, whereas cell concentration of 54 g/l with 46% P(3HB) was obtained with O_2 limitation [44].

Most processes for PHA production based on starch require the conversion of starch to easily convertible substrates, such as glucose by enzymatic or chemical hydrolysis [45, 46]. Alternatively, VFAs can be produced as fermentation substrates by acidogenesis [47]. The production of P(3HB-*co*-3HV) by *H. mediterranei* on extruded starch in a pH-stat fed-batch mode was recently described by Chen et al. (2006), where an exogenous source of α-amylases was used [45].

Rusendi and Sheppard (1995) described the enzymatic utilization of potato processing wastes for utilization as a substrate for P(3HB) production [48]. The amylase for starch hydrolysis was provided by barley malt. The hydrolysate, containing about 200 g/L glucose, was supplemented by other nutritional components needed for the cultivation of the production strain *Alcaligenes eutrophus*. In batch cultures, a concentration of 5 g/L P(3HB) was obtained, corresponding to 77% of cell dry mass [48].

Haas and colleagues used saccharified waste potato starch as a carbon source for P(3HB) production by *R. eutropha* NCIMB 11599 under phosphate-limited conditions [16]. The researchers achieved 179 g/L biomass, 94 g/L P(3HB), and reported the yield of total biomass from starch as 0.46 g/g, the yield for P(3HB) from starch as 0.22 g/g, and the volumetric P(3HB) productivity as 1.47 g/L/h. Residual maltose accumulated in the fed-batch reactor, but caused no noticeable inhibition [16].

4.6.2 Production of P(3HB) from Maple Sap

Maple sap, an abundant natural product especially in Canada, is rich in sucrose and thus may represent an ideal renewable feedstock for the production of a wide variety of value-added products. Maple sap has traditionally been used for the production of syrup, the most important non-timber forest product in Canada. Over the last 5 years, Canada has accounted for 84% of the world's production of maple syrup [26].

In 2007, Yezza and his colleagues had used maple sap as the sole carbon source for *Alcaligenes latus* in shake flasks and in batch fermentor (10 L working volume) for the production of P(3HB) [26]. *Alcaligenes latus* was chosen because it uses sucrose; the main sugar in maple sap, effectively for growth and for P(3HB) production, and it can be easily lysed to enhance the recovery of P(3HB). In shake flasks, they found that the biomass obtained from both the sap and sucrose were 4.4 ± 0.5 and 2.9 ± 0.3 g/L, and the P(3HB) contents were 77.6 ± 1.5 and 74.1 ± 2.0%, respectively. Subsequent batch fermentation (10 L sap) resulted in the formation of 4.2 ± 0.3 g/L biomass and a P(HB) content of 77.0 ± 2.6% [26].

4.6.3 Production of P(3HB) from Wheat-based Biorefinery

Koutinas et al. (2007) and Xu et al. (2010) demonstrated the potential of a novel feedstock derived from wheat for viable P(3HB) production by *C. necator* NCIMB11599 [17, 19]. Wheat was bioconverted into two feedstock streams, wheat hydrolysate (WH) and fungal extract (FE) that were rich in glucose and nitrogen, respectively. WH and FE were mixed in appropriate proportions to provide media with varying glucose (5–26 g/L) and free amino nitrogen (FAN) (0.1–1.2 g/L) concentrations for batch shake flask fermentations [17]. The fermentation study was then carried out in a 2L bioreactor to demonstrate that an efficient P(3HB) production by using the wheat-based biorefinery could be achieved in a bioreactor [19].

Results from shake flasks study showed that increasing FAN concentration resulted in higher microbial growth and less P(3HB) accumulation [17]. The consumption of various carbon sources (carbohydrates, amino acids, peptides) resulted in high growth yields (up to 1.07 g cells/g glucose) as related to glucose. Specific growth rates up to 0.16 h^{-1} were observed. Three WH with similar glucose (200–220 g/L) and varying FAN (0.3–1.48 g/L) concentrations were evaluated in fed-batch shake flask fermentations for *C. necator* growth and P(3HB) accumulation. The medium with the highest nitrogen concentration (WH3) gave the highest microbial biomass concentration (29.9 g/L), growth yield (0.28 g residual microbial biomass/g glucose), and P(3HB) yield (0.43 g/g glucose). WH2 gave the highest P(3HB) concentration (51.1 g/L) and content (70 wt.%) [17]. For the fed-batch experiment conducted in a 2L bioreactor, the consumption of amino acids and peptides derived from wheat gluten hydrolysis resulted in a high glucose to P(3HB) conversion yield of

0.47 g/g. The respective yield regarding the amount of wheat used for the production of enzymes and P(3HB) was around 0.3 g P(3HB)/g wheat, which corresponds to 82.8% of the maximum theoretical conversion yield. The productivity achieved was around 0.9 g/l/h [19].

4.7 POTENTIAL OF OIL PALM WASTE FOR P(3HB) PRODUCTION

Oil palm is the most important agricultural crop of Malaysia. Overall, the palm oil industry is the fourth largest contributor to the country's Gross National Income (GNI) accounting for about 8% or almost RM 50 billion of GNI. Globally, Malaysia is the second largest producer and the largest exporter of crude palm oil. In 2010, Malaysia exported 16.7 million tons of palm oil, worth 59.8 billion ringgit, a 5% increase compared to the export in 2009. To date, government support for downstream activities has been targeted at palm oil based products, such as oleochemicals and, more recently, at strengthening the role of the private sector in this industry as part of the Palm Oil National Key Economic Area (NKEA). At the same time, the palm oil industry generates significant amounts of biomass every year, which is mostly used as fertilizer in the plantations [49] .

There are six types of waste generated from oil palm industry and could be categorized into two groups; which are oil palm biomass and palm oil mill effluent (POME). In the plantations, oil palm fronds (OPF) are available throughout the year as they are regularly cut during harvesting of fresh fruit bunches (FFBs) and pruning of the palm trees. Additional fronds as well as oil palm trunks (OPT) become available in the plantations during the replanting of oil palm trees every 25 to 30 years.

In the mills, oil palm empty fruit bunches (OPEFBs) remain after the removal of the palm fruits from the fruit bunches. Mesocarp fiber (MF) and palm kernel shells (PKS) are recovered during the extraction of crude palm oil (CPO) and palm kernel oil (PKO), respectively. In addition, palm oil mill effluent (POME) accumulates as a liquid biomass at the mills.

With nearly 4.70 million hectares of planted land and 416 mills operating across the country in 2009, Malaysian oil palm industry is estimated to generate over 115 million tons of the total wet weight of oil palm biomass inclusive of OPEFB, OPF, and OPT, and 60 million tons of POME [50, 51]. Table 4.5 shows the estimated wet and dry weight of oil palm biomass generated from Malaysian oil palm industry in the year 2009.

Table 4.5 Estimated wet and dry weight of oil palm biomass generated from Malaysian oil palm industry in year 2009 [50, 51].

Types of oil palm biomass	*Wet weight* (millions tons)[a]	*Dry weight* (millions tons)[b]
Oil palm trunks	15.2	13.97
Oil palm frond	83	44.84
Empty fruit bunches	17.5	6.93
Oil palm mesocarp fiber and kernel shells	–	11.5
Palm oil mill effluent	60.3	–
Total	176	77.24

The purpose of a zero emission from palm oil industry incorporating the production of PHA from POME was extensively studied by Hassan and co-workers. Production of mixed organic acids from anaerobically treated POME has introduced it as a proven renewable and cheaper carbon source for PHAs production [52–57]. Recent development in this research field showed that poly(3-hydroxybutyrate-*co*-3-hydroxyvalerate) [P(3HB-*co*-3HV)] copolymer can be produced from a locally isolated bacterium, *Comamonas* sp. EB172, which was successfully isolated from a digester treating POME [55]. This bacterium exhibited PHA accumulation when organic acids from anaerobically treated POME were used as carbon sources. It was demonstrated that the strain has successfully converted the organic acids from anaerobically treated POME into P(3HB-*co*-3HV) copolymers with various fractions of poly(3-hydroxyvalerate) [P(3HV)] [57]. A two-stage cultivation process was adopted in this study, in which cells were first grown in the nutrient rich medium prior to inoculation in the second stage with nitrogen limiting conditions. Latest results showed that up to 14.6 g/L of cells could be obtained from 0.4 g/L inocula, which is better than previous report (9.6 g/L of cells from 1 g/L of inocula) [14]. However, moderate PHA accumulation (43 wt.%) and lower 3HV monomer incorporation (6 mol% of 3HV) were recorded compared to previous report i.e., 59 wt.% PHA content and 21 mol% of HV [57].

An alternative route to produce P(3HB) from oil palm biomass is by squeezing sugar juice from the petiole part of the OPF. Similar to sugar cane, the juice can then serve as feedstock for P(3HB) production. OPF can be obtained during plantation replanting, pruning of the oil palm tree, and harvesting of FFBs, with the latter accounting for the greatest share of volume. To get the fresh fruit bunches from the oil palm, usually 2 to 3 OPF are cut as the FFBs are compactly packed and hidden in the leaf axils. In order to cut off the fruit bunches and OPF on old, tall palms, curved knives fastened to bamboo poles are used. During the harvesting of fresh fruit bunches, this OPF is felled in between the inter rows of the oil palm plant [58]. The only reason to fell this OPF is to use it as decomposed fertilizer, as most of the oil palm plantation owners believed that all of the OPF contained high nutrients. However, based on the scientific findings, it showed that the petiole part contained lower crude protein and other nutrients, but higher sugars content compared to the leaflet [59].

As shown in Fig. 4.4, the OPF is approximately 2–3 meters long and weighs about 10 kg (wet weight). It consists of the petiole (the stem) and many long leaflets on either side of the stem. The top two-thirds of the frond contain most of the nutrients, while the basal (lower) third is rich in cellulosic materials and sugars, which are needed in the production of P(3HB). The collection of only the basal portion of the fronds for downstream uses has two key advantages: one-thirds of the desirable content for production of P(3HB) (contained in the basal portion of the frond) would be made available for downstream uses; at the same time two-thirds of the nutrients (contained in the remaining two-thirds of the frond) would remain in the plantations as fertilizer [82].

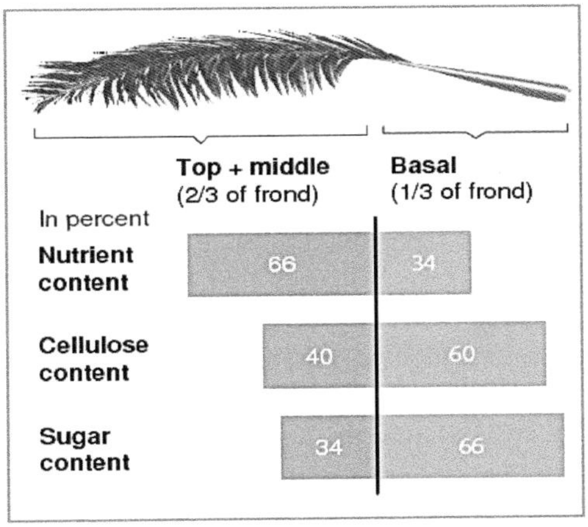

Figure 4.4 Picture of oil palm frond.

4.8 P(3HB) RECOVERY PROCESSES

In addition to the costs of maintaining pure cultures and the high costs of organic substrates, polymer recovery process is another factor that contributes to the high overall cost of PHA production. In the past two decades, several recovery processes have been investigated and studied in order to find an economic way to isolate and purify PHA. The methods of choice depend on factors, such as bacterial cell types, composition of growth medium, length of PHA granules, cell wall structure, incubation temperature and time, pH, pressure, economics of process, and the most important one is quality of PHA recovered [60] .

In most cases, bacterial biomasses are separated from substrate medium by centrifugation, filtration, or flocculation. Then, the biomasses are freeze dried (lyophilized). Basically, mild polar compounds, e.g., acetone and alcohols, solubilize non-PHA cellular materials, whereas PHA granules remain intact. Non-PHA cellular materials are nucleic acids, lipids, phospholipids, peptidoglycan, and proteinaceous materials. On the other hands, chloroform and other chlorinated hydrocarbons solubilize all PHAs. Therefore, both types of solvents are usually applied during recovery process. Finally, evaporation or precipitation with acetone or alcohol can be used to separate the dissolved polymer from the solvent [1].

According to Jacquel et al. (2008), several methods have been used as recovery processes for PHA [5]. Through the different research done by many research groups on the recovery of PHAs, lots of improvements have been done since the solvent extraction method used by [61]. The recovery by using solvents can lead to high purity and eliminate some endotoxins present in gram-negative bacteria. However, this method is also marked by high cost and the non-environmental friendly aspect of the use of some solvents. Other methods based on digestion of non-PHA cell material have also been envisaged, but by using hypochlorite, a degradation of the polymer was observed. Best results were obtained by combining hypochlorite with chloroform or surfactant treatment. Enzymatic digestion method was also introduced, but the use of expensive enzymes and complex process makes it economically unattractive. Mechanical cell disruption methods such as using bead mills and high pressure homogenization appear to be more economical. A clean recovery of P(3HB) have also been envisaged by the use of supercritical CO_2, but this method appears to be still expensive in comparison with other methods. Recently, new methods such as spontaneous liberation of P(3HB), dissolved air flotation, or air classification are being investigated and are probably promised to have much more success. The improvement of new extraction and purification methods leads to an optimal recovery of P(3HB), with high purity and recovery levels with low production cost [5].

4.9 PHYSICAL, THERMAL, AND MECHANICAL PROPERTIES OF P(3HB)

As mentioned previously, P(3HB) are accumulated intracellularly by a wide range of prokaryotes as carbon and energy reserves storage materials in the form of lipids, which are osmotically inert. These compounds are synthesized under conditions of nutrient imbalance, especially nitrogen, phosphorus, and oxygen, in the presence of excess carbon. In contrast to petroleum based plastics, P(3HB) can be degraded under the activation of enzymes by bacteria and fungi with no presence of the harmful products generated, such as water (H_2O) and carbon dioxide (CO_2). The rate of degradation depends on the soil and marine conditions, as well as the bacteria or fungi consortium [62]. The mineralization of P(3HB) into H_2O and CO_2 allows complete cycle of carbon. In addition, most P(3HB)-producing bacteria are able to degrade the polymer intracellularly. During intracellular degradation, the PHA depolymerase in the cell breaks down P(3HB) to give 3-hydroxybutyric acid. A dehydrogenase acts on the latter and oxidises it to acetylacetate and a β-ketothiolase acts on acetylacetate to break it down to acetyl-CoA. The *β*-ketothiolase enzyme plays an important role in both the biosynthetic and the biodegradation pathways. Under aerobic conditions, the acetyl-CoA enters the citric acid cycle and is oxidised to CO_2 [11].

The PHAs are formed as granular inclusion bodies in the cytoplasm. The diameter of P(3HB) granules varies from 0.2–0.7 μm covered with a membrane of 2–8 nm thickness. The molecular weight of PHB produced from wild-type bacteria is usually in the range of 10,000–3,000,000 Da with a polydispersity of around two [2, 22]. The large variation of the reported values is mainly due to the bacterial strain, substrate, growth rate, and temperature. Bacterially produced P(3HB) and other PHAs, however, have a sufficiently high molecular mass to have polymer characteristics that are similar to conventional plastics, such as polypropylene (see Table 4.6). Mechanical properties such as Young's modulus and tensile strength are close to that of polypropylene, though extension to break is markedly lower than that of polypropylene.

Table 4.6 Comparison between properties of polypropylene, P(3HB), and P(3HB-*co*-3HV).

Property	*Polypropylene*	*P(3HB)*	*P(3HB-co-3HV) (Biopol)*
Melting point (°C)	171–186	172–180	75–110
Glass transition temp. (°C)	–15	5–10	
Crystallinity (%)	65–70	65–80	less than P(3HB)
Density (g/cm³)	0.905–0.94	1.23–1.26	
Molecular weight (x 10^{-5})	2.2–7	1–8	
Flexural modulus (GPa)	1.7	3.5–4	
Tensile strength (MPa)	39	40	32
Extension to break (%)	400	6-8	250
UV resistance	poor	good	Good
Solvent resistance	good	poor	Poor
Oxygen permeability	good	poor	Good
Biodegradability	very poor	good	Good

Within the cell, P(3HB) exists in a fluid, amorphous state. However, after extraction from the cell with organic solvents, P(3HB) becomes highly crystalline and in this state is a stiff but brittle material [2]. Due to its brittleness, P(3HB) is not very stress resistant. Also, the relatively high melting temperature of P(3HB) (around 179°C) is close to the temperature where this polymer decomposes thermally, and thus limits the ability to process the homopolymer. Other properties of P(3HB) is that the decomposition temperature is higher than 205°C [22]. The densities of crystalline and amorphous P(3HB) are 1.26 and 1.18 g/cm, respectively.

4.10 GENERAL APPLICATION OF PHA/P(3HB)

The discovery that PHA can be processed by the traditional techniques used in the plastics industry such as injection molding, extrusion, blow molding, melt casting, and spinning, means that PHAs have the potential to become important source materials for the commodity plastics which are biodegradable. It must be pointed out that in high performance applications, such as high temperature or impact resistant plastics, these biodegradable materials could not replace conventional plastics. The homopolymer P(3HB) is itself quite brittle, a fact that reduces the range of its applications [63]. Nevertheless, nucleating agents, plasticizers, and other additives allow P(3HB) to be used in several processes, especially injection molding for pots, caps, and other single, hard pieces. Other uses, such as molds for metal casting, prosthetics, paper and granule coating, composites (starch, cellulose) are under development [63].

In general, the applications of bacterial PHAs have concentrated on three principal areas: medical and pharmaceutical, agricultural, and commodity packaging [64–66], as shown in Fig. 4.5). Based on the thermal and physical properties of PHA obtained, potential applications of PHA can be divided into two groups [14]. The first group refers to a polymer that is obtained from agricultural and industrial wastes. The products generated will be for agricultural purposes (plastic film for crop protection, encapsulation of seed, and encapsulation of fertilizer for slow release) [67], packaging (food) [68], and low-value products (stationery, comb, containers) [69]. The first consumer product made out of PHA was launched in 1990, by Wella AG. They tested their Sanara shampoo (products) with range of biodegradable shampoo bottles made of Biopol [83].

The second group is a polymer obtained from the pure substrate, such as glucose, fructose, and oils from plants. The products produced are probably in small quantities but for high value applications. To make PHA markets more competitive, some of the PHA families, particularly poly(4-hydroxybutyrate) [P(4HB)], poly(3-hydroxyhexanoate) [P(3HHx)], and poly(3-hydroxyoctanoate) [P(3HO)] were determined as suitable polymers to be used in medical applications. Owing to the fact that P(3HB) in the form of R-3-hydroxybutyric acids is a normal constituent of blood and present at concentration between 0.3 and 1.3 mM in the human bodies and also in the cell of envelope and eukaryotes, PHA is biocompatible and safe for medical applications [70, 71]. Several studies have been conducted in employment of PHA in medical applications. Medical devices (sutures), tissue repair, artificial organ construction, drug delivery systems are among potential applications of PHA [69, 70, 72].

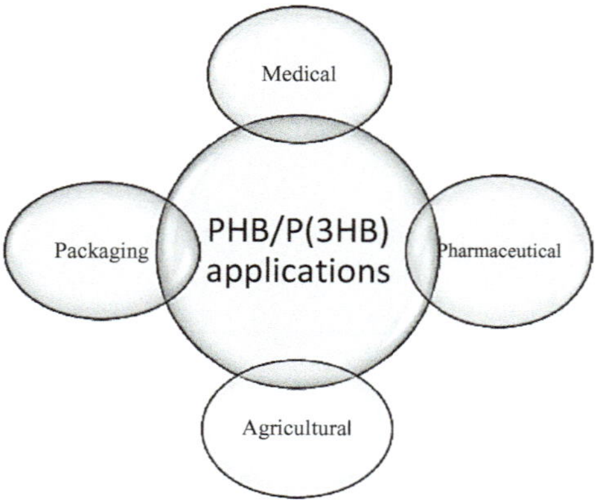

Figure 4.5: General applications of PHA/P(3HB).

In the last decades, the potential applications of PHAs and P(3HB) in medicine attracted much attention, due to the challenging combination of biomedical and biodegradable properties of P(3HB). The following paragraphs summarize its major applications in the medical industry.

1. Medical Devices

The perspective area of P(3HB) application is development of implanted medical devices for dental, craniomaxillofacial, orthopaedic, hernioplastic, and skin surgery. As shown in Fig. 4.6, a number of potential medical devices on the base of PHB: bioresorbable surgical sutures [73, 74], biodegradable screws and plates for cartilage and bone fixation, biodegradable membranes for periodontal treatment, surgical meshes with PHB coating for hernioplastic surgery [75], wound coverings [76] have been developed.

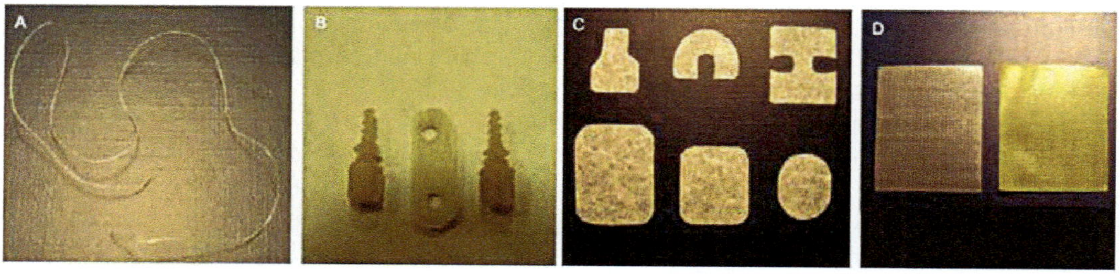

Figure 4.6 Medical devices on the base of PHB. (a) bioresorbable surgical suture; (b) biodegradable screws and plate for cartilage and bone fixation; (c) biodegradable membranes for periodontal treatment; (d) surgical meshes with PHB coating for hernioplastic surgery, pure (left) and loaded with antiplatelet drug, dipyridamole (right) [6].

2. Systems of Sustained Drug Delivery on the Base of P(3HB) Films

An improvement of medical devices on the base of biopolymers by encapsulating different drugs opens up the wide prospects in applications of these new devices with pharmacological activity in medicine. We have designed the novel systems for sustained delivery of an antiproliferative drug – dipyridamole (DP) and an anti-inflammatory drug – indomethacin. The kinetics of drug release from PHB films has been studied. The release occurs via two mechanisms, diffusion and degradation, operating simultaneously. Dipyridamole and indomethacin diffusion processes determine the rate of the release at the early stages of the contact of the system with the environment (the first 6–8 days). The coefficient of the release diffusion of a drug depends on its nature, the thickness of the PHB films containing the drug, the weight ratio of dipyridamole and indomethacin in polymer, and the molecular weight of PHB. Thus, it is possible to regulate the rate of drug release by changing of molecular weight of PHB, for example [75]. A number of other drugs have been also used for development of polymeric systems of sustained drug delivery: antibiotics (rifampicin, metronidazole, ciprofloxacin, levofloxacin), anti-inflammatory drugs (flurbiprofen, dexamethasone, prednisolone), and antitumor drugs (paclitaxel). Figure 4.7 shows an example of using PHB microspheres for the sustained delivery of dipyridamole for more than one month.

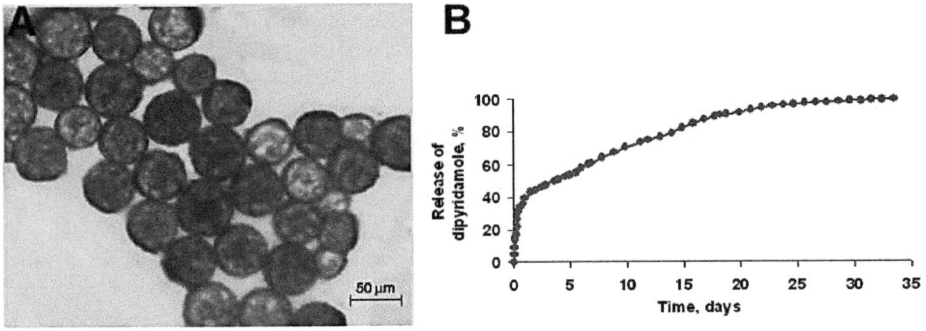

Figure 4.7 PHB microspheres for sustained delivery of drugs. (a) PHB microspheres (average diameter = 60 mkm, PHB Mw = 1000 kDa) loaded with dipyridamole (10% w/w); (b) Sustained delivery of dipyridamole from PHB micropsheres for more than 1 month [6].

3. Systems of Sustained Drug Delivery on the Base Microspheres and Microcapsules

Development of therapeutic systems of sustained drug delivery on the base of microspheres and microcapsules from biodegradable polymers is a new and promising trend in the modern pharmacology. We have developed systems of controlled dipyridamole release on the base of PHB microspheres. The coefficient of the release diffusion of DP extensively depends on diameter of microspheres. But it is possible to produce a system with prolonged uniform drug release that is important for producing therapeutic systems with adjusted drug dosing [6].

4. A Physiological Model on the Base of PHB Devices Loaded with Chemical Substances

Besides application of PHB for producing medical devices and systems of sustained drug delivery, PHB can be used for producing systems of sustained enzyme activators or inhibitors release for development physiological models. PHB with its minimal adverse inflammatory tissue reaction under implantation is a prospective tool in the design of novel physiological models of prolonged local enzyme activation or inhibition *in vivo* [6].

References

1. Punrattanasin, W. The utilization of activated sludge polyhydroxyalkanoates for the production of biodegradable plastics. Virginia Tech, 2001.
2. Madison, L.L. and G.W. Huisman. Metabolic engineering of poly (3-hydroxyalkanoates): From DNA to plastic. Microbiology and Molecular Biology Reviews 1999; 63(1): 21-53.
3. Sudesh, K., H. Abe and Y. Doi. Synthesis, structure and properties of polyhydroxyalkanoates: Biological polyesters. Progress in Polymer Science 2000; 25(10): 1503-1555.
4. Lee, S.Y. and S. Park. Fermentative production of short-chain-length PHAs. Biotechnology of Biopolymers–from Synthesis to Patents 2005; 1: 207-234.
5. Jacquel, N., C.-W. Lo, Y.-H. Wei, H.-S. Wu and S.S. Wang. Isolation and purification of bacterial poly (3-hydroxyalkanoates). Biochemical Engineering Journal 2008; 39: 15-27.
6. Bonartsev, A.P., V.L. Myshkina, D.A. Nikolaeva, E.K. Furina, T.A. Makhina, V.A. Livshits, A.P. Boskhomdzhiev, E.A. Ivanov, A.L. Iordanskii and G.A. Bonartseva. Biosynthesis, biodegradation, and application of poly (3-hydroxybutyrate) and its copolymers-natural polyesters produced by diazotrophic bacteria. Communicating Current Research and Educational Topics and Trends in Applied Microbiology 2007; 1: 295-307.
7. Doi, Y. Structure and properties of poly (3-hydroxybutyrate). Microbial Polyesters 1990.
8. Shen, L., J. Haufe and M.K. Patel. Product overview and market projection of emerging bio-based plastics PRO-BIP 2009. Report for European Polysaccharide Network of Excellence (EPNOE) and European Bioplastics 2009, Utrecht University, Utrecht: The Netherlands. p. 243.
9. Anderson, A.J. and E.A. Dawes. Occurrence, metabolism, metabolic role, and industrial uses of bacterial polyhydroxyalkanoates. Microbiological Reviews 1990; 54(4): 450-472.
10. Braunegg, G., G. Lefebvre and K.F. Genser. Polyhydroxyalkanoates, biopolyesters from renewable resources: Physiological and engineering aspects. Journal of Biotechnology 1998; 65(2-3): 127-161.
11. Philip, S., T. Keshavarz and I. Roy. Polyhydroxyalkanoates: Biodegradable polymers with a range of applications. Journal of Chemical Technology & Biotechnology: International Research in Process, Environmental & Clean Technology 2007; 82(3): 233-247.
12. Lee, S.Y., J.-i. Choi and H.H. Wong. Recent advances in polyhydroxyalkanoate production by bacterial fermentation: mini-review. International Journal of Biological Macromolecules 1999; 25(1-3): 31-36.
13. Reinecke, F. and A. Steinbüchel. Ralstonia eutropha strain H16 as model organism for PHA metabolism and for biotechnological production of technically interesting biopolymers. Journal of Molecular Microbiology and Biotechnology 2009; 16(1-2): 91-108.
14. Zakaria, M.R. Biosynthesis of Poly (3-hydroxybutyrate-co-3-hydroyvalerate) Copolymer from Organic Acids Using Comamonas Sp. EB172. PhD Thesis, Universiti Putra Malaysia, 2011.
15. Kim, B.S. and H.N. Chang. Control of glucose feeding using exit gas data and its application to the production of PHB from tapioca hydrolysate byAlcaligenes eutrophus. Biotechnology Techniques 1995; 9(5): 311-314.
16. Haas, R., B. Jin and F.T. Zepf. Production of poly (3-hydroxybutyrate) from waste potato starch. Bioscience, Biotechnology, and Biochemistry 2008; 72(1): 253-256.
17. Koutinas, A.A., Y. Xu, R. Wang and C. Webb. Polyhydroxybutyrate production from a novel feedstock derived from a wheat-based biorefinery. Enzyme and Microbial Technology 2007; 40(5): 1035-1044.
18. Zhang, S., O. Norrlow, J. Wawrzynczyk and E.S. Dey. Poly (3-hydroxybutyrate) biosynthesis in the biofilm of Alcaligenes eutrophus, using glucose enzymatically released from pulp fiber sludge. Applied and Environmental Microbiology 2004; 70(11): 6776-6782.
19. Xu, Y., R.H. Wang, A.A. Koutinas and C. Webb. Microbial biodegradable plastic production from a wheat-based biorefining strategy. Process Biochemistry 2010; 45(2): 153-163.
20. Ryu, H.W., S.K. Hahn, Y.K. Chang and H.N. Chang. Production of poly (3-hydroxybutyrate) by high cell density fed-batch culture of Alcaligenes eutrophus with phospate limitation. Biotechnology and Bioengineering 1997; 55(1): 28-32.
21. Poirier, Y., C. Nawrath and C. Somerville. Production of polyhydroxyalkanoates, a family of biodegradable plastics and elastomers, in bacteria and plants. Nature Biotechnology 1995; 13(2): 142.

22. Khanna, S. and A.K. Srivastava. Recent advances in microbial polyhydroxyalkanoates. Process Biochemistry 2005; 40(2): 607-619.
23. Kumar, T., M. Singh, H.J. Purohit and V.C. Kalia. Potential of *Bacillus* sp. to produce polyhydroxybutyrate from biowaste. Journal of Applied Microbiology 2009; 106(6): 2017-2023.
24. Cavalheiroo, J.M.B.T., M.C.M.D. de Almeida, C. Grandfils and M.M.R. da Fonseca. Poly (3-hydroxybutyrate) production by *Cupriavidus necator* using waste glycerol. Process Biochemistry 2009; 44(5): 509-515.
25. Kivanç, M., H. Bahar and M. Yilmaz. Production of poly (3-hydroxybutyrate) from molasses and peach pulp. Journal of Biotechnology 2008; 136: S405-S406.
26. Yezza, A., A. Halasz, W. Levadoux and J. Hawari. Production of poly-β-hydroxybutyrate (PHB) by *Alcaligenes latus* from maple sap. Applied Microbiology and Biotechnology 2007; 77(2): 269-274.
27. Omar, S., A. Rayes, A. Eqaab, I. Voß and A. Steinbüchel. Optimization of cell growth and poly (3-hydroxybutyrate) accumulation on date syrup by a *Bacillus megaterium* strain. Biotechnology Letters 2001; 23(14): 1119-1123.
28. Shimizu, H., S. Tamura, Y. Ishihara, S. Shioya and K.-I. Suga. Control of molecular weight distribution and mole fraction in poly (-D (−)-3-hydroxyalkanoate)(PHA) production by *Alcaligenes eutrophus*. *In*: Y. Doi and K. Fukuda (eds). Biodegradable Plastics and Polymers Studies: Studies in Polymer Science, Elsevier, 1994, pp. 365-372.
29. Lee, W.-H., M.N. Azizan and K. Sudesh. Effects of culture conditions on the composition of poly (3-hydroxybutyrate-co-4-hydroxybutyrate) synthesized by *Comamonas acidovorans*. Polymer Degradation and Stability 2004; 84(1): 129-134.
30. Loo, C.-Y. and K. Sudesh, Biosynthesis and native granule characteristics of poly (3-hydroxybutyrate-co-3-hydroxyvalerate) in Delftia acidovorans. International Journal of Biological Macromolecules 2007; 40(5): 466-471.
31. Chen J. and S. Wu S. Temperature Change on the Production of PHB by Ralstonia eutropha under Phosphorus-Limiting Conditions, 2004. Retrieved 10 September 2019. http://people.dyu.edu.tw/paper/9318416_c.pdf.
32. Bourque, D., Y. Pomerleau and D. Groleau. High-cell-density production of poly-β-hydroxybutyrate (PHB) from methanol by Methylobacterium extorquens: Production of high-molecular-mass PHB. Applied Microbiology and Biotechnology 1995; 44(3-4): 367-376.
33. Suzuki, T., T. Yamane and S. Shimizu. Mass production of poly-β-hydroxybutyric acid by fully automatic fed-batch culture of methylotroph. Applied Microbiology and Biotechnology 1986; 23(5): 322-329.
34. Dawes, E.A. and P.J. Senior. The role and regulation of energy reserve polymers in micro-organisms. Advances in Microbial Physiology 1973; 10: 135-266.
35. Braunegg, G., R. Bona and M. Koller. Sustainable polymer production. Polymer-plastics Technology and Engineering 2004; 43(6): 1779-1793.
36. Stanbury, P. and A. Whitaker. Principles of Fermentation Technology. Pergamon Press, 1984.
37. Kim, B.S. Production of poly(3-hydroxybutyrate) from inexpensive substrates. Enzyme and Microbial Technology 2000; 27: 774–777.
38. Reddy, C., R. Ghai and V.C. Kalia. Polyhydroxyalkanoates: An overview. Bioresource Technology 2003; 87(2): 137-146.
39. Koller, M., A. Atlić, M. Dias, A. Reiterer and G. Braunegg. Microbial PHA production from waste raw materials. *In*: G.Q. Chen (eds). Plastics from Bacteria. Microbiology Monographs, vol 14. Springer, Berlin, Heidelberg, 2010, pp. 85-119.
40. Suriyamongkol, P., R. Weselake, S. Narine, M. Moloney and S. Shah. Biotechnological approaches for the production of polyhydroxyalkanoates in microorganisms and plants–A review. Biotechnology Advances 2007; 25(2): 148-175.
41. Akiyama, M., T. Tsuge and Y. Doi. Environmental life cycle comparison of polyhydroxyalkanoates produced from renewable carbon resources by bacterial fermentation. Polymer Degradation and Stability 2003; 80(1): 183-194.
42. Petersen, K., P.V. Nielsen, G. Bertelsen, M. Lawther, M.B. Olsen, N.H. Nilsson and G. Mortensen. Potential of biobased materials for food packaging. Trends in Food Science & Technology 1999; 10(2): 52-68.

43. Guimaraes, W.V., et al. Ethanol production from starch by recombinant *Escherichia coli* containing integrated genes for ethanol production and plasmid genes for saccharification. Biotechnology Letters 1992; 14(5): 415-420.
44. Kim, B.S. and H.N. Chang. Production of poly (3-hydroxybutyrate) from starch by *Azotobacter chroococcum*. Biotechnology Letters 1998; 20(2): 109-112.
45. Chen, C.W., T.-M. Don and H.-F. Yen. Enzymatic extruded starch as a carbon source for the production of poly (3-hydroxybutyrate-co-3-hydroxyvalerate) by *Haloferax mediterranei*. Process Biochemistry 2006; 41(11): 2289-2296.
46. Huang, T.-Y., K.-J. Duan, S.-Y. Huang and C.W. Chen. Production of polyhydroxyalkanoates from inexpensive extruded rice bran and starch by *Haloferax mediterranei*. Journal of Industrial Microbiology and Biotechnology 2006; 33(8): 701-706.
47. Yu, J., Y. Si, W. Keung and R. Wong. Kinetics modeling of inhibition and utilization of mixed volatile fatty acids in the formation of polyhydroxyalkanoates by *Ralstonia eutropha*. Process Biochemistry 2002; 37(7): 731-738.
48. Rusendi, D. and J.D. Sheppard. Hydrolysis of potato processing waste for the production of poly-*β*-hydroxybutyrate. Bioresource Technology 1995; 54(2): 191-196.
49. Ng, W.P.Q., H.L. Lam, F.Y. Ng, M. Kamal and J.H.E. Lim. Waste-to-wealth: green potential from palm biomass in Malaysia. Journal of Cleaner Production 2012; 34: 57-65.
50. Zahari, M.A.K.M, S.S.S. Abdullah, A.M. Roslan, H. Ariffin, Y. Shirai, M.A. Hassan. Efficient utilization of oil palm frond for bio-based products and biorefinery. Journal of Cleaner Production 2014; 65: 252-260.
51. Ng, F.Y., F. K. Yew, Y. Basiron and K. Sundram. A renewable future driven with Malaysian palm oil-based green technology. Journal of Oil Palm, Environment and Health (JOPEH) 2012; 2: 1-7.
52. Hassan, M.A., Y. Shirai, H. Umeki, H. Yamazumi, S. Jin and S. Yamamoto. Acetic Acid Separation from Anaerobically Treated Palm Oil Mill Effluent by Ion Exchange Resins for the Production of Polyhydroxyaikanoate by *Alcaligenes eutrophus*. Bioscience, Biotechnology, and Biochemistry 1997; 61(9): 1465-1468.
53. Hassan, M.A., Y. Shirai, M. Inagaki, M.I. Abdul Karim, K. Nakanishi and K. Hashimoto. An economic analysis of the production of bacterial polyhydroxyalkanoates from palm oil mill effluent in Malaysia. Journal of Chemical Engineering Japan 1997; 30(4): 10-14.
54. Mumtaz, T., S. Abd-Aziz, N.A. Rahman, P.L. Yee, Y. Shirai and M.A.I. Hassan. Pilot-scale recovery of low molecular weight organic acids from anaerobically treated palm oil mill effluent (POME) with energy integrated system. African Journal of Biotechnology 2008; 7(21).
55. Zakaria, M.R., S. Abd-Aziz, H. Ariffin, A.R. NorAini, L.Y. Phang and M.A. Hassan, M.A. *Comamonas* sp. EB172 isolated from digester treating palm oil mill effluent as potential polyhydroxyalkanoate (PHA) producer. African Journal of Biotechnology 2008); 7: 4118-4121.
56. Zakaria, M.R., M. Tabatabaei, F.M. Ghazali, S. Abd-Aziz, Y. Shirai and M.A. Hassan. Polyhydroxyalkanoate production from anaerobically treated palm oil mill effluent by new bacterial strain *Comamonas* sp. EB172. World Journal of Microbiology and Biotechnology 2010; 26: 767-774.
57. Zakaria, M.R., H. Ariffin, N.A.M. Johar, S. Abd-Aziz, H. Nishida, Y. Shirai and M.A. Hassan. Biosynthesis and characterization of poly(3-hydroxybutyrate-co-3-hydroxyvalerate) a copolymer from wild-type *Comamonas* sp. EB172. Polymer Degradation and Stability 2010; 95: 1382–1386.
58. Islam, M., Nutritional evaluation and utilisation of oil palm (*Elaeis guineensis*) frond as feed for ruminants. Universiti Putra Malaysia, 1999.
59. IIslam, M., I. Dahlan, M.A. Rajion and Z.A. Jelan. Productivity and nutritive value of different fractions of oil palm (*Elaeis guineensis*) frond, 2000. Retrieved 10 September 2019. http://www.ajas.info/Editor/manuscript/upload/13-158.pdf.
60. Mohammadi, M., M.A. Hassan, Y. Shirai, H. Che Man, H. Ariffin, L.N. Yee, T. Mumtaz, M.L. Chong and L.Y. Phang. Separation and purification of polyhydroxyalkanoates from newly isolated *Comamonas* sp. EB172 by simple digestion with sodium hydroxide. Journal of Separation Science and Technology 2012; 47: 534–541.
61. Noel, B.J. Process for preparing poly-beta-hydroxybutyric acid. Google Patents, 1962.
62. Nishida, H. and Y. Tokiwa. Confirmation of colonization of degrading bacterium strain SC-17 on poly (3-hydroxybutyrate) cast film. Journal of Environmental Polymer Degradation 1995; 3(4): 187-197.

63. Nonato, R., P. Mantelatto and C. Rossell. Integrated production of biodegradable plastic, sugar and ethanol. Applied Microbiology and Biotechnology 2001; 57(1-2): 1-5.
64. Holmes, P. Applications of PHB-a microbially produced biodegradable thermoplastic. Physics in Technology 1985; 16(1): 32.
65. Huang, J.C., A.S. Shetty and M.S. Wang. Biodegradable plastics: a review. Advances in Polymer Technology 1990; 10(1): 23-30.
66. Lee, S.Y. Bacterial polyhydroxyalkanoates. Biotechnology and Bioengineering 1996; 49(1): 1-14.
67. Castro-Sowinski, S., S. Burdman, O. Matan and Y. Okon. Natural functions of bacterial polyhydroxyalkanoates. *In*: G.Q. Chen (eds). Plastics from Bacteria. Microbiology Monographs, vol 14. Springer, Berlin, Heidelberg, 2010, pp. 39-61.
68. SSiracusa, V., P. Rocculi, S. Romani and M.D. Rosa. Biodegradable polymers for food packaging: A review. Trends in Food Science & Technology 2008; 19(12): 634-643.
69. Chen, G.-Q. Plastics completely synthesized by bacteria: Polyhydroxyalkanoates. *In*: G.Q. Chen (eds). Plastics from Bacteria. Microbiology Monographs, vol 14. Springer, Berlin, Heidelberg, 2010, pp. 17-37.
70. Zinn, M., B. Witholt and T. Egli. Occurrence, synthesis and medical application of bacterial polyhydroxyalkanoate. Advanced Drug Delivery Reviews 2001; 53(1): 5-21.
71. Volova, T., E. Shishatskaya, V. Sevastianov, S. Efremov and O. Mogilnaya. Results of biomedical investigations of PHB and PHB/PHV fibers. Biochemical Engineering Journal 2003; 16(2): 125-133.
72. Wu, Q., Y. Wang and G.Q. Chen. Medical application of microbial biopolyesters polyhydroxyalkanoates. Artificial Cells, Blood Substitutes, and Biotechnology 2009; 37(1): 1-12.
73. Fedorov, M.B., G.A. Vikhoreva, N.R. Kil'deeva, A.N. Maslikova, G.A. Bonartseva and L.S. Gal'braikh. Modeling of surface modification of sewing thread. Fibre Chemistry 2005; 37(6): 441-446.
74. Rebrov, A.V., V.A. Dubinsky and Y.P. Nekrasov. Series B Structural Phenomena during Elastic Deformation of highly oriented poly (hydroxybutyrate). Vysokomolekuliarnye Soedineniia 2002; 44(2): 347-351.
75. Bonartsev, A.P., G.A. Bonartseva, T.K. Makhina, V.L. Myshking, E.S. Luchinina, V.A. Livshits, A.P. Boskhomdzhiev, V.S. Markin and A.L. Iordanski.. New poly (3-hydroxybutyrate)-based systems for controlled release of dipyridamole and indomethacin. Applied Biochemistry and Microbiology 2006; 42(6): 625-630.
76. Kil'deeva, N., G.A. Vikhoreva, L.S. Gal'braikh, A.V. Mironov, G.A. Bonartseva, P.A. Perminov, and A.N. Romashova. Preparation of biodegradable porous films for use as wound coverings. Applied Biochemistry and Microbiology 2006; 42(6): 631-635.
77. Wang, J. and Yu, H.-Q. Biosynthesis of polyhydroxybutyrate (PHB) and extracellular polymeric substances (EPS) by *Ralstonia eutropha* ATCC 17699 in batch cultures. Applied Microbiology and Biotechnology 2007; 75: 871-878.
78. [79] Lafferty, R.M., B. Korsatko, W. and Korsatko. (1988). Microbial production of poly-b-hydroxybutyric acid. *In*: Rehm, H.J. and G. Reed, G. (eds.). Biotechnology. VCH Publishers, New York, 1988, pp. 135-176.
79. [80] Choi, J and S.Y. Lee. Factors affecting the economics of polyhydroxyalkanoates production by bacterial fermentation. Applied Microbiology and Biotechnology 1999; 51: 13-21.
80. [81] Lunt, J. Manufacture and applications of PHBV polymers. Global Plastics Environmental Conference (GPEC) 2008. Retrieved 10 September 2019. https://www.gpec.ro/en/gpec-2008.
81. [82] Malaysian Innovation Agency (2011). National Biomass Strategy 2020: New wealth creation for Malaysia's palm oil industry, 2011. Retrieved 10 September 2019. https://www.cmtevents.com/MediaLibrary/BStgy2013RptAIM.pdf
82. [83] Min Zhang. Development of Polyhydroxybutyrate Based Blends for Compostable Packaging. PhD Thesis. Loughborough University, UK, 2010.
83. [84] Schlegel, H.G., R. Lafferty, I. Krauss. The isolation of mutants not accumulating poly-β-hydroxybutyric acid. Archiv für Mikrobiologie 1970; 71: 283-294.
84. [85] Park, C.-H. and V. Damodaran. Effect of alcohol feeding mode on the biosynthesis of poly (3-hydroxybutyrate-co-3-hyroxyvalerate). Biotechnology and Bioengineering 1994; 44(11): 1306-1314.

85. [86] Ramsay, B.A., K. Lomaliza, C. Chavarie, B. Dubé, P. Bataille and J.A. Ramsay, J.A. Production of poly-(β-hydroxybutyric-co-hydroxyvaleric) acids. Applied and Environmental Microbiology 1990; 56: 2093-2098.
86. [87] Bormann, E.J., M. Leißner and B. Beer. Growth-associated production poly(hydroxybutyrate) by Azobacter beijerinckii from organic nitrogen substrates. Applied Microbiology and Biotechnology 1998; 49: 84-88.

Chapter 5

Polyvinyl Alcohol and Polyvinyl Acetate

Mohammad S. Islam

School of Mechanical and Manufacturing Engineering
The University of New South Wales, Sydney, NSW 2052, Australia

5.1 POLYVINYL ALCOHOL

5.1.1 Introduction

Polyvinyl alcohol (PVA) was discovered by Hermann and Haehnel at Wacker Chemie in Germany in 1924 [1, 2], and around the same time by Staudinger [3]. PVA was the first fully synthetic stabilizer for colloidal systems [4].

Owing to its several characteristic properties, such as solubility in water, orientation characteristics, adhesive ability to a number of substrates, low toxicity, biodegradability, and biocompatibility, over one million metric tons of PVA is now produced in industry and consumed for a wide variety of applications every year [5, 6]. The basic monomer for PVA is vinyl acetate. The worldwide production capacity for vinyl acetate (VAM) was estimated to be close to 5,900 kilotons in 2009 with an actual production of around 5,500 kilotons [7]. About 30% of the VAM is converted to PVA [8]

The cost of PVA is USD 1.20 per kg [9]. PVA is a colorless, water-soluble polymer resin employed predominantly in the treating of textile materials and paper. It is the type of resin that most of us are likely to have encountered as a silky water-based white wood glue. Other than its use as an adhesive, it can be used as coatings for papers, textiles, and finishing material for leather. Another most interesting example of its use is its availability as a water-soluble yarn. Apart from adhesives and yarn, the water-solubility of PVA has resulted in a range of plastic films for packaging. It can also be used in fishing, where bait is put into PVA bags and then into the water. This allows a concentration of bait in the water while the bag itself dissolves. This same principle is exploited in laundry tablets, where a liquid detergent is enclosed within PVA capsules, which then dissolve in the machine. The crystalline structure of PVA allows it to polarize light, leading to applications for light filters, such as dichroic polarizers [9].

For Correspondance: Email: msislam@sydney.edu.au/saifulctp@yahoo.com; Phone: +61293512342

5.1.2 Raw Materials

PVA is unique as a polymer in the sense that it can be built up in polymerization reactions from single-unit precursor molecules known as monomers. It can be made by dissolving another polymer, polyvinyl acetate (PVAc), in an alcohol such as methanol, and treating it with an alkaline catalyst such as sodium hydroxide. The resulting hydrolysis or alcoholysis reaction removes the acetate groups from the PVAc molecules without disrupting their long-chain structure [10]. The chemical structure of the resulting vinyl alcohol repeating unit is given in Fig. 5.1.

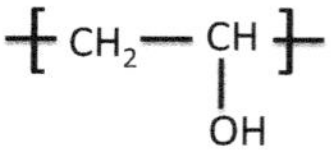

Figure 5.1 Chemical structure of vinyl alcohol repeating unit.

When the reaction is allowed to proceed to completion, the product becomes highly soluble in water and insoluble in practically all organic solvents. Incomplete removal of the acetate groups yields resins less soluble in water and more soluble in certain organic liquids.

There are more ways to prepare PVA than using VAM. A wide variety of vinyl esters can be polymerized via free radical polymerization to produce the PVA precursor similar to VAM [11]. Vinyl ethers can be homopolymerized via cationic polymerization using a Lewis acid catalyst. The homopolymer of vinyl ether with a bulky substituent, such as tert-butyl vinyl ether, can be easily hydrolyzed under acidic condition, subsequently to form PVA [12, 13]. Specifically, cationic polymerization of vinyl ether often gives relatively isotactic polymers, which consequently lead to isotactic PVA.

The worldwide production of PVAc and its copolymers containing more than 50 wt% of VAM reached about 2,300 kilotons by 2007, with an annual growth of about 3.4% [8]. Another major part of the produced VAM is converted to PVA. USA, Western Europe, China, and Japan are the major production regions. The major producers of VAM and VAM-based polymers include, but are not limited to, BP of the UK, Wacker Chemie of Germany, and Celanese Chemicals, Dow Chemical Corp., DuPont, and Millenium of the USA. In Asia, Kuraray, Nippon Gohsei, and Showa Denko of Japan; Dairen Chemical and Asian Acetyls of Taiwan; and Shanghai Petrochemical and Sichuan Vinylon Works of China produce VAM and their polymers. These regions are also the major markets for the different products obtained from PVAc and its derivatives [8].

5.1.3 Polymerization Process, Chemical Structure, and Production

Although PVA is one of the synthetic polymers derived from vinyl monomer units similar to polystyrene or polypropylene, it cannot generally be produced directly from vinyl alcohol, because vinyl alcohol itself is unstable and readily tautomerized into acetaldehyde. Therefore, it is usually prepared by the polymerization of a protected monomer, such as vinyl ester, followed by deprotection.

From the first breakthrough patent of the VAM preparation and polymerization into PVA by Herrmann et al., the production of PVA has progressed to a process of common polymeric materials [14, 15]. Nowadays, PVA in industrial applications is generally obtained from the saponification or hydrolysis of homopolymer of VAM. Although PVA is one of the oldest synthetic polymers, it still has been attracting much attention in the chemical, material, and medical fields because of a unique combination of its properties, including solubility in water, film orientation characteristics for the polarizer of a liquid crystal display, adhesive ability to a number of substrates, low toxicity, biodegradability, and biocompatibility [5, 16, 17]. These properties depend on degree of the hydrolysis as well as the primary structures of the original precursor of PVAc, such as head-totail regioselectivity, molecular weight, and tacticity. Therefore, from the viewpoint of synthetic chemistry, the control of the polymerization of VAM is still a challenging topic to improve the properties of PVA, and further contributes to the development of the PVA-based materials as well as the new polymerization systems for other protecting monomers.

PVA homopolymers along with a number of PVA-related materials are industrially produced and commercially available in many grades [5]. Copolymerization is a useful method for developing another polymeric material with different properties. Typically, the copolymerization of VAM with ethylene followed by hydrolysis results in ethylene-vinyl alcohol copolymer, which has excellent gas barrier properties with low permeability.

5.1.4 Properties

PVA exhibits many versatile physicochemical properties, such as viscosity film forming, emulsifying, dispersing power, tensile strength, and flexibilityflexibility, thermostability and tolerance towards solvents [18]. The most characteristic property of PVA is its solubility toward water, which depends both on the degree of polymerization and hydrolysis of PVAc [5]. Especially, the degree of hydrolysis significantly affects the water solubility, in which the original hydrophobic PVAc generally turns hydrophilic as the hydrolysis reaction proceeds due to the increasing affinity towards water with the hydroxyl groups. PVAc generally becomes water soluble over the range of 80% hydrolysis degree, in which the water solubility slightly decreases as the polymerization degree increases. The solubility in water also depends on the temperature. PVA with low degree of hydrolysis (80–85%) shows lower critical solution temperature-type phase diagram, in which the solubility becomes lower at a higher temperature [19].

The presence of too many adjacent hydroxyl groups in the consecutive side chain conclusively in turn result in water-insoluble polymers, because of the inter- and intramolecular hydrogen bondings to cause high crystallinity, such as in natural polymer cellulose, which has many hydroxyl groups. More specifically, over 90% hydrolysis degree, its water solubility gradually decreases after drying, and when the hydrolysis degree becomes over 98%, the polymer does not show water solubility anymore at the ambient temperature. In addition to that, thermal annealing or stretching for crystallization dramatically decreases the water solubility of PVA with high hydrolysis degree [5]. Consequently, a molded specimen or thread of pure PVA with quantitative hydrolysis degree and high crystallinity is hard to dissolve or sometimes insoluble in water after drying, similar to cellulose. Therefore, one has to tune the degree of hydrolysis depending on the water-soluble applications. In other words, only PVA with an adequate number of acetyl groups in the side chain (partially hydrolyzed PVAc) is accessible to many water-soluble applications.

Thus, PVA with different degrees of hydrolysis and polymerization are now commercially available depending on their applications. Generally, the fully hydrolyzed grade PVA has around 98–99% hydrolysis degree, with the degree of polymerization in the range of 200–3,000 [8].

5.1.5 Relationship between Structure and Properties

The chemical structure of PVA is relatively simple with a pendant hydroxyl group. The degrees of hydrolysis and polymerization affect the solubility of PVA in water [20]. It has been shown that PVA grades with high degrees of hydrolysis have low solubility in water. Residual hydrophobic acetate groups weaken the intra- and intermolecular hydrogen bonding of adjoining hydroxyl groups. The temperature must be raised well above 70°C for dissolution to occur. The presence of acetate groups also affects the ability of PVA to crystallize upon heat treatment. PVA grades containing high degrees of hydrolysis are more difficult to crystallize [21].

PVA is produced by free radical polymerization and subsequent hydrolysis, resulting in a fairly wide molecular weight distribution. A polydispersity index of 2 to 2.5 is common for most commercial grades. However, polydispersity indices of 5 are not uncommon. The molecular weight distribution is an important characteristic of PVA because it affects many of its properties, including crystallizability, adhesion, mechanical strength, and diffusivity [21].

It is important to note, that neighboring group effects affect the alkaline hydrolysis rate of the individual acetate groups. Acetate groups adjacent to alcohol groups are more readily hydrolysed than acetate groups having only other acetate groups in close proximity. As a consequence, partially hydrolysed PVA exhibits a block-like structure rather than a random structure and is best described as a (multi) block PVAc-co-PVA.

The degree of hydrolysis can be adjusted by the reaction time. Many common PVA resins have a degree of hydrolysis of about 88 mol%. These resins are soluble in cold water. Higher degree of hydrolysis leads to a reduced water solubility. A totally hydrolysed PVA is almost insoluble in cold water and can only be dissolved by boiling in water for an extended period of time. The insolubility is caused by the build-up of intramolecular hydrogen bonds, leading to a high degree of crystallisation not achieved by only partly hydrolysed PVA grades.

PVA subsequently can be acetalised with aldehydes, the most important of them being butyric aldehyde, leading to polyvinyl butyral (PVB) [8].

The presence of hydroxyl groups attached to the main chain in PVA has a number of significant effects. The first effect is that it makes the polymer hydrophilic. This will render PVA to be dissolved in water to a certain extent according to the degree of hydrolysis and the temperature. The formation of hydrogen bonding from the hydrolysis of PVA leads to a number of other effects, including decomposing unplasticized PVA below its flow temperature and making the tensile strength and toughness very high. The tensile properties will be greatly dependent on humidity; the higher the humidity, the more the water absorbed. Since water acts as a plasticizer, there will be a reduction in tensile strength, but an increase in elongation and tear strength. Figure 5.2 shows the relationship between tensile strength, percentage hydrolysis, and humidity [22].

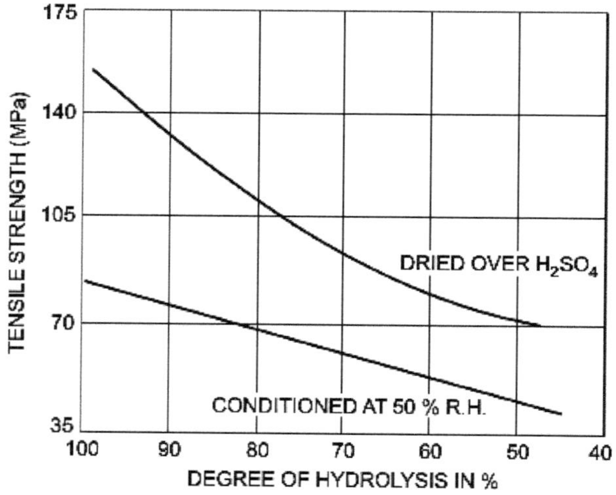

Figure 5.2 Relationship between tensile strength and degree of "hydrolysis" for unplasticized poly(vinyl alcohol film) [22].

5.1.6 Modifications according to Drawbacks

PVA is used in sizing agents that give greater strength to textile yarns and make paper more resistant to oils and greases. It is also employed as a component of adhesives and emulsifiers, as a water-soluble protective film, and as a starting material for the preparation of other resins. By reaction with butyraldehyde and formaldehyde, PVA can be made into the resins polyvinyl butyral (PVB) and polyvinyl formal (PVF). PVB, a tough, clear, adhesive, and water-resistant plastic film, is widely used in laminated safety glass, primarily for automobiles, while PVF is used in wire insulation.

PVA has high enough tensile strength and satisfactory flexibility. To improve deformability, PVA is usually plasticized by a variety of low molecular compounds, mostly containing polar groups [23], which associate with hydroxyl groups of PVA chain developing hydrogen bonds, thus reducing direct hydrogen bonding between macromolecules of PVA [23]. The comparatively high biodegradability in the environment may be one of the most important and desirable characteristics of PVA, because the polyvinyl-type polymer consisting of a carbon-carbon main chain is scarcely biodegradable [24]. This property of PVA has recently been reevaluated, and much effort has been made to produce PVA-based biodegradable polymeric materials having preferred physical and chemical properties for use in industrial and medical fields.

To improve mechanical or thermal properties, PVA is blended with different synthetic polymers, such as ethylenevinyl alcohol copolymer, poly(ethylene terephthalate), poly(vinyl chloride), polyurethanes, polyamides, polycarbonates, and others [25]. Natural polymers, such as starch, chitin, chitosan, lignin, or cellulose are also used [26].

Blends may be processed by conventional plastics technology to form various articles, which exhibit good balance of barrier and strength properties, low moisture absorptivity, and toughness/modulus combinations adequate for packaging uses [27].

The properties of PVA can also be modified by mixing with a nanofibrillated cellulose (NFC) fiber. Adding NFC to PVA resulted in an increase of tensile modulus of up to about three times of neat PVA (Fig. 5.3). Thermal analysis showed that the presence of NFC in PVA can serve as a nucleating agent, promoting the early onset of crystallization. However, at a higher NFC content, it also led to greater thermal degradation of PVA (Fig. 5.4). The PVA/NFC nanocomposites were sensitive to moisture content and dynamic mechanical analysis showed that, at room temperature, the storage modulus increased with decreasing moisture content. The solubility of carbon dioxide in the nanocomposites depended on their moisture content and decreased with the addition of NFC. Moreover, the desorption diffusivity became higher as the amount of NFC increased [28].

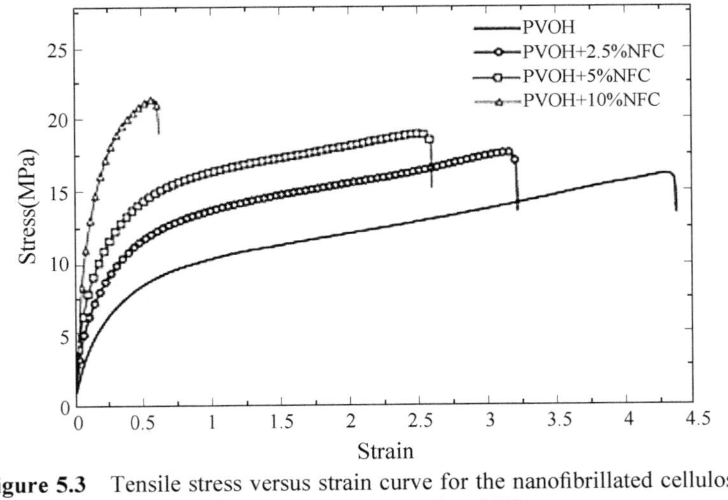

Figure 5.3 Tensile stress versus strain curve for the nanofibrillated cellulose, NFC, and PVA composites [28].

5.1.7 Industrial Applications

Being a water-soluble polymer with a strong adhesion property, PVA is used in paper manufacture, especially specialty papers, as a binder or primer, re-moistening agent, paper surface sizing, paper pigment coating, and paper internal sizing [29].

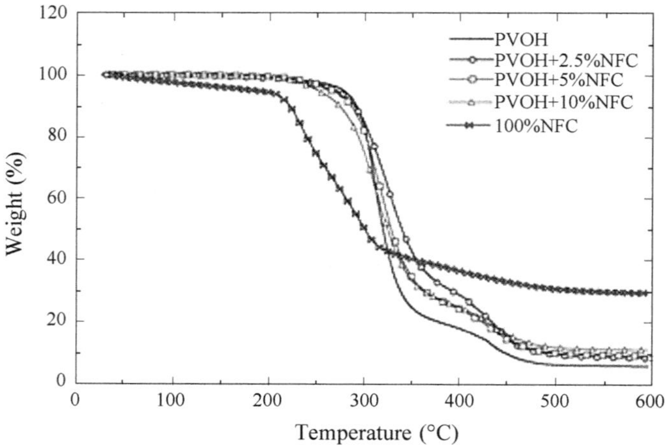

Figure 5.4 Thermogravimetric analysis curves for PVA/NFC composite samples [28].

PVA can be used as a de-characterizing agent of thermosetting resin, such as carbamide, formaldehyde resin, and melamine. It is mainly used as a binding agent for plywood, artificial board, and timber processing. When mixed with melamine-formaldehyde resin, the PVA size paste helps increase the viscosity and shortens both curing and cooling time with stronger initial cohesion.

PVA is biocompatible, nontoxic, noncarcinogenic, and has excellent chemical properties. Therefore, it is used in modern technologies for use in medicine and pharmacy [29]. The controlled drug release systems are produced from this material. They can have the form of soft contact lenses, active dressing, or tablets [30].

PVA is used with cement and mortar for efficient increase of their cohesion and fluidity, and reduction in the drying time for the concrete surface, thus increasing coating adaptability and preventing concrete cloth from chapping. Besides, due to its simple application method with desired effects, PVA is suitable for wall and ceiling decoration and tile facing.

It can also be used as an adhesive agent for profiles, such as prefabricated plasterboard and sound absorption board that are made of organic fibers or non-organic materials, on which highly cohesive and waterproof PVA series can be applied to make profiles through compression and cohesion.

One of the most traditional applications of PVA is found in fiber production, of which the first example is polyvinyl formal from PVA and formaldehyde developed in Japan by Sakurada et al. in 1939 and then Kurashiki Rayon Co., Ltd., started commercial production in 1950, which is often called as vinylon [31]. PVA fiber is not suited for clothing application. However, superiorly strong mechanical properties can be gained by orientation of polymer chains during drawing or stretching, owing to high crystallinity by pendent hydroxyl groups, which leads to high resistance to water and chemicals, including alkali and natural conditions. Therefore, PVA fiber is now composed of PVA homopolymer and the name of vinylon often indicates the PVA fiber.

PVA film is widely used for packaging, including a food packaging as barrier films and a water soluble and biodegradable film for packaging of detergents, water-soluble chemicals, agrochemicals, fishing, and dyes [5]. For food packaging film, the copolymer of vinyl alcohol and ethylene is used rather than PVA homopolymer in terms of water resistance, although its original gas permeability at dry condition is higher than that of PVA. As for the water-insoluble application, the water resistance of a dried PVA film increases with increasing molecular weights and degree of hydrolysis of the raw material polymers, which can be further improved by heat treating or annealing of the dried film over 100°C. Polyols, such as glycerol, ethylene glycol, di and tri ethylene glycol, and oligomer of ethylene glycol, can be used as plasticizers for PVA up to 30 wt%. PVA film as a water-soluble film is also used for laundry bags, water transfer printing, and embroidery applications. Nowadays, polarizer in liquid crystal display (LCD) is another major application of PVA films, which is the most essential component in the LCD panel, and LCD holds 95% consumption share of global polarizer production [32]. The polarizer film is prepared by stretching the PVA film, which is followed by doping of iodine molecule and tucking with triacetyl cellulose films. The market of PVA film for polarizer is oligopolized by Japanese two manufacturers, Kuraray Co. Ltd., and Nippon Synthetic Chemical Industry Co. Ltd., most of which are consumed by the polarizer manufacturers in Asia. The worldwide demand of PVA film for polarizer is now estimated to be over 200 million square meters. PVA is also used as the raw material in the manufacture of polyvinyl butyral resin, which is mainly used in automotive and architectural fields as a protective interlayer bonded between two panels of laminated glass [30].

Zhnng et al. [33] worked on improving the strength of PVA fbres by gel-spinning them with single-wall carbon nanotubes (SWNT). A homogenenus mixture of nanotubes, PVA dimethyl sulfoxide (DMSO), and water was prepared by stirring and sonication. The dispersion was extruded into the fiber via gel spinning. The modulus of the PVA/SWNT (3 wt%) composite fiber was reported by the authors to be 40% higher than that of the control PVA gel-spun fiber (Fig. 5.5). It was found that the SWNTs were well-dispersed with DMSO as the solvent [34].

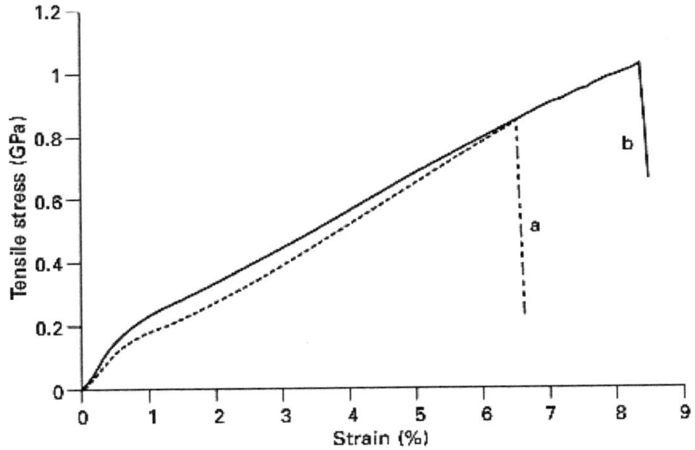

Figure 5.5 Stress–strain curves of (a) PVA and (b) PVA/SWNT gel-spun fibers [33].

The development of biodegradable packaging based on PVA film containing Isozyme incorporated with antimicrobial agents has been investigated by Conte et al. [35]. The active compound was sprayed along with a suitable bonding agent onto the surface of the crosslinked polymeric matrix. Active films show a high effectiveness against Micrococcus lysodeikticus cells. In particular, the rate of microorganism death increases as the amount of bound lysozyme increases. Several researchers have also evaluated the antimicrobial effectiveness of active films based on blends of PVA with natural and synthetic polymers and antimicrobial agents, such as carboxymethyl cellulose PVA films with clove oil [36], cassava bagasse/PVA and essential oils [37], and PVA/polypropylene with ethanolic extracts and essential oils [38].

5.1.8 Development and Future Trends

PVA is a synthetic polymer that has been used for the past 30 years in several medical and nonmedical devices. Multiple nonclinical and clinical studies have demonstrated that PVA is a synthetic alternative to native cartilage replacement, and it is readily available compared to cartilage transplantation, which has limited availability and disease transmission concerns. Several animal and clinical studies using PVA for cartilage, meniscus, embolization, and vitreous solutions demonstrate the biocompatibility and the safety related to this material. Follow-up periods of up to two years have been reported for animal and clinical studies, suggesting that PVA is stable and safe to use for medical devices. The biomechanical properties of PVA have also been investigated to better simulate the native tissue [39].

The PVA manufacturing process can be manipulated to generate the biomechanical properties desired. The thawing and freezing protocol, the addition of saline, crosslinking agents, and other materials, all play a role in the biomechanical properties of the end product. Many investigators have also reported the wear characteristics of PVA. The *in vivo* studies have determined that wear particles from PVA are less harmful than wear particles from metals and other polymers [39].

In the treatment of focal defects, implant devices of PVA cryogel for the replacement of cartilage do not require significant removal of healthy tissue. The device can also articulate directly against opposing cartilage with no apparent damage. Therefore, PVA cryogels have faster recovery times and require less surgical trauma. Patients that undergo PVA cryogel plug surgery for chondral defects exhibit full knee movement right after surgery, and the knee can withstand full loads after three weeks. It was also determined that the surgical insertion method and the implant site have an effect on the success rate of PVA implants for cartilage replacement *in vivo* [39].

There are some reports on dislocation and loosening of PVA implants following cartilage replacement surgery. Misplacement of these implants was the major reason for dislocation and loosening. Multiple implants were placed at the same site, touching each other and causing expulsion. PVA is a biologically compatible material that is stable *in vivo* (in both humans and animals) and has suitable biomechanical properties to be a promising material for future tissue replacement implants [40].

A biodegradable polymer does not generally sell simply because it is biodegradable. It must compete as a material on the basis of its own price/property characteristics, with biodegradability as an added bonus. Given the above advantageous properties of the PVA-based materials, potential markets fall into three main areas, depending upon the property to be exploited [41].

Applications which could make use of the controlled water solubility of PVA that include hospital applications, such as disposable infected laundry bags, biohazard bags, pathology sample bags, bedpans, urine bottle, and vomit bowl inserts. The packaging of powders which are dust producing and are then further processed in the aqueous phase is another possibility. Examples are toxic herbicides, pesticides, dyes, and pigments, where these materials are contained in a bag of cold water soluble PVA, avoiding operator contact.

PVA materials can now act as a more effective replacement for the ethylene vinyl alcohol polymers currently used as oxygen barrier layers for food packaging. The resistance of PVA to nonpolar organic solvents will also protect foodstuffs from secondary contamination by printing inks. Examples of potential food packaging uses are ketchup bottles, and enhanced shelf life packaging [42].

A variety of applications are anticipated where the driving force is increasing pressures from environmental awareness issues, high landfill charges, legislation limiting use of landfills and non-degradable products, and landfill directives. Such applications include film products made from hot water soluble PVA, floating mulch films or transparent greenhouse/cloche films and silage wrap, pallet wrap, fertiliser bags, and general industrial and consumer packaging. Strict segregation would be necessary to prevent the biodegradable products being mixed with recyclable products such as polyethylene. Knowledge of the product origin, service life, and end of life disposal method is necessary for applications where the biodegradability of the PVA is to be exploited.

5.1.9 Life Cycle Assessment

The commercial manufacturing of PVA is carried out by hydrolysis of PVAc derived from VAM [43, 44]. Different methods are applied in industry to polymerize VAM to PVAc, such as suspension, solution or emulsion polymerization, amongst which, solution polymerization is one of the most commonly used technologies since it is easy to control and it produces high-quality PVAc. Methanol solution polymerization with azobis-iso-butyronitrile (AIBN) as the initiator can be used assuming an efficient recycling/recovery system where waste VAM and methanol were separated, purified and recycled back into reactor [44]. The obtained PVAc from solution polymerization requires alcoholysis process to convert to PVA; generally alkaline alcoholysis is applied on an industrial scale [43]. Therefore, a continuous alkaline alcoholysis process with sodium hydroxide as the catalyst in methanol medium was used. Hydrolysis was followed by a drying process to separate the PVA gel from the methanol/methyl acetate solution. The solvent output from alcoholysis mainly including methanol, methyl acetate sodium acetate was assumed to be separated and recovered internally [4, 44], Methyl acetate is hydrolyzed into methanol and acetic acid through a cation exchange resin column [44]. Therefore, PVA was the only product from this process.

During the whole process, electricity and natural gas were assumed to be the only energy sources for heat and mechanical work. AIBN catalyst and chemicals added for solvent recovery were omitted from system boundary due to no data available [45].

5.1.10 Recycling (Method Choice, Details, Method, and Difficulties, Quality)

PVA along with polyacrylates (PAA) and starch are used in the textile industry for sizing cotton fibers. Sizing is a process where applied chemicals confer strength to the fiber and protect it during the weaving process. The PVA, PAA, and starch are removed from the cloth after weaving by washing it in hot water in a desizing operation, resulting in an aqueous effluent containing these chemicals. Some of these chemicals, particularly water soluble sizes such as PVA and PAA, can be recovered and reused in the process resulting in savings on expensive chemicals.

At the textile mill where the present case study was implemented, recovery of PVA was carried out on a pilot scale. An ultra-filtration process to reduce the amount of PVA and starch in the effluent followed by a closed-loop recycling operation was tested for 16 months in a pilot plant. The ultra-filtration membrane used in this process recovered PVA successfully. Starch is enzymatically solubilized prior to desizing and, therefore, cannot be recovered. The film forming characteristics of the PAA during testing were impaired by the formation of a calcium-polyacrylate complex.

The recycling of industrial materials could reduce their environmental impact and waste haulage fees and result in sustainable manufacturing. In this work, commercial PVA sponges are recycled into a macroporous carbon matrix to encapsulate size-tunable SnO_2 nanocrystals as anode materials for lithium-ion batteries through a scalable, flash-combustion method. The hydroxyl groups present copiously in the recycled PVA sponges guarantee a uniform chemical coupling with a tin(IV) citrate complex through intermolecular hydrogen bonds [46].

5.1.11 Conclusions

PVA has a simple chemical structure. It can be processed easily and has many potential uses, including adhesives, paper making, barrier film manufacturing, pharmaceutical, and biomedical applications. It is finding its rapid growth and expansion in the areas including blends, copolymers, and impact-modified products, which can create opportunities in the invention of unlimited novel produces.

5.2 POLYVINYL ACETATE

5.2.1 Introduction

Polyvinyl acetate (PVAc) and its related polymers are nothing new and the materials have been around for almost 100 years. It was discovered in Germany in 1912 by Dr. Fritz Klatte [47].

PVAc has been available commercially in the USA since the 1930s, although the growth was slow until the introduction of emulsions of PVA.

The price of PVAc is USD 2–3 per kg. Currently about 1.7 billion pounds of PVA resin is consumed per year. The adhesive industry is one of the most important outlets for PVA. The basic monomer for PVAc is vinyl acetate. The worldwide production capacity for vinyl acetate monomer (VAM) was estimated to be close to 5,900 kilotons in 2009 with an actual production of around 5,500 kilotons [7, 8]. About 50% of the VAM is converted to PVAc and vinyl acetate-containing polymers. Basically only ethylene and air are needed for the production of PVAc and its polymer analogous products, such as polyvinyl alcohol and polyvinyl butyrate.

Practical applications of polymers based on VAM have been known for over 80 years and products containing vinyl acetate-based polymers have been spreading into the environment for just as long. Wacker Chemie developed methods for the large scale production of vinyl acetate monomer (VAM) and also overcame synthetic limitations in the production of PVAc from VAM [48]. The resulting polymer PVAc was soon found to be suitable for use both as a binder and as a major component in adhesives [8].

It is the most widely used vinyl ester, is noted for its adhesion to substrates, and high cold flow. Polyvinyl acetate serves as the precursor for polyvinyl alcohol and, directly or indirectly, the polyvinyl acetals. It is insoluble in many organic solvents, but water sensitive. Polyvinyl acetate absorbs from 1 to 3% water, up to 8% on prolonged immersion [49]. US manufacturers currently sell polyvinyl acetate in emulsion form and typically are used in aqueous systems. Monomer residue has not been considered a problem in end-use products. Latexes or solutions of polyvinyl acetate that are essentially intermediates may contain residual vinyl acetate, essential emulsifiers, or initiators.

Polyvinyl acetate meeting certain specifications is permitted in stated food contact applications, such as packaging, coatings, and adhesives. Polyvinyl acetate with a minimum molecular weight of 2000 is permitted as a synthetic masticatory substance in chewing gum base [49]. Polyvinyl acetate can also be used as water-based (latex) paints, wood glue, white glue, paper adhesive during paper packaging converting, envelope adhesives, and wallpaper adhesives.

5.2.2 Raw Materials

Vinyl acetate monomer (VAM) is used as an intermediate in the manufacture of various copolymers and homopolymers, such as polyvinyl acetate, polyvinyl alcohol, ethylene-vinyl acetate, and ethylene-vinyl alcohol among others. Polyvinyl acetate accounted for the highest demand market share for vinyl acetate monomer over the past few years.

VAM is manufactured through catalytic conversion reaction of acetic acid and ethylene with oxygen. Due to higher reactivity of vinyl acetate monomer, a polymerization inhibitor is required in order to control the degree of polymerization. VAM can be produced starting from ethylene, which is converted to acetic acid via acetaldehyde by two sequential oxidation steps, the first step being the famous Wacker process [50]. Another way to produce acetic acid is based on a carbonylation of methanol in the so-called Monsanto process, which is the dominant technology for the production of acetic acid today [51]. Acetic acid is then converted to VAM by addition of ethylene to acetic acid in the gas phase using heterogeneous catalysts, usually based on palladium, cadmium, gold, and its alloys [52] supported on silica structures.

It should be pointed out that the raw materials for VAM and its related polymers are produced from fossil resources, mainly crude oil. It is possible to completely substitute the feedstock for these raw materials and switch to ethanol, which can be produced from renewable resources such as sugar cane, corn, or preferably straw, and other non-food parts of plants. Having that in mind, the whole production of PVAc, that nowadays is based on traditional fossil resources, could be switched to a renewable, sustainable, and CO_2-neutral production process based on bioethanol.

If the "vinyl acetate" circle can be closed by the important steps of biodegradation or hydrolysis and biodegradation of vinyl ester-based polymers back to carbon dioxide, then a truly sustainable material circle can be established [7].

5.2.3 Polymerization Process, Chemical Structure, and Production

Polyvinyl acetate (PVAc) is a rubbery synthetic resin which is prepared by the polymerization of vinyl acetate. Vinyl acetate is prepared from ethylene by reaction with oxygen and acetic acid over a palladium catalyst. Under the action of free-radical initiators, vinyl acetate monomers can be linked into long, branched polymers, in which the structure of the vinyl acetate repeating units is given in Fig. 5.6.

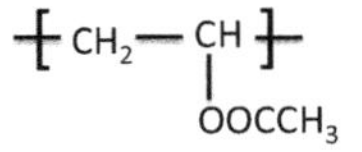

Figure 5.6 Structure of vinyl acetate repeating unit.

The monomer can be polymerized while dispersed in water to form a milky-white emulsion. This fluid can be processed directly into latex paints, in which the PVAc forms a strong, flexible, adherent film. It can also be made into a common household adhesive known as white glue or Elmer's glue. When employed in coatings or adhesives, PVAc is often partially hydrolyzed to a water-soluble polymer known as polyvinyl alcohol.

Vinyl acetate may be easily polymerised in bulk, solution, emulsion, and suspension. At conversions above 30%, chain transfer to polymer or monomer may occur. In the case of both polymer and monomer transfer, two mechanisms are possible, one at the tertiary carbon, the other at the acetate group.

The radical formed at either the tertiary carbon atom or at the acetate group will then initiate polymerisation and form branched structures. Since polyvinyl acetate is usually used in an emulsion form, the emulsion polymerisation process is commonly used. In a typical system, approximately equal quantities of vinyl acetate and water are stirred together in the presence of a suitable colloid-emulsifier system, such as polyvinyl alcohol and sodium lauryl sulphate, and a water-soluble initiator such as potassium persulphate. Polymerisation takes place over a period of about 4 hours at 70°C. The reaction is exothermic and provision must be made for cooling when the batch size exceeds a few liters. In order to achieve better control of the process and to obtain particles with a smaller particle size, part of the monomer is first polymerised and the rest, with some of the initiator, is then steadily added over a period of 3–4 hours. To minimise the hydrolysis of vinyl acetate or possible comonomers during polymerisation, it is necessary to control the pH throughout reaction. For this purpose, a buffer such as sodium acetate, is commonly employed.

5.2.4 Properties

PVAc is too soft and shows excessive 'cold flow' for use in moulded plastics. This is no doubt associated with the fact that the glass transition temperature of 28°C is a little above the usual ambient temperatures and in fact in many places at various times the glass temperature may be lower. It has a density of 1.19 g/cm^3 and a refractive index of 1.47. Commercial polymers are atactic and, since they do not crystallise, transparent (if free from emulsifier). They are successfully used in emulsion paints, as adhesives for textiles, paper and wood, as a sizing material and as a 'permanent starch'. A number of grades are supplied by manufacturers which differ in molecular weight and in the nature of comonomers (e.g., vinyl maleate) which are commonly used. The polymers are usually supplied as emulsions which also differ in the particle size, the sign of the charge on the particle, the pH of the aqueous phase, and in other details.

PVAc is a completely atactic, highly branched, noncrystalline thermoplastic, prepared by conventional free-radical polymerization. The homopolymer has a glass transition temperature around room temperature. It has good resistance to UV and oxidation, but is rather brittle below the T_g and very sticky above it. For this reason, it is usually blended with plasticizers to improve the flexibility. Due to the water sensitivity of the homopolymer, (unmodified) PVAc is generally viewed as inappropriate for exterior uses.

PVAcs are often used as adhesives. Water-based emulsions are, by far, the most common form in which polyvinyl acetates are used in the adhesive market. The adhesive properties are greatly affected by the molecular weight (distribution) and the compatibility of the vinyl acetate with other ingredients. The adhesive strength and moisture sensitivity also depend on the type and the amount of protective colloid (gelatin, natural gum, or cellulose derivatives) and on the functional additives, such as wetting agents and plasticizers.

PVAc emulsions are rather inexpensive and possess good adhesion to many porous substrates, such as wood and paper.

PVAc is not the only vinyl ester. Other vinyl esters have been synthesized, either by direct polymerization or by ester interchange between polyvinyl acetate and other esters/acids. Some known homologs are poly(vinyl butyrate), poly(vinyl propionate), and poly(vinyl formate). However, these vinyl ester polymers have found very little commercial use, probably due to their higher cost.

Of greater importance are the copolymers of PVAc. PVAc is a rather stiff homopolymer. For this reason, it is sometimes copolymerized with other monomers. For example, the combination of vinyl acetate and ethylene, called vinyl acetate ethylene (VAE), is a much softer polymer and finds applications in coatings and paints.

As an example of the concentration dependence of viscoelastic properties in Fig. 5.7 the shear creep compliance of PVAc is plotted vs. time for solutions of PVAc in diethyl phthalate, with indicated volume fractions of polymer, reduced to 40°C with the aid of the time temperature superposition principle [53]. From this figure, it becomes clear that the curves are parallel. We may conclude that the viscoelasticity may be shifted over the time axis to one curve, e.g., to the curve for $\varphi_2 = 1$, the pure polymer. In general it appears that viscoelastic properties measured at various concentrations may be reduced to one single curve at one concentration with the aid of a time-concentration superposition principle, which resembles the time-temperature superposition principle.

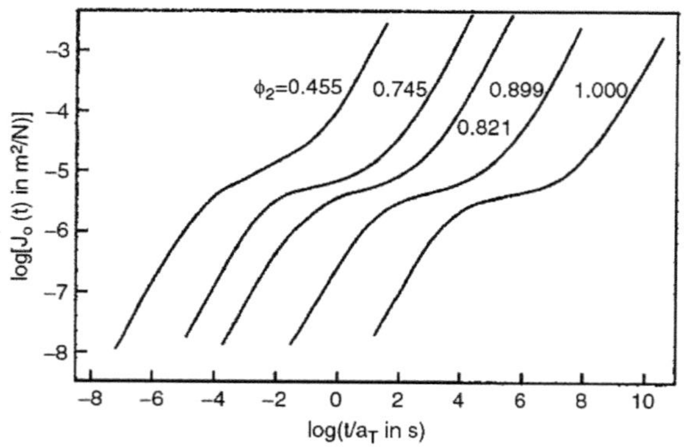

Figure 5.7 Shear creep compliance of PVAc, M=240 kg/mol, and four solutions in diethylphthalate with indicated values of the polymer volume fraction, φ_2, reduced to 40°C [53].

5.2.5 Relationship between Structure and Properties

The degree of polymerization of polyvinyl acetate typically is 100 to 5000 [54]. The ester groups of the polyvinyl acetate are sensitive for alkali and will slowly convert PVAc into polyvinyl alcohol and acetic acid. Under alkaline conditions, boron compounds, such as boric acid or borax causes the polymer to crosslink, forming toys, such as slime and flubber. A number of microorganisms, such as filamentous fungi, algae, yeasts, lichens, and bacteria have been shown to have the ability to degrade polyvinyl acetate [55].

In some instances, to achieve an acceptable balance between cost and film properties, blending of a low level of acrylic emulsion in a predominately PVAc paint is required. Generally, 100% PVAc binders have poor exterior paint performance properties including: adhesion, water resistance, dirt pickup resistance, efflorescence, and alkali resistance. Blending of a small amount of acrylic binder will upgrade the performance of PVAc paint in these properties. Also, block resistance is improved by modifying the PVAc binder with a small amount of acrylic binder. Conversely, some properties, such as scrub resistance may be degraded in the blended paints [56].

5.2.6 Modifications according to Drawbacks

Chemical modifications of PVAc are also of great importance. Among these, its conversion to polyvinyl alcohol (PVA) is the most important one. PVA cannot be derived from its monomer because the monomer is not stable. Instead, it is produced by hydrolyzing PVAc to PVA. The dry polymer is a hard, crystalline thermoplastic, with a T_g around 375 K. It is water soluble, resistant to solvents and oils, and has exceptionally good adhesion to cellulose and other hydrophilic surfaces.

Another important modification of PVAc is its derivatization with aldehydes to polyvinyl acetals. The two most important derivatives are polyvinyl formal and polyvinyl butyral. They are manufactured from PVA by reaction with butyraldehyde and formaldehyde [57].

All other polyvinyl esters are of limited use in the plastic industry. In most cases, they are copolymerized with other monomers. For example, vinyl stearate is sometimes copolymerized with vinyl chloride to reduce the melt viscosity, i.e., it acts as an internal lubricant and vinyl propionate is sometimes included in emulsion-paints, for example, it is copolymerized with vinyl acetate or vinyl acrylate to form copolymer latices for paints [57].

Sealants and caulks based on vinyl acetate homopolymers and copolymers are cheap low performance materials which are mostly used in residential applications for bathtub caulking, wall tile joints, and wall board joints. A major advantage of PVAc latex emulsions is their easy clean-up. The PVAc emulsions were designed for adhesive and coating use rather than for sealants. The homopolymers have little flexibility and are water sensitive. PVAc emulsions are made by polymerizing vinyl acetate monomer to make a homopolymer or with other monomers to make copolymers which can build flexibility into the polymer chain [57].

5.2.7 Industrial Applications

PVAc is used in a vast number of different applications [58]. The most common use of PVAc-based dispersions and dispersible polymer powders is in the construction and adhesives industry. The polymeric binders are used as additives to enhance the properties of tile adhesives, mortars, and self-levelling compounds. PVAc dispersions are also used in many other adhesive formulations.

Wood glues are often based on PVAc. It is also commonly used as a binder in the paper industry and as a binder in latex paints, although binders based on acrylics are far more common in paint technology. Almost one-third of the PVAc produced goes into binder and adhesive applications. As PVAc is approved by the Food and Drug Administration (FDA), it also has several uses in the food industry. PVAc can also be found in just about every chewing gum. It is a major component in the so-called gumbase, a mixture of different polymers that in combination with sugar, sweeteners, flavors, and other additives, make up a chewing gum [8].

PVAc is largely used in the gumbase industry for chewing gum, bubble gum, and xylitol. In medicals, it is applicable because it is a green and eco-friendly product. The applications of PVAc in industrial fields are realised as additives for anti shrink, anti fatigue, anti warping, adhesives, color enhancing, and strength enhancing. Meanwhile, PVAc is used in glass fiber reinforced polymer composites to improve the glossiness, anti-shrink, as well as the stress and strain. PVAc is also used in making auto head lamps, ferries, electronical plastics, pipes, and other composite materials. For the application in sheet and bulk moulded compounds, the PVAc is normally required in the form of liquid with styrene with a mixture ratio of 40 to 60. After dissolving well, the PVAc and styrene liquid is added into the unsaturated polyester at a ratio of 50 to 50.

PVAc is used as a key material in the adhesive industries for timber products, stationeries, plastic articles, and paper making. The PVAc has a range of viscosity, which is why it can be used in adhesives. But normally the applications in adhesives require the PVAc to be at a higher viscosity, related to a higher molecular weight.

PVAc is used in concrete and is generally added into the cement to improve waterproof, anti-shrink, and anti-fatigue properties of cement. This is an usual application in cement concrete, pillar, outside flowerpot, and sculptor.

PVAc is used in the coatings for wires. It has an amazing function in seizing the colors on the surfaces such as wires, bill boards, and advertisement boards. The theory of color seizing is that the PVAc has a viscosity which performs well in color stablizing. In its most important application, polyvinyl acetate serves as the film-forming ingredient in water-based paints.

PVAc as a low-profile additive is added to reduce shrinking of cured products which causes internal voids and surface quality. This additive absorbs some styrene in the early stages of curing. When the temperature is increased in the course of curing, the styrene eventually evaporates and consequently a counter pressure is formed which counter-balances the shrinking. The successful performance of low-profile additives depends essentially on the phase separation phenomena in the course of curing. The curing rate decreases with an increase of the molecular weight of the low-proflie additive, which causes the chain entanglement effect. The plasticizing effect is reduced with the increase of the molecular weight of the low-profile additive. Low-profile additives with higher molecular weight and lower content of additive seem to work better under low-temperature curing conditions [59].

Like natural fibers, PVAc as a synthetic water soluble biodegradable fiber has also been used for nonwoven applications. Thus, the target for biodegradable nonwovens is to replace synthetic fibers with biodegradable fibers in the disposable nonwovens, such as the wet laid pulp/polyester spun laced fabrics used mainly for industrial and professional wipe products and household and hygienic wipes, which are spun bonded or dry laid and then chemically or thermally bonded.

PVAc is used in the manufacture of dispersions for paints and binders and as a raw material for paints. It is also copolymerized with vinyl chloride and ethylene and to a lesser extent with acrylic esters. A substantial proportion of vinyl acetate is converted into polyvinyl alcohol by saponification or trans esterification of PVAc [60].

5.2.8 Development and Future Trends

A noble group of edible polymers was under enlargement, with the goal of allowing for the incorporation and/or controlled release of active compounds using nanotechnological solutions, such as nanoencapsulation and multifaceted systems. Nowadays, nanotechnologies are being used to enhance the nutritional features of food by means of nanoscale additives and nutrients and nanosized delivery systems for bioactive polymeric compounds. Nanocomposites perception represents a motivating route for creating new and innovative materials, also in the area of edible polymers. Materials with a large variety of properties have been realized, and even more are due to be realized. Micro- and nanoencapsulation of active compounds with edible polymer coatings may help to control their release under specific conditions, thus protecting them from moisture, heat, or other extreme conditions, and enhancing their stability and viability. Coating foods with nanolaminates involves either dipping them into a series of solutions containing substances that would be adsorbed to a food's surface or spraying substances onto the food surface. These nanolaminate coatings could be elaborated entirely from food-grade ingredients and could include various functional agents, such as antimicrobials, antibrowning agents, antioxidants, enzymes, flavorings, and colorant. In fact, the layer-by-layer electrodeposition technique could be used to coat highly hydrophilic food systems, such as fresh-cut fruits and vegetables, including further vitamins and antimicrobial agents [61]. The nanocomposite materials obtained by mixing natural, edible polymers, and sheets of crystalline solid layered offer a great variety of property profile. They were even able to compete, both in price and in performance, with synthetic polymeric materials. Consumer demands were driving research and development for alternatives to petroleum-based packaging materials, including those with recyclable or edible properties, as well as those materials made from renewable/sustainable agricultural products. Edible films, gels, or coatings were considered biopolymers with numerous desirable properties and may be made from a variety of materials, including polysaccharides, lipids, and proteins, alone or in combination with other components. Edible biopolymers also have been developed from other sources and applied to foods, including fungal exopolysaccharides (pullan) or fermentation by-products (polylactic acid) [61].

5.2.9 Recycling

Scientists have reviewed the chemical or feedstock recycling of PVAc. In this process, new components are formed, and these may be used as raw materials in other applications. Hence, waste is transformed into new products [62].

5.2.10 Conclusions

Polyvinyl acetate serves as the precursor for polyvinyl alcohol and, directly or indirectly, the polyvinyl acetals. In meeting certain specifications, it is permitted in stated food contact applications, such as packaging, coatings, and adhesives. Polyvinyl acetate can also be used as water-based (latex) paints, wood glue, white glue, paper adhesive during paper packaging converting, envelope adhesives, and wallpaper adhesives. It is finding rapid growth and expansion in the areas including blends and fiber reinforced composites.

References

1. Hermann, W. and W. Haehnel. Verfahren Zur Herstellung Von Polymerem Vinylalkohol. German Patent 1924, DE450286.
2. Hermann, W. and W. Haehnel. Alcohol Production. Canadian Patent 1926, 265172.
3. Staudinger, H. Über Hochpolymere Verbindungen. Chem Ber 1926; 59: 3019-3043.
4. Kim, N., E.D. Sudol, V.L. Dimonie, Md.S. El-Aasser. Poly(Vinyl Alcohol) stabilization of acrylic emulsion polymers using the miniemulsion approach. Macromolecules 2003; 36: 5573-5579.
5. Finch, C.A. Polyvinyl Alcohol: Developments. Wiley, 1992.
6. Kogure, K. and S. Hama. Nanomaterials for cosmetics. *In*: S. Kobayashi and K. Müllen (eds). Encyclopedia of Polymeric Nanomaterials. Springer, Berlin, Heidelberg, 2015, pp. 1349-1352.
7. Rieger, B., A. Künkel, G.W. Coates, R. Reichardt, E. Dinjus and T.A. Zevaco (eds). Synthetic Biodegradable Polymers. Springer Science & Business Media, Berlin, Germany, 2012.
8. Amann, M. and O. Minge. Biodegradability of poly (vinyl acetate) and related polymers. *In*: B. Rieger, A. Künkel, G.W. Coates, R. Reichardt, E. Dinjus and T.A. Zevaco (eds). Synthetic Biodegradable Polymers. Springer Science & Business Media, Berlin, Germany, 2012, pp. 137-172.
9. Lefteri, C. Materials for Design. ProQuest Ebook Central. 2014.

10. Umoren, S.A., I.B. Obot, A. Madhankumar and Z.M. Gasem. Effect of degree of hydrolysis of polyvinyl alcohol on the corrosion inhibition of steel: Theoretical and experimental studies. Journal of Adhesion Science and Technology 2015; 29(4): 271-295.
11. Moad, G. and D.H. Solomon. The Chemistry of Radical Polymerization. Elsevier, Sydney, 2005.
12. Murahashi, S., H. Yûcki, T. Sano, U. Yonemura, H. Tadokoro and Y. Chatani. Isotactic polyvinyl alcohol. Journal of Polymer Science 1962; 62(174).
13. Okamura, S., T. Kodama, and T. Higashimura. The cationic polymerization of t-butyl vinyl ether at low temperature and the conversion into polyvinyl alcohol of poly-t-butyl vinyl ether. Die Makromolekulare Chemie 1962; 53(1): 180-191.
14. Baum, E., W. Haehnel, and W.O. Herrmann. Esters of vinyl alcohol. DE Patents 1924; 483: 780.
15. Hachnel, W. and W. Herrmann. Vinyl alcohol. DE Patents 1924; 480: 866.
16. Drury, J.L. and D.J. Mooney. Hydrogels for tissue engineering: scaffold design variables and applications. Biomaterials 2003; 24(24): 4337-4351.
17. Jayasekara, R., I. Harding, I. Bowater, and G. Lonergan. Biodegradability of a selected range of polymers and polymer blends and standard methods for assessment of biodegradation. Journal of Polymers and the Environment 2005; 13(3): 231-251.
18. Halima, N.B. Poly (vinyl alcohol): Review of its promising applications and insights into biodegradation. RSC Advances 2016; 6(46): 39823-39832.
19. Nord, F., M. Bier and S.N. Timasheff. Investigations on proteins and polymers. IV. 1 Critical phenomena in polyvinyl alcohol-acetate copolymer solutions. Journal of the American Chemical Society 1951; 73(1): 289-293.
20. Finch, C.A. Polyvinyl Alcohol; Properties and Applications. John Wiley & Sons, 1973.
21. Hassan, C.M. and N.A. Peppas. Structure and applications of poly (vinyl alcohol) hydrogels produced by conventional crosslinking or by freezing/thawing methods. Biopolymers • PVA Hydrogels, Anionic Polymerisation Nanocomposites 2000; 153: 37-65.
22. Gilbert, M. Brydson's Plastics Materials, 8th Ed., Butterworth-Heinemann, 2016.
23. Jeļinska, N., M. Kalniņš, V. Tupureina, and A. Dzene. Poly (vinyl alcohol)/poly (vinyl acetate) blend films. Scientific Journal of Riga Technical University 2010; 21: 55-61.
24. Chiellini, E., A. Corti, S. D'Antone, R. Solaro. Biodegradation of poly (vinyl alcohol) based materials. Progress in Polymer Science 2003; 28(6): 963-1014.
25. Kulshreshtha, A.K. Handbook of Polymer Blends and Composites. iSmithers Rapra Publishing, 2002.
26. Jayasekara, R., I. Harding, I.C. Bowater, G.B.Y. Christie, G.T. Lonergan. Preparation, surface modification and characterisation of solution cast starch pva blended films. Polymer Testing 2004; 23(1): 17-27.
27. Gross, R.A. and B. Kalra. Biodegradable polymers for the environment. Science 2002; 297(5582): 803-807.
28. Satoh, K. Poly (vinyl alcohol)(PVA). Encyclopedia of Polymeric Nanomaterials 2015; 1734-1739.
29. Chen, Y., X. Cao, P.R. Changa and M.A. Huneault. Comparative study on the films of poly (vinyl alcohol)/pea starch nanocrystals and poly (vinyl alcohol)/native pea starch. Carbohydrate Polymers 2008; 73(1): 8-17.
30. Pajak, J., M. Ziemski and B. Nowak. Poly (vinyl alcohol)–biodegradable vinyl material. Chemik 2010; 64: 523-530.
31. Sakurada, I. Polyvinyl Alcohol Fibres. CRC Press, New York and Basel, 1985.
32. Land, E.H. Some aspects of the development of sheet polarizers. JOSA 1951; 41(12): 957-963.
33. Zhang, X., T. Liu, T.V. Sreekumar, S. Kumar, X. Hu and K. Smith. Gel spinning of PVA/SWNT composite fiber. Polymer 2004; 45(26): 8801-8807.
34. Alagirusamy, R. and A. Das, 2010. Technical Textile Yarns–Industrial and Medical Applications. Woodhead Publishing, Oxford, 2010.
35. Conte, A., G.G. Buonocore, A. Bevilacqua, M. Sinigaglia and M.A. Del Nobile. Immobilization of lysozyme on polyvinylalcohol films for active packaging applications. Journal of Food Protection 2006; 69(4): 866-887.

36. Muppalla, S.R., S.R. Kanatt, S.P Chawla and A. Sharma. Carboxymethyl cellulose-polyvinyl alcohol films with clove oil for active packaging of ground chicken meat. Food Packaging and Shelf Life 2014; 2(2): 51-58.
37. Debiagi, F., R.K.T. Kobayashi, G. Nakazato, L.A. Panagio and S. Mali. Biodegradable active packaging based on cassava bagasse, polyvinyl alcohol and essential oils. Industrial Crops and Products 2014; 52: 664-670.
38. Han, C.A., J. Wang, Y. Li, F. Lu and Y. Cui. Antimicrobial-coated polypropylene films with polyvinyl alcohol in packaging of fresh beef. Meat Science 2014; 96: 901-907.
39. Baker, M.I., S.P. Walsh, Z. Schwartz and B.D. Boyan. A review of polyvinyl alcohol and its uses in cartilage and orthopedic applications. Journal of Biomedical Materials Research Part B: Applied Biomaterials 2012; 100(5): 1451-1457.
40. Kobayashi, M. and H.S. Hyu. Development and evaluation of polyvinyl alcohol-hydrogels as an artificial atrticular cartilage for orthopedic implants. Materials 2010; 3(4): 2753-2771.
41. Middleton, J.C. and A.J. Tipton. Synthetic biodegradable polymers as orthopedic devices. Biomaterials 2000; 21(23): 2335-2346.
42. Massey, L.K. Permeability Properties of Plastics and Elastomers: A Guide to Packaging and Barrier Materials. William Andrew, New York, 2003.
43. Finch, C. Polyvinyl Alcohol-Developments. Wiley, New York, 1972.
44. Marten, F.L. Vinyl Alcohol Polymers. Kirk-Othmer Encyclopedia of Chemical Technology, 2000.
45. Guo, M. Life Cycle Assessment (LCA) of Light-Weight Eco-Composites. Springer Science & Business Media, 2013.
46. Ma, Y., C.-W. Tai, T. Gustafsson and K. Edström. Recycled poly (Vinyl Alcohol) sponge for carbon encapsulation of size-tunable tin dioxide nanocrystalline composites. ChemSusChem 2015; 8(12): 2084-2092.
47. Memon, R., M. Talib and A.S. Khan. A new approaches: Coating of a paper with biodegradable polymer. International Journal of Chemical and Analytical Science 2011; 2(5): 27-28.
48. Herrmann, W. and W. Haehnel. Polymerisation of Vinylester. Gerrman Patent 1925, DE490041.
49. Walker, B. and L. Burton. Polyvinyl acetate, alcohol, and derivatives, polystyrene, and acrylics. Patty's Toxicology, 2012.
50. Baerns, M., A. Behr, A. Brehm, J. Gmehling, K.-O. Hinrichsen, H. Hofmann, U. Onken, R. Palkovits and A. Renken. Technische Chemie. Wiely-VCH, Weinheim, 2013.
51. Malveda, M. and C. Funada. Acetic acid. Chemical economics handbook report. SRI Consulting, Englewood, 2010.
52. Weissermel, K. and H.J. Arpe. Industrial Organic Chemistry. John Wiley & Sons, Weinheim, 2008.
53. O¯yanagi, Y. and J.D. Ferry. Viscoelastic properties of polyvinyl acetates IV. Creep studies of plasticized fractions. Journal of Colloid and Interface Science 1966; 21(5): 547-559.
54. Rabasco, J.J., K.R. Lassila, R. Van Court Carr, and K.E. Minnich. Ink jet paper coatings containing polyvinyl alcohol-alkylated polyamine blends, Google Patents, 2002.
55. Cappitelli, F. and C. Sorlini. Microorganisms attack synthetic polymers in items representing our cultural heritage. Applied and Environmental Microbiology 2008; 74(3): 564-569.
56. Koleske, J.V. (ed.). Paint and Coating Testing Manual-Fifteenth Edition of the Gardner-Sward Handbook. ASTM International, 2012.
57. Polymer Database. Crow 2015. [Online]. http://polymerdatabase.com/polymer%20classes/Polyvinylester%20type.html. [Accessed 16 November 2016].
58. Chin H, T. Kälin and K. Yokose. Polyvinyl Acetate. Chemical Economics Handbook Report. SRI Consulting, Englewood, 2008.
59. Beheshti, M.H., H. Mehran and M. Vafaian, M. Evaluation of low-profile additives in the curing of unsaturated polyester resins at low temperatures. Iranian Polymer Journal 2006; 15(2): 143-154.
60. Yang, S.B., J.W. Kim and J.H. Yeum. Effect of saponification condition on the morphology and diameter of the electrospun poly(vinyl acetate) nanofibers for the fabrication of poly(vinyl alcohol) nanofiber mats. Polymers (Basel) 2016; 8(10): 376. doi: 10.3390/polym8100376. PMID: 30974653; PMCID: PMC6431977.

61. Shit, S.C., and P.M. Shah. Edible polymers: Challenges and opportunities. Journal of Polymers 2014; Article ID 427259, p. 13.
62. Blazevska-Gilev, J. and D. Spaseska. Remediation of poly (vinyl acetate) from waste products. Journal of the University of Chemical Technology and Metallurgy 2009; 44(2): 123-126.

Chapter 6

Starch and Starch-based Polymers

J. Fuchs* and H.-P. Heim

Institute of Material Engineering, Plastics Engineering
University of Kassel, Germany

6.1 INTRODUCTION

In connection with polymer engineering, renewable raw materials play an increasingly important role, having lost a great deal of importance in the course of the increased use of crude oil at the beginning of the last century. The use of biopolymers and bioplastics has several advantages: on the one hand, organic raw materials provide an almost inexhaustible supply of starting materials, such as wood or starch. Furthermore, these materials have an almost neutral CO_2 balance, with the exception of preparation and processing into products. During thermal recycling, at the end of the life of the respective product, the renewable raw materials theoretically release only as much CO_2 as the plant raw material previously absorbed during growth. The "bio-aspect" has been incorporated into the marketing strategies of companies for several years, which has enabled and even increased sales of these products, despite partly higher prices compared to comparable products made of fossil-based plastics [1–5].

Bioplastics are not only a supplement, but in some cases even an alternative to conventional plastics based on crude oil [6]. In the case of so-called "drop-in solutions", such as bio-polyethylene, the fossil-based plastic can be replaced one-to-one, in which a renewable raw material, such as cane sugar, is used as the raw material for the synthesis of the bioplastic. The resulting plastic has similar properties to its conventional counterpart [7, 8]. However, completely new plastics with individual property profiles, such as polylactic acid (PLA), can also be synthesized from renewable raw materials. Another possibility of using biopolymers is the direct use as a biogenic filler or as a natural reinforcing fiber. The advantage here is that the respective biopolymer is used directly without increased energy input, with the exception of extraction. The carbon footprint is particularly small, also in comparison to PLA, since different, energy-intensive processing steps are additionally required for its synthesis [9, 10]. For example, as high a proportion of native starch as a filler as possible in a blend with PLA is energy and cost efficient. However, the strong hydrophilicity of the starch on the one hand and the partly significantly reduced mechanical properties of the starch blend compared to unfilled PLA on the other hand pose a problem. The high moisture absorption of starch requires extensive drying prior to compounding with PLA in

**Corresponding author*: Email: j.fuchs@uni-kassel.de

order to prevent starch from being destructered by water, which acts as a plasticizer, and hydrolysis of the PLA. Furthermore, modifying the starch with suitable additives to improve the properties and adhesion to the thermoplastic matrix is a way of optimizing the mechanical properties and reducing the moisture absorption of starch blends [9–13].

6.2 STARCH

Besides cellulose, starch ($C_6H_{10}O_5$) is the most important renewable raw material in terms of quantity [14, 15]. Starch belongs to the group of polysaccharides, which in turn are assigned to the carbohydrate class. In plants containing chlorophyll, monosaccharides in the form of fructose (fructose-6-phosphate) are formed during photosynthesis (Fig. 6.1) [14, 16, 17].

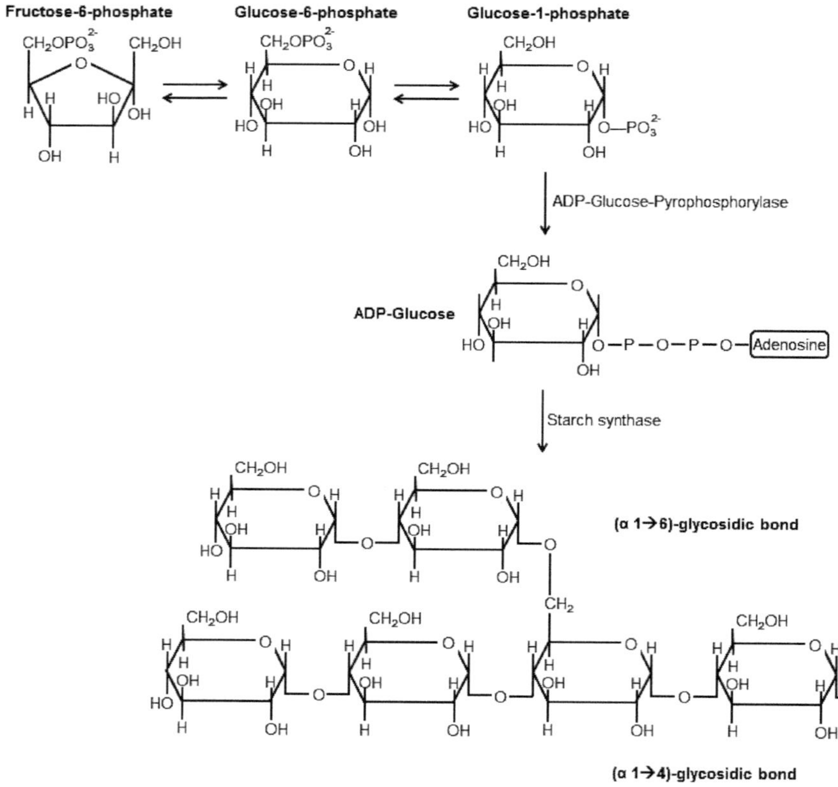

Figure 6.1 Steps in the biosynthesis of starch according to [14, 16–18].

From certain chain lengths, the (α-1→4)-connections are split, and the resulting chain residues are connected by means of a (α-1→6)-connection. Subsequently, the glucan chain (α-1,4-glucan) is further prolonged by the starch synthase until a (α-1→6)-branching occurs again. Glucose-1 -phosphate is then produced via a hexose phosphatisomerase with the intermediate glucose-6-phosphate and a phosphoglucomutase. With the help of adenosine triphosphate (ATP), glucose-1 -phosphate reacts within an ADP-glucose-pyrophosphorylase to ADP-glucose. The glucose activated by adenosine diphosphate is then transferred by the starch synthase to the glucan chain (α-1,4-glucan) [14, 16, 17].

Starch synthesis and storage mainly take place in the amyloplasts, a form of organelles found in plant sap, cytosol. Unlike the plastids involved in photosynthesis, amyloplasts do not contain chlorophyll. A distinction is made between transitory starch, i.e., temporary starch, and depot starch, i.e., permanently stored starch. The transitory starch is stored during the day, in sunshine, mainly in the chloroplasts of the leaves and is converted back into monosaccharides at night or in bad weather and transported away. The depot starch is stored in the amyloplasts, which are located in different areas (roots, tubers, seeds, etc.) depending on the plant species [16-20].

The resulting starch (depot starch) consists mainly of amylose and amylopectin, which in turn consist of a different number of anhydroglucose units (AGU), i.e., monomers [5].

Besides amylose and amylopectin, starch also consists of small amounts of proteins, phosphates, fats, and fatty acids. The quantities contained vary depending on the plant origin. Cereal starches from corn, wheat, etc. (0.6–1%) contain, in comparison with tuber starches from potatoes, tapioca etc. (0.05–0.1%), significantly higher proportions of fats and fatty acids. Similarly, proteins with a higher proportion are contained in cereal starches (0.25–0.5%) compared to tuber starches (0.06–0.1%). The different proportions have an influence on the solubility and color of the starch [21–26].

The two main components amylose and amylopectin differ in structure and properties. Amylose consists of linear, almost exclusively unbranched molecular chains, whereby the chain length can be very different depending on the plant origin [5, 20, 27]. The degree of polymerization of amylose is between 324 and 4920, whereby the chains have only a very small number of (α-1$\rightarrow$6)-branches [28, 29]. Amylose consists of about 99% (α-1$\rightarrow$4)-branches and has a relative molecular weight of 10^5–10^6 g/mol [30, 31]. In the case of amylose, a basic distinction can be made between A- and B-amylose (Fig. 6.2). A-amylose is mainly found in cereal starches and B-amylose mainly in tuber starches. Both consist of right- or left-handed, (anti)parallel double helices [25], which are constructed approximately hexagonal, whereby the A-amylose has a significantly higher packing density of the helices. The free spaces of the hexagonal arrangement of B-amylose are filled up by a significantly larger number of water molecules compared to A-amylose (Fig. 6.2) [27, 32, 33]. C-amylose, which consists of a mixture of A- and B-amylose, is usually found in beans. V-amylose is extremely rare in nature, but can be caused by heat treatment of starch under humid environmental conditions, for example. The different amylose types can be distinguished and detected by X-ray diffraction analysis [19, 34–37].

In contrast to amylose, amylopectin has a significantly higher number (approx. 5%) of branches (α-1$\rightarrow$6) of its molecular chains (Fig. 6.2). The molecular weight is about 10^7–10^9 g/mol [30, 31]. The degree of polymerization is between 9600 and 15900, whereby Takeda et al. [38] found out that there are always three fractions in amylopectin, which differ in their degree of polymerization. These are divided into a large (DP 13400–26500), a medium (DP 4400–8400), and a small fraction (DP 700–2100). The distribution of these fractions varies depending on the type of starch [38].

The structure of amylopectin has not yet been clearly clarified. For many years the model of Robin et al. [40] has established itself, which represents the branches as tufts distributed over the molecule. The tufts have short side chains, whereby two side chains each have a double helix, as with A- and B-amylose. These amylopectin trees (Fig. 6.2) form clusters and are thus distributed throughout the whole starch grain. The tufted branches represent the amorphous part, the double helices the crystalline part of amylopectin. Within the starch grain, the clusters form partly ordered and partly disordered areas. The mostly amorphous amylose is mainly found in the disordered cluster areas. Within the starch grain, the amylopectin content is largely responsible for the level of crystallinity [20, 27, 30, 31, 38–41].

The structure of the starch grain (Fig. 6.3) is determined by the arrangement of the two main components amylose and amylopectin. The single starch grain is built up in layers, whereby the core is formed by the so-called hilum, an amorphous region. Each growth ring consists of amorphous and semi-crystalline areas. The semi-crystalline regions are built up by alternating layers of amylose and amylopectin, each about 9 nm [42, 43] thick. Amylose molecules and disordered molecules of amylopectin are present in the amorphous region. The area consisting of a concentric semi-crystalline ring and an amorphous ring is called a growth ring. The growth rings have a layer thickness of several hundred nanometers [31, 43–52].

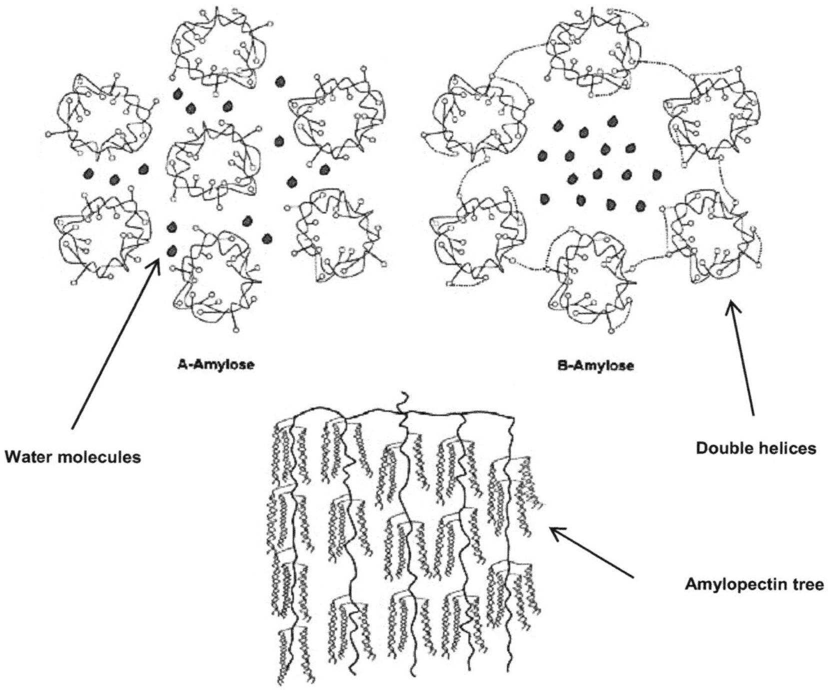

Figure 6.2 Models of A- and B-amylose (top) with bound water molecules, according to [32, 33] and model of amylopectin (bottom), according to [39].

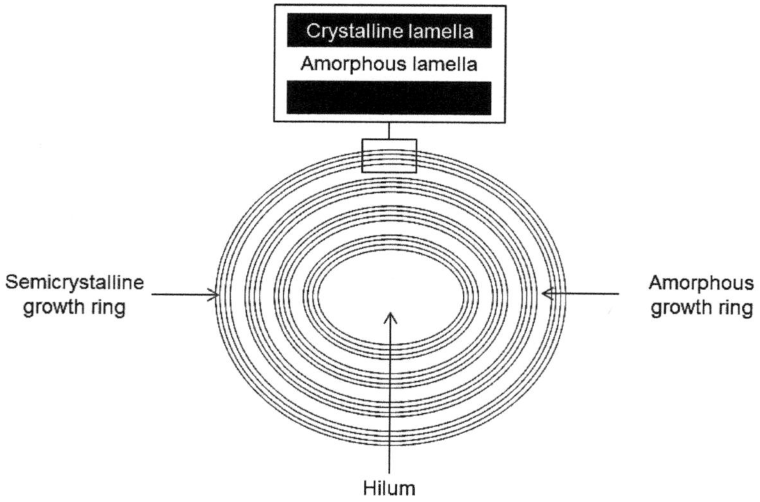

Figure 6.3 Structure of a starch grain, according to [23, 31, 53–55].

The composition of the starch grains from their main components amylose and amylopectin as well as their shape and size vary considerably depending on the type of starch [56, 57]. For native starch, for example, there are large variations in grain size, even within a starch type (Fig. 6.4). Cereal starches occur in an A and B type. Type A is lentil-shaped, whereas type B is spherical [58]. Potato starch grains are oval and have a diameter of up to 100 μm. Cereal starches often have small holes in their surface [58–60]. Compared to other types of starch, the grains of potato starch have a very smooth surface [45, 56, 57, 61].

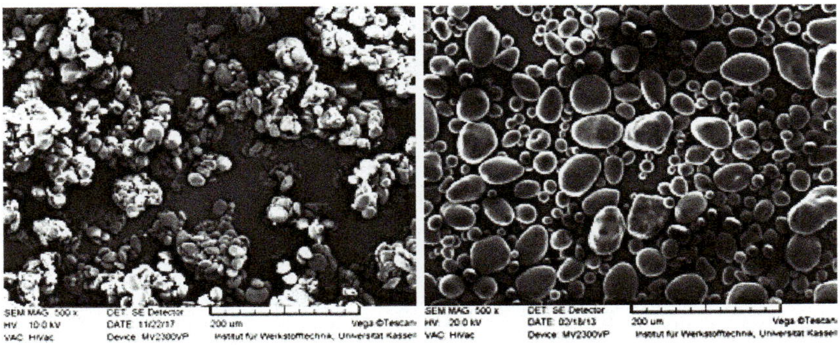

Figure 6.4 SEM images of native wheat starch (left) and native potato starch (right) at 500x magnification.

The degree of crystallinity of starch depends not only on the amylopectin content, but also on the water content, which can vary greatly depending on the variety and conditioning. Conventional starches contain 20–30% amylose and 70–80% amylopectin. Starches with a high amylopectin content (wax starches) or with a high amylose content (amylose starches) are also available, but are usually only used for special applications and not for the food industry [5, 23, 42, 56, 57, 61–69].

Various processes are available for extracting the starch from the plant. Basically, a distinction can be made between processes for obtaining tuber starch (potato, etc.) and seed starch (cereals, etc.), whereby only the process for extracting potato starch is described here. In the first step, the potatoes are pre-cleaned from foreign matter, such as soil and stones [66]. Afterwards, the tubers are washed in water basins, stirring constantly, whereby the peel is also partially removed by friction with each other. After this step, the potatoes are finely chopped by means of rapidly rotating saw blade grating, so that a mash is produced. By adding SO_2 to the wash water, possible oxidative degradation processes are to be reduced to a minimum. In separators, the starch and amniotic fluid are then separated from the pulp, the shell residues, which are used, for example, as animal feed. The separation of starch from the remaining amniotic fluid, which is mainly used as fertilizer, takes place in hydrocyclones [66]. The native starch is then first pre-conditioned to a moisture content of 40% with the aid of vacuum rotary filters and then dried in flash driers up to a moisture content of about 20% to enable storage. In addition to this large-scale process, other laboratory-scale extraction processes are known, in which, for example, dimethyl suloxide [70] or ammonia [71] are used as additives in the process water [20, 66, 67, 72, 73].

An iodine test can be performed to detect starch. The elementary iodine deposits in the helical structure of the starch and turns blue [74]. When starch grains treated in this way are examined under the microscope under polarised light, the characteristic "Maltese crosses" appear (Fig. 6.5) [75].

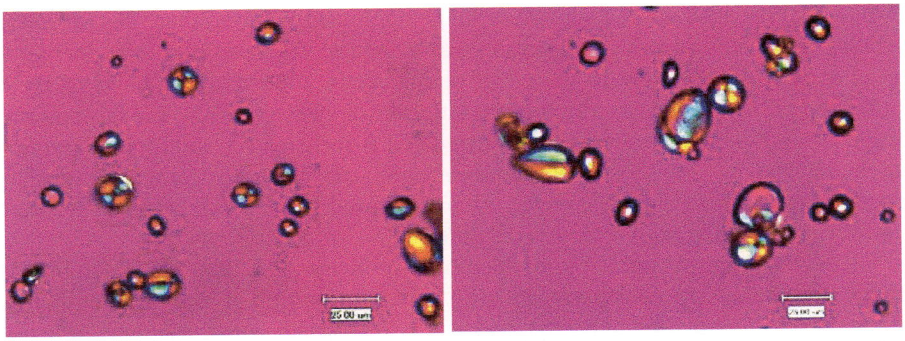

Figure 6.5 Microscopic images of native potato starch with characteristic "Maltese crosses" under polarized light at 40x magnification.

Native starch is basically hydrophilic and thus absorbs moisture from the environment very quickly in some cases. The speed and maximum amount of water absorption depends on the respective type of starch. An important property of starch, for example for applications in the food sector, is the so-called gelatinisation. Starch can swell very slowly (reversibly) in cold water, but rapid swelling of the starch grains starts in warm water at a temperature that varies depending on the type of starch. The swelling leads on the one hand to an increase in volume of the starch grains in which more water is stored, on the other hand it leads to a reduction in the ordered structure of the starch. This is accompanied by a change in the refraction of light intensity [76]. The maximum degree of swelling is determined by the degree of crystallinity of the starch; accordingly, a high amylopectin content (waxy starches) leads to an increased degree of swelling [77]. As already mentioned, the gelatinisation temperature is largely dependent on the respective type of starch, but it can also vary within starches of a plant origin. This is due to the conditions in the respective harvest year, i.e., the amount of rain, the quality of the soil, or the fertilisation at the point of cultivation, as all these factors have an influence on the composition of the starch [20, 76–92].

During gelatinisation, the viscosity of the mixture of water and starch increases continuously. This produces a paste-like starch pulp from which parts of the amylose diffuse out. This amylose then forms new crystalline structures, whereas amylopectin degrades further by breaking the hydrogen bonds between the molecular chains of glucose. The swelling of the starch continues until the individual starch grains burst open. When the temperature is further increased, a homogeneous mixture of amylose and amylopectin is produced [82, 87, 93–95].

When this mixture cools, the so-called retrogradation, i.e., the irreversible regression of the gelatinised starch suspension, begins when the specific start temperature is not reached. Water molecules that were previously bound to the free hydroxyl groups diffuse out. The molecular chains of amylose are predominantly bound to molecular aggregates, i.e., crystalline structures (Fig. 6.6). The lower the ambient temperature, the faster retrogradation takes place [96–101].

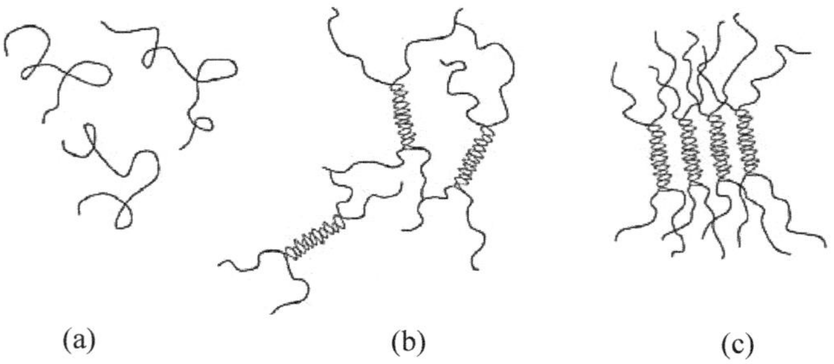

Figure 6.6 Retrogradation of amylose: Amylose in solution (a), amylose at the beginning of retrogradation (b) and molecular aggregates of amylose (c) according to [26].

The rate of retrogradation of amylopectin is significantly lower than that of amylose, so retrogradation takes considerably longer for wax starches, for example, than for amylopectin starches or normal starches due to their high amylopectin content. In addition to the ambient temperature, the pH value also has a significant influence on retrogradation. For this reason, various additives, such as diglycerides, are added to the starch suspension for controllable retrogradation. Targeted gelatinisation and subsequent retrogradation can be used, for example, to produce so-called swollen starches (Fig. 6.7), which are mainly used in the food sector. These starches are clearly different from their native form [102–106].

The biodegradation of starch is divided into primary and final degradation. First, the macromolecular chains are broken down. The enzyme α-amylase converts amylopectin and amylose into dextrins. Subsequently, the amylose is converted to maltose and the amylopectin to maltose and dextrose by the enzyme β-Amylase. The enzyme glycosidase then converts the maltose into glucose by hydrolysis. In addition to the enzymes mentioned, other microorganisms, such as certain bacterial strains of saccharolytic clostridia are involved in the degradation of starch. Temperature and ambient humidity are key factors influencing the rate of starch degradation [5, 107–111].

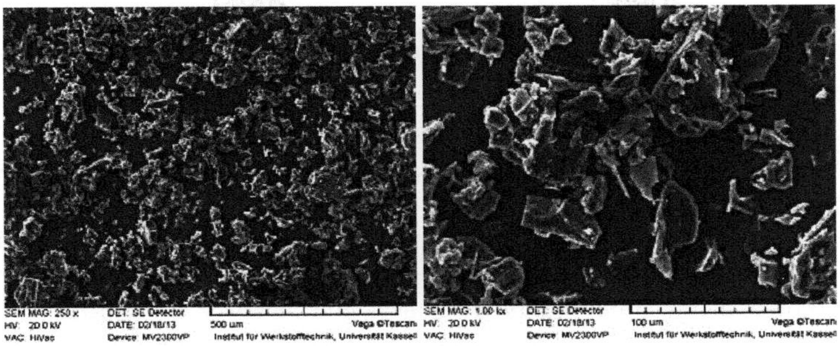

Figure 6.7 SEM images of swollen wheat starch at 250x (left) and 1000x magnification (right).

Starch continues to be mainly used in the food, paper, and corrugated board industries (Fig. 6.8) [67, 112]. Another important field of application for starch is the pharmaceutical industry. Starch serves here as a carrier substance for active ingredients in tablets. The starch offers the possibility of the targeted release (time, quantity, etc.) of the respective active ingredient. The German starch manufacturers process about 4 million tonnes of raw material annually into about 1.44 million tonnes of intermediate and end products [113]. Across Europe, approximately 8 million tonnes of starch are produced or extracted annually. This is about 13% of the world's annual production of native starch. More than 50% of the raw materials used are potatoes. The other starch raw materials worth mentioning are wheat and corn [113]. In the non-food sector, starch is not only used as an adhesive or additive in the production of paper and corrugated board, but also as a raw material for the production of biofuel, for example [20, 67, 73, 112, 113].

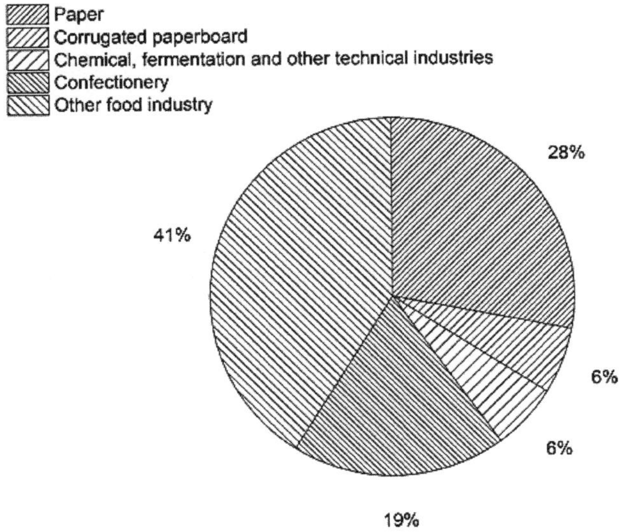

Figure 6.8 Starch consumption by application area in the Federal Republic of Germany in 2015, according to [113].

Another way of using native starch is to use starch as a raw material for the synthesis of bioplastics, such as PLA. Further information on this topic is also described in this book under Chapter 1.

6.3 STARCH MODIFICATION

In addition to the use of starch in its native form and as a raw material for the synthesis of bioplastics, starch is in most cases first modified before use in order to optimally adapt the properties for the respective area of application [114]. When modifying starch, a basic distinction is made between chemical and physical modification. An important physical modification is pre-gelatinisation (Fig. 6.9), in which native starch is heated to a temperature above the gelatinisation temperature and then rolled into a starch film [115, 116]. The powder produced after drying and grinding is mainly used in the food industry for the production of "instant products", such as pudding [114, 115, 117, 118].

Derivation is one way of chemically modifying native starch. By esterification or etherification, the hydroxyl groups of starch can be substituted by other functional groups [119]. If hydroxyl groups are substituted in the modification of starch, the "Degree of Substitution" (DS) is used as a measure of the effect of the modification. The maximum degree of substitution is three, due to the three hydroxyl groups within an AGU [120]. Etherification of starch is usually used to produce hydroxypropyl and hydroxyethyl starch (Fig. 6.9). Due to their high water-binding capacity and easy solubility in cold water, these starch ethers are often used as additives in the food, paper, and textile industries. In etherification, the hydroxyl groups can be (partially) replaced by ionic, hydrophobic, or neutral substituents [78]. Starch can be esterified with various acids, such as phosphoric, adipic, acetic, or fatty acids. In this way, for example, the gelatinisation temperature can be specifically reduced or hydrophobic groups can be incorporated into the starch molecule. Like starch ethers, starch esters can be used, among other things, as thickeners for foodstuffs [120–128].

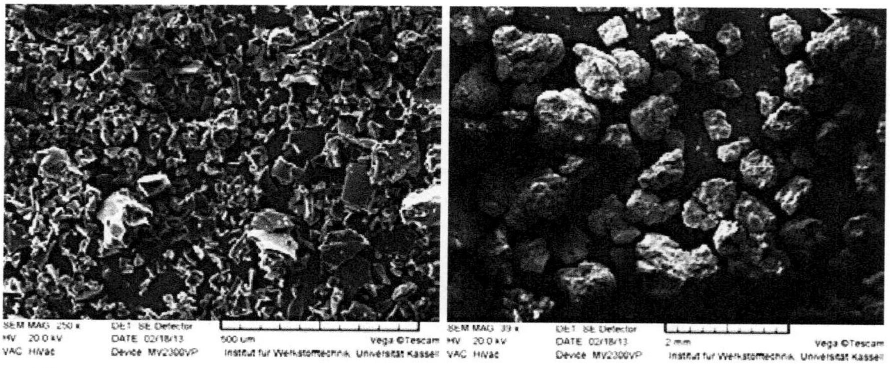

Figure 6.9 SEM images of pregelatinized starch at 250x magnification (left) and hydroxypropyl starch at 39x magnification (right).

In addition to esterification and etherification, the oxidation, cationization, and grafting of starch offer further possibilities for chemical modification [124]. Oxidation is used to reduce the molecular chain length of starch, thereby reducing the viscosity of the starch and increasing the maximum starch content in a product. During the process, carboxyl and/or aldehyde groups are formed within the starch. In the oxidation of starch, a distinction is made, among other things, between wet and dry oxidation [129]. For example, hydrogen peroxide or sodium hypochloride are used as oxidizing agents and various metals, such as copper or iron are used as catalysts [130]. The exact course of the reaction is still being discussed, with the description by Floor et al. [131] receiving the broadest approval so far. Nowadays, various methods are used to activate the reaction, such as the use of microwaves [129, 130, 132–134].

Reagents, such as glycidyltrimethylammonium chlorides, are often used in the cationization of starch [135]. Cationic starches are mainly used in the paper and corrugated board industry, where they serve as binders or auxiliary materials to enlarge the surface [136]. As with oxidation for the modification of native starch, various wet and dry processes are also available for cationization [135–139].

Starch can be used as a carrier material for the grafting of copolymers. In this form, the modified starch can serve as hydrogel, flocculant, or ion exchanger [140]. A frequent application is the grafting of starch-polyacrylonitrile, which is used in this form as a highly absorbent component, for example in diapers [141–145].

Further possibilities for modification of native starch, such as treatment with ultrasound, gamma radiation, etc. are described in [116, 145–148].

In the past, epichlorohydrin (EPI) was mainly used in the modification of native starch by a crosslinking reaction. The first patent application for this reaction dates back to the 1920s [149]. In addition to epichlorohydrin, various other bi- and multifunctional crosslinkers, such as phosphorus oxychloride [150], sodium trimetaphosphate [151], or sodium tripolyphosphate [152] have been and continue to be used. As in esterification and etherification, hydroxyl groups are substituted by other functional groups. In addition, however, individual molecular chains of starch are inter- and intramolecularly crosslinked. In addition to the actual crosslinker, a suitable catalyst, such as sodium hydroxide, is required to start the reaction. Crosslinked starches may exhibit higher resistance to low pH values, higher temperatures, and shear rates, among other things [153]. Paschall et al. [66] and Kartha and Srivastava [154, 155] describe the reaction of starch with epichlorohydrin in three steps (Fig. 6.10). During the reaction, the hydroxide ions of the dissolved sodium hydroxide form a compound with the hydrogen atoms of the hydroxyl groups of the starch by splitting off water (1). The oxygen ions (O–) of starch cause a ring opening of the epoxy group of epichlorohydrin and then form a new epoxy group with it. During this reaction sodium chloride splits off (2). The reaction described above is repeated on the newly formed epo-oxide ring and thus leads to a linkage of two starch molecules (3). Thus, all 100–3000 AGUs of starch are created by such cross-linking points [153]. The cross-linking points are mainly located in the amorphous areas, i.e., between the clusters, of amylopectin [156]. In addition to crosslinked starch, polymerized epichlorhydrin is also produced during the reaction by the reaction of EPI with itself. [154, 155, 157–164].

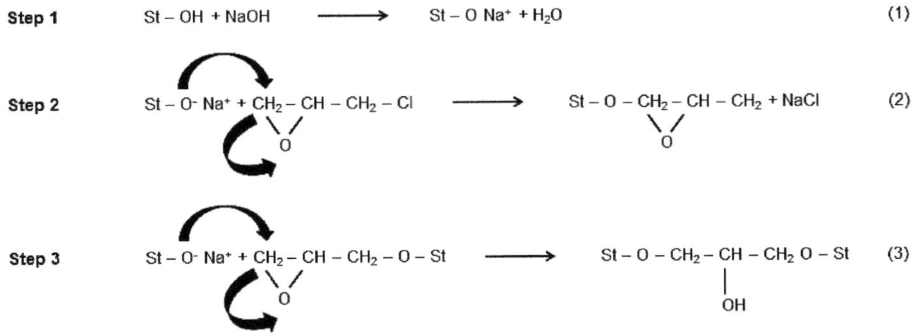

Figure 6.10 Three-step reaction of starch with epichlorohydrin, according to [66, 154, 155].

After the crosslinking reaction, the starch suspension must be neutralized and matured to remove unreacted residues and the resulting salts. Starch crosslinked with epichlorohydrin is strengthened by further covalent bonds in addition to hydrogen bonds. When heating crosslinked starch in water, the grains remain intact due to the covalent bonds, although the hydrogen bonds are significantly reduced or partly broken up [165, 166]. The crosslinking of native starch therefore always leads to hydrophobicity of the polysaccharide. Crosslinked starch is used, for example, as an absorbent of heavy metals in waste water [167]. Starches crosslinked by epichlorohydrin and products containing them are not suitable for consumption, as the crosslinker has a very high toxicity [168–172]. In addition to epichlorohydrin, other less toxic or even completely harmless crosslinkers, such as polycarboxylic acid, citric acid, malonic acid, or genipine are also available [168, 168, 173–185].

Another possibility of crosslinking native starch with safer reagents are glycidyl ethers, i.e., epichlorohydrin derivatives [186]. The reaction is similar to that of epichlorohydrin with native starch. Studies till date point to high possible degrees of substitution (DS) in starch [185, 187–189].

The possibilities described above for chemical and physical modification of starch relate primarily to areas of application outside plastics technology and especially outside the use in a blend with thermoplastics. Some of these reagents and methods have also earlier been and still are used to modify starch on plastics processing machines. In particular, the cross-linking with the toxic epichlorohydrin is to be mentioned here.

6.4 STARCH BLENDS

Starch can be incorporated into thermoplastics to produce blends. In addition to native starch, thermoplastic or modified starch can also be used. For the production of thermoplastic starch (TPS), co-rotating twin-screw extruders are generally used [190, 191]. During the process, the native starch is thermomechanically destructured. In order to reinforce the effect of destructuring and to make the TPS thermoplastically processable, further additives are required in addition to a sufficiently high rotation speed as well as shear and temperature effects. The focus is on a softening effect. Reagents, such as sorbitol, glycerine, urea, or water are used accordingly [190, 192, 193]. The energy input required for starch disintegration depends on the plant origin of the starch and is approx. 650 kJ/kg for potato starch, approx. 380 kJ/kg for corn starch, and approx. 435 kJ/kg for wheat starch [194]. During the destructuring process, the hydrogen bonds and thus the crystallite structures of the semi-crystalline starch are partially or completely broken up, resulting in a predominantly amorphous, pasty starch gel [191, 194–201].

Thermoplastic starch can be used as a foamed extrudate without any further plastic components. A moisture content of more than 10% of the starch before processing and a sufficiently high temperature (>100°C) is required to ensure foaming at the exit of the material from the die of the extruder [191]. As cuttings, the foamed TPS is used as so-called "Loose-Fills" (packaging chips) (Fig. 6.11). Another possibility in the production of thermoplastic starch is to prevent foaming by using a starch with a lower moisture content or pre-drying the starch. As a result, the thermoplastic starch can be granulated like conventional thermoplastics at the end of the preparation process and then further processed, for example in the injection molding process (Fig. 6.11). In addition to the production of thermoplastic starch in granular form or as a foamed extrudate, the TPS can also be blended with thermoplastics directly after the destructuring process. Low-melting thermoplastics, such as polyethylene or PLA, are often used here [192, 199, 202–209].

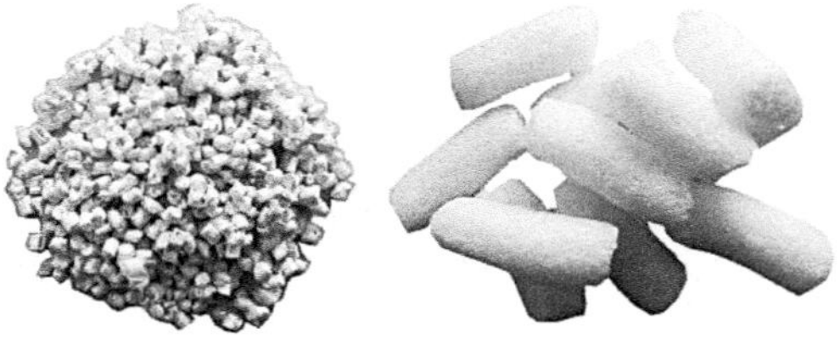

Figure 6.11 Thermoplastic starch as granulate material (left) and as foamed packaging chips (right).

In the production of PLA-TPS blends (Fig. 6.12), hydrolysis of the PLA is accepted by the use of moist starch, since a reduction in the impact strength and tensile strength of the compound during subsequent application, particularly as a film material, plays a minor role compared to the desired increase in elongation (boosted by plasticizers) [210, 210, 211].

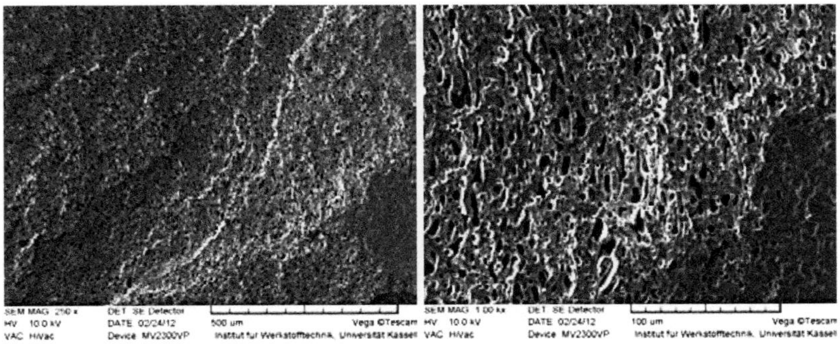

Figure 6.12 SEM images of a PLA-TPS blend with completely destructured and plasticized starch at 250x (left) and 1000x magnification (right).

The properties of the blends vary greatly depending on the starch content and the moisture content of the starch before compounding and the moisture content of the compounds before or during the test [212–214].

Due to the low mechanical properties, the high moisture absorption and thus the solubility of the material in water, the areas of application of PLA-TPS blends are severely limited. Applications for these blends include compostable organic waste bags and mulch films for agriculture, which are used, for example, in asparagus cultivation [215–218].

In blends of thermoplastics with native starch, starch serves as a pure filler. In contrast to the production of blends with thermoplastic starch, the aim here is in most cases to avoid starch granule disintegration. For this reason, as in conventional compounding processes, for example, for the production of fiber-reinforced plastics, the thermoplastic is first melted and then the starch is added via side metering. Blends with native starch and PLA are very rarely used compared to TPS-PLA blends. The reason for this are the already brittle properties of pure PLA, which are further enhanced by the addition of granular starch, especially in large quantities [191, 195, 219, 219–221].

Analogous to the addition of conventional fillers to thermoplastics, the modulus of elasticity increases to a certain degree, whereas the strength and elongation at break are reduced compared to unfilled matrix plastics. One reason for this is the reduction of the load-bearing cross-sectional area of the PLA matrix by introducing the native starch as filler [195]. One of the most important factors influencing the mechanical properties and the moisture absorption of the starch blends is the adhesion of the starch granules to the respective matrix plastic. In the case of nonpolar thermoplastics, such as polyolefins, coupling agents are often used to ensure sufficient bonding of the polar starch to the matrix (Fig. 6.13) [222–227].

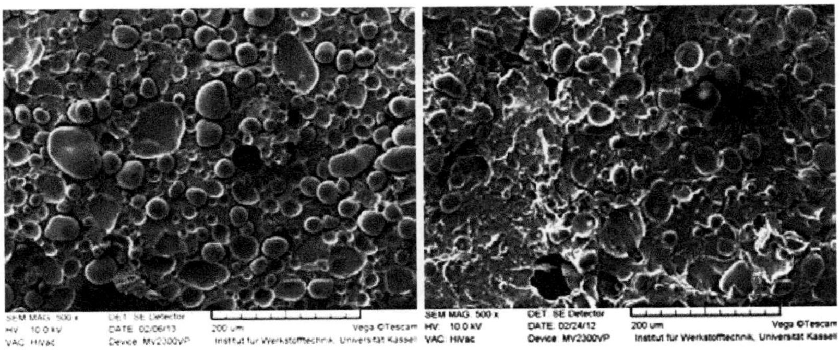

Figure 6.13 SEM images of a PP starch blend without coupling agent (left) and a PP starch blend with coupling agent (right) at 500x magnification.

Even with the use of polar PLA as matrix plastic, a satisfactory bonding of the starch is partially not given (Fig. 6.14). One reason for this is often an insufficient or even complete lack of drying of the starch before the compounding process. Drying the starch before processing prevents on the one hand hydrolysis of the PLA and on the other hand a disintegration of the starch so that it remains in its original grain shape and grain size even in the blend. Furthermore, drying prevents the formation of pores within the PLA [228], which can also lead to poorer bonding of the starch to the matrix. The starch can be dried in conventional convection ovens or special drying equipment for powders. Studies by Kovács and Tábi [226] have shown that a sufficiently high temperature is required to completely dry the native starch. For example, drying moist starch at 60°C results in a constant sample weight after about 48 hours, but complete drying at this temperature is not possible because water molecules are still bound in the starch. Temperatures above 100°C are required to completely dry the starch. However, no temperature close to the decomposition temperature of the starch should be selected for a longer period of time [224, 229–231].

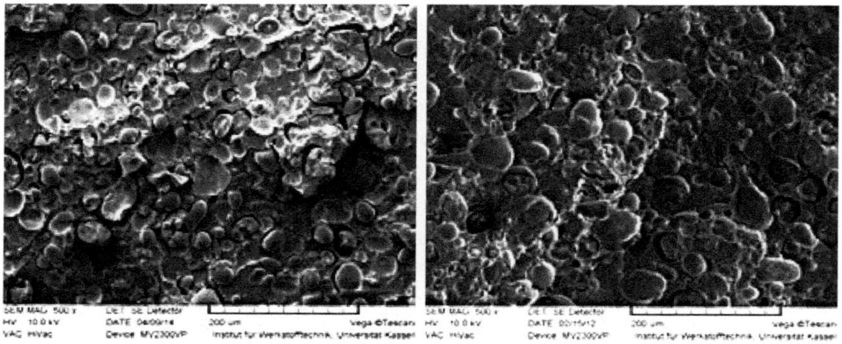

Figure 6.14 SEM images of a PLA starch blend with moist starch (left) and a PLA starch blend with pre-dried starch (right) at 500x magnification.

The products Mater-Bi from Novamont and Ecovio from BASF are the best-known commercially available starch blends. The mechanical properties of PLA starch blends vary greatly depending on their composition, for example the starch content, the starch type (granular/TPS), or the PLA type [232–234].

In the injection molding process, even small amounts of starch lead to accelerated crystallization of the PLA, whereby this nucleation effect is higher with conventional fillers, such as talcum powder [231].

As already mentioned, blends with granular starch have rarely been used up till now. Nevertheless, these materials offer great potential and their usability could also be proven on the basis of sample components (Fig. 6.15).

The maximum moisture absorption and the speed of moisture absorption of the PLA starch blend increase with increasing starch content [226]. Moisture absorption is higher for blends with thermoplastic starch than for blends with granular starch due to complete destructuring (Fig. 6.12 and Fig. 6.14). Moist storage of samples or components made of PLA blends with thermoplastic or granular starch results in a reduction of the mechanical properties [235, 236].

To reduce the moisture absorption of starch blends with PLA and thus also to improve the mechanical properties, the aforementioned complete drying of the starch before processing can be carried out. This is associated with increased energy and time expenditure. Another starting point is the use of modified starch produced in an upstream process (batch process) during compounding. However, it should be noted that the starch in this two-stage process is exposed to a double thermal stress first through modification and then through compounding. Furthermore, starch can be modified within the compounding process. This is known as reactive extrusion or reactive blending. In contrast to conventional compounding, the blending partners are not only mixed physically, but the extruder also serves as a chemical reactor for actively changing the chemical structure of a component or even all blending partners. Compared to the two-stage preparation process, the single-stage reactive extrusion saves time and energy. Nevertheless, pre-drying of the starch to achieve good bonding of the filler to the blending partner and thus maximum mechanical properties should not be omitted [12, 237–241].

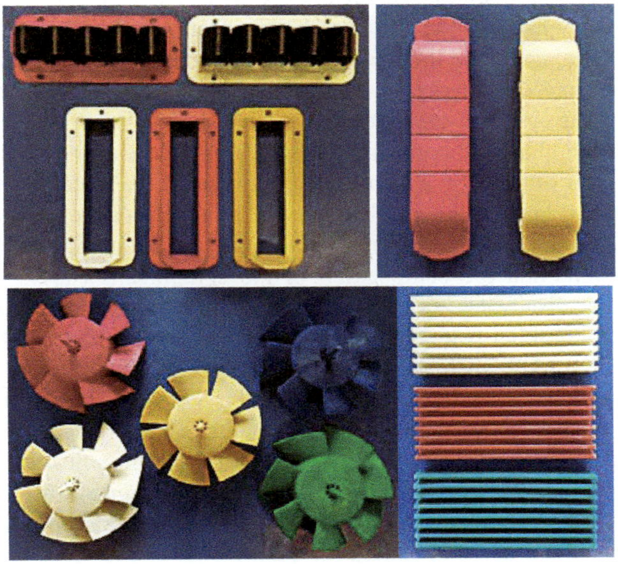

Figure 6.15 Sample components made of starch blends with granular starch: cable holder (top left), chair clamp (top right), and fan impeller (bottom left) made of PLA starch and ribbed profile (bottom right) made of Bio-PE with starch.

The mechanisms involved in reactive extrusion have been described in detail by Michaeli et al. [238]. The heat transport, the flow behavior, and the reaction as such are the most important, alternating factors in this complex process (Fig. 6.16). These factors are additionally influenced by the respective material and machine properties [238, 242].

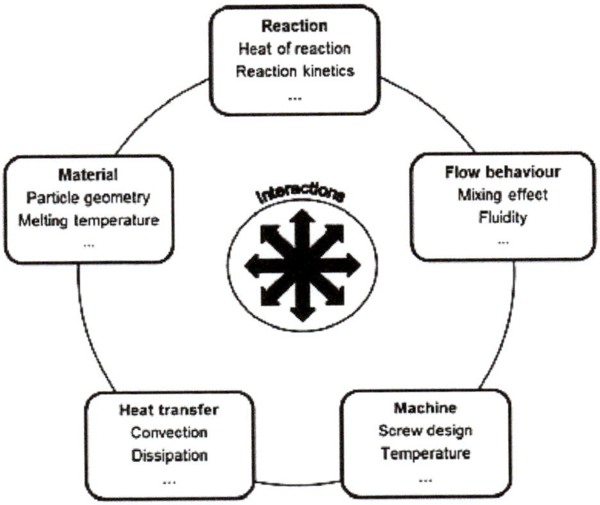

Figure 6.16 Interacting factors in reactive extrusion according to [238].

Various programs, such as “Morex” (RWTH Aachen), “SIGMA” (University of Paderborn) or “Akro-Co-Twin-Screw” (University of Akron, Ohio) are known for simulating reactive extrusion processes and for designing the screw of the respective extruder [243]. Although these programs provide information on all required parameters and physical properties, they are only partially suitable for simulating the behavior of blends with biogenic fillers. The particle geometry plays a decisive role in the flow behavior or the mixing effect. Due to the variety of shapes and sizes within a starch fraction, the modelling of such materials is very complex [244–246].

The single-stage preparation process for the production of blends of PLA and modified starch offers some advantages over the familiar two-stage processes. In addition to the previously mentioned savings in time and energy, smaller quantities can also be produced quickly with this process. Furthermore, the only one-time thermal load on the starch and the possibility of quick and easy modification of the glare components to adapt the properties for subsequent application are to be mentioned. This is offset by the short residence time of the material in the extruder, which does cause a low thermal load on the starch, but requires a high reaction speed with the respective additives. The high demands on the dosing of the individual components, both in terms of the number of dosing points and the composition of the components (liquid, powdered, etc.), are also to be mentioned as disadvantages. Last but not least, the often low throughput and unwanted degradation and/or crosslinking reactions represent further disadvantages [247–254].

Various processes are known for reactive extrusion to modify starch. For the most part, not native starch is modified, but thermoplastic starch is first produced in the extruder and then modified within this process. The production of thermoplastic starch is already understood by some working groups as reactive extrusion. As with the batch processes, the modification aims to improve the binding of starch to the matrix, reduce moisture absorption, or improve the mechanical properties of the compounds. Analogous to the modification options, the following modifications can be made within the reactive extrusion of starch, for example [250, 255–260]:

▪ Esterification or etherification	▪ Thermal or catalytic degradation
▪ Oxidation	▪ Phosphorylation
▪ Cationization	▪ Silylation
▪ Coupling	▪ Cross-linking
▪ Grafting	▪ ...

For the adhesion of starch with PLA in reactive extrusion, isocyanates [227, 255, 261], peroxides [262, 263], or maleic anhydride [264] are used. As in batch processes, amonium chlorides are mainly used for the cationization of starch [136, 265]. Starch is crosslinked primarily with epichlorohydrin and thermoplastic starch. In addition to the aforementioned sodium hydroxide, various metal soaps and amines can be used as catalysts [165, 266–269].

6.4.1 Mechanical Properties

The following figures show selected mechanical properties of various starch blends and pure PLA. Looking at the tensile strength of the materials, it becomes clear that the blend with thermoplastic starch as well as the blend with moist starch have the lowest values in comparison (Fig. 6.17). In the case of these materials and in particular with TPS, (partial) destructuring of the starch occurs during the compounding process. This results in a reduction in tensile strength. The unfilled PLA, on the other hand, shows, with approx. 62 MPa, the highest tensile strength. The blend with modified starch is also at a high level, with approx. 58 MPa. In terms of Young’s modulus, the blends with native and modified starch show significantly higher values compared to the unfilled PLA and the blend with TPS and PLA.

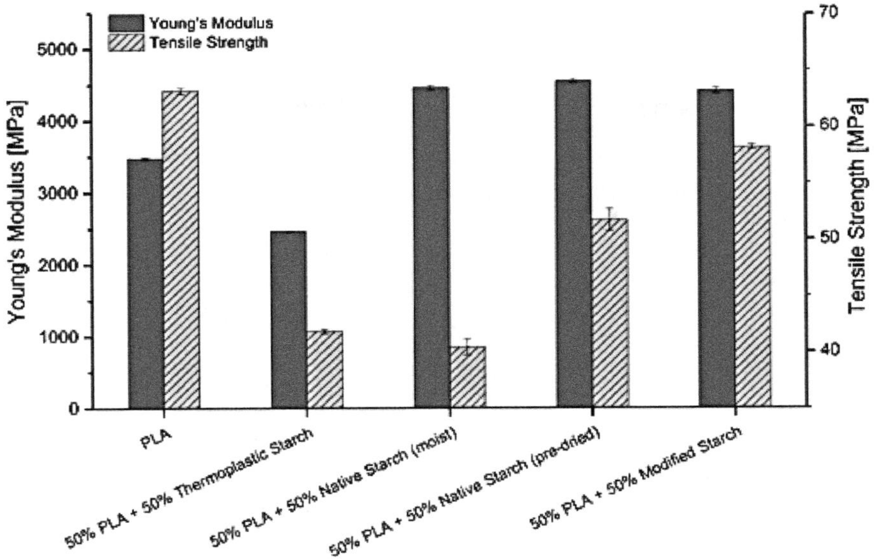

Figure 6.17 Young's modulus and tensile strength of selected starch blends and pure PLA [270].

The unfilled PLA shows the highest values for charpy impact strength and tensile elongation at break (Fig. 6.18). The blend with modified starch has an impact strength of approx. 10 kJ/m², which is higher than the achieved value of the compound with dried starch and thus at a high level. The blends with moist and pre-dried native starch show the lowest impact strength and elongation at break compared to the other materials investigated. Due to the complete destructuring of the starch in the PLA-TPS blend, this material exhibits a relatively high elongation at break in comparison. Basically, blends with granular starch, whose original granular shape was retained during compounding, have a significantly lower elongation at break compared to blends with thermoplastic starch.

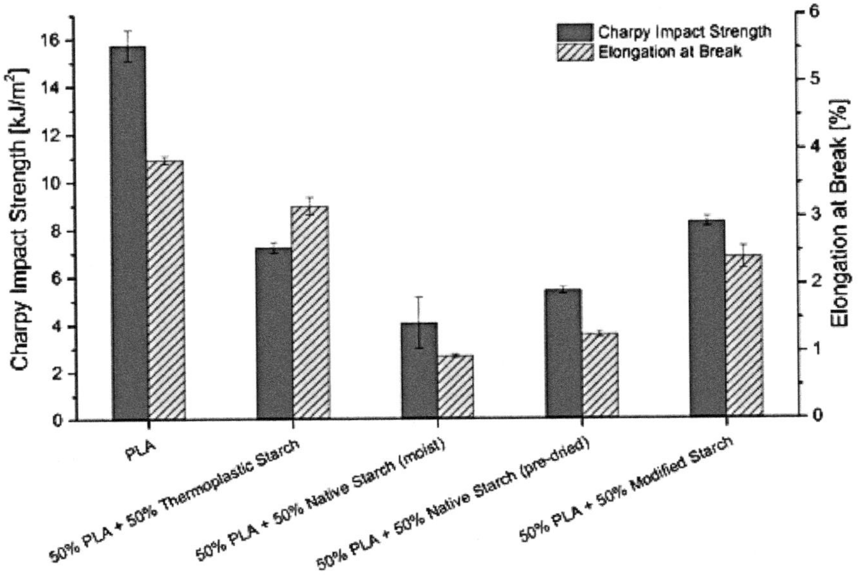

Figure 6.18 Charpy impact strength and elongation at break of selected starch blends and pure PLA [270].

A comparison of the stress-strain curves in the tensile test reveals clear differences between the materials (Fig. 6.19). The unfilled PLA shows the highest stress and elongation. The blend with PLA and TPS also achieves comparable elongation values. The blends with moist and pre-dried native starch show a very brittle failure. The maximum tension of the blend with dried native starch is significantly higher than the tension of the PLA blend with TPS, whereas the maximum tension of the blend with moist native starch is in the range of the PLA-TPS compound. The blend with modified starch achieves significantly higher stresses.

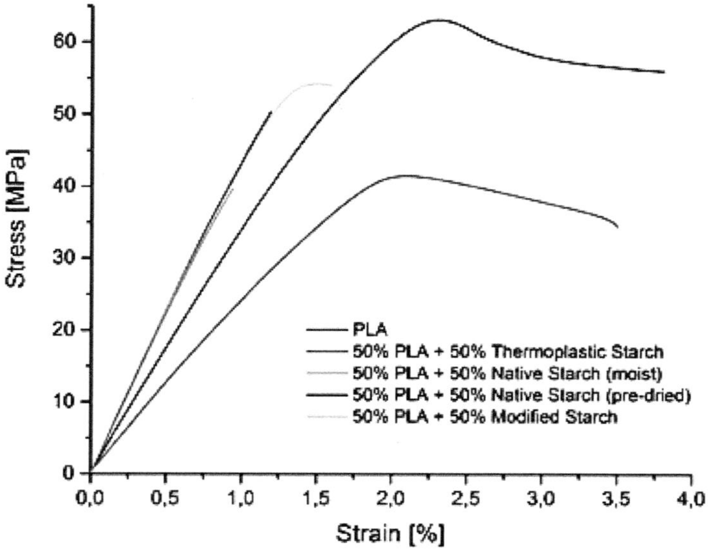

Figure 6.19 Stress-strain-diagram of selected starch blends and pure PLA [270].

6.5 CONCLUSION

Starch is a versatile raw material that is available in large quantities and at a relatively low price. In the field of plastics technology, starch can not only be used for the synthesis of bioplastics, such as PLA, but also as a filler in blends. Thermoplastic starch in particular is already used in large quantities for various products. By a suitable modification of the native starch, the properties can be adapted to the respective application area. When granular starch is used as a filler, drying the starch before compounding is crucial for the high mechanical properties of the blends. Drying mainly improves the bonding of starch granules to the thermoplastic matrix. With an additional modification, for example with suitable crosslinkers, it is possible to further improve the connection and thus the mechanical properties. In this form, blends with granular starch can also be used in part for technical applications.

References

1. Yamamoto, M. and A. Schneller. Industrielle perspektiven von biopolymeren. Chemie Ingenieur Technik 2010; 82: 1245-1249.
2. Feldmann, J. Mit Kunststoffen Ressourcen Schonen. Kunststoffe 2007; 97: 72-77.
3. Käb, H. Eine Logische Entwicklung. Kunststoffe 2009; 99: 12-19.
4. Fachagentur Nachwachsende Rohstoffe e.V. (FNR). Biokunststoffe: Pflanzen, Rohstoffe, Produkte, 2013.

5. Endres, H.-J. and A. Siebert-Raths. Technische Biopolymere: Rahmenbedingungen, Marktsituation, Herstellung, Aufbau und Eigenschaften. Hanser, München, 2009.
6. Reimer, V., A. Künkel and S. Philipp. Sinn oder Unsinn von Bio. Kunststoffe 2008; 98: 32-36.
7. Fachagentur Nachwachsende Rohstoffe e.V. (FNR). Hintergrundinformationen zu Biokunststoffen, 2015.
8. Käb, H. Biokunststoffe. Kunststoffe 2010; 100: 172-177.
9. Heim, H.-P. (ed.). Specialized Injection Molding Techniques: 5 Injection Molding of BioBased Plastics Polymers and Composites. Elsevier, Oxford, 2016.
10. Fuchs, J., M. Feldmann and H.-P. Heim. Natürlich und fest anbinden. Kunststoffe 2013; 103: 100-103.
11. Funke, U., W. Bergthaller and M.G. Lindhauer. Processing and characterization of biodegradable products based on starch. Polymer Degradation and Stability 1998; 59: 293-296.
12. Xie, F., L. Yu, H. Liu and L. Chen. Starch modification using reactive extrusion. Starch-Stärke 2006; 58: 131-139.
13. Wang, H., X. Sun and P. Seib. Effects of stach moisture on properties of wheat starch/poly(lactic acid) blend con-taining methylene-diphenyl diisocyanate. Journal of Polymers and the Environment 2002; 10: 133-138.
14. Martin, C. and A.M. Smith. Starch biosynthesis. The Plant Cell 1995; 7: 971-985.
15. Abo, S. -El-Fetoh, H.M.A. Al-Sayed and N.M.N. Nabih. Physicochemical properties of starch extracted from different sources and their application in pudding and white sauce. World Journal of Dairy & Food Sciences 2010; 5: 173-182.
16. Heldt, H.-W. and B. Piechulla. Pflanzenbiochemie, Fifth., Überarb. Aufl. Springer Spektrum, Berlin U.A., 2015.
17. Tegge, G. (ed.). Stärke und Stärkederivate, Third., Vollst. Überarb. Aufl. Behr, Hamburg, 2004.
18. Emes, M.J., C.G. Bowsher, C. Hedley, M.M. Burrell, E.S.F. Scrase-Field and I.J. Tetlow. Starch synthesis and carbon partitioning in developing endosperm. Journal of Experimental Botany 2003; 54: 569-575.
19. Belitz, H.-D., W. Grosch and P. Schieberle. Lebensmittelchemie, 6th Ed. Springer-Verlag, Berlin Heidelberg, 2008.
20. Galliard, T. (ed.). Starch: Properties and Potential. Elsevier, London U.A., 1987.
21. Swinkels, J.J.M. Composition and properties of commercial native starches. Starch–Stärke 1985; 37(1): 1-5.
22. Ratnayake, W.S., R. Hoover and T. Warkentin. Pea starch: Composition, structure and properties—a review. Starch–Stärke 2002; 54(6): 217-234.
23. Pérez, S. and E. Bertoft. The molecular structures of starch components and their contribution to the architecture of starch granules: A comprehensive review. Starch–Stärke 2010; 62: 389-420.
24. Yusuph, M., R.F. Tester, R. Ansell and C.E. Snape. Composition and properties of starches extracted from tubers of different potato varieties grown under the same environmental conditions. Food Chemistry 2003; 82: 283-289.
25. Tester, R.F., J. Karkalas and X. Qi. Starch-composition, fine structure and architecture. Journal of Cereal Science 2004; 39: 151-165.
26. Rein, H., R. Wansorra and K.J. Steffens. Stärke als komplexbildner–neue aspekte eines bekannten hilfsstoffes. Pharmazie in Unserer Zeit 1999; 28: 257-265.
27. Zobel, H.F. Molecules to granules: A comprehensive starch review. Starch–Stärke 1988; 40: 44-50.
28. Yoshimoto, Y., T. Takenouchib and Y. Takeda. Molecular structure and some physicochemical properties of waxy and low-amylose barley starches. Carbohydrate Polymers 2002; 47: 159-167.
29. Yoshimoto, Y., J. Tashiro, T. Takenouchi and Y. Takeda. Molecular structure and some physicochemical properties of high-amylose barley starches. Cereal Chemistry 2000; 77: 279-285.
30. Mua, J.P. and D.S. Jackson. Fine structure of corn amylose and amylopectin fractions with various molecular weights. Journal of Agricultural and Food Chemistry 1997; 45(10): 3840-3847.
31. Buléon, A., P. Colonna, V. Planchot and S. Ball. Starch granules: Structure and biosynthesis. International Journal of Biological Macromolecules 1998; 23: 85-112.
32. Hsien-Chih, H.W. and A. Sarko. The double-helical molecular structure of crystalline a-amylose. Carbohydrate Research 1978; 61: 27-40.

33. Hsein-Chih, H.W. and A. Sarko. The double-helical molecular structure of crystalline b-amylose. Carbohydrate Research 1978; 61: 7-25.
34. Popov, D., A. Buléon, M. Burghammer, H. Chanzy, N. Montesanti, J.-L. Putaux, G. Potocki-Véronèse and C. Riekel. Crystal structure of a-amylose: A revisit from synchrotron microdiffraction analysis of single crystals. Macromolecules 2009; 42: 1167-1174.
35. Buléon, A., F. Duprat, F. Booy and H. Chanzy. Single crystals of amylose with a low degree of polymerization. Carbohydrate Polymers 1984; 4: 161-173.
36. Zobel, H.F. Starch crystal transformations and their industrial importance. Starch–Stärke 1988; 40: 1-7.
37. Hoover, R. Composition, molecular structure, and physicochemical properties of tuber and root starches: A review. Carbohydrate Polymers 2001; 45: 253-267.
38. Takeda, Y., S. Shibahara and I. Hanashiro. Examination of the structure of amylopectin molecules by fluorescent labeling. Carbohydrate Research 2003; 338: 471-475.
39. Imberty, A., A. Buléon, V. Tran and S. Péerez. Recent advances in knowledge of starch structure. Starch–Stärke 1991; 43: 375-384.
40. Robin, P.J., C. Mercier, F. Duprat, R. Charbonnigre and A. Guilbot. Lintnerized starches. Chromatographic and enzymatic studies of insoluble residues from acid hydrolysis of various cereal starches, particularly waxy maize starch. Starch–Stärke 1975; 27: 36-45.
41. Bello-Pérez, L.A., O. Paredes-López, P. Roger and P. Colonna. Amylopectin-properties and fine structure. Food Chemistry 1996; 56: 171-176.
42. Jenkins, P.J., R.E. Cameron and A.M. Donald. A universal feature in the structure of starch granules from different botanical sources. Starch–Stärke 1993; 45: 417-420.
43. Baker, A.A., M.J. Miles and W. Helbert. Internal structure of the starch granule revealed by AFM. Carbohydrate Research 2001; 330: 249-256.
44. Pilling, E. and A.M. Smith. Growth ring formation in the starch granules of potato tubers. Plant Physiology 2003; 132: 365-371.
45. Gallant, D.J., B. Bouchet and P.M. Baldwin. Microscopy of starch: Evidence of a new level of granule organization. Carbohydrate Polymers 1997; 32: 177-191.
46. Sevenou, O., S.E. Hill, I.A. Farhat and J.R. Mitchell. Organisation of the external region of the starch granule as determined by infrared spectroscopy. International Journal of Biological Macromolecules 2002; 31: 79-85.
47. Putaux, J.-L., S. Molina-Boisseau, T. Momaur and A. Dufresne. Platelet nanocrystals resulting from the disruption of waxy maize starch granules by acid hydrolysis. Biomacromolecules 2003; 4: 1198-1202.
48. Vermeylen, R., B. Goderis, H. Reynaers and J.A. Delcour. Amylopectin molecular structure reflected in macromolecular organization of granular starch. Biomacromolecules 2004; 5: 1775-1786.
49. Waigh, T.A., K.L. Kato, A.M. Donald, M.J. Gidley, C.J. Clarke and C. Riekel. Side-chain liquid-crystalline model for starch. Starch–Stärke 2000; 52: 450-460.
50. Jeon, J.-S., N. Ryoo, T.-R. Hahn, H. Walia and Y. Nakamura. Starch biosynthesis in cereal endosperm, Plant Physiology and Biochemistry PPB 2010; 48: 383-392.
51. Angellier-Coussy, H., J.-L. Putaux, S. Molina-Boisseau, A. Dufresne, E. Bertoft and S. Perez. The molecular structure of waxy maize starch nanocrystals. Carbohydrate Research 2009; 344: 1558-1566.
52. Jenkins, P.J. and A.M. Donald. The influence of amylose on starch granule structure. International Journal of Biological Macromolecules 1995; 17: 315-321.
53. Buléon, A., G. Véronèse and J.-L. Putaux. Self-association and crystallization of amylose. Australian Journal of Chemistry 2007; 60: 706-718.
54. Ball, S., H.-P. Guan, M. James, A. Myers, P. Keeling, G. Mouille, A. Buléon, P. Colonna and J. Preiss. From glycogen to amylopectin: A model for the biogenesis of the plant starch granule. Cell 1996; 86: 349-352.
55. Jenkins, P.J., R.E. Comerson, A.M. Donald, W. Bras, G.E. Derbyshire, G.R. Mant and A.J. Ryan. In situ simultaneous small and wide angle x-ray scattering: A new technique to study starch gelatinization. The Journal of Polymer Science Part B: Polymer Physics 1994; 32: 1579-1583.
56. Jane, J.-L., T. Kasemsuwan, S. Leas, H. Zobel and J.F. Robyt. Anthology of starch granule morphology by scanning electron microscopy. Starch–Stärke 1994; 46: 121-129.

57. Gallant, D.J. and B. Bouchet. Ultrastructure of maize starch granules. A review, Food Structure 1986; 5: 141-155.
58. Fannon, J.E., J.M. Shull and J.N. BeMiller. Interior channels of starch granules. Cereal Chemistry 1993; 70: 611-613.
59. Huber, K.C. and J.N. BeMiller. Channels of maize and sorghum starch granules. Carbohydrate Polymers 2000; 41: 269-276.
60. Baldwin, P.M., J. Adler, M.C. Davies and C.D. Melia. Holes in starch granules: Confocal, SEM and light microscopy studies of starch granule structure. Starch–Stärke 1994; 46: 341-346.
61. Alcázar-Alay, S.C. and M.A.A. Meireles. Physicochemical properties, modifications and applications of starches from different botanical sources. Food Science and Technology (Campinas) 2015; 35: 215-236.
62. Singh, N., J. Singh, L. Kaur, N.S. Sodhi and B.S. Gill. Morphological, thermal and rheological properties of starches from different botanical sources. Food Chemistry 2003; 81: 219-231.
63. Srichuwong, S., T.C. Sunarti, T. Mishima, N. Isono and M. Hisamatsu. Starches from different botanical sources II: Contribution of starch structure to swelling and pasting properties. Carbohydrate Polymers 2005; 62: 25-34.
64. Oates, C.G. Towards an understanding of starch granule structure and hydrolysis. Trends in Food Science & Technology 1997; 8: 375-382.
65. Liu, Q., R. Tarn, D. Lynch and N. Skjodt. Physicochemical properties of dry matter and starch from potatoes grown in Canada. Food Chemistry 2007; 105: 897-907.
66. Paschall, E.F., J.N. BeMiller and R.L. Whistler (eds.). Starch: Chemistry and Technology, 2nd Ed. Academic Press, Orlando, 1984.
67. Ellis, R.P., M.P. Cochrane, M.F.B. Dale, C.M. Duffus, A. Lynn, I.M. Morrison, R.D.M. Prentice, J.S. Swanston and S.A. Tiller. Starch production and industrial use. Journal of the Science of Food and Agriculture 1998; 77: 289-311.
68. Visser, R.G.F., Suurs, C. J. M. Luc, P.M. Bruinenberg, I. Bleeker and E. Jacobsen. Comparison between amylose-free and amylose containing potato starches. Starch–Stärke 1997; 49: 438-443.
69. Li, J.H., T. Vasanthan, B. Rossnagel and R. Hoover. Starch from hull-less barley: I. Granule morphology, composition and amylopectin structure. Food Chemistry 2001; 74: 395-405.
70. Carpita, N.C. and J. Kanabus. Extraction of starch by dimethyl sulfoxide and quantitation by enzymatic assay. Analytical Biochemistry 1987; 161: 132-139.
71. Moorthy, S.N. Extraction of starches from tuber crops using ammonia. Carbohydrate Polymers 1991; 16: 391-398.
72. Schenck F.W. (ed.). Starch Hydrolysis Products: Worldwide Technology, Production, and Applications. VCH, Weinheim U.A., 1992.
73. Bergthaller, W., W. Witt and H.-P. Goldau. Potato starch technology. Starch–Stärke 1999; 51: 235-242.
74. Saenger, W. The structure of the blue starch-iodine complex. Naturwissenschaften 1984; 71: 31-36.
75. Burchard, W. and K. Balser (eds.). Polysaccharide: Eigenschaften und Nutzung; eine Einführung. Springer, Berlin U.A., 1985.
76. Liu, H., J. Lelievre and W. Ayoung-Chee. A study of starch gelatinization using differential scanning calorimetry, X-ray, and birefringence measurements. Carbohydrate Research 1991; 210: 79-87.
77. Hoover, R. and H. Manuel. The effect of heat–moisture treatment on the structure and physicochemical properties of normal maize, waxy maize, dull waxy maize and amylomaize V starches. Journal of Cereal Science 1996; 23: 153-162.
78. Fink, H.-P., H. Ebeling and W. Vorwerg. Technologien der Cellulose- und Stärkeverarbeitung. Chemie Ingenieur Technik 2009; 81: 1757-1766.
79. Biliaderis, C.G., T.J. Maurice and J.R. Vose. Starch gelatinization phenomena studied by differential scanning calorimetry. Journal of Food Science 1980; 45: 1669-1674.
80. Krueger, B.R., C.A. Knutson, G.E. Inglett and C.E. Walker. A differential scanning calorimetry study on the effect of annealing on gelatinization behavior of corn starch. Journal of Food Science 1987; 52: 715-718.
81. Lund, D. Influence of time, temperature, moisture, ingredients, and processing conditions on starch gelatinization. Critical Reviews in Food Science and Nutrition 1984; 20: 249-273.

82. Alvani, K., X. Qi, R.F. Tester and C.E. Snape. Physico-chemical properties of potato starches. Food Chemistry 2011; 125: 958-965.
83. Cooke, D. and M.J. Gidley. Loss of crystalline and molecular order during starch gelatinisation: Origin of the enthalpic transition. Carbohydrate Research 1992; 227: 103-112.
84. Karim, A.A., L.C. Toon, V.P.L. Lee, W.Y. Ong, A. Fazilah and T. Noda. Effects of phosphorus contents on the gelatinization and retrogradation of potato starch. Journal of Food Science 2007; 72: C132-8.
85. Nakamura, Y., A. Sato and B.O. Juliano. Short-chain-length distribution in debranched rice starches differing in gelatinization temperature or cooked rice hardness. Starch–Stärke 2006; 58: 155-160.
86. Tester, R.F. and W.R. Morrison. Swelling and gelatinization of cereal starches. I. Effects of amylopectin, amylose, and lipids. Cereal Chemistry 1990; 67: 551-557.
87. Gunaratne, A. Effect of heat-moisture treatment on the structure and physicochemical properties of tuber and root starches. Carbohydrate Polymers 2002; 49: 425-437.
88. Hoover, R. and T. Vasanthan. Effect of heat-moisture treatment on the structure and physicochemical properties of cereal, legume, and tuber starches. Carbohydrate Research 1994; 252: 33-53.
89. Fredriksson, H., J. Silverio, R. Andersson, A.-C. Eliasson and P. Åman. The influence of amylose and amylopectin characteristics on gelatinization and retrogradation properties of different starches. Carbohydrate Polymers 1998; 35: 119-134.
90. Asaoka, M., K. Okuno and H. Fuwa. Effect of environmental temperature at the milky stage on amylose content and fine structure of amylopectin of waxy and nonwaxy endosperm starches of rice (*Oryza sativa* L.). Agricultural and Biological Chemistry 2014; 49: 373-379.
91. Zeng, M., C.F. Morris, I.L. Batey and C.W. Wrigley. Sources of variation for starch gelatinization, pasting, and gelation properties in wheat. Cereal Chemistry 1997; 74: 63-71.
92. Evans, I.D. and D.R. Haisman. The effect of solutes on the gelatinization temperature range of potato starch. Starch–Stärke 1982; 34: 224-231.
93. Blennow, A., M. Hansen, A. Schulz, K. Jørgensen, A.M. Donald and J. Sanderson. The molecular deposition of transgenically modified starch in the starch granule as imaged by functional microscopy. Journal of Structural Biology 2003; 143: 229-241.
94. Tester, R. Annealing of maize starch. Carbohydrate Polymers 2000; 42: 287-299.
95. Larsson, I. and A.-C. Eliasson. Annealing of starch at an intermediate water content. Starch–Stärke 1991; 43: 227-231.
96. Goñi, O., M.I. Escribano and C. Merodio. Gelatinization and retrogradation of native starch from cherimoya fruit during ripening, using differential scanning calorimetry. LWT—Food Science and Technology 2008; 41: 303-310.
97. Gudmundsson, M. Retrogradation of starch and the role of its components. Thermochimica Acta 1994; 246: 329-341.
98. Iturriaga, L.B., B. Lopez de Mishima and M.C. Añon. A study of the retrogradation process in five argentine rice starches. LWT—Food Science and Technology 2010; 43: 670-674.
99. Karim, A. Methods for the study of starch retrogradation. Food Chemistry 2000; 71: 9-36.
100. Liu, H., L. Yu, L. Chen and L. Li. Retrogradation of corn starch after thermal treatment at different temperatures. Carbohydrate Polymers 2007; 69: 756-762.
101. Tian, Y., Y. Li, X. Xu and Z. Jin. Starch retrogradation studied by thermogravimetric analysis (TGA). Carbohydrate Polymers 2011; 84: 1165-1168.
102. Aguirre, J.F., C.A. Osella, C.R. Carrara, H.D. Sánchez and M.D.P. Buera. Effect of storage temperature on starch retrogradation of bread staling. Starch–Stärke 2011; 63: 587-593.
103. Hyang Aee, L., K. Nam Hie and K. Nishinari. DSC and rheological studies of the effects of sucrose on the gelatinization and retrogradation of acorn starch. Thermochimica Acta 1998; 322: 39-46.
104. Takaya, T. Thermal studies on the gelatinisation and retrogradation of heat-moisture treated starch. Carbohydrate Polymers 2000; 41: 97-100.
105. Tako, M., Y. Tamaki, T. Konishi, K. Shibanuma, I. Hanashiro and Y. Takeda. Gelatinization and retrogradation characteristics of wheat (*Rosella*) starch. Food Research International 2008; 41: 797-802.
106. Zhou, X., B.-K. Baik, R. Wang and S.-T. Lim. Retrogradation of waxy and normal corn starch gels by temperature cycling. Journal of Cereal Science 2010; 51: 57-65.

107. Bernfeld, P. Enzymes of starch degradation and synthesis. Advances in Enzymology and Related Areas of Molecular Biology 2006; 12: 379-428.

108. Lu, Y., J.P. Gehan and T.D. Sharkey. Daylength and circadian effects on starch degradation and maltose metabolism. Plant Physiology 2005; 138: 2280-2291.

109. Smith, A.M., S.C. Zeeman and S.M. Smith. Starch degradation. Annual Review of Plant Biology 2005; 56: 73-98.

110. Bastioli, C. (ed.). Handbook of Biodegradable Polymers, 2nd Ed. Smithers Rapra, Shropshire, England, 2014.

111. Bergmann, F.W., J.-I. Abe and S. Hizukuri. Selection of microorganisms which produce raw-starch degrading enzymes. Applied Microbiology and Biotechnology 1988; 27: 443-446.

112. Chen, C.-J., Y.-C. Shen and A.-I. Yeh. Physico-chemical characteristics of media-milled corn starch. Journal of Agricultural and Food Chemistry 2010; 58: 9083-9091.

113. Information on http://www.staerkeverband.de/html/zahlen.html.

114. Kim, H.R., A.-M. Hermansson and C.E. Eriksson. Structural characteristics of hydroxypropyl potato starch granules depending on their molar substitution. Starch–Stärke 1992; 44: 111-116.

115. Meuser, F., N. Gimmler and J. Oeding. Systemanalytische betrachtung der derivatisierung von stärke mit einem kochextruder als reaktor. Starch–Stärke 1990; 42: 330-336.

116. Lawal, M.V., M.A. Odeniyi and O.A. Itiola. Material and rheological properties of native, acetylated, and pregelatinized forms of corn, cassava, and sweet potato starches. Starch–Stärke 2015; 67: 964-975.

117. Singh, J., L. Kaur and N. Singh. Effect of acetylation on some properties of corn and potato starches. Starch–Stärke 2004; 56: 586-601.

118. Tänzer, W. Biologisch abbaubare Polymere: 11 Tabellen, Dt. Verl. für Grundstoffindustrie. Stuttgart, 2000.

119. Singh, J., L. Kaur and O.J. McCarthy. Factors influencing the physico-chemical, morphological, thermal and rheological properties of some chemically modified starches for food applications—A review. Food Hydrocolloids 2007; 21: 1-22.

120. Tessler, M.M. and R.L. Billmers. Preparation of starch esters. Journal of Environmental Polymer Degradation 1996; 4: 85-89.

121. BeMiller, J.N. Starch modification: Challenges and prospects. Starch–Stärke 1997; 49: 127-131.

122. Choi, S.G. and W.L. Kerr. Water mobility and textural properties of native and hydroxypropylated wheat starch gels. Carbohydrate Polymers 2003; 51: 1-8.

123. Hermansson, A.-M. and K. Svegmark. Developments in the understanding of starch functionality. Trends in Food Science & Technology 1996; 7: 345-353.

124. Kaur, B., F. Ariffin, R. Bhat and A.A. Karim. Progress in starch modification in the last decade. Food Hydrocolloids 2012; 26: 398-404.

125. Fang, J. The preparation and characterisation of a series of chemically modified potato starches. Carbohydrate Polymers 2002; 47: 245-252.

126. Whistler, R.L., M.A. Madson, J. Zhao and J.R. Daniel. Surface derivatization of corn starch granules. Cereal Chemistry 1998; 75: 72-74.

127. Jeon, Y.-S., A.V. Lowell and R.A. Gross. Studies of starch esterification: Reactions with alkenylsuccinates in aqueous slurry systems. Starch–Stärke 1999; 51: 90-93.

128. Bien, F., B. Wiege and S. Warwel. Hydrophobic modification of starch by alkali-catalyzed addition of 1,2-epoxyalkanes. Starch–Stärke 2001; 53: 555.

129. Tomasik, P. and C.H. Schilling. Chemical modification of starch. Advances in Carbohydrate Chemistry and Biochemistry 2004; 59: 175-403.

130. Tolvanen, P., P. Mäki-Arvela, A.B. Sorokin, T. Salmi and D.Y. Murzin. Kinetics of starch oxidation using hydrogen peroxide as an environmentally friendly oxidant and an iron complex as a catalyst. Chemical Engineering Journal 2009; 154: 52-59.

131. Floor, M., K.M. Schenk, A.P.G. Kieboom and H. van Bekkum. Oxidation of maltodextrins and starch by the system tungstate-hydrogen peroxide. Starch–Stärke 1989; 41: 303-309.

132. Lukasiewicz, M., S. Bednarz and A. Ptaszek. Environmental friendly polysaccharide modification-microwave-assisted oxidation of starch. Starch–Stärke 2011; 63: 268-273.

133. Chong, W.T., U. Uthumporn, A.A. Karim and L.H. Cheng. The influence of ultrasound on the degree of oxidation of hypochlorite-oxidized corn starch. LWT—Food Science and Technology 2013; 50: 439-443.
134. Boruch, M. Transformations of potato starch during oxidation with hypochlorite. Starch–Stärke 1985; 37: 91-98.
135. Bendoraitiene, J., R. Kavaliauskaite, R. Klimaviciute and A. Zemaitaitis. Peculiarities of starch cationization with glycidyltrimethylammonium chloride. Starch–Stärke 2006; 58: 623-631.
136. Radosta, S., W. Vorwerg, A. Ebert, A.H. Begli, D. Grülc and M. Wastyn. Properties of low-substituted cationic starch derivatives prepared by different derivatisation processes. Starch–Stärke 2004; 56: 277-287.
137. Hellwig, G., D. Bischoff and A. Rubo. Production of cationic starch ethers using an improved dry process. Starch–Stärke 1992; 44: 69-74.
138. Kavaliauskaite, R., R. Klimaviciute and A. Zemaitaitis. Factors influencing production of cationic starches. Carbohydrate Polymers 2008; 73: 665-675.
139. Maurer, H.W. and R.L. Kearney. Opportunities and challenges for starch in the paper industry. Starch–Stärke 1998; 50: 396-402.
140. Athawale, V.D. and S.C. Rathi. Graft polymerization: starch as a model substrate. Journal of Macromolecular Science, Part C: Polymer Reviews 1999; 39: 445-480.
141. Brockway, C.E. and P.A. Seaberg. Grafting of polyacrylonitrile to granular corn starch. Journal of Polymer Science Part A-1: Polymer Chemistry 1967; 5: 1313-1326.
142. Chan, W.-C. and C.-Y. Chiang. Flocculation of clay suspensions with water-insoluble starch grafting acrylamide/sodium allylsulfonated copolymer powder. Journal of Applied Polymer Science 1995; 58: 1721-1726.
143. Labet, M., W. Thielemans and A. Dufresne. Polymer grafting onto starch nanocrystals. Biomacromolecules 2007; 8: 2916-2927.
144. Pourjavadi, A. and M.J. Zohuriaan-Mehr. Modification of carbohydrate polymers via grafting in air. 1. ceric-induced synthesis of starch-g-polyacrylonitrile in presence and absence of oxygen. Starch–Stärke 2002; 54: 140-147.
145. Salimi, K., M. Topuzogullari, S. Dincer, H.M. Aydin and E. Piskin. Microwave-assisted green approach for graft copolymerization of l -lactic acid onto starch, Journal of Applied Polymer Science 2016; 133: 42937-42944.
146. Hu, H., W. Liu, J. Shi, Z. Huang, Y. Zhang, A. Huang, M. Yang, X. Qin and F. Shen. Structure and functional properties of octenyl succinic anhydride modified starch prepared by a non-conventional technology. Starch–Stärke 2016; 68: 151-159.
147. Radosta, S., B. Kiessler, W. Vorwerg and T. Brenner. Molecular composition of surface sizing starch prepared using oxidation, enzymatic hydrolysis and ultrasonic treatment methods. Starch–Stärke 2016; 68: 541-548.
148. Sujka, M., K. Cieśla and J. Jamroz. Structure and selected functional properties of gamma-irradiated potato starch. Starch–Stärke 2015; 67: 1002-1010.
149. Kuniak, L. and R.H. Marchessault. Study of the crosslinking reaction between epichlorohydrin and starch. Starch–Stärke 1972; 24: 110-116.
150. Reddy, I. and P.A. Seib. Modified waxy wheat starch compared to modified waxy corn starch. Journal of Cereal Science 2000; 31: 25-39.
151. Woo, K. and P.A. Seib. Cross-linking of wheat starch and hydroxypropylated wheat starch in alkaline slurry with sodium trimetaphosphate. Carbohydrate Polymers 1997; 33: 263-271.
152. Woo, K.S. and P.A. Seib. Cross-linked resistant starch: Preparation and properties. Cereal Chemistry 2002; 79: 819-825.
153. Ayoub, A.S. and S.S.H. Rizvi. An overview on the technology of cross-linking of starch for nonfood applications. Journal of Plastic Film & Sheeting 2009; 25: 25-45.
154. Kartha, K.P.R. and H.C. Srivastava. Reaction of epichlorhydrin with carbohydrate polymers. Part I. starch reaction kinetics. Starch–Stärke 1985; 37: 270-276.
155. Kartha, K.P.R. and H.C. Srivastava. Reaction of epichlorhydrin with carbohydrate polymers. Part II. starch reaction mechanism and physicochemical properties of modified starch. Starch–Stärke 1985; 37: 297-306.

156. Wood, L.F. and C. Mercier. Molecular structure of unmodified and chemically modified manioc starches. Carbohydrate Research 1978; 61: 53-66.

157. Bhattacharya, A. (ed.). Polymer Grafting and Crosslinking. Wiley & Sons, Hoboken NY, 2009.

158. Kim, M. and S.-J. Lee. Characteristics of crosslinked potato starch and starch-filled linear low-density polyethylene films. Carbohydrate Polymers 2002; 50: 331-337.

159. Liu, H., L. Ramsden and H. Corke. Physical properties of cross-linked and acetylated normal and waxy rice starch. Starch–Stärke 1999; 51: 249-252.

160. Šimkovic, I., J.A. Laszlo and A.R. Thompson. Preparation of a weakly basic ion exchanger by crosslinking starch with epichlorohydrin in the presence of NH4OH11Names are necessary to report factually on available data; however, the USDA neither guarantees nor warrants the standard of the product, and the use of the name by USDA implies no approval of the product to the exclusion of others that may also be suitable. Carbohydrate Polymers 1996; 30: 25-30.

161. Sreedhar, B., D.K. Chattopadhyay, M.S.H. Karunakar and A.R.K. Sastry. Thermal and surface characterization of plasticized starch polyvinyl alcohol blends crosslinked with epichlorohydrin. Journal of Applied Polymer Science 2006; 101: 25-34.

162. Jyothi, A.N., S.N. Moorthy and K.N. Rajasekharan. Effect of cross-linking with epichlorohydrin on the properties of cassava (*Manihot esculenta* Crantz) starch. Starch–Stärke 2006; 58: 292-299.

163. Hamdi, G., G. Ponchel and D. Duchêne. Formulation of epichlorohydrin cross-linked starch microspheres. Journal of Microencapsulation 2001; 18: 373-383.

164. Morin-Crini, N. and G. Crini. Environmental applications of water-insoluble β-cyclodextrin–epichlorohydrin polymers. Progress in Polymer Science 2013; 38: 344-368.

165. Finch, C.A. Modified starches: Properties and uses. *In*: O.B. Wurzburg (ed.). CRC Press, Boca Raton, Florida. 1986. pp. vi + 277, ISBN 0-8493-5964-3. British Polymer Journal 1989; 21: 87-88.

166. Ačkar, Đ., J. Babić, D. Šubarić, M. Kopjar and B. Miličević. Isolation of starch from two wheat varieties and their modification with epichlorohydrin. Carbohydrate Polymers 2010; 81: 76-82.

167. Kim, B.S. and S.-T. Lim. Removal of heavy metal ions from water by cross-linked carboxymethyl corn starch. Carbohydrate Polymers 1999; 39: 217-223.

168. Reddy, N. and Y. Yang. Citric acid cross-linking of starch films. Food Chemistry 2010; 118: 702-711.

169. Trimble, E., Modified starches in foods. Journal of Consumer Studies and Home Economics 1983; 7: 247-260.

170. Shen, L., H. Xu, L. Kong and Y. Yang. Non-toxic crosslinking of starch using polycarboxylic acids: Kinetic study and quantitative correlation of mechanical properties and crosslinking degrees. Journal of Polymers and the Environment 2015; 23: 588-594.

171. Perocco, P., P. Rocchi, A.M. Ferreri and A. Capucci. Toxic, DNA-damaging and mutagenic activity of epichlorohydrin on human cells cultured in vitro. Tumori 1983; 69: 191-194.

172. Marks, T.A., F.S. Gerling and R.E. Staples. Teratogenic evaluation of epichlorohydrin in the mouse and rat and glycidol in the mouse. Journal of Toxicology and Environmental Health 1982; 9: 87-96.

173. Carmona-Garcia, R., M.M. Sanchez-Rivera, G. Méndez-Montealvo, B. Garza-Montoya and L.A. Bello-Pérez. Effect of the cross-linked reagent type on some morphological, physicochemical and functional characteristics of banana starch (*Musa paradisiaca*). Carbohydrate Polymers 2009; 76: 117-122.

174. Delval, F., G. Crini, S. Bertini, C. Filiatre and G. Torri. Preparation, characterization and sorption properties of crosslinked starch-based exchangers. Carbohydrate Polymers 2005; 60: 67-75.

175. Heinze, T. and A. Koschella. Carboxymethyl ethers of cellulose and starch—A review. Macromolecular Symposia 2005; 223: 13-40.

176. Kittipongpatana, O.S. and N. Kittipongpatana. Physicochemical, in vitro digestibility and functional properties of carboxymethyl rice starch cross-linked with epichlorohydrin. Food Chemistry 2013; 141: 1438-1444.

177. Šimkovic, I. One-step quaternization/crosslinking of starch with 3-chloro-2-hydroxypropylammonium chloride/epichlorohydrin in the presence of NH_4OH. Carbohydrate Polymers 1996; 31: 47-51.

178. Hirsch, J.B. and J.L. Kokini. Understanding the mechanism of cross-linking agents (POCl 3 STMP, and EPI) through swelling behavior and pasting properties of cross-linked waxy maize starches. Cereal Chemistry 2002; 79: 102-107.

179. Okoli, C.P., G.O. Adewuyi, Q. Zhang, G. Zhu, C. Wang and Q. Guo. Aqueous scavenging of polycyclic aromatic hydrocarbons using epichlorohydrin, 1,6-hexamethylene diisocyanate and 4,4-methylene diphenyl diisocyanate modified starch: Pollution remediation approach. Arabian Journal of Chemistry 2015; 1-14. http://dx.doi.org/10.1016/j.arabjc.2015.06.00.

180. Truhaut, R., B. Coquet, X. Fouillet, D. Galland, D. Guyot, D. Long and J.L. Rouaud. Two-year oral toxicity and multigeneration studies in rats on two chemically modified maize starches. Food and Cosmetics Toxicology 1979; 17: 11-17.

181. de Groot, A.P., H.P. Til, V.J. Feron, H.C. Dreef-van der Meulen and M.I. Willems. Two-year feeding and multigeneration studies in rats on five chemically modified starches. Food and Cosmetics Toxicology 1974; 12: 651-663.

182. Baran, E.T., J.F. Mano and R.L. Reis. Starch-chitosan hydrogels prepared by reductive alkylation cross-linking. Journal of Materials Science: Materials in Medicine 2004; 15: 759-765.

183. Crini, G. Recent developments in polysaccharide-based materials used as adsorbents in wastewater treatment. Progress in Polymer Science 2005; 30: 38-70.

184. Xu, H., L. Shen, L. Xu and Y. Yang. Controlled delivery of hollow corn protein nanoparticles via non-toxic crosslinking: In vivo and drug loading study. Biomedical Microdevices 2015; 17: 8.

185. Ghosh, T., Dastidar and A.N. Netravali. 'Green' crosslinking of native starches with malonic acid and their properties. Carbohydrate Polymers 2012; 90: 1620-1628.

186. Huijbrechts, A.A.M.L., J. Huang, H.A. Schols, B. van Lagen, G.M. Visser, C.G. Boeriu and E.J.R. Sudhölter. 1-Allyloxy-2-hydroxy-propyl-starch: Synthesis and characterization, Journal of Polymer Science Part A: Polymer Chemistry 2007; 45: 2734-2744.

187. Halley, P.J. (ed.). Starch Polymers: From Genetic Engineering to Green Applications. Elsevier, Amsterdam U.A., 2014.

188. Franssen, M.C.R. and C.G. Boeriu. Chemically modified starch; allyl- and epoxy-starch derivatives, *In*: P.J. Halley (ed.). Starch Polymers: From Genetic Engineering to Green Applications. Elsevier, Amsterdam U.A., 2014, pp. 145-184.

189. Duanmu, J., E.K. Gamstedt and A. Rosling. Synthesis and preparation of crosslinked allylglycidyl ether-modified starch-wood fibre composites. Starch–Stärke 2007; 59: 523-532.

190. Fritz, H.G., W. Aichholzer, T. Seidenstücker and B. Widmann. Abbaubare polymerwerkstoffe auf der basis nachwachsender rohstoffe—möglichkeiten und grenzen. Starch–Stärke 1995; 47: 475-491.

191. Potente, H., V. Schöppner and A. Rücker. Verarbeitung von kartoffelstärke auf kunststoff-verarbeitungsmaschinen. Starch–Stärke 1991; 43: 231-235.

192. Li, H. and M.A. Huneault. Comparison of sorbitol and glycerol as plasticizers for thermoplastic starch in TPS/PLA blends. Journal of Applied Polymer Science 2011; 119: 2439-2448.

193. Della Valle, G., J. Tayeb and J.P. Melcion. Relationship of extrusion variables with pressure and temperature during twin screw extrusion cooking of starch. Journal of Food Engineering 1987; 6: 423-444.

194. Endres, H.-J., H. Kammerstetter and M. Hobelsberger. Plastification behaviour of different native starches. Starch–Stärke 1994; 46: 474-480.

195. Fritz, H.-G. and B. Widmann. Der einsatz von stärke bei der modifizierung synthetischer kunststoffe. Starch–Stärke 1993; 45: 314-322.

196. Aichholzer, W. and H.-G. Fritz. Charakterisierung der Stärkedestrukturierung bei der Aufbereitung von bioabbaubaren polymerwerkstoffen. Starch–Stärke 1996; 48: 434-444.

197. Averous, L. and N. Boquillon. Biocomposites based on plasticized starch: Thermal and mechanical behaviours. Carbohydrate Polymers 2004; 56: 111-122.

198. Avérous, L. Biodegradable multiphase systems based on plasticized starch: A review, Journal of Macromolecular Science. Part C: Polymer Reviews 2004; 44: 231-274.

199. Avérous, L. and P.J. Halley. Biocomposites based on plasticized starch. Biofuels, Bioproducts and Biorefining 2009; 3: 329-343.

200. Wiedmann, W. and E. Strobel. Compounding of thermoplastic starch with twin-screw extruders. Starch–Stärke 1991; 43: 138-145.

201. Pushpadass, H.A., A. Kumar, D.S. Jackson, R.L. Wehling, J.J. Dumais and M.A. Hanna. Macromolecular changes in extruded starch-films plasticized with glycerol, water and stearic acid. Starch–Stärke 2009; 61: 256-266.

202. Altskar, A., R. Andersson, A. Boldizar, K. Koch, M. Stading, M. Rigdahl and M. Thunwall. Some effects of processing on the molecular structure and morphology of thermoplastic starch. Carbohydrate Polymers 2008; 71: 591-597.
203. Bhatnagar, S. and M.A. Hanna. Properties of extruded starch-based plastic foam. Industrial Crops and Products 1995; 4: 71-77.
204. Gross, R.A. and B. Kalra. Biodegradable polymers for the environment. Science 2002; 297: 803-807.
205. Janssen, L.P.B.M. and L. Moscicki (eds). Thermoplastic Starch: A Green Material for Various Industries. Wiley-VCH, Weinheim, 2009.
206. Tatarka, P.D. and R.L. Cunningham. Properties of protective loose-fill foams. Journal of Applied Polymer Science 1998; 67: 1157-1176.
207. St-Pierre, N., B.D. Favis, B.A. Ramsay, J.A. Ramsay and H. Verhoogt. Processing and characterization of thermoplastic starch/polyethylene blends. Polymer 1997; 38: 647-655.
208. Rodriguez-Gonzalez, F.J., B.A. Ramsay and B.D. Favis. High performance LDPE/thermoplastic starch blends: A sustainable alternative to pure polyethylene. Polymer 2003; 44: 1517-1526.
209. Huneault, M.A. and H. Li. Morphology and properties of compatibilized polylactide/thermoplastic starch blends. Polymer 2007; 48: 270-280.
210. Wang, N., J. Yu, P.R. Chang and X. Ma. Influence of formamide and water on the properties of thermoplastic starch/poly(lactic acid) blends. Carbohydrate Polymers 2008; 71: 109-118.
211. Garlotta, D. A literature review of poly(Lactic Acid). Journal of Polymers and the Environment 2001; 9: 63-84.
212. Huneault, M.A. and H. Li. Preparation and properties of extruded thermoplastic starch/polymer blends. Journal of Applied Polymer Science 2012; 126: E96-E108.
213. Sangeetha, V.H., H. Deka, T.O. Varghese and S.K. Nayak. State of the art and future prospectives of poly(Lactic Acid) based blends and composites. Polymer Composites 2018; 39: 81-101.
214. Yu, L., K. Dean and L. Li. Polymer blends and composites from renewable resources. Progress in Polymer Science 2006; 31: 576-602.
215. Tabi, T. and J.G. Kovacs. Examination of injection moulded thermoplastic maize starch. Express Polymer Letters 2007; 1(12): 804-809.
216. Stepto, R.F.T. Thermoplastic starch. Macromolecular Symposia 2009; 279: 163-168.
217. Stepto, R.F.T. The processing of starch as a thermoplastic. Macromolecular Symposia 2003; 201: 203-212.
218. Rodriguez-Gonzalez, F.J., B.A. Ramsay and B.D. Favis. Rheological and thermal properties of thermoplastic starch with high glycerol content. Carbohydrate Polymers 2004; 58: 139-147.
219. Jacobsen, S. and H.G. Fritz. Filling of poly(Lactic Acid) with native starch. Polymer Engineering & Science 1996; 36: 2799-2804.
220. Rahmat, A.R., W.A.W.A. Rahman, L.T. Sin and A.A. Yussuf. Approaches to improve compatibility of starch filled polymer system: A review, Materials Science and Engineering: C 2009; 29: 2370-2377.
221. Shah, P.B., S. Bandopadhyay and J.R. Bellare. Environmentally degradable starch filled low density polyethylene. Polymer Degradation and Stability 1995; 47: 165-173.
222. Zhang, J.-F. and X. Sun. Mechanical properties of poly(lactic acid)/starch composites compatibilized by maleic anhydride. Biomacromolecules 2004; 5: 1446-1451.
223. Wang, H., X. Sun and P. Seib. Properties of poly(lactic acid) blends with various starches as affected by physical aging. Journal of Applied Polymer Science 2003; 90: 3683-3689.
224. Ke, T. and X. Sun. Effects of moisture content and heat treatment on the physical properties of starch and poly(Lactic Acid) blends. Journal of Applied Polymer Science 2001; 81: 3069-3082.
225. Ke, T., S.X. Sun and P. Seib. Blending of poly(Lactic Acid) and starches containing varying amylose content. Journal of Applied Polymer Science 2003; 89: 3639-3646.
226. Kovács, J.G. and T. Tábi. Examination of starch preprocess drying and water absorption of injection-molded starch-filled poly(Lactic Acid) products. Polymer Engineering & Science 2011; 51: 843-850.
227. Wang, H., X. Sun and P. Seib. Strengthening blends of poly(Lactic Acid) and starch with methylenediphenyl diisocyanate. Journal of Applied Polymer Science 2001; 82: 1761-1767.
228. Natarajan, L., J. New, A. Dasari, S. Yu and M.A. Manan. Surface morphology of electrospun PLA fibers: Mechanisms of pore formation. RSC Adv 2014; 4: 44082-44088.

229. Ke, T. and X. Sun. Physical properties of poly(Lactic Acid) and starch composites with various blending ratios. Cereal Chemistry 2000; 77: 761-768.

230. Chandra, R. Biodegradable polymers. Progress in Polymer Science 1998; 23: 1273-1335.

231. Ke, T. and X. Sun. Melting behavior and crystallization kinetics of starch and poly(lactic acid) composites. Journal of Applied Polymer Science 2003; 89: 1203-1210.

232. Lu, D.R., C.M. Xiao and S.J. Xu. Starch-based completely biodegradable polymer materials. Express Polymer Letters 2009; 3: 366-375.

233. Lörcks, J. Properties and applications of compostable starch-based plastic material. Polymer Degradation and Stability 1998; 59: 245-249.

234. Bastioli, C. Properties and applications of Mater-Bi starch-based materials. Polymer Degradation and Stability 1998; 59: 263-272.

235. Yew, G.H., A.M. Mohd Yusof, Z.A. Mohd Ishak and U.S. Ishiaku. Water absorption and enzymatic degradation of poly(Lactic Acid)/rice starch composites. Polymer Degradation and Stability 2005; 90: 488-500.

236. Gáspár, M., Z. Benkő, G. Dogossy, K. Réczey and T. Czigány. Reducing water absorption in compostable starch-based plastics. Polymer Degradation and Stability 2005; 90: 563-569.

237. Della Valle, G., P. Colonna and J. Tayeb. Use of a twin-screw extruder as a chemical reactor for starch cationization. Starch–Stärke 1991; 43: 300-307.

238. Michaeli, W., A. Grefenstein and U. Berghaus. Twin-Screw extruders for reactive extrusion. Polymer Engineering & Science 1995; 35: 1485-1504.

239. Raquez, J.-M., R. Narayan and P. Dubois. Recent advances in reactive extrusion processing of biodegradable polymer-based compositions. Macromolecular Materials and Engineering 2008; 293: 447-470.

240. Tzoganakis, C. Reactive extrusion of polymers: A review. Advances in Polymer Technology 1989; 9: 321-330.

241. Wang, N., J. Yu and X. Ma. Preparation and characterization of thermoplastic starch/PLA blends by one-step reactive extrusion. Polymer International 2007; 56: 1440-1447.

242. Michaeli, W. and A. Grefenstein. Engineering analysis and design of twin-screw extruders for reactive extrusion. Advances in Polymer Technology 1995; 14: 263-276.

243. Emin, M.A., Modeling extrusion processes. *In*: S. Bakalis, K. Knoerzer, P.J. Fryer (eds). Modeling Food Processing Operations. Woodhead Publisher, Cambridge UK, 2015, pp. 235-253.

244. Garber, B.W., F. Hsieh and H.E. Huff. Influence of particle size on the twin-screw extrusion of corn meal. Cereal Chemistry 1997; 74: 656-661.

245. Li, M., J. Hasjim, F. Xie, P.J. Halley and R.G. Gilbert. Shear degradation of molecular, crystalline, and granular structures of starch during extrusion. Starch–Stärke 2014; 66: 595-605.

246. Okechukwu, P.E. and M.A. Rao. Influence of granule size on viscosity of cornstarch suspension. Journal of Texture Studies 1995; 26: 501-516.

247. Xanthos, M. Reactive extrusion: Principles and practice a monograph with 92 illustrations and 17 tables. Hanser Publishers, Munich, Germany, 1992.

248. Grefenstein, A. Reaktive Extrusion und Aufbereitung: Maschinentechnik und Verfahren, Hanser, München, 1996.

249. Moad, G. The synthesis of polyolefin graft copolymers by reactive extrusion. Progress in Polymer Science 1999; 24: 81-142.

250. Moad, G. Chemical modification of starch by reactive extrusion. Progress in Polymer Science 2011; 36: 218-237.

251. Wang, L., R.L. Shogren and J.L. Willett. Preparation of starch succinates by reactive extrusion. Starch–Stärke 1997; 49: 116-120.

252. Benham, J.L. (ed.). Chemical reactions on polymers: Developed from a symposium sponsored by the Div. of Polymer Chemistr, Inc., and the Div. of Polymeric Materials: Science and Engineering at the 192th Meeting of the American Chemical Society, Anaheim, California, September 7-12, 1986. ACS American Chemical Society, Washington DC, 1988.

253. Al-Malaika, S. Reactive modifiers for polymers, *In*: J.L. Benham (ed.). Chemical Reactions on Polymers: Developed from a Symposium Sponsored by the Div. of Polymer Chemistr, Inc., and the Div. of Polymeric Materials: Science and Engineering at the 192th Meeting of the American Chemical Society, Anaheim, California, September 7–12, 1986. ACS American Chemical Society, Washington DC, 1988, pp. 409-425.

254. Janssen, L.P.B.M. Reactive Extrusion Systems. Marcel Dekker, New York, 2004.

255. Jun, C.L. Reactive blending of biodegradable polymers: PLA and starch. Journal of Polymers and the Environment 2000; 8: 33-37.

256. Kalambur, S. and S.S.H. Rizvi. An overview of starch-based plastic blends from reactive extrusion. Journal of Plastic Film & Sheeting 2016; 22: 39-58.

257. Miladinov, V.D. and M.A. Hanna. Starch esterification by reactive extrusion. Industrial Crops and Products 2000; 11: 51-57.

258. Murúa-Pagola, B., C.I. Beristain-Guevara and F. Martínez-Bustos. Preparation of starch derivatives using reactive extrusion and evaluation of modified starches as shell materials for encapsulation of flavoring agents by spray drying. Journal of Food Engineering 2009; 91: 380-386.

259. Raquez, J.-M., Y. Nabar, M. Srinivasan, B.-Y. Shin, R. Narayan and P. Dubois. Maleated thermoplastic starch by reactive extrusion. Carbohydrate Polymers 2008; 74: 159-169.

260. Demirgöz, D., C. Elvira, J.F. Mano, A.M. Cunha, E. Piskin and R.L. Reis. Chemical modification of starch based biodegradable polymeric blends: Effects on water uptake, degradation behaviour and mechanical properties. Polymer Degradation and Stability 2000; 70: 161-170.

261. Ohkita, T. and S.-H. Lee. Effect of aliphatic isocyanates (HDI and LDI) as coupling agents on the properties of eco-composites from biodegradable polymers and corn starch. Journal of Adhesion Science and Technology 2004; 18: 905-924.

262. Schwach, E., J.-L. Six and L. Avérous. Biodegradable blends based on starch and poly(Lactic Acid): comparison of different strategies and estimate of compatibilization. Journal of Polymers and the Environment 2008; 16: 286-297.

263. Tanrattanakul, V. and W. Chumeka. Effect of potassium persulfate on graft copolymerization and mechanical properties of cassava starch/natural rubber foams. Journal of Applied Polymer Science 2010; 116: 93-105.

264. Raquez, J.-M., Y. Nabar, R. Narayan and P. Dubois. In situ compatibilization of maleated thermoplastic starch/polyester melt-blends by reactive extrusion. Polymer Engineering & Science 2008; 48: 1747-1754.

265. Carr, M.E. Preparation of cationic starch containing quaternary ammonium substituents by reactive twin-screw extrusion processing. Journal of Applied Polymer Science 1994; 54: 1855-1861.

266. Ayoub, A. and S.S.H. Rizvi. Properties of supercritical fluid extrusion-based crosslinked starch extrudates. Journal of Applied Polymer Science 2008; 107: 3663-3671.

267. Patel, S., R.A. Venditti, J.J. Pawlak, A. Ayoub and S.S.H. Rizvi. Development of cross-linked starch microcellular foam by solvent exchange and reactive supercritical fluid extrusion. Journal of Applied Polymer Science 2009; 111: 2917-2929.

268. Chen, L., S.H. Imam, S.H. Gordon and R.V. Greene. Starch- polyvinyl alcohol crosslinked film—performance and biodegradation. Journal of Environmental Polymer Degradation 1997; 5: 111-117.

269. O'Brien, S., Y.-J. Wang, C. Vervaet and J.P. Remon. Starch phosphates prepared by reactive extrusion as a sustained release agent. Carbohydrate Polymers 2009; 76: 557-566.

270. Johannes, F. Blends aus Stärke und PLA: Prozessintegrierte Trocknung und Vernetzung–Struktur–Eigenschaften. Kassel University Press, GmbH, Kassel, 2018.

Chapter 7

Chemistry of Cellulose

M.D.H. Beg*, K. Najwa and J.O. Akindoyo

Faculty of Chemical and Natural Resources Engineering
Universiti Malaysia Pahang Lebuhraya Tun Razak
Gambang 26300, Kuantan, Malaysia

7.1 INTRODUCTION

Among the natural organic polymers, cellulose is the most abundant member, which exists as cell wall components of a main plant. It carries immense importance because it is a biodegradable organic resource which is renewable. Due to these, studies on cellulose have been on for over 20 decades, but there is still need for further research to fully elucidate its chemical synthesis, biosynthesis, regiospecific substitution reactions, crystal structure, interrelationships between the structure and function of its derivatives, etc. Specifically, synthesis of cellulose is highly important, but also a severely difficult problem to solve. Enzymatic synthesis of cellulose with the help of cellulase came as a welcome development, however the approach seems not to meet the modern molecular design of function-specific cellulose derivatives. This is because it does not support regiospecific introduction of functional groups into the hydroxyl group of interest in the repeating cellulose pyranose units. The available functional cellulose derivatives include cellulose esters and liquid crystalline ethers, ethers with chiral recognition features, sulfated anticoagulant cellulose, antitumor branched cellulose, etc. [1]. Despite all these, certain areas still remain unclear, such as structure-property relationship, the most actively functional derivatives with respect to the substituted positions (2, 3- or 6-). For molecular design of advanced cellulose based materials, it is pertinent to use an approach which gives room for preparation of cellulose derivatives such that functional groups can be easily incorporated into either of the 2,3,6-hydroxyl groups of the repeating cellulose pyranose units.

Cellulose occupies a large percentage (35–50%) of lignocellulosic materials and it has been identified as a potential renewable source of bio-based chemicals and biofuels. Several important industrial fuels and chemicals, such as ketones, carboxylic acids, hydrocarbons, and ethanol can be obtained from the hydrolysis of cellulose followed by fermentation of saccharides [2, 3]. It has been noted however that production of chemical from cellulose is sometimes not cost effective compared to those obtained from petrochemicals and they are sometimes even more expensive than those obtained from other renewable sources, such as starch. This is as a result of the relatively

**Corresponding author*: Email: dhbeg@yahoo.com

low yield of saccharide obtained in the hydrolysis of cellulose based on the high resistance and stability of cellulose towards chemical attack [4]. Cellulose is highly hydrophilic due to presence of a large number of –OH groups in the cellulose structure and these are responsible for the several hydrogen bonds within the cellulose structure. The hydrogen bonds in combination with van der waals forces cause some of the cellulose molecules to align together in a well ordered crystalline manner. The high crystallinity (65–75%) of cellulose means that pretreatment is necessary prior to cellulose hydrolysis [3].

Several researches have been carried out on pretreatment of cellulosic materials in order to improve the yield of fermentable saccharide [5–7]. Different approaches have also been investigated, such as alkali [8], combined ionic liquid and microwave [9], super- or sub- critical solvents [10, 11], other organic or ionic liquids [12]. Nevertheless, most of these pretreatment methods involve the use of organic solvent and other chemicals which may significantly add to the production cost and as well lead to environmental pollution [13, 14]. Report shows that cellulose is unstable after certain chemical modification and can thereafter be easily hydrolysed [15]. In a particular research, it was reported that carboxymethyl cellulose with low degree of substitution is more easily hydrolysable compared to original cellulose [16]. Likewise, a phenomenon known as reactive tendering, often used in the textile industry, in which reactive dyes are covalently bonded with cellulose, was reported to increase the hydrolysis of by β-1,4-glycosidic bonds [17].

In another vein, cellulose has been recognized as a suitable replacement for traditional synthetic fibers in reinforced composite materials [18]. Literature review revealed that synthetic polymers have played very important roles in different applications [19, 20]. Nevertheless, environmental benign natural polymers have gained much research interest in the past few decades due to depletion of petroleum and fossil fuel, as well as increasing environmental awareness and legislations [21]. The inherent properties of natural polymers, such as low cost, renewability, availability, and biodegradability, had led to their progressive emergence as suitable substitutes for petroleum based materials [21, 22].

Cellulose is the most readily abundant, biodegradable, and renewable natural material resource on earth [23, 24], and has been perceived to be the best alternative to meet the present energy demands. Cellulose was first discovered and isolated in 1838, and it has since been extensively studied [25]. Over the years, cellulose has been studied for structural analysis [26, 27], biosynthesis [28], chemical modification [29], application in different fields [30], and cellulosic materials regeneration [31, 32].Cellulosic polymers have proven to possess desirable properties comparable to conventional synthetic polymers. Effective use of these biorenewable materials can drastically reduce the persisting dependence on petroleum based materials [33]. In fact, cellulosic composites are currently one of the fastest growing fields in the composite industry based on their high specific strength and modulus, low density, and cheap cost compared to synthetic fiber based composites [34]. Nevertheless, the inherent hydrophilic properties of cellulosic fibers often limit their applications due to poor moisture and chemical resistance [35]. After a prolonged period, these shortcomings could affect the overall performance of the composite material [18]. The recent progresses on cellulose functional materials [36], surface modification of cellulose [37, 38], nanocellulose [39, 40], and new solvents for cellulosic materials regeneration [41], are available in literature.

These days, the major uses of cellulose are for cardboards, tissues, explosives, textiles, papers, membranes, and materials for construction. Wider extensive applications of cellulose seem difficult due to its insolubility in conventional solvents as well as its lack of characteristic features native to synthetic polymers. Based on these, chemical modification of cellulose is necessary not only to increase the yield of saccharides in cellulose hydrolysis, but also very important to impart hydrophobicity to the cellulose so as to enhance its wider application in composite materials. Chemical modification of cellulose is therefore currently being focused on for the improvement of cellulose resistance towards abrasion and heat, increased oil and water repellence, antibacterial properties, and improved mechanical strength [24].

7.2 STRUCTURE AND PROPERTIES OF CELLULOSE

In the field of polysaccharides, the structure of cellulose is one of the most simple but unique. The structure therefore has a significant complex impact on the nature of the chemical reactions of the polymer. Furthermore, the structure determines the macroscopic features of the polymer. To provide a clear description of the complex structure of cellulose, it is often convenient to specify three structural categories, such as (i) molecular level of single macromolecules; (ii) supramolecular level of mutual packing and arrangement of the macromolecules; (iii) morphological level based on architecture of the complex structure and the corresponding porous system.

7.2.1 The Basic Molecular Structure

Cellulose is the most plenteous available polysaccharide on earth. It is a polymer of glucose which is connected via a linear β-1,4 glycosidic bond. The basic structural unit of cellulose is made up of anhydroglucose unit (AGU), as depicted in Fig. 7.1.

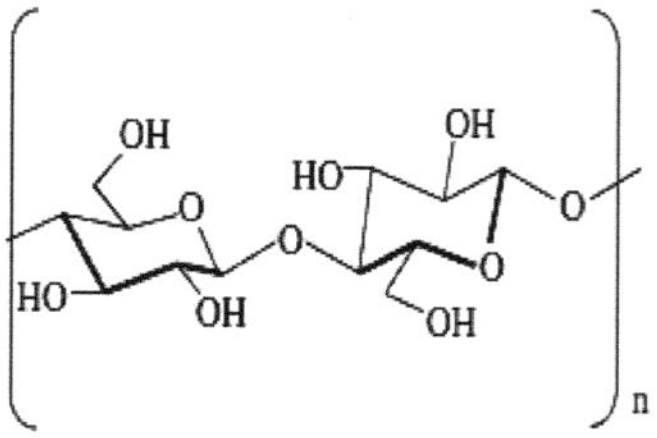

Figure 7.1 Structural unit of cellulose.

Alternate positioning of functional groups, such as –OH and –H, below or above the anhydroglucose unit plane is being determined by the β-configuration of the cellulose. The degree of polymerization of (DP) is determined from the total number of repeating structural units. The DP of any particular type of cellulose varies based on factors, such as material treatment, and origin. Specifically, cellulose obtained from wood pulp has a DP of about 300–1700 [42] whereas, bacterial cellulose and cellulose obtained from plants, such as cotton, exhibits higher DP of around 800–10,000 [42]. On the other hand, lower DP values (250–500) are characteristic of cellulose obtained from regenerate fibers as well as cellulose within the microcrystalline group [43, 44]. In cellulose the AGU are connected via β-1,4-glycosidic bonds, making it a 1,4-β-D-glucan. The AGU rings are in the 4C1 configuration such that the –OH groups, –CH_2OH groups, and the glycosidic bonds, are equatorial with respect to the mean planes. The supramolecular structure of cellulose is a factor of the molecular structure and this determines its major physical and chemical properties. Adjacent chains in the AGU are oriented at an angle of 180° to each other in their mean planes, forming a dimmer structural unit known as cellobiose. Therefore, the repeating unit in cellulose is anhydrocellobiose and the DP per molecule is twice the number of its repeating units.

7.2.2 The Supramolecular Structure

Based on the spatial conformation and chemical constitution of cellulose, the chains have a great tendency to form aggregates of well-ordered structural entities. The molecular structure of cellulose shows that it predominantly comprises of hydroxyl groups which engage in intermolecular hydrogen bonding, which are responsible for interchain cohesion. Knowing well that the arrangement of macromolecules in cellulosic fibers is not uniform all through the structure, it can be said that the cellulose structure comprises of regions of low order (amorphous) and regions of high order (crystalline). The relative amount of polymer in regions of the high order (crystalline regions) is called degree of crystallinity, and this depends on factors such as origin as well as pretreatment of the cellulosic material.

7.2.3 The Gross Morphological Structure

The morphology of cellulose can be described as a highly organized architecture of fibrillar entities. In native cellulose, a group of fibrillar entities which are organized in layer of different fibrillar texture can be found. On the other hand, regenerated cellulose and films usually exhibits fibrillar network morphology of different tightness. However, products of the regenerated cellulose often reveal a skin-core morphological structure. Cellulose is a nontoxic, odorless, and colorless solid polymer which is insoluble in water and conventional organic solvents. It is however swellable in several polar protic and aprotic liquids. Based on the strong adhesion between cellulose macromolecules through intermolecular forces, notably hydrogen bonds, it is difficult to transform cellulose into the molten state. Due to the fibrillar morphology, supramolecular arrangement, and molecular structure of cellulose, it is a typical fiber polymer which is most often processed and used in the fabric form. Availability of cellulose in nature is abundant as various plant fibers as well as deliginified woody plants (wood pulp). Dimensions of some cellulose fibers obtained from different plants are included in Table 7.1. Dimensions of the cellulose fibers can be cut into sizes by different milling procedures. For example, wet milling in paper-makers beater, majorly apply a fibrillation action but can also decrease the fiber length depending on the technique used.

Table 7.1 Dimensions of some cellulose fibers obtained from different plants.

Material source	*Fiber width* (μm)	*Fiber length* (μm)
Eucalypt	20	850
Wheat straw	15	1410
Spruce	31	3400
Bamboo	14	2700
Bagasse	20	1700
Pine	25	3100
Cotton linters	19	9000
Beech	21	120

7.2.4 Mechanical Properties of Cellulose

Based on the strong macromolecule adhesion, linear polymer chain, fibrillar morphology, and high supramolecular order, cellulose can be referred to as an ideal fibrous polymer. Therefore, discussions on mechanical properties of cellulose are most often based on criteria applicable to textile end uses, such as elongation at break or breaking strength. Imperfection at the different structural levels of cellulose often lead to lower experimental breaking strength of cellulose compared to the theoretically obtained ultimate tensile strength and also varies depending on fiber structure. Oven dried cellulose is highly hygroscopic and it is relatively brittle. In wet state, the tenacity of cellulose may become reduced or increased with respect to the two negating effects of softening action on the one hand, as well as a more even stress distribution on the other hand. This is the reason for lower tenacity of dry cotton fibers compared to wet samples. Air dried cellulose exhibits high resistance to mechanical impact, such that before the microscopic and macroscopic structure can be destroyed, a very large amount of mechanical energy needs to be expended. However, continuous freezing or oven-drying of the cellulosic material can lead to significant reduction in toughness and enhance mechanical breakdowns as well.

7.2.5 Chemical Stability and Environmental Properties of Cellulose

Cellulose, like several other organic polymers, is inert and may be combined with different types of construction materials. Cellulose is stable in water of either neutral or slightly alkaline pH. It is also stable in vast majority of organic liquids with different polar nature, but may also swell in these liquids at varying extent. Cellulose is, however vulnerable to several degradation routes with mechanism as follows:

(i) Hydrolytic degradation in aqueous acid and other non-aqueous media through the glucosidic bond cleavage;
(ii) Oxidation process through different pathways leading to incorporation of carboxyl and carbonyl groups, followed by fragmentation of chains through cleavage;
(iii) Hydrolytic cleavage of glucosidic bonds either in water or moist environment through the action of various cellulolytic enzymes, including bacterial and fungal;
(iv) Thermal degradation of cellulose at temperature >180–200°C.

Cellulose exhibits a physicochemical distinctiveness based on its strong sorption capacity for several group of substances. One of the most important examples is the sorption of water vapor. At a relatively low humidity, strong chemisorption of water molecules via interaction with the –OH groups of the polymer occurs, followed by a subsequent multilayer sorption (at medium relative humidity), and then the condensation of free water at relatively high humidity through capillary action. In effect, cellulose and various cellulose materials are the preferred polymers from the environmental point of view. This is because they can be conveniently returned to nature by ordinary rotting. Furthermore, cellulose bears no toxicity towards living organisms, including human beings.

7.3 GENERAL MODIFICATION OF CELLULOSE

Based on the various beneficial features of cellulose, modification techniques to improve the original properties or to incorporate new functionalities to cellulose have been severally investigated. These have enhanced development of the science and technology of cellulose. Chemical treatments have been specifically placed at the forefront of the field of cellulose modification with particular aims, which may be grouped as follows: (i) to modify cellulosic materials in order to exhibit specific features; (ii) to provide laboratory scale characterization of cellulosic materials; (iii) to tailor the nature of cellulose for scientific interests. However, due to the many unsolved and unattended areas of cellulose science, modification is often used to investigate the fundamental areas of cellulose science. This includes molecular dynamics of cellulose when in solutions, molecular interaction of cellulose with other materials, and cellulosic solid-state structure, such as chain conformation, amorphous and crystal structures, as well as the hydrogen bonding patterns.

The conventional approaches to cellulose modification are etherifications and esterifications mainly at the hydroxyl groups of cellulose. Therefore, most of the cellulose derivatives which are soluble in water and organic solvents are usually prepared via these modification techniques. Other possible approaches for chemical modification of cellulose are radical and ionic grafting, oxidation, deoxyhalogenation, and acetylation. Cellulosic materials of wood and cotton pulp origin contain carboxyl and aldehyde groups in relatively small amounts depending on the pulp purity. These groups are also sometimes the target points for chemical modification of cellulose.

One convenient way of incorporating new physical and chemical functionalities into cellulose is through graft modification. The three common approaches under this technique include "grafting onto", which involves the interaction between functional groups of two entirely different polymers. The second approach is "grafting from", which involves the initiation of vinyl monomers polymerization by a particular polymer with functional groups. The third approach is "grafting through", which is the copolymerization of macromonomers [45]. Among these three approaches, "grafting onto" and "grafting from" are the most commonly used for producing cellulose graft copolymers. Due to the low activity of macromolecular reactions, the "grafting onto" approach often exhibits low reaction efficiency [46]. Thus "grafting from" is often employed because there is the possibility to easily create reactive sites unto the main chain either via chemical treatment or irradiation, and a subsequent addition of monomers to produce the graft copolymer. There have been several investigations on the "grafting from" approach, one of which involves the direct growth of polymer grafts from the cellulose backbone via an ordinary process of free radical polymerization. Generation of radicals along the cellulose backbone may be achieved in the presence of some chemical initiators [47], or via the application of irradiation [48]. However, due to some shortcomings, the widespread application of cellulose graft polymers obtained through this grafting is limited [24]. In a bid to overcome this challenge, cationic and anionic controlled polymerization technique has been tried [49]. However, this approach also has its inherent shortcomings, thereby necessitating improved modification techniques [24].

7.3.1 Chemical Treatments

Cellulose fibers have been chemically modified through different approaches and with the help of different modifiers [50, 51]. The first series of modification is aimed at improving the decay resistance and dimensional stability, to improve the intra- and inter-bondability of the fibers. Some of the previous studies include the introduction of aldehyde and carboxyl groups unto the fiber surface. Chemical modification of cellulose has also been reported to improve the optical properties of the fiber. For example, acetylation process with acetic anhydride was observed to cause significant improvement in strength of the cellulose material without any observable side effect. Furthermore, it was reported that acetylation was efficient at eliminating chromophore groups, thereby improving the optical properties of the cellulosic materials [52]. Another strategy often employed for modification of cellulose is carboxymethylation reaction. This is particularly employed for modification of cellulose during the papermaking process.

7.3.2 Enzymatic Treatments

Treatment of cellulose with the help of enzymes is often conducted with cellulase, an enzyme complex which functions in a synergistic manner. They are grouped as endo-1-4-β-D-glucanases (endoglucanases) whose function is to create new terminal groups through the breakage of glycosidic cellulose chains; exo-1,4-β-D-glucanases (cellobiohydrolases) whose function is to convert the created cellulose terminal groups into cellobiose; and 1,4-β-D-glucosidases which hydrolyses cellobiose to glucose. Endoglucanases act haphazardly at the amorphous regions of cellulose and cellulose derivatives, and in the process hydrolysing the glucosidic β-(1,4) bonds. Endoglucanases and cellobiohydrolases function by reducing the cellulose molecular weight, leading to decreased viscosity [53]. There have been series of research on the enzymatic modification of cellulose fibers for improved properties [54]. In a particular research, a mixture of oxidized enzymes was used to pre-treat chemical-mechanical pulps in a bid to improve their physical, optical, and mechanical properties [55]. In another research aimed at improving pulp strength, cellulases obtained from Trichoderma reesei were applied to pine kraft pulp which was not bleached. Different dosage of the enzyme was incorporated and it was reported that the kind and amount of enzyme added determines to a very large extent the pulp viscosity [56].

7.3.3 Biological Treatments

In recent years, there have been progressive efforts towards improvement of cellulose fiber surface through bio modification of cellulose fibres as an alternative to traditional modification methods.

This approach is considered to be environmentally benign, energy saving, and offers less fiber damage compared to conventional modification methods. Specifically, this approach is attractive for mechanical and chemomechanical pulps of relatively low strength. In some previous research, it was reported that biological treatment of cellulose was significant on the fiber wall structure, in the form of observable internal fibrillation leading to increased fiber internal bonding [57, 58]. This further resulted into significant increases in strength of the modified fiber.

7.3.4 Hemicellulose Treatment

Another approach for the modification of cellulose fibers was proposed, which involves the use of carboxymethyl cellulose (CMC) and xylan for improvement of paper strength [59]. For this treatment, chemical modification of cellulose was carried out via irreversible attachment of hardwood xylan and CMC to chemical and recycled fibers. It was observed that both xylan and CMC treatments improved the properties of the paper to a similar extent and this was achieved through increased water retention and presence of more carboxylix groups.

7.4 APPLICATIONS OF CELLULOSE

7.4.1 Adsorbents

Removal of toxic anions (such as AsO_4^{3-}, CrO_4^{2-}, AsO_3^-, F^-) and toxic ions (such as Hg^{2+}, Cu^{2-}, Pb^{2+}, Cd^{2+}) from waste and water bodies have become so important recently in an effort to reduce the problems associated with ecological and industrial wastes. Adsorption is known to be an efficient, effective, and cost effective way of water decontamination as well as analytical separation [60]. These days, there is increased attention on adsorbents obtained from naturally occurring materials, mainly due to large availability, cheap cost, and the ease of introducing specific functional groups through chemical modification [24]. Cellulose, a naturally occurring polymer having many hydroxyl groups in its chain is a suitable material for production of adsorbents for toxic ions removal. Usually, surface modification may be used to introduce functional groups which are capable of chelating toxic ions unto the cellulosic material. Fabrication and properties of cellulosic adsorbents used for toxic ions removal have been extensively reviewed in literature [61, 62].

7.4.2 Antimicrobial

It is usually a matter of necessity for materials intended to be used for applications, such as medical items, military items, household, food packaging and sanitary materials, to exhibit some antibacterial properties [63]. Incorporation of antibacterial properties unto materials surfaces may be achieved through covalent bonding or through non-covalent interactions between the material surfaces and antibacterial species. Immobilization of antibacterial agents on the material surface through covalent bonding could result into durable, efficient, and effective antimicrobial features [24]. Surface grafting has been successfully used for the generation of antibacterial surfaces using various monomers with different functional groups, such as phenol derivatives [64], N-halamines [65], antibiotics [66], quaternary ammonium and phosphonium salts [67, 68], and polypeptide mimics [69].

7.4.3 Resistance to Protein Adsorption

Cellulose and cellulose derivatives have been largely employed for biomedical applications. However, the biocompatibility, most especially blood compatibility of raw cellulose, is insufficient and therefore needs to be enhanced prior to use. It is well known that protein adsorption on surface of biomaterials is the first stage of many subsequent undesirable bio-response and bio-reactions, and so on [70, 71]. Surface modification of cellulosic materials for good biocompatibility properties has been well reported [72–74]. It is reported that surfaces of modified cellulosic materials exhibit highly desirable resistance to platelet and protein adsorption [24].

7.4.4 Micelles and Drug Release

Modified cellulose with specially designed architecture has been produced which exhibits the properties of the cellulose backbone as well as side chains. Stimuli-responsive features of the modified cellulose may either be tailored via the cellulose backbone or otherwise the side chain's chemical structure. The modified cellulose can be designed into single [12, 75], or multi stimuli [76–78] responsive micelles in specific solvents based on the nature of modification, and this possesses the potential to be used for gene delivery, drug carriers, as well as controlled release [75, 79, 80]. In order to imitate the physiological environment, a lot of research has been focused on production of functional materials which is sensitive to pH and temperature [76, 77]. The fabricated material was reported to exhibit dual stimuli-responsive features. In recent years, aqueous solutions of polymer based stimuli-responsive nanomaterial are of particular interest because of the various potential applications [81]. Different properties and structure of polymer based stimuli-responsive materials can be modified by altering the parameters of its molecular structure or the environmental conditions, such as temperature, photochemical processes, ionic strength, pH, and light, etc. [82–86].

Based on this, and several other potential applications of nanomaterials, the following section of this chapter will focus on cellulose nanocrystals. It is known that cellulose microfibrils are made up of a combination of well ordered (crystalline) and disordered (amorphous) regions. The crystalline portions of cellulose are cellulose nanocrystals [87]. The crystalline structure found in cellulose comprises about 65–95% and this can be extracted from a wide variety of sources [88]. A few terms have been used for cellulose nanomaterials, such as microcrystals, whiskers, nanocrystals, nanoparticles, microcrystallites, or nanofibers. In this chapter the term cellulose nanocrystals (CNC) is used.

7.5 CELLULOSE NANOCRYSTALS

Cellulose nanocrystals CNCs can be extracted from plants such as jute, cotton, ramie, and wood, and also from biomass residues, such as agricultural crops. In addition, they can also be produced by bacteria and, in rare cases, found in sea creatures such as tunicates [89]. The first colloidal suspension of cellulose was isolated in the 1950s via sulfuric acid-catalysed degradation of cellulose fibers [90]. Meanwhile, Marchessault et al. found that a colloidal suspensions of cellulose nanocrystals formed nematic liquid crystalline alignment after optimisation was done with an acid hydrolysis process [91]. Dried suspension shows the presence of aggregates of needle-shaped particles. Subsequently, improvements in the mechanical properties of nanocomposites by cellulose nanocrystal reinforcement were also found. Indeed, it is a good candidate as reinforcing filler for composite in many other applications such as barrier films, transparent films, flexible displays, biomedical implants, separation membranes, supercapacitors, and templates for electronic components [92].

Concurrently, microcrystalline cellulose (MCC) was also commercially produced as a result of hydrochloric acid-assisted degradation of cellulose fibers from wood pulps [93]. Since it is chemically inactive, stable, and physiologically inert with attractive binding properties, it has applications in various fields. It is used in the food processing industry as a fat replacer and texturizing agent, in the pharmaceutical industry as a tablet binder, and as an additive in paper manufacturing, and has composite applications as well [94].

Recent findings have proposed CNCs to be amongst the most promising new nanoscale building blocks for next generation biomaterials and engineering applications due to their renewable origins, combined with low toxicity, and unique physical properties. As shown in Fig. 7.2, the schematic diagram of cellulose sources and structure of its microfibrils consist of regions that are highly ordered (crystalline) and regions that are more disordered (amorphous) [92].

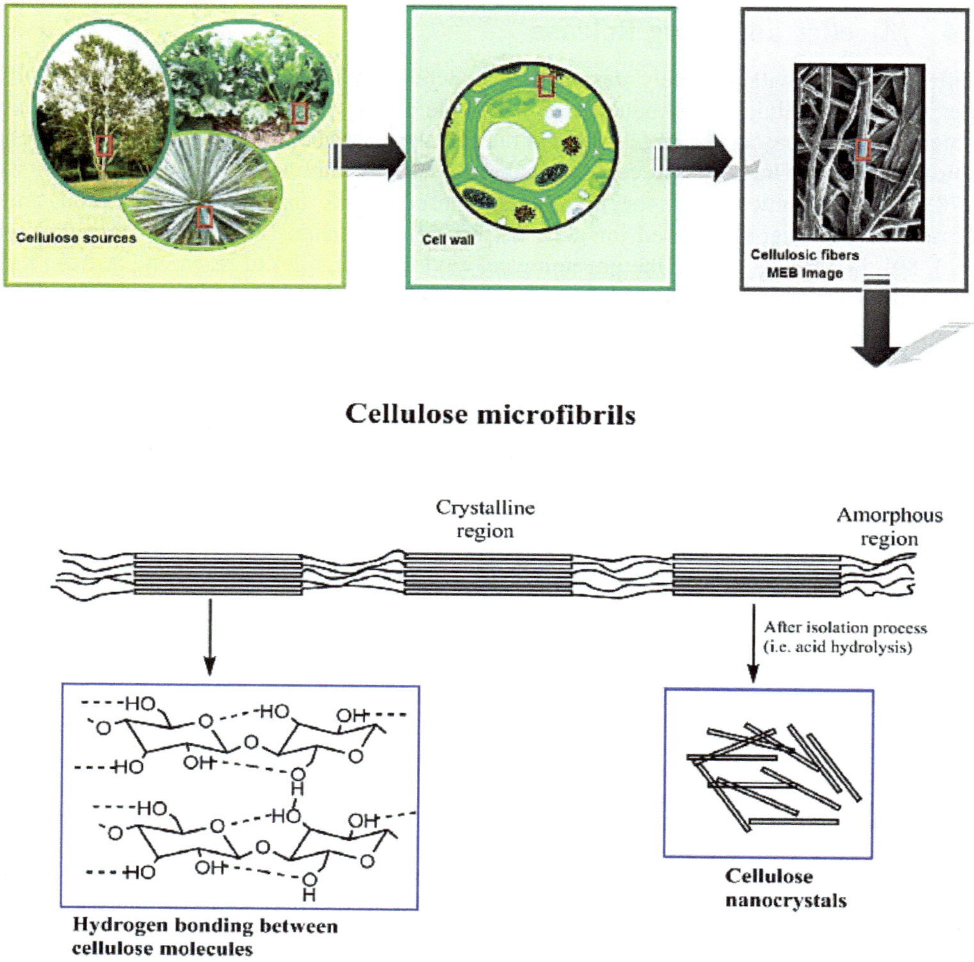

Figure 7.2 Schematic diagram of cellulose structure and cellulose microfibrils [39].

7.5.1 Isolation of Cellulose Nanocrystals

Cellulose isolation from cellulose source materials occurs in two phases. The first phase involves purification and homogenisation pretreatments of the source material. In this stage, the pretreatments for wood and plants consist of removal of matrix materials such as hemicellulose and lignin completely or partially, as well as separating the individual complete fibers. Thus, it can react consistently during subsequent treatments. The second phase is the isolation of microfibrillar and/or crystalline components from the "purified" cellulose materials. Crystalline fibrils of cellulose can be liberated via many extraction methods. Generally, an isolation method to obtain CNC can be divided into three which are chemical/biological, for instance acid hydrolysis and enzymatic treatments, mechanical treatment such as ultra-sonication [95], homogenization, refining, grinding, and cryocrushing [96]. The third method can be a combination of these methods, such as enzymatic treatment with acid hydrolysis.

The most common method reported by other researchers to isolate CNC is acid hydrolysis using sulfuric acid. To obtain CNC, the concentration of sulfuric acid required in the hydrolysis reaction is approximately 65% (wt.), and reaction temperature can reach upto 70°C meanwhile hydrolysis time can vary from 30 minutes to overnight depending on the temperature [97]. In this process, amorphous and paracrystalline regions are preferentially hydrolysed. The crystalline parts of the cellulose have a higher resistance to acid attack and remain undamaged [98, 99]. Another typical acid used in hydrolysis process is hydrochloric acid. It can also generate similar shape of CNC as sulfuric acid, but it has limited dispersion ability because the suspension tends to flocculate [100]. Whereas, in sulfuric acid hydrolysis it becomes a hydrolysing agent which introduces sulfate charges on the surface of CNCs and prevents agglomeration, thus promoting good dispersion in water [101]. Other types of acids for instance, phosphoric [102], formic [103], and hydrobromic [104] acids have also been used for isolation of CNC. However the use of acid in extraction process has a few disadvantages such as corrosivity, is environmentally hazardous, associated with a high energy demand, and can cause cellulose degradation [95].

Furthermore, there are a few shapes of cellulose nanocrystals that can be produced, such as rods and spheres [105]. A range of CNC with different dimensions (aspect ratio) and morphology could be obtained depending on the source and the isolation methods used, as displayed in Table 7.2. The common cellulose sources used to isolate CNC, especially at the laboratory scale, is cotton due to its high cellulose content which often results into a higher yield of CNC [106].

Table 7.2 Source of cellulose and isolation method to produce CNC and CNF and range of its size.

Source of cellulose	*Isolation method*	*Size*		*References*
		***L* (nm)**	***w* (nm)**	
Wood pulp	Acid hydrolysis	100–300	3–5	[100, 107]
	Enzymatic hydrolysis	100–1800	30–80	[108]
	Chemical-mechanical	65	15	[109]
Cotton	Acid hydrolysis	70–160	15	[110]
	Microbial hydrolysis	120	40	[111]
Tunicates	Acid hydrolysis	1000–10000	3–20	[94, 112]
Bacteria	Acid hydrolysis	250–1000	16–54	[113]
	Enzymatic hydrolysis	100–300	10–15	[114]
Agriculture crops (coconut husk fibers, rice straw, banana stem, etc.)	Acid hydrolysis	177–200	5–7	[115]
	Chemical-mechanical	–	5–40	[116]
	Mechanical	–	3.5–60	[117]

7.5.2 CNC as Potential Reinforcing Filler

For the last 15 years, nanocellulose has been studied as a reinforcing element in nanocomposites [88]. One of the most desirable features of CNC is the hydroxyl groups on the surface of the nanocrystal. The abundance of –OH groups favors the formation of hydrogen bonding, causing the cellulose chains to assemble into highly ordered structures. These reactive hydroxyl groups will interact with the functional groups of other molecules, and are valuable for tailoring the functional properties of CNC [118]. Moreover, the physical properties of CNC are superior to other engineering materials (Table 7.3) and indeed, these features represent compelling criteria as advanced reinforcing or functional fillers for polymer nanocomposites.

Moreover, CNC has lower density with high modulus [119]. A good aspect ratio (length/diameter) value also contributes to its reinforcing abilities, which enables a critical length for stress transfer from the matrix to the reinforcing phase [120]. Furthermore, CNCs are biologically renewable, sustainable, low cost, combustible, nontoxic, and have biodegradable properties [121].

Table 7.3 Properties of various materials compared to cellulose.

Type of material	*Modulus* (GPa)	*Specific modulus* (GPa cm^3/g)	*Density* (g/cm^3)	*References*
Aluminium	62	23	2.7	[122]
Steel	200	25	7.8	[123, 124]
Glass Fiber	73	28	2.6	[125]
Cellulose nanocrystal	110–220	138	1.6	[92]

Many researchers have demonstrated the potential of CNC by incorporating it into various polymer matrices (polypropylene [126], polyethylene [127], polyvinyl alcohol [128], poly(lactic acid) [129]. The nanocomposites obtained demonstrated improvements in mechanical properties. In particular, CNC also display an outstanding potential in increasing the composite material properties at low concentrations [130]. CNC are also unique in their capacity to maintain the transparency of the host material [131].

7.5.3 Modification of Cellulose Nanocrystals

7.5.3.1 Acid Hydrolysis

Conversion of lignocellulosic biomass (mostly plant) to monosaccharides required chemical pre-treatments to break down the complex structure, which involved acid hydrolysis, especially for cellulose derystallisation. Concentrated acid hydrolysis process is a two-step process; concentrated acid decrystallises cellulose to less crystallized oligosaccharides, followed by less concentrated and higher reaction temperature for converting decrystallised oligosaccharides to monosaccharides [132]. Advantages of concentrated acid hydrolysis process are higher conversion from polysaccharides to monosaccharides with minimum formation of reaction by-products with careful control of reaction conditions, no limitations of biomass resources, and long history of commercial trials. Main disadvantages of this process are higher usages of acids than dilute acid hydrolysis process and sensitivity to moisture content of biomass raw materials [133]. Choice of acid will determine the properties of nanocellulose. The typical acid that has been used to isolate cellulose nanocrystal is sulfuric acid [134–136]. Recently phosphoric acid is used to increase the thermal stability of CNC [137] so that it can be used in high temperature post-processing method. The resulting CNC hydrolysed by those acids have a negative surface charge due to sulfate and phosphate groups on the surface of CNC. This provides electrostatic stabilisation of CNC suspension [[137, 138]. The presence of these ionisable groups on the surface of CNC can be considered important structure-wise as well as for the reactivity of CNC.

7.5.3.2 Oxidation

Oxidation reactions are performed on cellulose to introduce acid or aldehyde functionalities [133]. The most common oxidation reactions are 2,2,6,6-tetramethylpiperidine-1-oxy (TEMPO) oxidation reaction. In TEMPO oxidation the main reaction corresponded to selective oxidation on surface primary hydroxyl groups into carboxylic groups, as shown in Fig. 7.3 [132]. This reaction introduced negative charges on the interface of crystalline domains, thus improving dispersion stability of crystallites. These carboxylated CNCs will prevent flocculate and sedimentation. Another advantage of TEMPO-oxidation method is its capability to produce high oxidized yield up to 90% [40]. Further modification of TEMPO-oxidized CNC is also being used as a precursor for further functionalization, such as hydrophobic CNC [136]. Due to various benefits of this, CNC modification has been utilised to make CNC a good reinforcing filler in many polymer composites, such as polystyrene [137], polypropylene [138], and hydrogels [139].

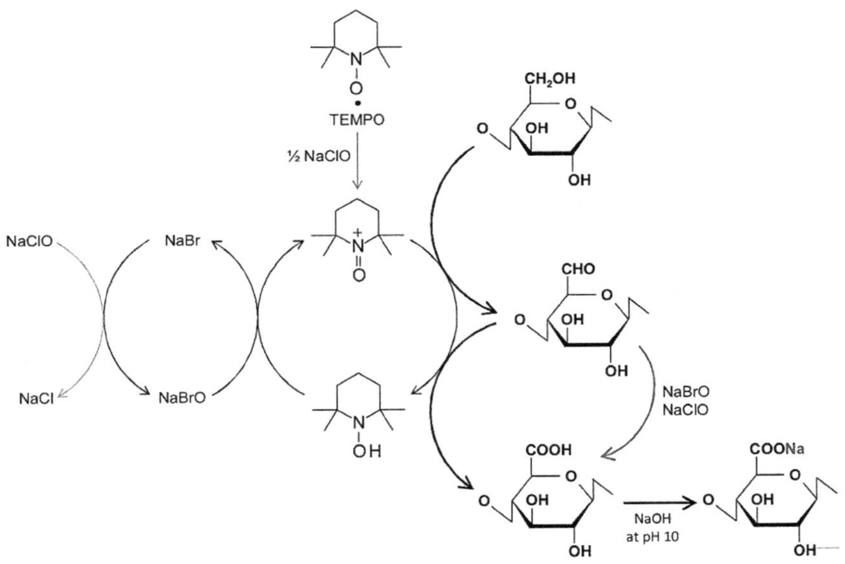

Figure 7.3 Oxidation of cellulose by TEMPO-mediated oxidation (with permission from Elsevier) [135].

7.5.3.3 Amination and Amidation

Typically, amidation of CNC will use CNC produced via oxidation technique as a starting material. The method involves activation of the carboxylic acid moieties on the CNCs through formation of the N-hydroxysuccinimidyl ester for amination, followed by reaction with a primary amine to form the amide product [133]. This amide-CNC has been a good intermediate precursor to be incorporated further with other substances, such as fullerene, to become better photochemical stable sensitizers [140, 141].

Meanwhile, in Sirviö et al. [142], oxidation process by sodium periodate was reported for CNC precursor in order to produce amino-modified CNC with the goal to produce adjustable hydrophobic CNC, where the oxidation process is used. The route preparation of amino cellulose is shown in Fig. 7.4. With the same aim of hydrophobic properties of CNC, this method has been widely used, for instance drug delivery composite [143, 144]. This method is also useful to improve cationic efficiency of CNC, which is low with typical etherification reaction [145].

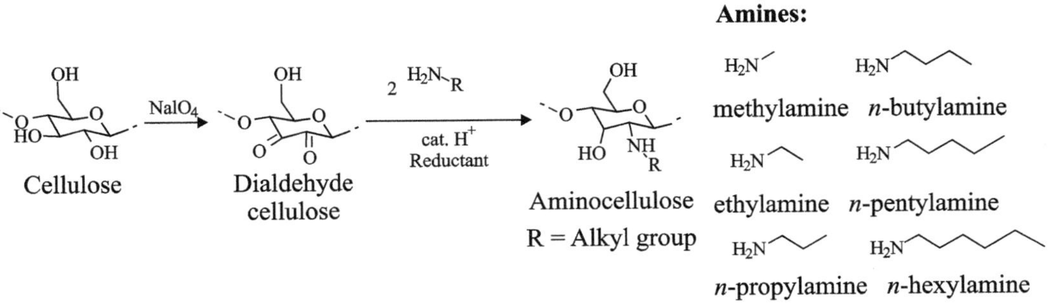

Figure 7.4 Preparation of amino cellulose (with permission from Elsevier) [142].

7.5.3.4 Esterification

Sulfation and phosphorylation that occur in hydrolysis process are examples of esterification reactions. The other examples of simultaneous esterification and acid hydrolysis in the literature are the production of acetylated and butyrated cellulose nanocrystals by Fisher esterification, using mixed acetic or butyric acid and hydrochloric acid [146]. The resulting surface of modified CNCs increase in hydrophobicity and have a good dispersion in organic solution. Meanwhile, in Boujemaoui et al. [134], different functional organic acids also have been carried out, such as combination of hydrochloric acid with hydrobromic acid, and managed to reduce the reaction time significantly while retaining the desired functionality. The schematic diagram of CNC modified by esterification is illustrated in Fig. 7.5.

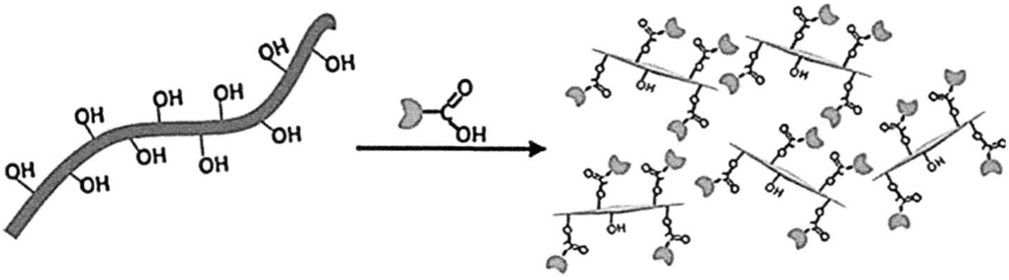

Figure 7.5 Schematic diagram of cellulose nanocrystals modified by esterification reaction (with permission from Elsevier) [134].

7.5.3.5 Etherification

To cationize the surface of the cellulose the most common etherification of cellulose nanocrystals appears to be the application of glycidyltrimethylammonium chloride (GTMAC) or derivatives [145, 147, 148]. Cationically modified CNC have hydrophobic properties and lead to compatibility with non-polar solvents. As shown in Fig. 7.6, a nucleophilic reaction occurs between the alkali activated hydroxyl group of CNC and epoxy group of GTMAC. The main etherification reaction is also accompanied by the alkaline hydrolysis.

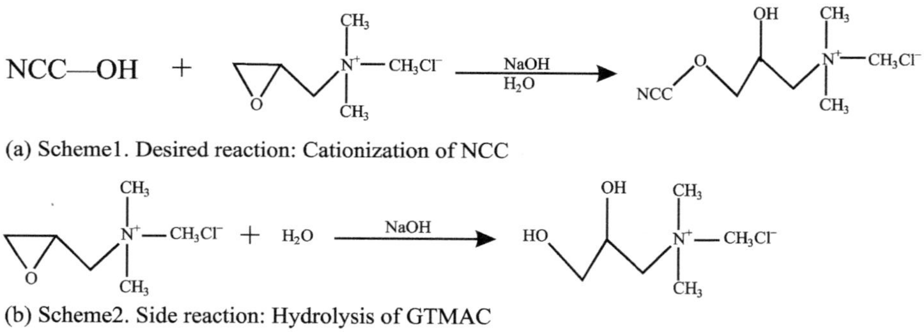

Figure 7.6 Cationic modification of CNC using GTMAC (with permission from Elsevier) [149].

7.6 CONCLUSION

As a summary, it should be noted that the main features of cellulose as a polymer ensues from its hydrophilicity in combinations with several other salient features, such as nontoxicity, desirable mechanical properties, potential as a good adsorbent, and safe after use disposability.

References

1. Hon, D.N.-S. and N. Shiraishi (ed.). Wood and Cellulosic Chemistry, Revised, and Expanded, 2nd Ed. CRC Press, 2000.

2. Ferreira, S., N. Gil, J. Queiroz, A. Duarte and F. Domingues. An evaluation of the potential of Acacia dealbata as raw material for bioethanol production. Bioresource Technology 2011; 102(7): 4766-4773.
3. Jiang, X., J. Gu, X. Tian, Y. Li and D. Huang. Modification of cellulose for high glucose generation. Bioresource Technology 2012; 104: 473-479.
4. Alvira, P., E. Tomás-Pejó, M. Ballesteros and M. Negro. Pretreatment technologies for an efficient bioethanol production process based on enzymatic hydrolysis: a review. Bioresource Technology 2010; 101(13): 4851-4861.
5. Gupta, R., Y.P. Khasa and R.C. Kuhad. Evaluation of pretreatment methods in improving the enzymatic saccharification of cellulosic materials. Carbohydrate Polymers 2011; 84(3): 1103-1109.
6. Ingram, T., K. Wörmeyer, J.C.I. Lima, V. Bockemühl, G. Antranikian, G. Brunner and I. Smirnova. Comparison of different pretreatment methods for lignocellulosic materials. Part I: Conversion of rye straw to valuable products. Bioresource Technology 2011; 102(8): 5221-5228.
7. Zhang, J., X. Ma, J. Yu, X. Zhang and T. Tan. The effects of four different pretreatments on enzymatic hydrolysis of sweet sorghum bagasse. Bioresource Technology 2011; 102(6): 4585-4589.
8. Wu, L., M. Arakane, M. Ike, M. Wada, T. Takai, M. Gau, and K. Tokuyasu. Low temperature alkali pretreatment for improving enzymatic digestibility of sweet sorghum bagasse for ethanol production. Bioresource Technology 2011; 102(7): 4793-4799.
9. Ha, S.H., N.L. Mai, G. An and Y.-M. Koo. Microwave-assisted pretreatment of cellulose in ionic liquid for accelerated enzymatic hydrolysis. Bioresource Technology 2011; 102(2): 1214-1219.
10. Ju, Y.-H., L.-H. Huynh, N.S. Kasim, T.-J. Guo, J.-H. Wang and A.E. Fazary. Analysis of soluble and insoluble fractions of alkali and subcritical water treated sugarcane bagasse. Carbohydrate Polymers 2011; 83(2): 591-599.
11. Kim, K.H. and J. Hong. Supercritical CO_2 pretreatment of lignocellulose enhances enzymatic cellulose hydrolysis. Bioresource Technology 2001; 77(2): 139-144.
12. Wang, K., H.Y. Yang, F. Xu and R.C. Sun. Structural comparison and enhanced enzymatic hydrolysis of the cellulosic preparation from *Populus tomentosa* Carr., by different cellulose-soluble solvent systems. Bioresource Technology 2011; 102(6): 4524-4529.
13. Akindoyo, J.O., Md. D.H. Beg, S. Ghazali and Md. R. Islam. Effects of poly (dimethyl siloxane) on the water absorption and natural degradation of poly (lactic acid)/oil-palm empty-fruit-bunch fiber biocomposites. Journal of Applied Polymer Science 2015; 132(45): 42784-42793.
14. Akindoyo, J.O., Md. D.H. Beg, S. Ghazali, Md. R. Islam and A. Al Mamun. Preparation and characterization of poly (lactic acid)-based composites reinforced with poly dimethyl siloxane/ ultrasound-treated oil palm empty fruit bunch. Polymer-Plastics Technology and Engineering 2015; 54(13): 1321-1333.
15. Karst, D. and Y. Yang. Effect of structure of large aromatic molecules grafted onto cellulose on hydrolysis of the glycosidic linkages. Macromolecular Chemistry and Physics 2007; 208(7): 784-791.
16. Borsa, J., I. Tanczos and I. Rusznak. Acid hydrolysis of carboxymethylcellulose of low degree of substitution. Colloid and Polymer Science 1990; 268(7): 649-657.
17. Karst, D.T., Y. Yang and G. Tanaka. An explanation of increased hydrolysis of the β-(1, 4)-glycosidic linkages of grafted cellulose using molecular modeling. Polymer 2006; 47(18): 6464-6471.
18. Thakur, M.K., R.K. Gupta and V.K. Thakur. Surface modification of cellulose using silane coupling agent. Carbohydrate Polymers 2014; 111: 849-855.
19. Zhou, D., R. Zhou, C. Chen, W.-A. Yee, J. Kong, G. Ding and X. Lu. Non-volatile polymer electrolyte based on poly (propylene carbonate), ionic liquid, and lithium perchlorate for electrochromic devices. The Journal of Physical Chemistry B 2013; 117(25): 7783-7789.
20. Ding, G., C.M. Cho, C. Chen, D. Zhou, X. Wang, A.Y.X. Tan, J. Xu, X. Lu. Black-to-transmissive electrochromism of azulene-based donor–acceptor copolymers complemented by poly (4-styrene sulfonic acid)-doped poly (3, 4-ethylenedioxythiophene). Organic Electronics 2013; 14(11): 2748-2755.
21. Thakur, V., A. Singha and M. Thakur. Natural cellulosic polymers as potential reinforcement in composites: physicochemical and mechanical studies. Advances in Polymer Technology 2013; 32(S1): E427-E435.
22. Basta, A.H., H. El-Saied, O. El-Hadi and C. El-Dewiny. Evaluation of rice straw-based hydrogels for purification of wastewater. Polymer-Plastics Technology and Engineering 2013; 52(11): 1074-1080.

23. Mishra, S., G.U. Rani and G. Sen. Microwave initiated synthesis and application of polyacrylic acid grafted carboxymethyl cellulose. Carbohydrate Polymers 2012; 87(3): 2255-2262.
24. Kang, H., R. Liu and Y. Huang. Graft modification of cellulose: methods, properties and applications. Polymer 2015; 70: A1-A16.
25. Krässig, H.A. Cellulose: Structure, Accessibility and Reactivity. Gordon and Breach Publishers, Tacony Street, Philadelphia, PA, 1993.
26. Yu, H., R. Liu, L. Qiu and Y. Huang. Composition of the cell wall in the stem and leaf sheath of wheat straw. Journal of Applied Polymer Science 2007; 104(2): 1236-1240.
27. Yu, H., R. Liu, D. Shen, Z. Wu and Y. Huang. Arrangement of cellulose microfibrils in the wheat straw cell wall. Carbohydrate Polymers 2008; 72(1): 122-127.
28. Sani, A. and Y. Dahman. Improvements in the production of bacterial synthesized biocellulose nanofibres using different culture methods. Journal of Chemical Technology and Biotechnology 2010; 85(2): 151-164.
29. Roy, D., M. Semsarilar, J.T. Guthrie and S. Perrier. Cellulose modification by polymer grafting: A review. Chemical Society Reviews 2009; 38(7): 2046-2064.
30. Chen, P., S.Y. Cho and H.-J. Jin. Modification and applications of bacterial celluloses in polymer science. Macromolecular Research 2010; 18(4): 309-320.
31. Pinkert, A., K.N. Marsh, S. Pang and M.P. Staiger.. Ionic liquids and their interaction with cellulose. Chemical Reviews 2009; 109(12): 6712-6728.
32. Cai, J., L. Zhang, J. Zhou, H. Qi, H. Chen, T. Kondo, X. Chen and B. Chu. Multifilament fibers based on dissolution of cellulose in NaOH/urea aqueous solution: Structure and properties. Advanced Materials 2007; 19(6): 821-825.
33. Rani, G.U., S. Mishra, G. Sen and U. Jha. Polyacrylamide grafted agar: Synthesis and applications of conventional and microwave assisted technique. Carbohydrate Polymers 2012; 90(2): 784-791.
34. Abdul Khalil, H.P.S., M. Jawaid, P. Firoozian, M. Amjad, E. Zainudin and M.T. Paridah. Tensile, electrical conductivity, and morphological properties of carbon black–filled epoxy composites. International Journal of Polymer Analysis and Characterization 2013; 18(5): 329-338.
35. Thakur, V.K., M.K. Thakur and R.K. Gupta. Graft copolymers from natural polymers using free radical polymerization. International Journal of Polymer Analysis and Characterization 2013; 18(7): 495-503.
36. Fox, S.C., B. Li, D. Xu and K.J. Edgar. Regioselective esterification and etherification of cellulose: A review. Biomacromolecules 2011; 12(6): 1956-1972.
37. Malmström, E. and A. Carlmark. Controlled grafting of cellulose fibres–an outlook beyond paper and cardboard. Polymer Chemistry 2012; 3(7): 1702-1713.
38. Carlmark, A., E. Larsson and E. Malmström. Grafting of cellulose by ring-opening polymerisation–A review. European Polymer Journal 2012; 48(10): 1646-1659.
39. Lavoine, N., I. Desloges, A. Dufresne and J. Bras. Microfibrillated cellulose–Its barrier properties and applications in cellulosic materials: A review. Carbohydrate Polymers 2012; 90(2): 735-764.
40. Isogai, A., T. Saito and H. Fukuzumi. TEMPO-oxidized cellulose nanofibers. Nanoscale 2011; 3(1): 71-85.
41. Wang, H., G. Gurau and R.D. Rogers. Ionic liquid processing of cellulose. Chemical Society Reviews 2012; 41(4): 1519-1537.
42. Coseri, S. Cellulose: To depolymerize… or not to?. Biotechnology Advances 2017; 35(2): 251-266.
43. Song, Y., J. Zhang, X. Zhang and T. Tan. The correlation between cellulose allomorphs (I and II) and conversion after removal of hemicellulose and lignin of lignocellulose. Bioresource Technology 2015; 193: 164-170.
44. Gupta, V., P. Carrott, R. Singh, M. Chaudhary and S. Kushwaha. Cellulose: A review as natural, modified and activated carbon adsorbent. Bioresource Technology 2016; 216: 1066-1076.
45. Odian, G. Principles of Polymerization, 4th Ed. John Wiley & Sons, Inc., Hoboken, New Jersey, 2004.
46. Hansson, S., V. Trouillet, T. Tischer, A.S. Goldmann, A. Carlmark, C. Barner-Kowollik, and E. Malmström. Grafting efficiency of synthetic polymers onto biomaterials: A comparative study of grafting-from versus grafting-to. Biomacromolecules 2012; 14(1): 64-74.
47. Littunen, K., U. Hippi, L.-S. Johansson, M. Österberg, T. Tammelin, J. Laine and J. Seppälä. Free radical graft copolymerization of nanofibrillated cellulose with acrylic monomers. Carbohydrate Polymers 2011; 84(3): 1039-1047.

48. Prasad, G., K. Prasad, R. Meena and A. Siddhanta.. Facile preparation of Chaetomorpha antennina based porous polysaccharide–PMMA hybrid material by radical polymerization under microwave irradiation. Journal of Materials Science 2009; 44(15): 4062-4068.
49. Feit, B.A., A. Bar-Nun, M. Lahav and A. Zilkha. Anionic graft polymerization of vinyl monomers on cellulose and polyvinyl alcohol. Journal of Applied Polymer Science 1964; 8(4): 1869-1888.
50. Papadopoulos, A.N. Chemical modification of pine wood with propionic anhydride: Effect on decay resistance and sorption of water vapour. BioResources 2006; 1(1): 67-74.
51. Saito, T. and A. Isogai. Ion-exchange behavior of carboxylate groups in fibrous cellulose oxidized by the TEMPO-mediated system. Carbohydrate Polymers 2005; 61(2): 183-190.
52. Mirshokraie, S.A., A. Abdulkhani, A.A. Enayati and A.J. Latibari. Evaluation of mechanical and optical properties of modified bagasse chemi-mechanical pulp through acetylation in liquid phase. Iranian Polymer Journal 2005; 14(11): 982-988.
53. Ogeda, T.L. and D.F. Petri. Hidrólise enzimática de biomassa. Química Nova 2010; 33(7): 1549-1558.
54. Gandini, A. and D. Pasquini. The impact of cellulose fibre surface modification on some physico-chemical properties of the ensuing papers. Industrial Crops and Products 2012; 35(1): 15-21.
55. Boeva, R.K., E. Petkova, N. Georgieva, L. Yotova and I. Spiridonov. Utilization of a chemical-mechanical pulp with improved properties from poplar wood in the composition of packing papers. BioResources 2007; 2(1): 34-40.
56. Pere, J., M. Siika-aho, J. Buchert and L. Viikari. Effects of purified trichoderma reesei cellulases on the fiber properties of kraft pulp. Tappi Journal (USA), 1995; 78: 71-78.
57. Yang, Q., H. Zhan, S. Wang, S. Fu and K. Li. Modification of eucalyptus CTMP fibres with white-rot fungus Trameteshirsute–Effects on fibre morphology and paper physical strengths. Bioresource Technology 2008; 99(17): 8118-8124.
58. Yang, Q., H. Zhan, S. Wang, S. Fu and K. Li. Bio-modification of eucalyptus chemithermo-mechanical pulp with different white-rot fungi. BioResources 2007; 2(4): 682-692.
59. Arndt, T. and R. Zelm. New nanotechnology-produced fibre compounds in papermaking applications—Review and first own experiences. Internationale Papierwirtschaft 2008 9: 59-63.
60. Crini, G. Recent developments in polysaccharide-based materials used as adsorbents in wastewater treatment. Progress in Polymer Science 2005; 30(1): 38-70.
61. O'Connell, D.W., C. Birkinshaw and T.F. O'Dwyer. Heavy metal adsorbents prepared from the modification of cellulose: A review. Bioresource Technology 2008; 99(15): 6709-6724.
62. Wojnárovits, L., C.M. Földváry and E. Takács. Radiation-induced grafting of cellulose for adsorption of hazardous water pollutants: A review. Radiation Physics and Chemistry 2010; 79(8): 848-862.
63. Lee, S.B., R.R. Koepsel, S.W. Morley, K. Matyjaszewski, Y. Sun and A.J. Russell. Permanent, non-leaching antibacterial surfaces. 1. Synthesis by atom transfer radical polymerization. Biomacromolecules 2004; 5(3): 877-882.
64. Patel, M.V., S.A. Patel, A. Ray and R.M. Patel. Antimicrobial activity on the copolymers of 2, 4-dichlorophenyl methacrylate with methyl methacrylate: Synthesis and characterization. Journal of Polymer Science Part A: Polymer Chemistry 2004; 42(20): 5227-5234.
65. Sun, Y. and G. Sun. Durable and regenerable antimicrobial textile materials prepared by a continuous grafting process. Journal of Applied Polymer Science 2002; 84(8): 1592-1599.
66. Soykan, C., Ş. Güven and R. Coşkun. 2-[(5-methylisoxazol-3-yl) amino]-2-oxo-ethyl methacrylate with glycidyl methacrylate copolymers: Synthesis, thermal properties, monomer reactivity ratios, and antimicrobial activity. Journal of Polymer Science Part A: Polymer Chemistry 2005; 43(13): 2901-2911.
67. Jin, D., S. Fu, F. Zhang, X. Lu and P. Han. Preparation of silica gel grafting bisquarternary phosphonium and its antibacterial activities. Asian Journal of Chemistry 2012; 24(1): 107-111.
68. Kenawy, E.-R., S. Worley and R. Broughton, The chemistry and applications of antimicrobial polymers: a state-of-the-art review. Biomacromolecules, 2007. 8(5): 1359-1384.
69. Arnt, L., J.R. Rennie, S. Linser, R. Willumeit and G.N. Tew. Membrane activity of biomimetic facially amphiphilic antibiotics. The Journal of Physical Chemistry B 2006; 110(8): 3527-3532.
70. Seo, J.-H., R. Matsuno, T. Konno, M. Takai and K. Ishihara. Surface tethering of phosphorylcholine groups onto poly (dimethylsiloxane) through swelling–deswelling methods with phospholipids moiety containing ABA-type block copolymers. Biomaterials 2008; 29(10): 1367-1376.

71. Lee, J.H., Y.M. Ju and D.M. Kim. Platelet adhesion onto segmented polyurethane film surfaces modified by addition and crosslinking of PEO-containing block copolymers. Biomaterials 2000; 21(7): 683-691.
72. Zhang, J., J. Yuan, Y. Yuan, J. Shen and S. Lin. Chemical modification of cellulose membranes with sulfo ammonium zwitterionic vinyl monomer to improve hemocompatibility. Colloids and Surfaces B: Biointerfaces 2003; 30(3): 249-257.
73. Ishihara, K., N. Nakabayashi, K. Fukumoto and J. Aoki. Improvement of blood compatibility on cellulose dialysis membrane I. Grafting of 2-methacryloyloxyethyl phosphorylcholine on to a cellulose membrane surface. Biomaterials 1992; 13(3): 145-149.
74. Ishihara, K., R. Takayama, N. Nakabayashi, K. Fukumoto and J. Aoki. Improvement of blood compatibility on cellulose dialysis membrane: 2. Blood compatibility of phospholipid polymer grafted cellulose membrane. Biomaterials 1992; 13(4): 235-239.
75. Yan, Q., J. Yuan, F. Zhang, X. Sui, X. Xie, Y. Yin, S. Wang and Y. Wei. Cellulose-based dual graft molecular brushes as potential drug nanocarriers: stimulus-responsive micelles, self-assembled phase transition behavior, and tunable crystalline morphologies. Biomacromolecules 2009; 10(8): 2033-2042.
76. Ma, L., H. Kang, R. Liu and Y. Huang. Smart assembly behaviors of hydroxypropylcellulose-graft-poly (4-vinyl pyridine) copolymers in aqueous solution by thermo and pH stimuli. Langmuir 2010; 26(23): 18519-18525.
77. Ma, L., R. Liu, J. Tan, D. Wang, X. Jin, H. Kang, M. Wu and Y. Huang. Self-assembly and dual-stimuli sensitivities of hydroxypropylcellulose-graft-poly (N, N-dimethyl aminoethyl methacrylate) copolymers in aqueous solution. Langmuir 2010; 26(11): 8697-8703.
78. Jin, X., H. Kang, R. Liu and Y. Huang. Regulation of the thermal sensitivity of hydroxypropyl cellulose by poly (N-isopropylacryamide) side chains. Carbohydrate Polymers 2013; 95(1): 155-160.
79. Wang, D., J. Tan, H. Kang, L. Ma, X. Jin, R. Liu and Y. Huang. Synthesis, self-assembly and drug release behaviors of pH-responsive copolymers ethyl cellulose-graft-PDEAEMA through ATRP. Carbohydrate Polymers 2011; 84(1): 195-202.
80. Tan, J., Y. Li, R. Liu, H. Kang, D. Wang, L. Ma, W. Liu, M. Wu and Y. Huang. Micellization and sustained drug release behavior of EC-g-PPEGMA amphiphilic copolymers. Carbohydrate Polymers 2010; 81(2): 213-218.
81. Zhong, J.-F., X.-S. Chai and S.-Y. Fu. Homogeneous grafting poly (methyl methacrylate) on cellulose by atom transfer radical polymerization. Carbohydrate Polymers 2012; 87(2): 1869-1873.
82. Aznar, E., R. Casasús, B. García-Acosta, M.D. Marcos, R. Martínez-Máñez, F. Sancenón, J. Soto and P. Amorós. Photochemical and chemical two-channel control of functional nanogated hybrid architectures. Advanced Materials 2007; 19(17): 2228-2231.
83. Black, A.L., J.M. Lenhardt and S.L. Craig. From molecular mechanochemistry to stress-responsive materials. Journal of Materials Chemistry 2011; 21(6): 1655-1663.
84. Ercole, F., T.P. Davis and R.A. Evans. Photo-responsive systems and biomaterials: photochromic polymers, light-triggered self-assembly, surface modification, fluorescence modulation and beyond. Polymer Chemistry 2010; 1(1): 37-54.
85. Jiang, X., C.A. Lavender, J.W. Woodcock and B. Zhao. Multiple micellization and dissociation transitions of thermo-and light-sensitive poly (ethylene oxide)-b-poly (ethoxytri (ethylene glycol) acrylate-co-o-nitrobenzyl acrylate) in water. Macromolecules 2008; 41(7): 2632-2643.
86. Schumers, J.M., C.A. Fustin and J.F. Gohy. Light-responsive block copolymers. Macromolecular Rapid Communications 2010; 31(18): 1588-1607.
87. Rahimi, M. and R. Behrooz. Effect of cellulose characteristic and hydrolyze conditions on morphology and size of nanocrystal cellulose extracted from wheat straw. International Journal of Polymeric Materials 2011; 60(8): 529-541.
88. Favier, V., H. Chanzy and J. Cavaille. Polymer nanocomposites reinforced by cellulose whiskers. Macromolecules 1995; 28(18): 6365-6367.
89. Brännvall, E. Aspects on Strenght Delivery and Higher Utilisation of the Strength Potential of Kraft Pulp Fibres. 2007.
90. Rånby, B. and R. Marchessault. Inductive effects in the hydrolysis of cellulose chains. Journal of Polymer Science 1959; 36(130): 561-564.

91. Marchessault, R., F. Morehead and N. Walter. Liquid crystal systems from fibrillar polysaccharides. Nature 1959; 184(4686): 632-633.

92. Moon, R.J., A. Martini, J. Nairn, J. Simonsen and J. Youngblood. Cellulose nanomaterials review: structure, properties and nanocomposites. Chemical Society Reviews 2011; 40(7): 3941-3994.

93. Battista, O.A. Hydrolysis and crystallization of cellulose. Industrial & Engineering Chemistry 1950; 42(3): 502-507.

94. Habibi, Y., T. Heim and R. Douillard. AC electric field-assisted assembly and alignment of cellulose nanocrystals. Journal of Polymer Science Part B: Polymer Physics 2008; 46(14): 1430-1436.

95. Li, W., J. Yue and S. Liu. Preparation of nanocrystalline cellulose via ultrasound and its reinforcement capability for poly (vinyl alcohol) composites. Ultrasonics Sonochemistry 2012; 19(3): 479-485.

96. Alemdar, A. and M. Sain. Isolation and characterization of nanofibers from agricultural residues–Wheat straw and soy hulls. Bioresource Technology 2008; 99(6): 1664-1671.

97. Habibi, Y., L.A. Lucia and O.J. Rojas. Cellulose nanocrystals: chemistry, self-assembly, and applications. Chemical Reviews 2010; 110(6): 3479-3500.

98. Saxena, I.M. and R.M. Brown, Cellulose biosynthesis: current views and evolving concepts. Annals of Botany 2005; 96(1): 9-21.

99. Ruiz, M.M., J.Y. Cavaillé, A. Dufresne, J.F. Gérard and C. Graillat. Processing and characterization of new thermoset nanocomposites based on cellulose whiskers. Composite Interfaces 2000; 7(2): 117-131.

100. Araki, J., M. Wada, S. Kuga and T. Okano. Flow properties of microcrystalline cellulose suspension prepared by acid treatment of native cellulose. Colloids and Surfaces A: Physicochemical and Engineering Aspects 1998; 142(1): 75-82.

101. Revol, J.-F., H. Bradford, J. Giasson, R. Marchessault and D. Gray. Helicoidal self-ordering of cellulose microfibrils in aqueous suspension. International Journal of Biological Macromolecules 1992; 14(3): 170-172.

102. Camarero Espinosa, S., T. Kuhnt, E.J. Foster and C. Weder. Isolation of thermally stable cellulose nanocrystals by phosphoric acid hydrolysis. Biomacromolecules 2013; 14(4): 1223-1230.

103. Sun, Y., L. Lin, C. Pang, H. Deng, H. Peng, J. Li, B. He and S. Liu. Hydrolysis of cotton fiber cellulose in formic acid. Energy & Fuels 2007; 21(4): 2386-2389.

104. Lee, S.-Y., D.J. Mohan, I.-A. Kang, G.-H. Doh, S. Lee and S.O. Han. Nanocellulose reinforced PVA composite films: Effects of acid treatment and filler loading. Fibers and Polymers 2009; 10(1): 77-82.

105. Lu, P. and Y.-L. Hsieh. Preparation and properties of cellulose nanocrystals: rods, spheres, and network. Carbohydrate Polymers 2010; 82(2): 329-336.

106. Jonoobi, M., R. Oladi, Y. Davoudpour, K. Oksman, A. Dufresne, Y. Hamzeh and R. Davoodi. Different preparation methods and properties of nanostructured cellulose from various natural resources and residues: A review. Cellulose 2015; 22(2): 935-969.

107. Beck-Candanedo, S., M. Roman and D.G. Gray. Effect of reaction conditions on the properties and behavior of wood cellulose nanocrystal suspensions. Biomacromolecules 2005; 6(2): 1048-1054.

108. Filson, P.B., B.E. Dawson-Andoh and D. Schwegler-Berry. Enzymatic-mediated production of cellulose nanocrystals from recycled pulp. Green Chemistry 2009; 11(11): 1808-1814.

109. Filson, P.B. and B.E. Dawson-Andoh. Sono-chemical preparation of cellulose nanocrystals from lignocellulose derived materials. Bioresource Technology 2009; 100(7): 2259-2264.

110. Yue, Y., C. Zhou, A.D. French, G. Xia, G. Han, Q. Wang and Q. Wu. Comparative properties of cellulose nano-crystals from native and mercerized cotton fibers. Cellulose 2012; 19(4): 1173-1187.

111. Satyamurthy, P., P. Jain, R.H. Balasubramanya and N. Vigneshwaran. Preparation and characterization of cellulose nanowhiskers from cotton fibres by controlled microbial hydrolysis. Carbohydrate Polymers 2011; 83(1): 122-129.

112. Iwamoto, S., W. Kai, A. Isogai and T. Iwata. Elastic modulus of single cellulose microfibrils from tunicate measured by atomic force microscopy. Biomacromolecules 2009; 10(9): 2571-2576.

113. Hirai, A., O. Inui, F. Horii and M. Tsuji. Phase separation behavior in aqueous suspensions of bacterial cellulose nanocrystals prepared by sulfuric acid treatment. Langmuir 2008; 25(1): 497-502.

114. George, J., K. Ramana and A. Bawa. Bacterial cellulose nanocrystals exhibiting high thermal stability and their polymer nanocomposites. International Journal of Biological Macromolecules 2011; 48(1): 50-57.

115. Rosa, M.F., E.S. Medeiros, J.A. Malmonge, K.S. Gregorski, D.F. Wood, L.H.C. Mattoso, G. Glenn, W.J. Orts and S.H. Imam. Cellulose nanowhiskers from coconut husk fibers: Effect of preparation conditions on their thermal and morphological behavior. Carbohydrate Polymers 2010; 81(1): 83-92.
116. Jonoobi, M., A. Khazaeian, P. Md. Tahir, S.S. Azry and K. Oksman. Characteristics of cellulose nanofibers isolated from rubberwood and empty fruit bunches of oil palm using chemo-mechanical process. Cellulose 2011; 18(4): 1085-1095.
117. Hassan, M.L., A.P. Mathew, E.A. Hassan, N.A. El-Wakil and K. Oksman. Nanofibers from bagasse and rice straw: process optimization and properties. Wood Science and Technology 2012; 46(1-3): 193-205.
118. Salas, C., T. Nypelö, C. Rodriguez-Abreu, C. Carrillo and O.J. Rojas. Nanocellulose properties and applications in colloids and interfaces. Current Opinion in Colloid & Interface Science 2014; 19(5): 383-396.
119. Samir, M.A.S.A., F. Alloin, J.-Y. Sanchez and A. Dufresne. Cellulose nanocrystals reinforced poly (oxyethylene). Polymer 2004; 45(12): 4149-4157.
120. Eichhorn, S.J., A. Dufresne, M. Aranguren, N.E. Marcovich, J.R. Capadona, S.J. Rowan, C. Weder, W. Thielemans, M. Roman, S. Renneckar, W. Gindl, S. Veigel, J. Keckes, H. Yano, K. Abe, M. Nogi, A.N. Nakagaito, A. Mangalam, J. Simonsen, A.S. Benight, A. Bismarck, L.A. Berglund and T. Peijs. Review: current international research into cellulose nanofibres and nanocomposites. Journal of Materials Science 2010; 45(1): 1-33.
121. Kalia, S., A. Dufresne, B.M. Cherian, B. Kaith, L. Avérous, J. Njuguna and E. Nassiopoulos. Cellulose-based bio-and nanocomposites: A review. International Journal of Polymer Science 2011; ID 837875: 35.
122. Noble, B., S. Harris and K. Dinsdale. The elastic modulus of aluminium-lithium alloys. Journal of Materials Science 1982; 17(2): 461-468.
123. Shackelford, J.F., Y.-H. Han, S. Kim and S.-H. Kwon. CRC Materials Science and Engineering Handbook, 4th Ed. CRC Press, Boca Raton, 2016.
124. Soroushian, P. and K.-B. Choi, Steel mechanical properties at different strain rates. Journal of Structural Engineering 1987; 113(4): 663-672.
125. Wambua, P., J. Ivens and I. Verpoest. Natural fibres: Can they replace glass in fibre reinforced plastics? Composites Science and Technology 2003; 63(9): 1259-1264.
126. Ljungberg, N., C. Bonini, F. Bortolussi, C. Boisson, L. Heux and J.Y. Cavaillé. New nanocomposite materials reinforced with cellulose whiskers in atactic polypropylene: effect of surface and dispersion characteristics. Biomacromolecules 2005; 6(5): 2732-2739.
127. de Menezes, A.J., G. Siqueira, A.A. Curvelo and A. Dufresne. Extrusion and characterization of functionalized cellulose whiskers reinforced polyethylene nanocomposites. Polymer 2009; 50(19): 4552-4563.
128. Roohani, M., Y. Habibi, N.M. Belgacem, G. Ebrahim, A.N. Karimi and A. Dufresne. Cellulose whiskers reinforced polyvinyl alcohol copolymers nanocomposites. European Polymer Journal 2008; 44(8): 2489-2498.
129. Oksman, K., A.P. Mathew, D. Bondeson and I. Kvien. Manufacturing process of cellulose whiskers/ polylactic acid nanocomposites. Composites Science and Technology 2006; 66(15): 2776-2784.
130. Pandey, J.K., A.N. Nakagaito and H. Takagi. Fabrication and applications of cellulose nanoparticle-based polymer composites. Polymer Engineering & Science 2013; 53(1): 1-8.
131. Pan, H., L. Song, L. Ma and Y. Hu. Transparent epoxy acrylate resin nanocomposites reinforced with cellulose nanocrystals. Industrial & Engineering Chemistry Research 2012; 51(50): 16326-16332.
132. Montanari, S., M. Roumani, L. Heux and M.R. Vignon. Topochemistry of carboxylated cellulose nanocrystals resulting from TEMPO-mediated oxidation. Macromolecules 2005; 38(5): 1665-1671.
133. Eyley, S. and W. Thielemans. Surface modification of cellulose nanocrystals. Nanoscale 2014; 6(14): 7764-7779.
134. Boujemaoui, A., S. Mongkhontreerat, E. Malmströmand, A. Carlmark. Preparation and characterization of functionalized cellulose nanocrystals. Carbohydrate Polymers 2015; 115: 457-464.
135. Rohaizu, R. and W. Wanrosli. Sono-assisted TEMPO oxidation of oil palm lignocellulosic biomass for isolation of nanocrystalline cellulose. Ultrasonics Sonochemistry 2017; 34: 631-639.
136. Salajková, M., L.A. Berglund and Q. Zhou. Hydrophobic cellulose nanocrystals modified with quaternary ammonium salts. Journal of Materials Chemistry 2012; 22(37): 19798-19805.

137. Fujisawa, S., T. Ikeuchi, M. Takeuchi, T. Saito and A. Isogai. Superior reinforcement effect of TEMPO-oxidized cellulose nanofibrils in polystyrene matrix: optical, thermal, and mechanical studies. Biomacromolecules 2012; 13(7): 2188-2194.
138. Nagalakshmaiah, M., N. El Kissi and A. Dufresne. Ionic compatibilization of cellulose nanocrystals with quaternary ammonium salt and their melt extrusion with polypropylene. ACS Applied Materials & Interfaces 2016; 8(13): 8755-8764.
139. Yang, J., C.-R. Han, X.-M. Zhang, F. Xu and R.-C. Sun . Cellulose nanocrystals mechanical reinforcement in composite hydrogels with multiple cross-links: correlations between dissipation properties and deformation mechanisms. Macromolecules 2014; 47(12): 4077-4086.
140. Herreros-López, A., M. Carini, T. Da Ros, T. Carofiglio, C. Marega, V. La Parola, V. Rapozzi, L.E. Xodo, Ali A. Alshatwi, C. Hadad and M. Prato. Nanocrystalline cellulose-fullerene: Novel conjugates. Carbohydrate Polymers 2017; 164: 92-101.
141. Javanbakht, T., W. Raphael and J.R. Tavares. Physicochemical properties of cellulose nanocrystals treated by photo-initiated chemical vapour deposition (PICVD). The Canadian Journal of Chemical Engineering 2016; 94(6): 1135-1139.
142. Sirviö, J.A., M. Visanko, O. Laitinen, A. Ämmälä and H. Liimatainen. Amino-modified cellulose nanocrystals with adjustable hydrophobicity from combined regioselective oxidation and reductive amination. Carbohydrate Polymers 2016; 136: 581-587.
143. Gwon, J.-G., H.-J. Cho, S.-J. Chun, S. Lee, Q. Wu and S.-Y. Lee. Physiochemical, optical and mechanical properties of poly (lactic acid) nanocomposites filled with toluene diisocyanate grafted cellulose nanocrystals. RSC Advances 2016; 6(12): 9438-9445.
144. Ooi, S.Y., I. Ahmad and M.C.I.M. Amin. Cellulose nanocrystals extracted from rice husks as a reinforcing material in gelatin hydrogels for use in controlled drug delivery systems. Industrial Crops and Products 2016; 93: 227-234.
145. Tian, X., D. Yan, Q. Lu and X. Jiang. Cationic surface modification of nanocrystalline cellulose as reinforcements for preparation of the chitosan-based nanocomposite films. Cellulose 2017; 24(1): 163-174.
146. Braun, B. and J.R. Dorgan. Single-step method for the isolation and surface functionalization of cellulosic nanowhiskers. Biomacromolecules 2008; 10(2): 334-341.
147. Zoppe, J.O., X. Xu, C. Känel, P. Orsolini, G. Siqueira, P. Tingaut, T. Zimmermann and H.-A. Klok. Effect of surface charge on surface-initiated atom transfer radical polymerization from cellulose nanocrystals in aqueous media. Biomacromolecules 2016; 17(4): 1404-1413.
148. Courtenay, J.C., M.A. Johns, J.L. Scott and R.I. Sharma. Surface modified cellulose scaffolds for tissue engineering. Cellulose 2017; 24(1): 253-267.
149. Zaman, M., H. Xiao, F. Chibante and Y. Ni. Synthesis and characterization of cationically modified nanocrystalline cellulose. Carbohydrate Polymers 2012; 89(1): 163-170.

Chapter 8

Chitin and Chitosan and Their Polymers

Md. Saifur Rahaman[1], Jahid M.M. Islam[2,3], Md. Serajum Manir[2], Md. Rabiul Islam[4] and Mubarak A. Khan[2,5*]

[1]Institute of Nuclear Science and Technology, Atomic Energy Research Establishment, Bangladesh Atomic Energy Commission, Savar, Dhaka, Bangladesh

[2]Institute of Radiation and Polymer Technology, Atomic Energy Research Establishment, Bangladesh Atomic Energy Commission, Savar, Dhaka, Bangladesh

[3]Monash University Malaysia, Bandar Sunway, Selangor, 47500, Malaysia

[4]Radioactivity Testing and Monitoring Laboratory, Mongla Port Area, Bagerhat, Bangladesh

[5]Bangladgesh Jute Mills Corporation, Adamjee Court Annex-1, 115–120 Motijheel, Dhaka

8.1 INTRODUCTION

Chitin, a linear polysaccharide composed of (1-4)-linked 2-acetamido-2-deoxy-β-D-glucopyranose units, is the second prevalent form of natural polymerized carbon after cellulose (Fig. 8.1). It is categorized as a cellulose derivative, in spite of the fact that it does not appear in organisms producing cellulose. Its structure is similar to cellulose, but at the C2 position, it has an acetamide group ($NHCOCH_3$) [1].

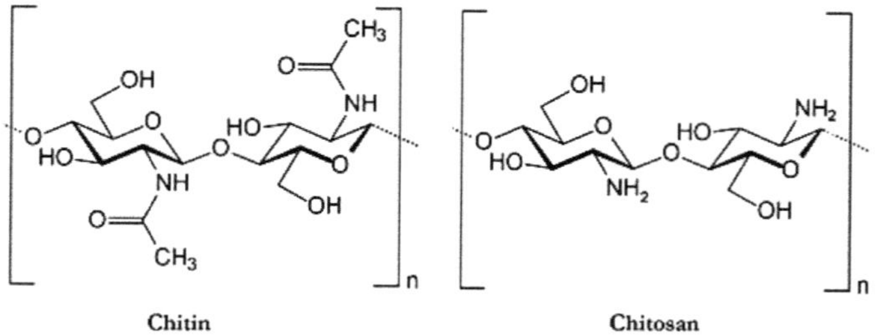

Figure 8.1 Structure of Chitin and Chitosan.

*Corresponding author: Email: makhan.inst@gmail.com

Chitin occurrs as a structural polysaccharide in the outer skeleton of animals belonging to the phylum Arthropoda (animals with an outer skeleton) and a component of the cell walls of certain fungi and algae. It is also produced by a number of other living organisms in the lower plant and animal kingdoms (Fig. 8.2) (the radulae of molluscs, cephalopod beaks, and the scales of fish and lissamphibians), serving in many functions where reinforcement and strength are required. In contrast, chitosan is much less abundant in nature than chitin, and has so far been found only in the cell walls of certain fungi. The derivative of chitin with degree of deacetylation of approximately 50 % is known as "chitosan" and it is soluble in aqueous acidic solutions. Chitosan is therefore a copolymer that comprises of *N*-acetyl-D-glucosamine and deacetylated D-glucosamine units (Fig. 8.1). The structure of chitosan obtained by complete deacetylation of chitin [1–4].

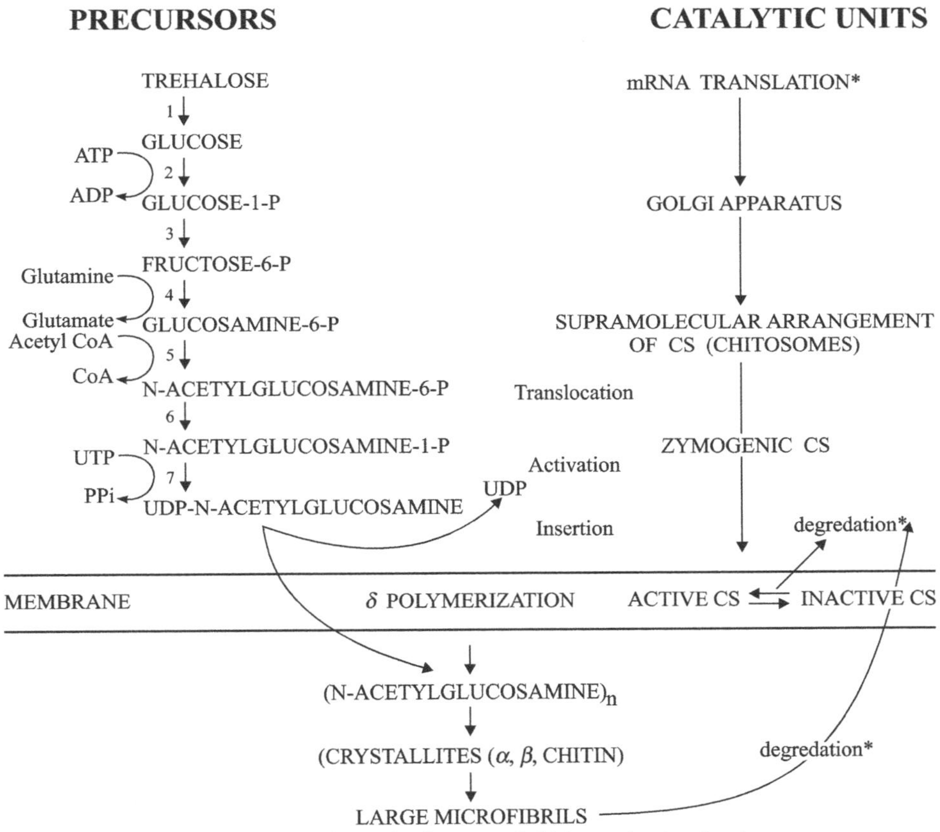

Figure 8.2 Schematic diagram of chitin production in vivo.

Every year, molluscs, crustaceans, insects, fungus, algae, and related organisms approximately produce 10 billion ton of chitin [5]. Chitin is bio-renewable, environmentally friendly, biocompatible, biodegradable, and biofunctional, and is beneficial as a chelating agent, water treatment additive, drug carrier, biodegradable pressure-sensitive adhesive tape, wound-healing agent in membranes, and has other advantages for several important applications in medicine and biochemistry. Due to these advantages, much attention is paid to this characteristic biomaterial. However, nowadays, chitin is not vastly employed by the pharmaceutical industry. Due to its weak solubility, it has unique applications. Chitin is insoluble in common organic solvents and diluted aqueous solvents because it is highly hydrophobic due to the highly expanded hydrogen-bonded semi-crystalline structure of chitin. Its derivative, chitosan, is prepared by deacetylation and depolymerization of native chitin, (partial) deacetylation of chitin in the solid state under alkaline conditions (concentrated NaOH), or enzymatic hydrolysis in the presence of a chitin deacetylase. [2–4].

8.2 RAW MATERIALS

Fishing is one the most important sectors, particularly in the Asian countries including, China, Japan, India, and Thailand. Asia plays a leading role in shrimp farming, accounting for almost 80% of world shrimp production [6–7]. Globally, over 6 to 8 million tons of waste shrimp, lobster, and crab shells are produced each year, out of which over 1.5 million tons is produced in Southeast Asia alone [8]. Shrimp landed by large trawlers are however, deheaded at sea or supplied to processing industries. Heads are usually removed in peeling sheds near the landing or at packing plants. Generally, shrimp is exported in frozen form without exoskeleton. About 45–48% by weight of shrimp raw material is discarded as waste depending on the species. The shrimp waste is composed mainly of protein (40%), minerals (35%), and chitin (14–30%), and is very rich in carotenoid pigments, mainly Astaxanthin [9]. In majority of the developing countries, the shell waste is often dumped directly into landfills or in the sea. This disposal for raw material is costly, for example, it costs up to USD 150/ton in Australia, and hence the raw material suppliers are likely to offer these products even at low costs, which are expected to offer an advantage for chitosan manufacturers. In Europe, according to the statistic estimation of Food and Agriculture Organization, FAOSTAT, over 750,000 tons of crustacean shell waste is generated every year. The waste shells serve as growth media for pathogenic bacteria and can cause significant health and environmental hazards and surface pollution in coastal regions. To avoid these problems, manufacturers used to burn the shell waste, which again is a costly activity and has many environmental disadvantages. Thus, these manufacturers tend to sell this waste to chitin and chitosan manufacturers.

8.3 PRODUCTION

A variety of procedures have been developed and proposed over the years for preparation of pure chitosan. Most of the processes are based on the chemical processes for industrial production of chitosan from crustacean shell waste. For chitin production, the raw materials most abundantly available are the shells of crab, shrimp, and prawn (69–70%) [10–12]. As chitin is associated with other constituents, harsh treatments are required to remove them from chitinous material to prepare chitin and then chitosan on a large scale. Proteins are removed from ground shells by treating them with either sodium hydroxide or by digestion with proteolytic enzymes, such as papain, pepsin, trypsin, and pronase [13]. Minerals, such as calcium carbonate and calcium phosphate, are extracted with hydrochloric acid. Pigments, such as melanin and carotenoids, are eliminated with 0.02% potassium permanganate at 60°C or hydrogen peroxide or sodium hypochlorite. Conversion of chitin to chitosan is generally achieved by hydrolysis of acetamide groups of chitin. This is normally conducted by severe alkaline hydrolysis treatment due to the resistance of such groups imposed by the trans-arrangement of the C2–C3 substituent in the sugar ring [14]. Thermal treatments of chitin under strong aqueous alkali are usually needed to give partially deacetylated chitin (degree of acetylation, DA <30%), regarded as chitosan. Usually, this process is achieved by treatment with concentrated sodium or potassium hydroxide solution (40–50%) at 100°C or higher to remove some or all the acetyl groups from the polymer [11, 15–16]. This process, called deacetylation, releases amine groups (NH_2) and gives the chitosan a cationic characteristic. This is especially interesting in an acid environment, where the majority of polysaccharides are usually neutral or negatively charged. The deacetylation process is carried out either at room temperature (homogeneous deacetylation) or at elevated temperature (heterogeneous deacetylation), depending on the nature of the final product desired. However, the latter is preferred for industrial purposes. In some cases, the deacetylation reaction is carried out in the presence of thiophenol as a scavenger of oxygen or under N_2 atmosphere to prevent chain degradation that invariably occurs due to peeling reaction under strong alkaline conditions [17].

One other method of preparing chitosan of improved purity is to dissolve the materials in an acid (e.g., acetic acid) and filter to remove extraneous materials. The clarified product is then lyophilized to give a water-soluble chitosonium acid salt or precipitated with NaOH, washed, and dried to give a product in the free amine form [11]. Recent advances in fermentation technologies suggest that the cultivation of selected fungi can provide an alternative source of chitin and chitosan. The amount of these polysaccharides depends on the fungi species and culture conditions. Fungal mycelia are relatively consistent in composition and are not associated with inorganic materials; therefore, no demineralization treatment is required to recover fungal chitosan [18–27]. The use of biomass from fungi has demonstrated great advantages, such as independence of seasonal factor, wide scale production, simultaneous extraction of chitin and chitosan, simple and cheap extraction process resulting in reduction in time and cost required for production, and also absence of proteins contamination, mainly the proteins that could cause allergy reactions in individuals with shellfish allergies [18, 28–32]. However, to optimize the production of chitin and chitosan from fungi, complex or synthetics cultures media, which are expensive, are usually used. It has become necessary to obtain economic culture media that promote the growth of fungi and stimulate the production of the polymers. Recently, microbiological processes were used for chitin and chitosan production by Cunninghamellaelegans grown by submerse fermentation in economic culture medium, with yam bean (*Pachyrhizus erosus* L. Urban) as substrate. The main characteristic of yam bean is the simple manipulation and low nutrition requirements when compared with other similar cultures, and tuberous roots yields are up to 60 ton/ha [33].

8.4 LIMITATION OF CHITOSAN AND MODIFICATIONS

Chitosan suffers from poor solubility in water at a physiological pH of 7.4, which is a major drawback for drug formulations, limiting its use as absorption enhancer in, for example, nasal or peroral delivery systems [34]. Indeed, chitosan is only soluble in acidic solutions of pH below 6.5, required to insure the protonation of the primary amine. In such cases, the presence of positive charges on the chitosan skeleton increases the repulsion between the different polymer chains, facilitating their solubilization. Another limitation of chitosan for the preparation of sustained release systems arises from its rapidly adsorbing water and higher swelling degree in aqueous environments, leading to fast drug release [35]. As far as organic solvents are concerned, chitosan is slightly soluble in dimethyl sulfoxide (DMSO) and p-toluene sulfonic acid [36]. This poor solubility is a limitation for the processing of chitosan and is also a brake in its chemical modification. In order to overcome these problems, a number of chemically modified chitosan derivatives have been synthesized.

In order to improve the solubility or impart new properties to chitosan, modification of the chitosan chains, generally by either grafting of small molecules or polymer chains onto the chitosan backbone or by alkylation, sulfation, thiolation, carboxymethylation, etc. of the amino groups, has been investigated. Chitosan chains possess three attractive reactive sites for chemical modification: two hydroxyl groups (primary or secondary) at C(3) and C(6), respectively and one primary amine at C(2). The site of modification is dictated by the desired application of the final chitosan derivative.

8.4.1 Modification with Functional Groups

The three reactive sites of chitosan impart great significance because the fundamental skeleton of chitosan and its original outstanding physicochemical and biochemical properties as a biomaterial are retained after modification, while achieving desirable structures and functions particularly suitable for desired biomedical applications. Important chitosan derivatives include sulfated chitosan, trimethyl chitosan, thiolated chitosan, hydroxyalkyl chitosan, carboxyalkyl chitosan, phosphorylated chitosan, and cyclodextrin-linked chitosan.

8.4.1.1 Modification by Sulfate

Derivatization of chitosan via the chemical modification of the amino or hydroxyl group with a sulfate group has been of increasing interest, mainly because of the structural similarity of the modified chitosan to heparin, which displays anticoagulant, antisclerotic, and antiviral activity, and bioaffinity to growth factors [37–38]. Various reagents, such as concentrated sulfuric acid, oleum, chloro sulfonic acid–sulfuric acid, sulfurtrioxide, sulfurtrioxide/pyridine, sulfur trioxide/sulfur dioxide, sulfurtrioxide/trimethylamine, and chlorosulfonic acid have been most commonly used in the sulfation of chitosan (Fig. 8.3) [39–41]. These synthetic processes have been performed under both homogeneous and heterogeneous conditions in solvent media, such as tetrahydrofuran, formic acid dimethylformamide (DMF), and DMF–dichloroacetic acid [42]. It has been reported that increasing the sulfur content in chitosan increased its anticoagulant activity. For example, 6-O-carboxymethyl chitosan N-sulfate exhibited 23% of the activity of heparin, O-sulfated form of 6-O-carboxymethyl chitosan exhibited 45% of the activity of heparin, and N-Carboxymethyl chitosan 3,6-disulfate exhibited anticoagulant heparin-like activity [43–45]. Synthesis and comparative bioactivity of bone morphogenetic protein-2 (BMP-2) studies of N-Sulfate, 6-O-sulfated, and 2-N,6-O-sulfated chitosans showed that the 6-O-sulfated substitution primarily increased the bioactivity of BMP-2, while 2-N-sulfate acted as a subsidiary group, which provided less activity, and a low dose of 2-N, 6-O-sulfated chitosan markedly enhanced the activity of alkaline phosphatase (ALP) and the mineralization that was induced by BMP- 2 [46]. Affinity to many kinds of growth factors has made the sulfated chitosan a promising component for building of scaffolds or delivery systems for regeneration and tissue repair. Amphiphilic N-octyl-O-sulfate chitosan can form micelles and has been used as a delivery system for paclitaxel, which is a water-insoluble drug [47].

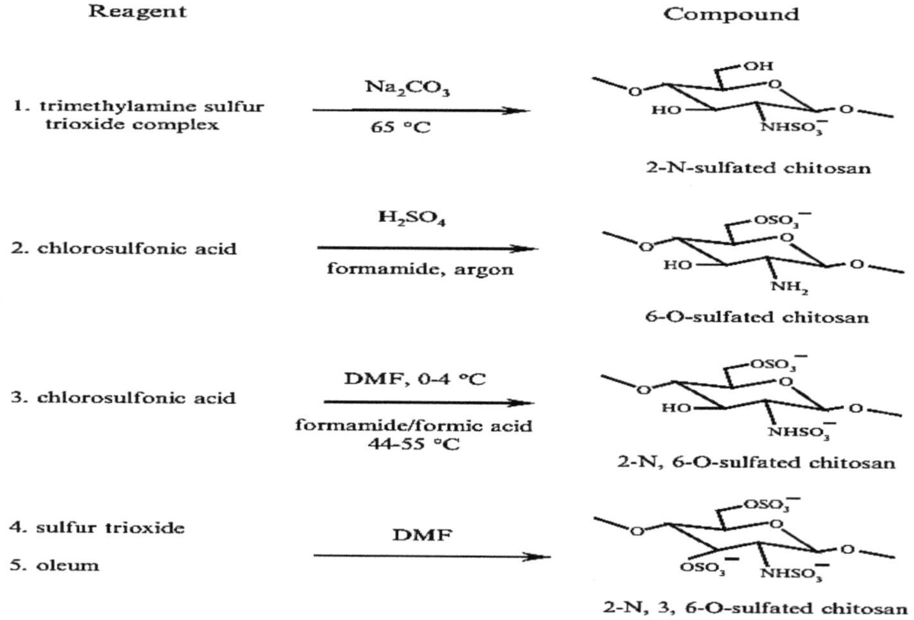

Figure 8.3 Reaction scheme for the sulfation of chitosan [48].

8.4.1.2 Modified by Methylation (Quaternization)

Protonated form of chitosan in acid conditions has been proven to exhibit bio adhesion and permeation properties by transiently opening the tight junctions in cell membrane, which is of great significance in drug or gene delivery [49]. Development of trimethyl chitosan (TMC) is an effort to boost the positive charge density, resulting in an enhanced solubility of chitosan over a broader range of pH. Methylation of chitosan can be achieved via the reaction of methyl iodide and amino groups (Fig. 8.4) [50]. This methylation or quaternization is based on the nucleophilic alkylation of the primary amino group at the C-2 position of chitosan [51]. The physicochemical and biological properties of TMC are strongly determined by the degree of quaternization (DQ), which is controlled by the time and route of the reaction. Properties such as, permeation enhancement has led the use of TMC as a delivery vector for proteins and genes. For example, TMC nanoparticles were fabricated by ionic crosslinking with tripoly phosphate (TPP) and used as a delivery system for ovalbumin [52]. The influence of DQ on the property of protein-loaded TMC/TPP nanoparticles was investigated by Chen et al. [48]. They reported that a lower DQ leads to an increased particle size and a slower release rate of protein.

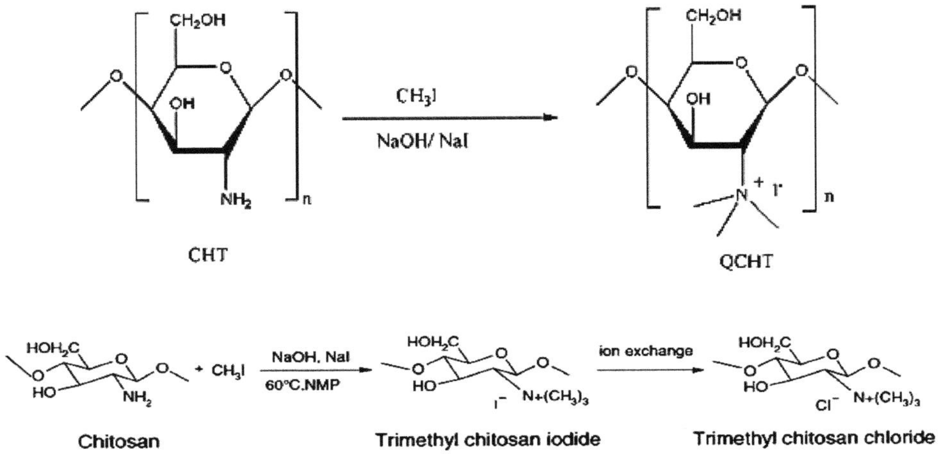

Figure 8.4 Synthesis of quaternized chitosan [50, 53–54].

TMC with high Mw were prepared and explored for the role of DQ in terms of cytotoxicity and gene transfection efficiency [55]. It was found that the TMC showed some toxicity under a long-term cell culture and an increase of the DQ above 12% lowers the gene transfection efficiency. Poly(N-isopropylacrylamide) (PNIPAAm) can further improve the gene delivery efficiency and biocompatibility of TMC [56]. The TMC-g-PNIPAAm/DNA can form compact particles above the lower critical solution temperature (LCST) of PNIPAAm as a result of collapse of the PNIPAAm chains. After cellular uptake of the particles, the culture was maintained below the LCST for a period of time to facilitate the unpacking of DNA from the complexes. By this strategy, the gene transfection efficiency was significantly improved, while the overall good cytocompatibility was maintained. Other studies have shown that modifying TMC with PEG and folate can further enhance the colloidal stability and cellular uptake of the TMC/gene particles [57–58].

8.4.1.3 Modified by Carboxymethylation

Carboxymethylation is an outstanding process to fabricate modified chitosans with high water solubility along with numerous chemical, physical, and biological properties, including high viscosity, gel formation, film formation, biodegradability, nontoxicity, biocompatibility, and antibacterial and antifungal activities. These excellent features of carboxymethylated chitosans are attractive for being used in biomedical, pharmaceutical, and cosmetic applications [59–60]. Both N-carboxymethyl and O-carboxymethyl chitosan derivatives have been prepared under different reaction conditions using monochloroacetic acid in the presence of NaOH to achieve the N-versus-O selectivity (Fig. 8.5) [61–62].

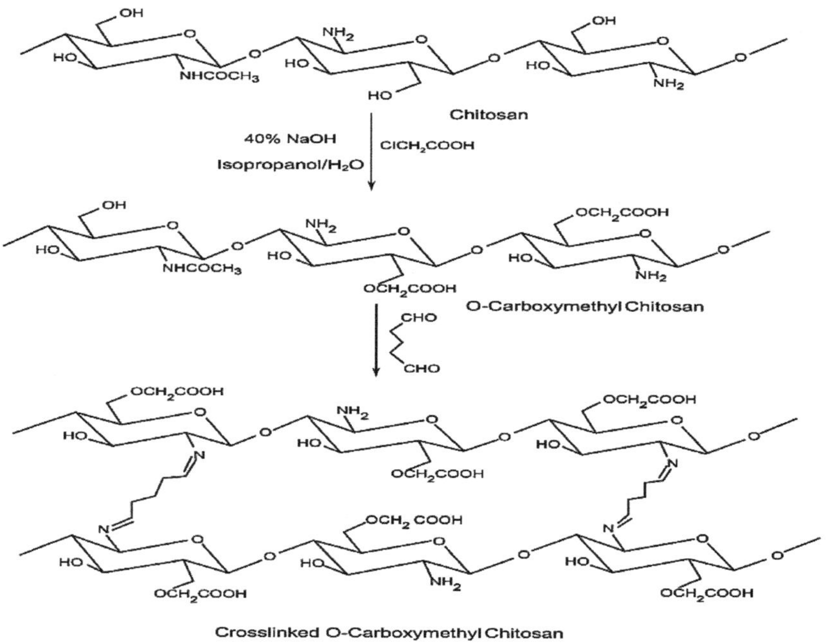

Figure 8.5 The reaction scheme for the carboxymethylation of chitosan [63].

N,O-carboxymethyl chitosan (N,O–CC) is a chitosan derivative, having carboxymethyl substituents at some of both the amine and the 6-hydroxyl sites of its glucosamine units, and can be easily prepared using chitosan, sodium hydroxide, and isopropanol with chloroacetic acid [64]. N,O–CC is used in the development of wound dressing materials and various functional hydrogels, including superporous or pH sensitive hydrogels for protein drug delivery [65–66]. Another promising carboxymethyl chitosan derivative called N-succinyl chitosan has been reported to be obtained by introducing succinyl groups at the N-terminal of the glucosamine units of chitosan [67]. Various macromolecule–antitumor drug conjugates are reported by researchers from the synthesis process of N-succinyl chitosan and drug conjugation [68–69].

8.4.1.4 Modified by Thiolation

Thiolated chitosans can be synthesized by covalently coupling thiol groups onto chitosan. They exhibit *in situ* gel-forming properties owing to the formation of inter- and/or intra-molecular disulfide bonds at pH values above 5 [70]. The thiolation reaction of chitosan involves amide bond formation between the carboxylic acid group of the thiol reagent (Fig. 8.6) and the primary amino group of chitosan, mediated by a water-soluble carbodiimide [71]. Thiolated chitosans can also be prepared by modifying chitosan with 2-iminothiolane or isopropyl-S-acetyl thioacetimidate. Along with a sulfhydryl group, a cationic moiety is also introduced in the form of an amidine substructure, resulting in a chitosan–4-thio-butylamidine conjugate (chitosan–TBA) or chitosan–thioethylamidine conjugate (chitosan–TEA) [72]. The mucoadhesion properties of chitosan are reported to be enhanced by the formation of disulfide bonds with cysteine-rich subdomains of mucus glycoproteins, and these bonds are stronger than non-covalent bonds [73–74]. Chitosan–TBA conjugates provide 140-fold stronger mucoadhesion than unmodified polymers [70, 75]. Thiolated chitosan based on chitosan–thioglycolic acid (chitosan–TGA) can be formulated using oxidizing agents, such as hydrogen peroxide (H_2O_2), sodium periodate ($NaIO_4$), ammonium persulfate ($(NH_4)_2S_2O_8$), and sodium hypochlorite (NaOCl). Chitosan-TGA has been proven to reduce the reverse volume phase time and increase the dynamic viscosity [76].

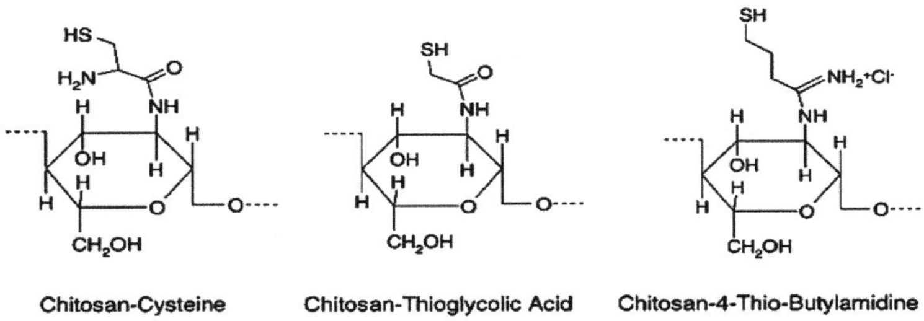

Figure 8.6 Substructures of thiolated chitosan derivatives

8.4.1.5 Modification by Aldehydes

Aldehydes can be used to modify chitosan moiety. Traditionally, chitosan is crosslinked by glutaraldehyde. The crosslinking process involves formation of imine bonds between the amino groups on chitosan chains and bifunctional glutaraldehyde crosslinker [77]. The glutaraldehydecrosslinked chitosan gels can be reformed into membranes, beads, tablets, and nano- or microspheres. The reaction temperature and concentration of glutaraldehyde were used to control the degree of crosslinking in chitosan-based microspheres [78–80]. Crosslinked chitosan microspheres have been investigated for potential use in drug delivery system [81–83]. The studies on drug release rates demonstrate that drug release rates may be manipulated by controlling the degree of crosslinking of the chitosan hydrogels or microspheres. Crosslinked chitosan microspheres reportedly release hydroquinone faster compared to their uncrosslinked counterparts [84].

8.4.2 Ionic Modification

The ionic interactions between the positively charged amino groups on chitosan and either small anionic molecules, such as sulfates, citrates, and phosphates, or some metal ions, have been successfully used in the preparation of chitosan beads or hydrogels [85–86]. Metal ions form coordinate-covalent bonds with chitosan, whereas small anionic molecules bind to the protonated amino groups on chitosan via electrostatic attractions [87]. In the ionotropic gelation method,

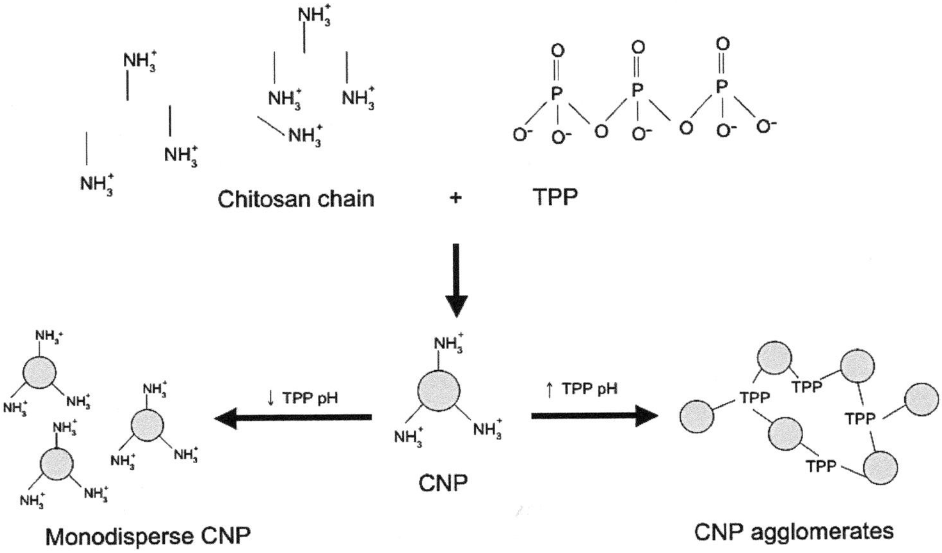

Figure 8.7 TPP-Chitosan synthesis [93]

chitosan dissolved in aqueous acidic solution is added drop-wise with constant stirring to TPP (Tripolyphosphate), food additive solution to form spherical particles. Changing the pH value of the curing agent, TPP, from basic to acidic, the ionic-crosslinking density of chitosan beads can be improved (Fig. 8.7) [88–89]. A new approach for preparing chitosan nanoparticles based on the ionotropic gelation method by aluminium monostearate in lactic acid solution has been developed lately (Fig. 8.8) [90]. Ionically crosslinked chitosan nanoparticles have been reported to show drug delivery actions. Anticancer or protein drugs, such as doxorubicin or insulin, can be effectively entrapped into the chitosan nanoparticles during ionotropic gelation process [91–92].

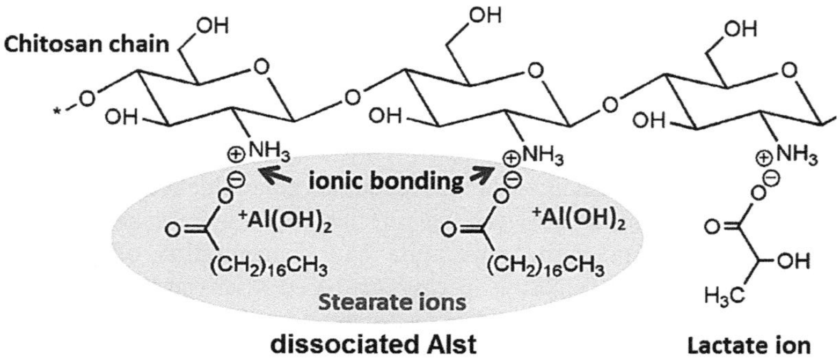

Figure 8.8 Ionic bonding between ammonium group of chitosan and the acid side chain (stearate molecule) [90]

8.4.3 Modification by Specific Ligands

Many attempts have been made to functionalize chitosan with specific ligands, in order to induce desirable interactions with cells and mediate specific cell responses and behavior. Combination with extra cellular matrix (ECM) proteins, such as collagen, fibronectin, and laminin is an attractive strategy for chitosan modification. In addition, implication of sugar moieties in cell signaling mechanisms and in recognition processes indicates them as another key component for chitosan modification.

8.4.3.1 Peptides

Direct modification by proteins has several limitations, such as low cell binding efficiency, immunogenicity and easy denaturation. However, the drawbacks are overcome by using small peptide sequences derived from these ECMs, since the short peptides are more stable against heat treatment, sterilization, and variation in pH. Arginine–glycine–aspartate (RGD) sequence or RGD-containing short peptides are by far the most extensively used ECM-derived biomolecules [94]. Various approaches have been developed to produce the RGD containing peptides on chitosan. Photochemical techniques based on phenyl-azido chemistry have been used by for grafting of RGDS peptide on the surface of chitosan film (Fig. 8.9) [95]. In another strategy, chitosan reacted with 2-iminothiolane to generate a SH-chitosan derivative, and subsequently, RGDSGGC was introduced by disulfide bond linkage with the aid of dimethyl sulfoxide [96]. This chitosan–RGDSGGC conjugate shows excellent adhesion and proliferation for chondrocytes and fibroblasts. A99a (ALRGDN) (avb3 integrin-binding peptide) and AG73 (RKRLQVQLSIRT) (syndecan-binding peptide)—these two kinds of peptide-conjugated chitosan membranes were prepared by binding A99a and AG73 to maleimidobenzoyloxy–chitosan [97]. Both bioactive membranes promoted cell attachment according to the cell-type specificity, with different mechanisms: AG73–chitosan membrane interacts with proteoglycan, whereas the A99a–chitosan membrane recognizes an integrin receptor. YIGSR is another laminin-derived peptide, such as A993 and

AG73, and has been regarded as capable of enhancing neural outgrowth [98]. When YIGSR-modified chitosan/hydroxyapatite tubes were implanted as bridge grafts into the sciatic nerve of Sprague-Dawley rats, enhancement of the nerve regeneration and sprouting from the proximal nerve stump occurred [98]. Glycine spacers were introduced to yield the CGGYIGSR peptide, to further enhance the bioactivity of the YIGSR sequence [99]. Bae et al. (2002) illustrated that EPDIM (an ECM protein, derived from big-h3), promoted keratinocyte adhesion, migration, and proliferation by integrin-mediated interactions [100].

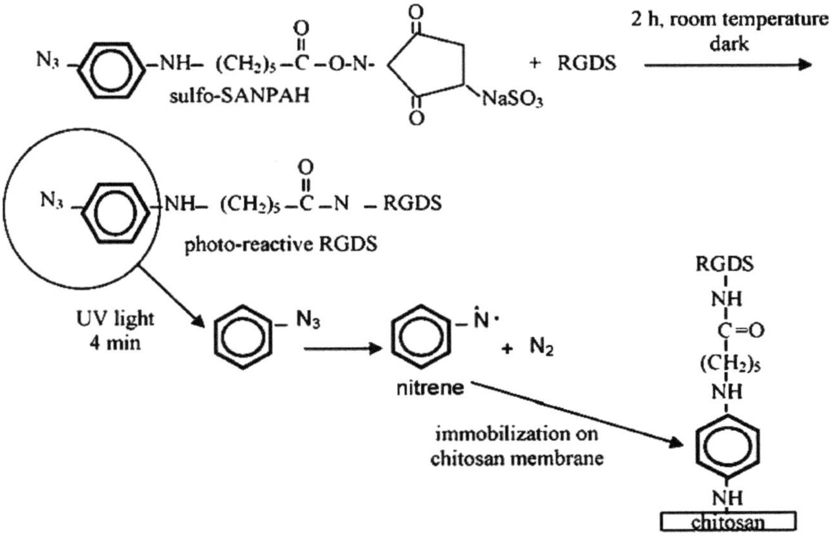

Figure 8.9 Production of nitrene groups in phenyl-azidoderivatized RGDS by UV irradiation and immobilization on chitosan surface [95].

8.4.3.2 Sugars

Galactose is the most popular ligand used to modify biomaterials to enhance hepatocyte adhesion as well as liver-specific functions [101]. Galactosylated chitosan (GC) is synthesized from chitosan and lactobionic acid bearing galactose ligands by using an active ester intermediate, 1-ethyl-3-(3-dimethylaminopropyl) carbodiimide (EDC) [102]. Other methods reported etherization of chitosan and galactose in THF using $BF_3.OEt_2$ as the catalyst [103]. An electrospinning technique was utilized by Feng et al. (2009) to fabricate a GC-based nanofibrous scaffold, in which rat primary hepatocytes formed 3D flat aggregates with high mechanical stability and excellent bioactivity [104]. A multilayer radial-flow bioreactor was also developed by the nanofibrous scaffold, which showed exciting properties such as, short-term support of patients with hepatic failure [105]. As a gene vector, GC or its copolymers (such as GC-g-PEG or GC-g-dextran) can form complexes with DNA and exhibit hepatocyte-targeted transfection [106–107]. Donati et al. (2005) conjugated lactose, which is a disaccharide consisting of galactose and glucose, to a highly deacetylated chitosan via a reductive N-alkylation, resulting in chondrocyte aggregation, formation of nodules of high dimensions, and synthesis of aggrecans and type II collagen [108]. Tan et al. (2008) reported the blending of lactose–chitosan with heparin for chondrocyte culture, resulting in better cell adhesion, proliferation, and GAG secretion [109]. By a thiourea reaction between the isothiocyanate group and the amine groups, a mannose moiety can be introduced onto chitosan or chitosan-g-polyethyleneimine (PEI) molecules (Fig. 8.10) [110–111]. The mannose-bearing polymers exhibit enhanced macrophage-specific bioactivity both as a gene delivery system and as a vaccine. Li et al. (2000) researched on linking of L-fucose to chitosan, which can act as a somatic agglutinin to induce bacteria aggregation [112]. Specific recognition and binding of the L-fucose moiety with PA-P lectin on Pseudomonas aeruginosa surface is the reason for the enhanced antimicrobial activity of chitosan–L-fucose.

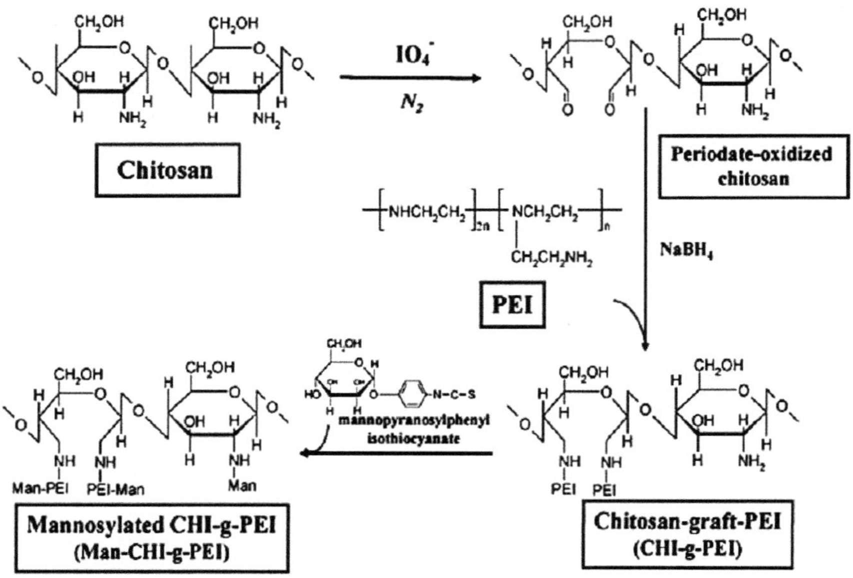

Figure 8.10 Reaction scheme for synthesis of mannosylated chitosan-g-PEI [111].

8.4.4 Modification by Synthetic Macromolecules

Synthetic biopolymers display reasonable mechanical properties and degradation behaviors, making them suitable for some biological applications. However, hydrophobicity and poor cell affinity, as well as lack of biological responses are the disadvantages of synthetic biopolymers [113]. To generate biomaterials with a balance between the physicochemical and biological properties and to overcome the disadvantages, combining synthetic biopolymers with natural biopolymers, such as chitosan, by blending or coating is a simple but effective strategy. PCL, PLA, PGA, and PLGA are the most widely used synthetic biodegradable polymers. Decomposition of chitosan before melt flow can be overcome by using a common solvent for the solution blending of synthetic biopolymers and chitosan. Wan et al. (2008) used hexafluoro-2-propanol as the common solvent and sodium chloride to fabricate a chitosan/PCL scaffold, possessing well wet state compressive properties, dimension stability [114]. Formic acid/acetone mixture (70:30 vol%) was used by Malheiro et al. (2010) as a common solvent to obtain a fibrous chitosan/PCL hybrid scaffold by a simple wet spinning method [115]. Mechanical and biological properties can be tuned by optimizing the processing parameters. Cruze et al. (2009) synthesized semi-interpenetrating polymer networks (semi-IPNs) to prepare porous scaffold by simultaneous precipitation of chitosan/PCL blend, followed by physical crosslinking of chitosan, and finally melt processing and leaching out [116]. The resulting semi-IPNs scaffold exhibited good mechanical properties and high porosity. Xu et al. (2009) used electro spinning to prepare chitosan/PLA blend micro-/ nano-fibers from a trifluoro acetic acid mixture solution. They reported the increase of PLA content, the fiber diameter was enlarged, with a fine morphology [117].

8.5 APPLICATIONS

Bioactivities, biocompatibility, biodegradability, low-allergenicity, and non-toxicity have given chitosan and its derivatives recognition as versatile biomaterials with superior physical properties, such as high surface area, porosity, tensile strength, and conductivity. Additionally, they can be easily molded into different shapes and forms (fibers, sponges, films, beads, gels, powder, and solutions) [118–119]. Some of the potential biomedical applications of are briefly discussed below (Fig. 8.11).

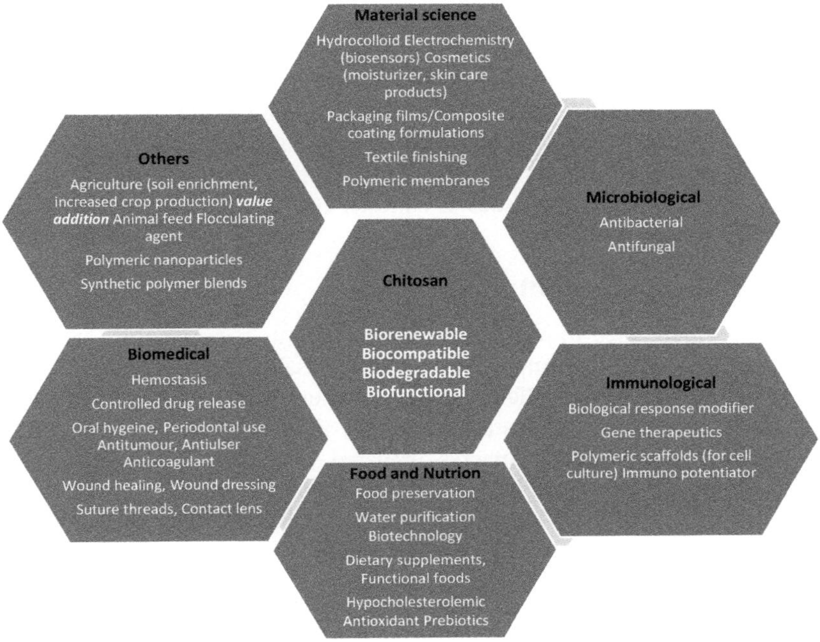

Figure 8.11 Potential applications of Chitosan and its derivatives.

8.5.1 Tissue Engineering

Chitosan-based biomaterials have become a potential target in developing tissue engineering lately. Tissue engineering describes the development of tissues that mimic and improve biological functions by using a combination of cells, engineering materials, and suitable biochemical factors. This field includes the repair or replacement of part of or whole tissues (for example, bone, blood vessels, cartilage, bladder, muscle, and skin). Chitosan derivatives provide certain structural and mechanical properties for suitable functioning for the repaired tissues (Fig. 8.12) [120].

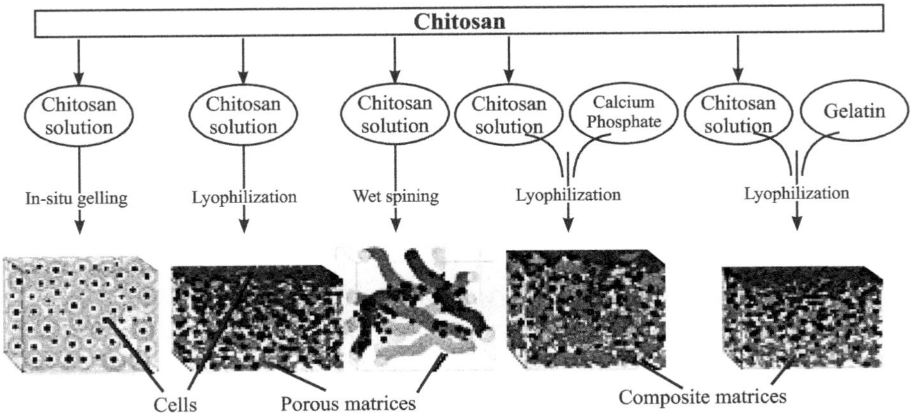

Figure 8.12 Use of Chitosan in tissue engineering [127].

For tissue engineering and quick transdermal curing, N-methacryloyl chitosan hydrogel acts as building material which facilitates quick and cost-effective construction of microgels with distinctive amino groups [121]. Chitosan/tricalcium phosphate demonstrates potential activity in enhancing osteogenesis and vascularization and repair of bone defects in conjunction with mesenchymal stem cells [122]. For cartilage tissue engineering, silk reinforced chitosan microparticles have been found to be promising [123]. The repair of the damaged spinal cord is hindered by the loss of spinal cord tissue and cavity formation. Chitosan scaffolds are reported to enhance neural stem cell differentiation into neurons, oligodendrocytes, and astrocytes, and hence are promising for the repair of damaged spinal cords [124]. A chitosan hollow tube employed for regeneration of the injured rodent transected sciatic nerve showed activities similiar to autologous nerve graft repair [125]. Chitosan/bioactive glass nanoparticles scaffolds can be applied in bone regeneration because they possesses the shape memory characteristics of chitosan and the biomineralization activity [126].

8.5.2 Drug Delivery System

Chitosan has been widely used in pharmaceutical industry as a carrier for drug delivery in both implantable as well as injectable formulae through oral, nasal, and ocular routes in different forms, such as tablets, microspheres, micelles, vaccines, nucleic acids, hydrogels, nanoparticles, and conjugates. Chitosan and its derivatives are commonly used as an excipient in tablet formulation for oral medication. Chitosan microspheres prepared by complexation between the cationic chitosan in addition to the anionic (Fig. 8.13) compounds such as tripolyphosphate or alginates, have been extensively investigated for controlled release of drugs and vaccines through oral and nasal delivery [128–129].

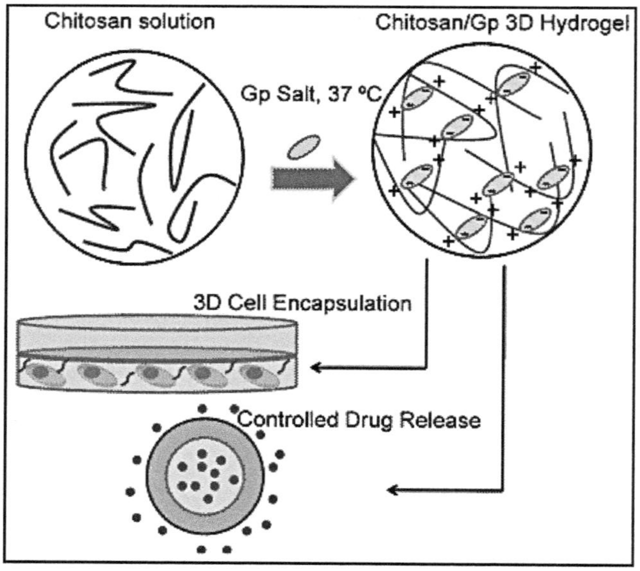

Figure 8.13 Chitosan/ glycerophosphate for drug delivery action [130].

For targeting different intracellular compartments, glycol chitosan nanogels may be potentially important as drug delivery vectors [131]. Glipizide was developed by polyionic complexation technique using chitosan and xanthan gum for the preparation of beads with controlled-release, floating, and mucoadhesive properties. The beads displayed good bioadhesive strength, comparable floating capacity in the gastric fluids, and pH-dependent swelling kinetics. Altering the chitosan to xanthan gum ratio did not affect the drug release [132].

The colon is a significant drug delivery target due to the long transit time and thus a prolonged drug absorption time. Progesterone having an abbreviated half-life, much first-pass metabolism, and low oral bioavailability must have a suitable delivery system. An oral Zn-pectinate/chitosan multi-particulate system allowing increased oral bioavailability of progesterone as well as progesterone residence time in plasma has been found to be a potential candidate for colonic-specific progesterone delivery [133].

In cancer treatment, nanoparticles of chitosan derivatives conjugated with antitumor agents to form a good partner in drug delivery. Due to predominant distribution into the cancer cells and a progressive release of the free drug from the conjugates, they manifest reduced side effects compared with the original drugs [134]. N-stearoyl O-butylglyceryl chitosan has been used as shell material for microcapsules loaded with fish oil, demonstrating sustained release of fish oil having higher thermo-stability [135].

8.5.3 Wound Dressing Materials

Chitosan and its derivatives have long been considered potential candidates for being used in wound care because of the properties such as biodegradability, biocompatibility, antimicrobial activity, and low immunogenicity [136]. Besides these properties, they offer 3D tissue growth matrix, stimulate cell proliferation, and activate macrophage activity [137]. Chitosan promotes activity of polymorphonuclear leukocytes, macrophages and fibroblasts essential for the wound healing and bleeding control (Fig. 8.14) [138]. Chitosan slowly degrades into N-acetyl-β-d-glucosamine that stimulates fibroblast proliferation, aids regular collagen deposition, accelerating the healing progress and preventing scar formation [139].

As wound dressing materials, nanofibrous and adhesive-based chitosan have been developed of late [140]. The adhesive-based wound dressing from chitosan, used in surgery, shows strong sealing strength, and does not require sutures or staples. It can effectively stop bleeding from blood vessels along with air leakage from the lung [141].

For patients seeking plastic surgery, skin grafting, and endoscopic sinus surgery, wound healing is a prime concern, and chitosan based wound dressing have been found to be suitable in this regard [142–143]. There are a number of chitosan-based wound dressings available in the market at present, in the form of non-wovens, composites, nanofibers, sponges, and films [144–145]. These wound dressing bandages showed efficacious hemostasis in penetrating limb trauma, emergency bleeding control capability, and all of them offer an antibacterial barrier.

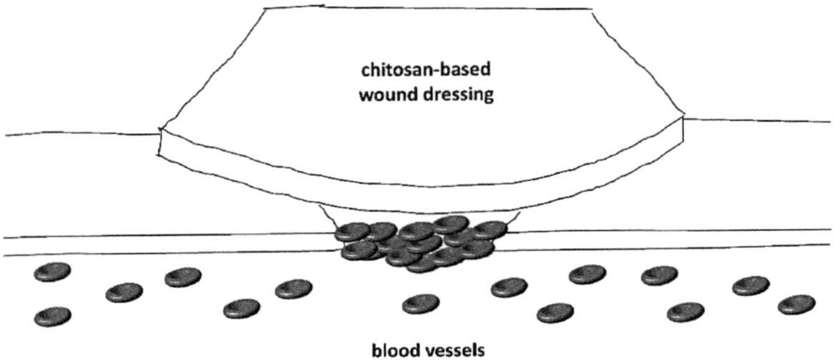

Figure 18.14 A diagrammatic presentation of how the chitosan-based wound dressing works.

Curcumin-encapsulated bioglass-chitosan, which is promising for wound healing applications, displayed higher 1,1-diphenyl-2-picrylhydrazyl and superoxide free radical quenching activities compared with unmodified curcumin, antibacterial activity against Staphylococcus aureus, and reduction in tumor necrosis factor- α production [146].

8.5.4 Blood Anticoagulants

Chitosan derivatives exhibit not only hemostatic effect, but also anticoagulant properties as well [147]. Sulfonated chitosan derivatives offer potential possibility of their application as anticoagulant because of their heparin like structure (Fig. 8.15) [148–149]. It has been reported that, with an increase in sulfur content in chitosan, the anticoagulant activity increases. N-Carboxymethyl chitosan 3,6-disulfonate of low molecular weight exhibited heparin-like anticoagulant activity and posed no adverse effects on the cellular structures when added to blood [45]. This may be useful in storing blood.

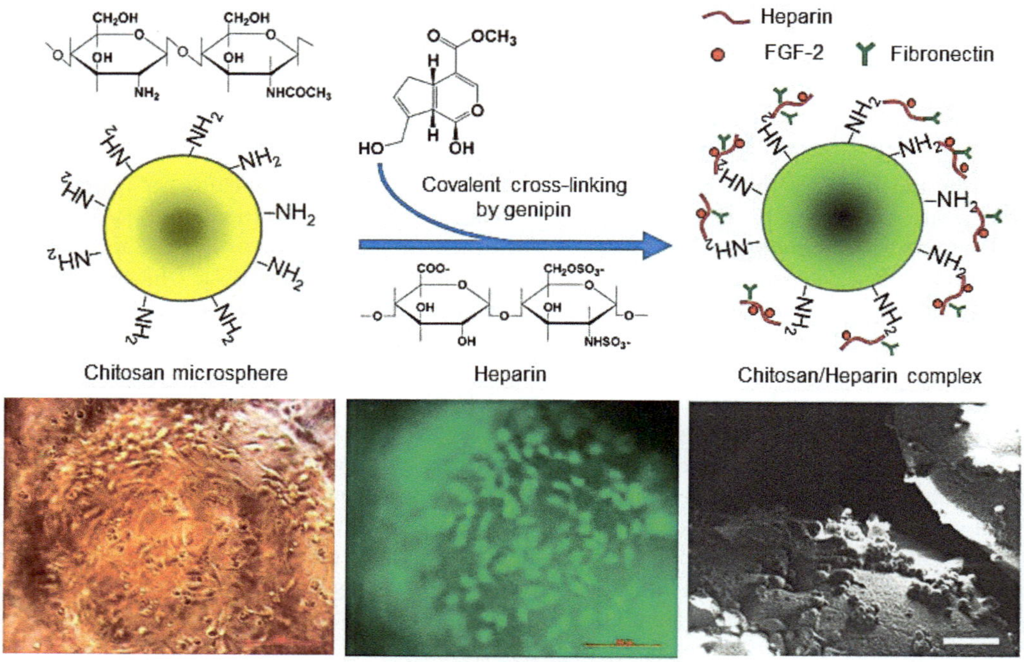

Figure 8.15 Heparin like blood anticoagulant preparation from chitosan [150].

8.5.5 Obesity Treatment

Chitosan has been considered and used as a dietary supplement or nutraceutical for containing serum cholesterol and controlling obesity for ages. Chitosan helps controlling obesity by physically filling the stomach with a feeling of satiety, by inhibiting pancreatic lipase activity, by binding fats, and by precipitating fats [151]. Chitosan also interferes with emulsification of neutral lipids such as cholesterol and other sterols by binding them with hydrophobic interaction [152]. Larger and better-controlled clinical trials have proven the worth of chitosan on body weight controlling. However, serum cholesterol appeared to be ineffective [153–154].

8.5.6 Treatment of Age Related Dysfunctions

Many researchers have explored the potential usefulness of chitosan and their derivatives in forestalling and therapy of aging-associated diseases (Fig. 8.16) [155]. The pathophysiological roles played by oxidative stress, oxidation of low density lipoprotein, increase of tissue stiffness, conformational changes of protein, aging-associated, and chronic inflammation have been reviewed [155].

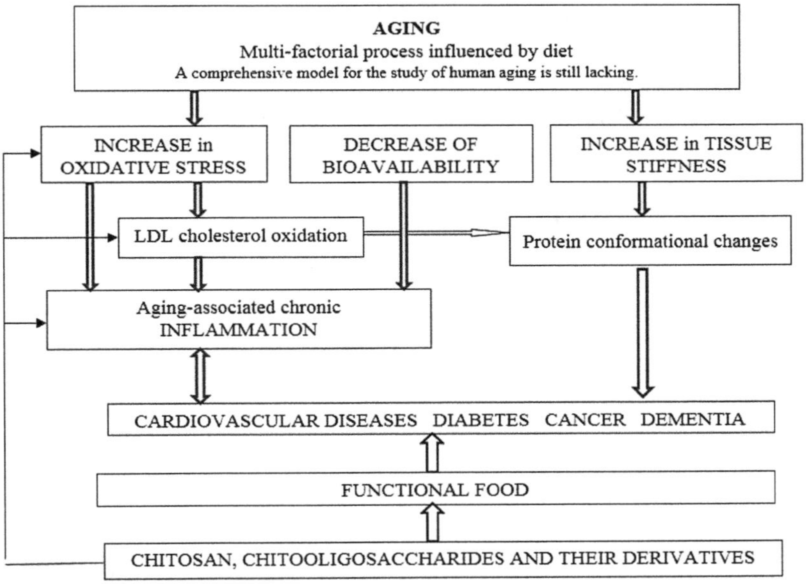

Figure 8.16 The potential of Chitosan in age related dysfunction.

8.5.7 Mucosal Immunity Enhancer

Chitosan-based formulations represent the most promising candidates for adjuvant and delivery systems for mucosal vaccines, as they have been found to coordinate both humoral and cellular responses [156].

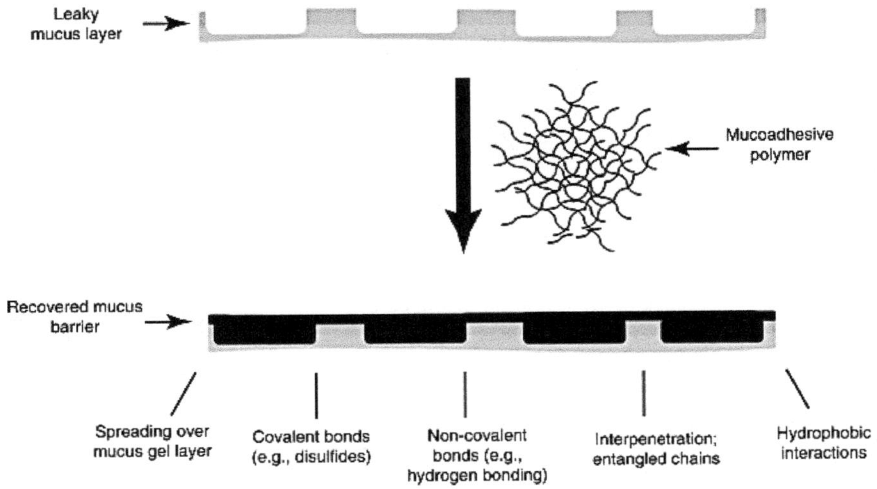

Figure 8.17 Chitosan for the treatment of dry X syndrome.

Numerous studies have demonstrated the potent stimulation of cellular responses, positive effects on mucosal adjuvanticity, and activation of the innate immune system (Fig. 8.17) [157–159].

Nasal administration of Bacillus anthracis protective antigen adsorbed on chitosan nanoparticles into mice produced elevated serum titers of antibodies against protective antigen. This activity also advocates in favor of chitosan nanoparticles to act in concert with a cell activator to affect nasal immunization [160].

8.5.8 Dry Mouth Syndrome Treatment

Low pH (acidic) in the body is the reason for almost all oral degenerative diseases, including tooth decay. People with a dry mouth syndrome experience mouth acidity caused by saliva thickening or drying up because of dehydration. Chitosan-thioglycolic-mercaptonicotinamide conjugates manifested improved swelling and cohesive characteristics compared with unmodified chitosan, and were promising for therapy of dry mouth syndrome, in which the necessary conditions are mucoadhesiveness and lubrication of the mucosa [161].

8.5.9 Gene Silencing

For the treatment of type 2 diabetes mellitus, glucosamine-based polymer chitosan has been proven to be effective as a cationic polymer-based *in vitro* delivery system for GLP-1, DPP-IV resistant GLP-1 analogs, and siRNA targeting DPP-IV mRNA (T2DM) [162]. Chitosan/interfering RNA nanoparticles in the form of food and taken up by larvae of Anopheles gambiae (vector of the primary African malaria vector) and Aedes aegypti (vector of the dengue and yellow fever) represented a methodology that can be applied to many insects and pests [163]. An usual mechanism of gene silencing activity of chitosan derivatives is given in the following figure (Fig. 8.18).

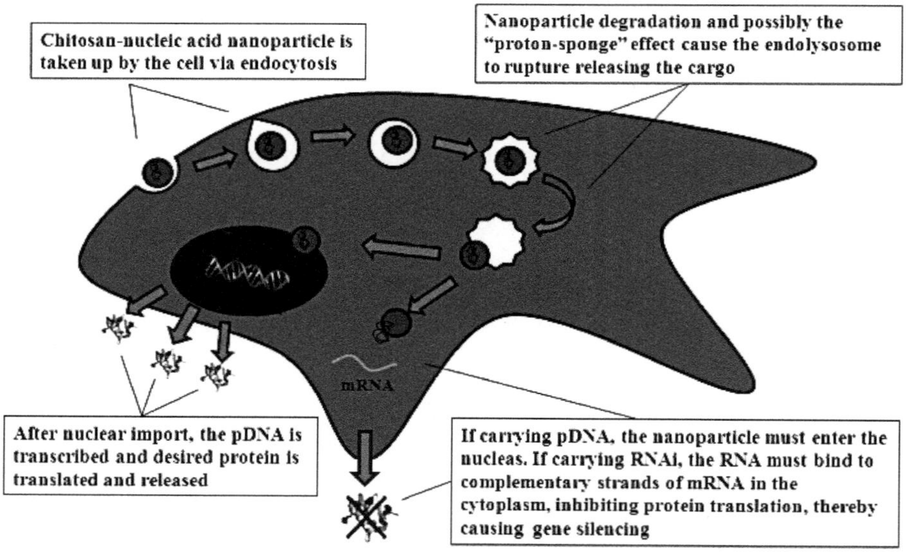

Figure 8.18 Chitosan mediated gene silencing [164].

8.6 DEVELOPMENT AND FUTURE TRENDS OF CHITIN AND CHITOSAN AND THEIR POLYMER

The majority of the applications are still in the developing stage. Major obstacles for large-scale technical applications of high quality chitin/chitosan are the high prices of the polymers and the difficulties in preparing uniformly reproducible charges in bulk quantities from various marine organisms. Issues for future research are:

- Development of economically competitive production processes
- Development of technical processes for the preparation of chitin derivatives and composite materials
- Investigations of the biodegradation of chemically modified chitosan

The situation of a rapidly rising demand for chitin and chitin-based products may arise in the foreseeable future. Care must be taken in the exploration of natural resources, and concern about ecological questions could arise once chitin is produced from crabs and other marine organisms that are caught for the purpose of chitin production only. Therefore, the planning of an eventually large-scale utilization of chitin must also include thoughts about a possible decrease of biodiversity and shifts in the ecological balance in the marine environment.

References

1. Batista, A.C.L., M.C.F. Silva, J.B. Batista, A.E. Nascimento and G.M. Campos-Takaki. Eco-friendly chitosan production by syncephalastrumracemosum and application to the removal of acid orange 7 (AO7) from wastewaters. Molecules 2013; 18: 7646-7660.
2. Pillai, C.K.S., W. Paul and C.P. Sharma. Chitin and chitosan polymers: Chemistry, solubility and fiber formation. Progress in Polymer Science 2009; 34: 641-678.
3. Cohen, E. Chitin biochemistry: Synthesis and inhibition. Annual Review of Entomology 1987; 32: 71-93.
4. Cohen, E. Chitin biochemistry: Synthesis, hydrolysis and inhibition. Advances in Insect Physiology 2010; 38: 5-74.
5. Sakthivel, D., N. Vijayakumar and V. Anandan. Extraction of chitin and chitosan from mangrove crab *sesarma plicatum* from Thengaithittu estuary Pondicherry southeast coast of India, International Journal of Pharmacy and Pharmaceutical Research, Human Journals, 2015; 4(1): 12-24.
6. Fuchs, J. and J.M. Martin. Impact of tropical shrimp aquaculture on the environment in Asia and the Pacific. Shrimp Culture 1999; 12(4): 9-13.
7. Rosenberry, B. World shrimp farming. Shrimp News International, San Diego, California, USA. 1998; 328 p.
8. Food and Agriculture Organization of the United Nations. The State of World Fisheries and Aquaculture. Food and Agricultural Organization 2014.
9. Britton, G. Proceedings of the 11th international symposium on carotenoids, Leiden, The Netherlands. Pure and Applied Chemistry 1997; 69: 2027–2173.
10. Kurita, K. Chitin and chitosan: Functional biopolymers from marine crustaceans. Marine Biotechnology 2006; 8(3): 203-226.
11. No, H.K. and S.P. Meyers. Preparation of chitin and chitosan. *In*: R.A.A. Muzzarelli and M.G. Peter (eds). Chitin Handbook. European Chitin Society, Grottammare, Italy, 1997, pp. 475-489.
12. Hirano, S. Chitin biotechnology applications. Biotechnology Annual Review 1996; 2: 237-258.
13. Sandford, P.A. Chitosan—commercial uses and potential applications. *In*: G.S. Brack, T. Anthonsen and P. Sandford (eds). Chitin and Chitosan. Elsevier, New York, NY, USA, 1989; pp. 51-69.
14. Horton, D. and D.R. Lineback. N-deacetylation, chitosan from chitin. *In*: R.L. Whistler and M.L. Wolfson (eds). Methods in Carbohydrate Chemistry. Academic Press, New York, NY, USA, 1965, p. 403.
15. Bough, W.A., W.L. Salter and A.C.M. Wu and B.E. Perkins. Influence of manufacturing variables on the characteristics and effectiveness of chitosan products. I. Chemical composition, viscosity, and molecular weight distribution of chitosan products. Biotechnology and Bioengineering 1978; 20(12): 1931-1943.
16. No, H.K., K.S. Lee and S.P. Meyers. Correlation between physicochemical characteristics and binding capacities of chitosan products. Journal of Food Science 2000; 65(7): 1134-1137.
17. Dung, P.L., M. Milas, M. Rinaudo and J. Desbrières. Water soluble derivatives obtained by controlled chemical modifications of chitosan. Carbohydrate Polymers 1994; 24(3): 209-214.
18. Andrade, V.S., B. de B. Neto, K. Fukushima and G.M. Campos-Takaki. Effect of medium components and time of cultivation on chitin production by Mucorcircinelloides (Mucorjavanicus IFO 4570)–A factorial study. Revista Iberoamericana de Micologia 2003; 20(4): 149-153.
19. Chandy, T. and C.P. Sharma. Chitosan—As a biomaterial. Biomaterials, Artificial Cells and Artificial Organs 1990; 18(1): 1-24.

20. Feofilova, E.P., D.V. Nemtsev, V.M. Tereshina and V.P. Kozlov. Polyaminosaccharides of mycelial fungi: New biotechnological use and practical implications (review). Applied Biochemistry and Microbiology 1996; 32(5): 437-445.
21. Hu, K.-J., J.-L. Hu, K.-P. Ho and K.-W. Yeung. Screening of fungi for chitosan producers, and copper adsorption capacity of fungal chitosan and chitosanaceous materials. Carbohydrate Polymers 2004; 58(1): 45-52.
22. Rinaudo, M., P. Dung Le, C. Gey and M. Milas. Substituent distribution on O,N-carboxymethyl chitosans by 1H and 13C N.M.R. International Journal of Biological Macromolecules 1992; 14(3): 122-128.
23. Nemtsev, S.V., O.Y. Zueva, M.R. Khismatullin, A.I. Albulov and V.P. Varlamov. Isolation of chitin and chitosan from honeybees. Applied Biochemistry and Microbiology 2004; 40(1): 39-43.
24. Pochanavanich, P. and W. Suntornsuk. Fungal chitosan production and its characterization. Letters in Applied Microbiology 2002; 35(1): 17-21.
25. Synowiecki, J. and N.A. Al-Khateeb. Production, properties, and some new applications of chitin and its derivatives. Critical Reviews in Food Science and Nutrition 2003; 43(2): 145-171.
26. Tan, S.C., T.K. Tan, S.M. Wong and E. Khor. The chitosan yield of zygomycetes at their optimum harvesting time. Carbohydrate Polymers 1996; 30(4): 239-242.
27. Teng, W.L., E. Khor, T.K. Tan, L.Y. Lim and S.C. Tan. Concurrent production of chitin from shrimp shells and fungi. Carbohydrate Research 2001; 332(3): 305-316.
28. Andrade, V.S., B.B. Neto, W. Souza and G.M. Campos-Takaki. A factorial design analysis of chitin production by Cunninghamellaelegans. Canadian Journal of Microbiology 2000; 46(11): 1042-1045.
29. Campos-Takaki, G.M. The fungal versatility on the copolymers chitin and chitosan production. *In*: P.K. Dutta (eds). Chitin and Chitosan Opportunities and Challenges. SSM: International Publication, Midnapore, India, 2005, pp. 69-94.
30. Amorim, R.V.S., W.M. Ledingham, K. Fukushima and G.M. Campos-Takaki. Screening of chitin deacetylase from Mucoralean strains (*Zygomycetes*) and its relationship to cell growth rate. Journal of Industrial Microbiology and Biotechnology 2005; 32(1): 19-23.
31. Benjakul, S. and P. Sophanodora. Chitosan production from carapace and shell of black tiger shrimp. ASEAN Food Journal 1993; 8: 145-148.
32. Nadarajah, K., J. Kader, M. Mazmira and D.C. Paul. Production of chitosan by fungi. Pakistan Journal of Biological Science 2001; 4: 263-265.
33. Stamford, T.C.M., T.L.M. Stamford, N.P. Stamford, B. de B. Neto and G.M. Campos-Takaki. Growth of Cunninghamellaelegans UCP 542 and production of chitin and chitosan using yam bean medium. Electronic Journal of Biotechnology 2007; 10(1): 61-68.
34. Bhattarai, N., J.J. Gunn and M. Zhang. Chitosan-based hydrogels for controlled, localized drug delivery. Advanced Drug Delivery Reviews 2010; 62: 83-99.
35. Park, J.H., G. Saravanakumar, K. Kim and I.C. Kwon. Targeted delivery of low molecular drugs using chitosan and its derivatives. Advanced Drug Delivery Reviews 2010; 62(1): 28-41.
36. Mourya, V.K. and N.N. Inamdar. Trimethyl chitosan and its applications in drug delivery. Journal of Materials Science: Materials in Medicine 2009; 20: 1057-1080.
37. Desai, U.R. New antithrombin-based anticoagulants. Medicinal Research Reviews 2004; 24(2): 151-181.
38. Jayakumar, R., N. Nwe, S. Tokura and H. Tamura. Sulfated chitin and chitosan as novel biomaterials. International Journal of Biological Macromolecules 2007; 40: 175-181.
39. Vikhoreva, G., A.G. Bannikov, P. Stolbushkina, A. Panov, N. Drozd, V. Makarov, V. Varlamov and L. Galbraikh, L. Preparation and anticoagulant activity of a low-molecular-weight sulfated chitosan. Carbohydrate Polymers 2005; 62: 327-332.
40. Je, J.Y., P.J. Park and S.K. Kim. Prolyl endopeptidase inhibitory activity of chitosan sulfates with different degree of deacetylation. Carbohydrate Polymers 2005; 60: 553-556.
41. Can, Z., Q.N. Ping, H. Zhang and J. Shen. Preparation of N-alkyl-O-sulfate chitosan derivatives and micellar solubilization of taxol. Carbohydrate Polymers 2003; 54: 137-141.
42. Xing, R.E., S. Liu, H. Yu, Z. Guo, Z. Li and P. Li. Preparation of high-molecular weight and high sulfate content chitosans and their potential antioxidant activity in vitro. Carbohydrate Polymers 2005; 61: 148-154.

43. Horton, D. and E.K. Just. Preparation from chitin of (1-4)-2-amino-2-deoxy-beta-D-glucopyranuronan and its 2-sulfoamino analog having blood anticoagulant properties. Carbohydrate Research 1973; 29: 173-179.
44. Whistler, R.L. and M. Kosik. Anticoagulant activity of oxidized and N-sulfated and O-sulfated chitosan. Archives of Biochemistry and Biophysics 1971; 142(1): 106-110.
45. Muzzarelli, R.A.A., F. Tanfani, M. Emanuelli, D.P. Pace, E. Chiurazzi and M. Piani. Sulfated N-carboxymethylchitosans as blood anticoagulants. *In*: R. Muzzarelli, C. Jeuniaux and G.W. Gooday (eds). Chitin in Nature and Technology. Plenum Press, New York, 1986, pp. 469-476.
46. Zhou, H.J., J.C. Qian, J. Wang, W. Yao, C. Liu, J. Chen and X. Cao. Enhanced bioactivity of bone morphogenetic protein-2 with low dose of 2-N, 6-O-sulfated chitosan in vitro and in vivo. Biomaterials 2009; 30: 1715-1724.
47. Qu, G.W., Z. Yao, C. Zhang, X.L. Wu and Q.E. Ping. PEG conjugated N-octyl-O-sulfate chitosan micelles for delivery of paclitaxel: In vitro characterization and in vivo evaluation. European Journal of Pharmaceutical Sciences 2009; 37(2): 98-105.
48. Chen, M-C., F.-L. Mi, Z.-X. Liao and H.-W. Sung. Chitosan: Its applications in drug-eluting devices in chitosan for biomaterials I. Advances in Polymer Sciences 2011; 243: 185-230.
49. Lehr, C.M., J.A. Bouwstra, E.H. Schacht and H.E. Junginger. In vitro evaluation of mucoadhesive properties of chitosan and some other natural polymers. International Journal of Pharmaceutics 1992; 78: 43-48.
50. Mourya, V.K. and N.N. Inamdar. Trimethyl chitosan and its applications in drug delivery. Journal of Materials Science: Materials in Medicine 2009; 20: 1057-1080.
51. Domard, A., M. Rinaudo and C. Terrassin, C. New method for the quaternization of chitosan. International Journal of Biological Macromolecules 1986; 8: 105-107.
52. Amidi, M., S.G. Romeijn, G. Borchard, H.E. Junginger, W.E. Hennink and W. Jiskoot. Preparation and characterization of protein-loaded N-trimethyl chitosan nanoparticles as nasal delivery system. Journal of Controlled Release 2006; 111: 107-116.
53. Bernkop-Schnurch, A., M. Hornof and D. Guggi. Thiolatedchitosans. European Journal of Pharmaceutics and Biopharmaceutics 2004; 57: 9-17.
54. Tharun, J., Y. Hwang, R. Roshan, S. Ahn, A.K. Cherian and P. Dae-Won. A novel approach of utilizing quaternized chitosan as a catalyst for the eco-friendly cycloaddition of epoxides with CO_2. Catalysis Science & Technology 2012; 2: 1674-1680.
55. Mao, Z.W., L. Ma, Y. Jiang, M. Yan, C.Y. Gao and J.C. Shen. N,N,N-Trimethylchitosan chloride as a gene vector: Synthesis and application. Macromolecular Bioscience 2007; 7(2): 855-863.
56. Mao, Z.W., L. Ma, J. Yan, M. Yan, C.Y. Gao and J.C. Shen. The gene transfection efficiency of thermo-responsive N,N,N-trimethyl chitosan chloride-g-poly(N-isopropylacrylamide) copolymer, Biomaterials 2007; 28(30): 4488-4500.
57. Germershaus, O., S.R. Mao, J. Sitterberg, U. Bakowsky and T. Kissel. Gene delivery using chitosan, trimethyl chitosan or polyethylenglycol-graft-trimethyl chitosan block copolymers: Establishment of structure-activity relationships in vitro. Journal of Controlled Release 2008; 125(2): 145-154.
58. Zheng, Y., Z. Cai, X.R. Song, B. Yu, Y.Q. Bi, Q.H. Chen, D. Zhao, J.P. Xu and S.X. Hou. Receptor mediated gene delivery by folate conjugated N-trimethyl chitosan in vitro. International Journal of Pharmaceutics 2009; 382: 262-269.
59. Muzzarelli, R.A.A. Carboxymethylated chitins and chitosans. Carbohydrate Polymers 1988; 8: 1-21.
60. Pavlov, G.M., E.V. Korneeva, S.E. Harding and G.A. Vichoreva. Dilute solution properties of carboxymethyl chitins in high ionic-strength solvent. Polymer 1998; 39: 6951-6961.
61. Jayakumar, R., M. Prabaharan, S.V. Nair, S. Tokura and N. Selvamurugan. Novel carboxymethyl derivatives of chitin and chitosan materials and their biomedical applications. Progress in Materials Science 2010; 55: 675-709.
62. Kim, C.H. and K.S. Choi. Synthesis and properties of carboxyalkyl chitosan derivatives. Journal of Industrial and Engineering Chemistry 1998; 4: 19-25.
63. Sarkar, K., M. Debnath and P.P. Kundu. Recyclable crosslinked O-carboxymethyl chitosan for removal of cationic dye from aqueous solutions. Hydrology: Current Research 2012; 3: 4.

64. Hayes, E.R. N, O-carboxymethyl chitosan and preparative methods therefor. US Patent 1986; 4: 619, 995.
65. Lin, Y.H., H.F. Liang, C. Hung and C. Ketal. Physically crosslinked alginate/N.O-carboxymethyl chitosan hydrogels with calcium for oral delivery of protein drugs. Biomaterials 2005; 26: 2105-2113.
66. Yin, L.C., L.K. Fei, F.Y. Cui, C. Tang and C. Yin. Superporous hydrogels containing poly(acrylic acidco-acrylamide)/ O-carboxymethyl chitosan interpenetrating polymer networks. Biomaterials 2007; 28: 1258-1266.
67. Kato, Y., H. Onishi and Y. Machida. Depolymerization of N-succinyl-chitosan by hydrochloric acid. Carbohydrate Research 2002; 337: 561-562.
68. Kato, Y., H. Onishi and Y. Machida. N-succinyl-chitosan as a drug carrier: Water-insoluble and water soluble conjugates. Biomaterials, 2004; 25: 907-915.
69. Zhu, A., T. Chen, L. Yuan, W. Hao and L. Ping. Synthesis and characterization of N-succinylchitosan and its self-assembly of nanospheres. Carbohydrate Polymers 2006; 66: 274-279.
70. Kast, C.E. and A. Bernkop-Schnurch. Thiolated polymers–Thiomers: development and in vitro evaluation of chitosan-thioglycolic acid conjugates. Biomaterials 2001; 22: 2345-2352.
71. Kast, C.E. and A. Bernkop-Schnurch. Polymer-cysteamine conjugates: new mucoadhesive excipients for drug delivery. International Journal of Pharmaceutics 2002; 234: 91-99.
72. Kafedjiiski, K., A.H. Krauland, M.H. Hoffer and A. Bernkop-Schnurch. Synthesis and in vitro evaluation of a novel thiolated chitosan. Biomaterials 2005; 26: 819-826.
73. Hassan, E.E. and J.M. Gallo. A simple rheological method for the in vitro assessment of mucin -polymer bioadhesive bond strength. Pharmaceutical Research 1990; 7: 491-495.
74. Leitner, V.M., G.F. Walker and A. Bernkop-Schnurch. Thiolated polymers: evidence for the formation of disulphide bonds with mucus glycoproteins. European Journal of Pharmaceutics and Biopharmaceutics 2003; 56: 207-214.
75. Bernkop-Schnurch, A., M. Hornof and T. Zoidl. Thiolated polymers-thiomers: synthesis and in vitro evaluation of chitosan-2 -iminothiolane conjugates. International Journal of Pharmaceutics 2003; 260: 229-237.
76. Sakloetsakun, D., J.M. Hombach and A. Bernkop-Schnurch. In situ gelling properties of chito-santhioglycolic acid conjugate in the presence of oxidizing agents. Biomaterials 2009; 30: 6151-6157.
77. Peppas, N.A., P. Bures, W. Leobandung and H. Ichikawa. Hydrogels in pharmaceutical formulations, European Journal of Pharmaceutics and Biopharmaceutics 2000; 50: 27-46.
78. Mi, F.L., C.Y. Kuan, S.S. Shyu, S-T. Lee and S-F. Chang. The study of gelation kinetics and chainrelaxation properties of glutaraldehyde-cross-linked chitosan gel and their effects on microspheres preparation and drug release. Carbohydrate Polymers 2000; 41: 389-396.
79. Gupta, K.C. and F.H. Jabrail. Effects of degree of deacetylation and cross-linking on physical characteristics, swelling and release behavior of chitosan microspheres. Carbohydrate Polymers 2006; 66: 43-54.
80. Arguelles-Monal, W., F.M. Goycoolea, C. Peniche and I. Higuera-Ciapara, I. Rheological study of the chitosan glutaraldehyde chemical gel system. Polymer Gels and Networks 1998; 6: 429-440.
81. Hassan, E.E., R.C. Parish and J.M. Gallo. Optimized formulation of magnetic chitosan microspheres containing the anticancer agent, oxantrazole. Pharmaceutical Research 1992; 9: 390-397.
82. Chung, T.W., S.Y. Lin, D.Z. Liu, Y.C. Tyan and J.S. Yang. Sustained release of 5-FU from poloxamer gels interpenetrated by crosslinking chitosan network. International Journal of Pharmaceutics 2009; 382: 39-44.
83. Gupta, K.C. and F.H. Jabrail. Glutaraldehyde and glyoxal cross-linked chitosan microspheres for controlled delivery of centchroman. Carbohydrate Research 2006; 341: 744-756.
84. Dini, E., S. Alexandridou and C. Kiparissides. Synthesis and characterization of cross-linked chitosan microspheres for drug delivery applications. Journal of Microencapsulation 2003; 20: 375-385.
85. Shu, X.Z. and K.J Zhu. Controlled drug release properties of ionically cross-linked chitosan beads: The influence of anion structure. International Journal of Pharmaceutics 2002; 233: 217-225.
86. Dambies, L., T. Vincent, A. Domard and G. Eric. Preparation of chitosan gel beads by ionotropic molybdate gelation. Biomacromolecules 2001; 2: 1198-1205.

87. Brack, H.P., S.A. Tirmizi and W.M. Risen. A spectroscopic and viscometric study of the metal ioninduced gelation of the biopolymer chitosan. Polymer 1997; 38: 2351-2362.

88. Mi, F.L., S.-S. Shyu, T.-Bi. Wong, S.-F. Jang, S.-T. Lee and K.-T. Lu. Chitosan-polyelectrolyte complexation for the preparation of gel beads and controlled release of anticancer drug. II. Effect of pH-dependent ionic crosslinking or interpolymer complex using tripolyphosphate or polyphosphate as reagent. Journal of Applied Polymer Science 1999; 74: 1093-1107.

89. Bodmeier, R., K.H. Oh and Y. Pramar. Preparation and evaluation of drug-containing chitosan beads. Drug Dev Ind Pharm 1989; 15: 1475-1494.

90. Yodkhum, K. and T. Phaechamud. Hydrophobic chitosan sponges modified by aluminum monostearate and dehydrothermal treatment as sustained drug delivery system. Materials Science and Engineering C 2014; 42: 715-725.

91. Janes, K.A., M.P. Fresneau, A. Marazuela, A. Fabra and M.J. Alonso. Chitosan nanoparticles as delivery systems for doxorubicin. Journal of Controlled Release 2001; 73: 255-267.

92. Fernandez-Urrusuno, R., P. Calvo, C. Remunan-Lopez and V-J. Luis. Enhancement of nasal absorption of insulin using chitosan nanoparticles. Pharmaceutical Research 1999: 16: 1576-1581.

93. Masarudin, M.J., M.C. Suzanne, J.E. Benny, R.P. Don and P.J. Pigram. Factors determining the stability, size distribution, and cellular accumulation of small, monodisperse chitosan nanoparticles as candidate vectors for anticancer drug delivery: Application to the passive encapsulation of (14C)-doxorubicin. Nanotechnology, Science and Applications, 2015; 8: 67-80.

94. Hersel, U., C. Dahmen and H. Kessler. RGD modified polymers: Biomaterials for stimulated cell adhesion and beyond. Biomaterials 2003; 24(2): 4385-4415.

95. Karakeçili, A.G. and M. Gümüşderelioğlu. Physico-chemical and thermodynamic aspects of fibroblastic attachment on RGDS-modified chitosan membranes. Colloids and Surfaces B: Biointerfaces 2008; 61(2): 216-223.

96. Masuko, T., N. Iwasaki, S. Yamane, T. Funakoshi, T. Majima, A. Minami, N. Ohsuga, T. Ohta and S.I. Nishimura. Chitosan–RGDSGGC conjugate as a scaffold material for musculoskeletal tissue engineering. Biomaterials 2005; 26(26): 5339-5347.

97. Mochizuki, M., Y. Kadoya, Y. Wakabayashi, K. Kato, I. Okazaki, M. Yamada, T. Sato, N. Sakairi, N. Nishi and M. Nomizu. Laminin-1 peptide-conjugated chitosan membranes as a novel approach for cell engineering. FASEB J. 2003; 17: 875-877.

98. Itoh, S., A. Matsuda, H. Kobayashi, S. Ichinose, K. Shinomiya and J. Tanaka. Effects of a laminin peptide (YIGSR) immobilized on crab-tendon chitosan tubes on nerve regeneration. Journal of Biomedical Materials Research Part B 2005; 73B: 375-382.

99. Wang, W., S. Itoh, A. Matsuda, T. Aizawa, M. Demura, S. Ichinose, K. Shinomiya and J. Tanaka. Enhanced nerve regeneration through a bilayered chitosan tube: The effect of introduction of glycine spacer into the CYIGSR sequence, Journal of Biomedical Materials Research Part A 2008; 85A(4): 919-928.

100. Bae, J.S., S.H. Lee, J.E. Kim, J.Y. Choi, R.W. Park, J.Y. Park, H.S. Park, Y.S. Sohn, D.S. Lee, E.B. Lee and I.S. Kim. Betaig-h3 supports keratinocyte adhesion, migration, and proliferation through alpha3beta1 integrin. Biochemical and Biophysical Research Communications 2002; 294: 940-948.

101. Chung, T.-W., Y.-F. Lu, S.-S. Wang, Y.-S. Lin and S.-H. Chu. Growth of human endothelial cells on photochemically grafted Gly-Arg-Gly-Asp (GRGD) chitosans. Biomaterials 2002; 23: 4803-4809.

102. Kim, T.H., I.K. Park, J.W. Nah, Y.J. Choi and C.S. Cho. Galactosylated chitosan/DNA nanoparticles prepared using water-soluble chitosan as a gene carrier. Biomaterials 2004; 25(17): 3783-3792.

103. Song, B.F., W. Zhang, R. Peng, J. Huang, T. Me, Y. Li, Q. Jiang and R. Gao. Synthesis and cell activity of novel galactosylated chitosan as a gene carrier. Colloids and Surfaces B: Biointerfaces 2009; 70: 181-186.

104. Feng, Z.Q., X.H. Chu, N.P. Huang, T. Wang, Y.C. Wang, X.L. Shi, Y.T. Ding and Z.Z. Gu. The effect of nanofibrous galactosylated chitosan scaffolds on the formation of rat primary hepatocyte aggregates and the maintenance of liver function. Biomaterials 2009; 30: 2753-2763.

105. Chu, X.H., X.L. Shi, Z.Q. Feng, J.Y. Gu, H.Y. Xu, Y. Zhang, Z.Z. Gu and Y.T. Ding. In vitro evaluation of a multi-layer radial-flow bioreactor based on galactosylated chitosan nanofiber scaffolds. Biomaterials 2009; 30: 4533-4538.

106. Park, I.K., T.H. Kim, Y.H. Park, B.A. Shin, E.S. Choi, E.H. Chowdhury, T. Akaike and Cho, C.S. Galactosylated chitosan-graft-poly(ethylene glycol) as hepatocyte-targeting DNA carrier. Journal of Controlled Release 2001; 76(3): 349-362.

107. Park, Y.K., Y.H. Park, B.A. Shin, E.S. Choi, Y.R. Park, T. Akaike and C.S. Cho. Galactosylated chitosan-graft-dextran as hepatocyte-targeting DNA carrier. Journal of Controlled Release 2000; 69(1): 97-108.

108. Donati, I., S. Stredanska, G. Silvestrini, A. Vetere, P. Marcon, E. Marsich, P. Mozetic, A. Gamini, S. Paoletti and F. Vittur. The aggregation of pig articular chondrocyte and synthesis ofextracellular matrix by a lactose-modified chitosan. Biomaterials 2005; 26: 987–998.

109. Tan, H.P., L.H. Lao, J.D. Wu, Y.H. Gong and C.Y. Gao. Biomimetic modification of chitosan with covalently grafted lactose and blended heparin for improvement of in vitro cellular interaction, Polymer for Advanced Technology 2008; 19: 15-23.

110. Jiang, H.L., M.L. Kang, J.S. Quan, S.G. Kang, T. Akaike, H.S. Yoo and C.S. Cho. The potential of mannosylated chitosan microspheres to target macrophage mannose receptors in an adjuvant-delivery system for intranasal immunization. Biomaterials 2008; 29(12): 1931-1939.

111. Jiang, H.L., Y.K. Kim, R. Arote, D. Jere, J.S. Quan, J.H. Yu, Y.J. Choi, J.W. Nah, M.H. Cho and C.S. Cho. Mannosylated chitosan-graft-polyethylenimine as a gene carrier for Raw 264.7 cell targeting. International Journal of Pharmaceutics 2009; 375: 133-139.

112. Li, X.B., Y. Tushima, M. Morimoto, H. Saimoto, Y. Okamoto, S. Minami and Y. Shigemasa. Biological activity of chitosan–sugar hybrids: specific interaction with lectin. Polymers for Advanced Technologies 2000; 11(4): 176-179.

113. Ratner, B.R., A.S. Hoffman, F.J. Schoen and J.E. Lemons. Biomaterials Science, 2nd Ed. Academic, New York, 1996.

114. Wan, Y., H. Wu, X.Y. Cao and S. Dalai. Compressive mechanical properties and biodegradability of porous poly(caprolactone)/chitosan scaffolds. Polymer Degradation and Stability 2008; 93(10): 1736-1741.

115. Malheiro, V.N., S.G. Caridade, N.M. Alves and J.F. Mano. New poly(ε-caprolactone)/chitosan blend fibers for tissue engineering applications. Acta Biomaterialia 2010; 6(2): 418-428.

116. Cruz, D.M.G., D.F. Coutinho, J.F. Mano, J.L.G. Ribelles and M.S. Sanchez. Physical interactions in macroporous scaffolds based on poly(ε-caprolactone)/chitosan semi-interpenetrating polymer networks. Polymer 2009; 50: 2058.

117. Xu, J., J.H. Zhang, W.Q. Gao, H.W. Liang, H.Y. Wang and J.F. Li. Large-scale synthesis of singlecrystal alpha manganese sesquioxide nanowires via solid-state reaction. Materials Letters 2009; 63(8): 658-663.

118. Cheung, R.C.F., T.B. Ng, J.H. Wong and W.Y. Chan. Chitosan: An update on potential biomedical and pharmaceutical applications. Marine Drugs 2015; 13(8): 5156-5186.

119. Shukla, S.K., A.K. Mishra, O.A. Arotiba and B.B. Mamba. Chitosan-based nanomaterials: A state-ofthe-art review. International Journal of Biological Macromolecules 2013; 59: 46-58.

120. Kim, I.Y., S.J. Seo, H.S. Moon, M.K. Yoo, I.Y. Park, B.C. Kim and C.S. Cho. Chitosan and its derivatives for tissue engineering applications. Biotechnology Advances 2008; 26: 1-21.

121. Li, B., L. Wang, F. Xu, X. Gang, U. Demirci, D. Wei, Y. Li, Y. Feng, D. Jia and Y. Zhou. Hydrosoluble, UV-crosslinkable and injectable chitosan for patterned cell-laden microgel and rapid transdermal curing hydrogel in vivo. Acta Biomaterialia 2015; 22: 59-69.

122. Yang, L., Q. Wang, L. Peng, H. Yue and Z. Zhang. Vascularization of repaired limb bone defects using chitosan-β-tricalcium phosphate composite as a tissue engineering bone scaffold. Molecular Medicine Reports 2015; 12: 2343-2347.

123. Chameettachal, S., S. Murab, R. Vaid, S. Midha and S. Ghosh. Effect of visco-elastic silk-chitosan microcomposite scaffolds on matrix deposition and biomechanical functionality for cartilage tissue engineering. Journal of Tissue Engineering and Regenerative Medicine 2017; 11(4): 1212-1229.

124. Jian, R., Y. Yixu, L. Sheyu, S. Jianhong, Y. Yaohua, S. Xing, H. Qingfeng, L. Xiaojian, Z. Lei, Z. Yan, X. Fangling, G. Huasong and G. Yilu. Repair of spinal cord injury by chitosan scaffold with glioma ECM and SB216763 implantation in adult rats. Journal of Biomedical Materials Research Part A 2015; 103(10): 3259-3272.

125. Shapira, Y., M. Tolmasov, M. Nissan, E. Reider, A. Koren, T. Biron, Y. Bitan, M. Livnat, G. Ronchi, S. Geuna and S. Rochkind. Comparison of results between chitosan hollow tube and autologous nerve graft in reconstruction of peripheral nerve defect: An experimental study. Microsurgery 2016; 36(8): 664-671.
126. Correia, C.O., A.J. Leite and J.F. Mano. Chitosan/bioactive glass nanoparticles scaffolds with shape memory properties. Carbohydrate Polymer 2015; 123: 39-45.
127. Camelia, E.O. Applications of functionalized chitosan. Scientific Study & Research 2007; VIII(3): 227-256.
128. Jiang, H.L., I.K. Park, N.R. Shin, S.G. Kang, H.S. Yoo, S.I. Kim, S.B. Suh, T. Akaike and C.S. Cho. In vitro study of the immune stimulating activity of an atrophic rhinitis vaccine associated to chitosan microspheres. European Journal of Pharmaceutics and Biopharmaceutics 2004; 58: 471-476.
129. Thanou, M., J.C. Verhoef and H.E. Junginger. Oral drug absorption enhancement by chitosan and its derivatives. Advanced Drug Delivery Reviews 2001; 52: 117-126.
130. Tahrir, F.G., F. Ganji and T.M. Ahooyi. Injectable thermosensitive chitosan/glycerophosphate-based hydrogels for tissue engineering and drug delivery applications: A review. Recent Patents on Drug Delivery & Formulations 2015; 9(2): 107-120.
131. Pereira, P., S.S. Pedrosa, J.M. Wymant, E. Sayers, A. Correia, M. Vilanova, A.T. Jones and F.M. Gama. siRNA inhibition of endocytic pathways to characterize the cellular uptake mechanisms of folatefunctionalized glycol chitosan nanogels. Molecular Pharmaceutics 2015; 2: 1970-1979.
132. Kulkarni, N., P. Wakte and J. Naik. Development of floating chitosan-xanthan beads for oral controlled release of glipizide. International Journal of Pharmaceutical Investigation 2015; 129; 5: 73-80.
133. Gadalla, H.H., G.M. Soliman, F.A. Mohammed and A.M. El-Sayed. Development and in vitro/in vivo evaluation of Zn-pectinate microparticles reinforced with chitosan for the colonic delivery of progesterone. Drug Delivery 2015; 8: 1-14.
134. Kato, Y., H. Onishi and Y. Machida. Contribution of chitosan and its derivatives to cancer chemotherapy. In Vivo 2005; 19: 301-310.
135. Chatterjee, S. and Z.M.A. Judeh. Encapsulation of fish oil with N-stearoylO-butylglyceryl chitosan using membrane and ultrasonic emulsification processes. Carbohydrate Polymer 2015; 123: 432-442.
136. Rhoades, J. and S. Roller. Antimicrobial actions of degraded and native chitosan against spoilage organisms in laboratory media and foods. Applied and Environmental Microbiology 2000; 66: 80-86.
137. Jayasree, R.S., K. Rathinam and C.P. Sharma. Development of artificial skin (Template) and influence of different types of sterilization procedures on wound healing pattern in rabbits and guinea pigs. Journal of Biomaterials Applications 1995; 10: 144-162.
138. Ueno, H., T. Mori and T. Fujinaga. Topical formulations and wound healing applications of chitosan. Advanced Drug Delivery Reviews 2001; 52: 105-115.
139. Muzzarelli, R.A., M. Mattioli-Belmonte, A. Pugnaloni and G. Biagini. Biochemistry, histology and clinical uses of chitins and chitosans in wound healing. EXS 1999; 87: 251-264.
140. Azuma, K., R. Izumi, T. Osaki, S. Ifuku, M. Morimoto, H. Saimoto, S. Minami and Y. Okamoto. Chitin, chitosan, and its derivatives for wound healing: Old and new materials. Journal of Functional Biomaterials 2015; 6: 104-142.
141. Ishihara, M., K. Obara, S. Nakamura, M. Fujita, K. Masuoka, Y. Kanatani, B. Takase, H. Hattori, Y. Morimoto, M. Ishihara, T. Maehara and M. Kikuchi. Chitosan hydrogel as a drug delivery carrier to control angiogenesis. Journal of Artificial Organs 2006; 9: 8-16.
142. Azad, A.K., N. Sermsintham, S. Chandrkrachang and W.F. Stevens. Chitosan membrane as a woundhealing dressing: Characterization and clinical application. Journal of Biomedical Materials Research – Part B: Applied Biomaterials 2004; 69: 216-222.
143. Valentine, R., T. Athanasiadis, S. Moratti, L. Hanton, S. Robinson and P.J. Wormald. The efficacy of a novel chitosan gel on hemostasis and wound healing after endoscopic sinus surgery. American Journal of Rhinology & Allergy 2010; 24: 70-75.
144. Bennett, B.L., L.F. Littlejohn, B.S. Kheirabadi, F.K. Butler, R.S. Kotwal, M.A. Dubick and J.A. Bailey. Management of external hemorrhage in tactical combat casualty care: Chitosan-based hemostatic gauze dressings—TCCC guidelines-change 13-05. Special Operations Medical Journal 2014; 14: 40-57.
145. Hatamabadi, H.R., F.A. Zarchi, H. Kariman, A.A. Dolatabadi, A. Tabatabaey and A. Amini. Celox-coated gauze for the treatment of civilian penetrating trauma: A randomized clinical trial. Trauma Monthly 2015; 20(1): e23862.

146. Jebahi, S., M. Saoudi, L. Farhat, H. Oudadesse, T. Rebai, A. Kabir, A. El-Feki and H. Keskes. Effect of novel curcumin-encapsulated chitosan-bioglass drug on bone and skin repair after gamma radiation: Experimental study on a Wistar rat model. Cell Biochemistry and Function 2015; 33: 150-159.
147. Dutkiewicz, J., L. Szosland, M. Kucharska, L. Judkiewicz and R. Ciszewski. Structure-bioactivity relationship of chitin derivatives—Part I: The effect of solid chitin derivatives on blood coagulation. Journal of Bioactive and Compatible Polymers 1990; 5: 293.
148. Ruocco, N., S. Costantini, S. Guariniello and M. Costantini. Polysaccharides from themarine environment with pharmacological, cosmeceutical and nutraceutical potential. Molecules 2016; 21(5): 551.
149. Muzzarelli, R.A.A., F. Tanfani and M. Emanuelli. Sulfated N-(carboxymethyl)chitosans: Novel blood anticoagulants. Carbohydrate Research 1984; 126: 225.
150. Nolan, B.S., C. Frances, W.L. Steven, D.G. Chirag and H.C. Cheul. Heparin crosslinked chitosan microspheres for the delivery of neural stem cells and growth factors for central nervous system repair. Acta Biomaterialia 2013; 9(6): 6834-6843.
151. Heber, D. Herbal preparations for obesity: Are they useful? Primary Care. 2003; 30: 441–463.
152. Ylitalo, R., S. Lehtinen, E. Wuolijoki, P. Ylitalo and T. Lehtimäki. Cholesterol-lowering properties and safety of chitosan. Arzneimittel Forschung. 2002; 52: 1-7.
153. Mhurchu, C.N., S.D. Poppitt, A.T. McGill, F.E. Leahy, D.A. Bennett, R.B. Lin, D. Ormrod, L. Ward, C. Strik and A. Rodgers. The effect of the dietary supplement, chitosan, on body weight: A randomised controlled trial in 250 overweight and obese adults. International Journal of Obesity and Related Metabolic Disorders 2004; 28: 1149-1156.
154. Egras, A.M., W.R. Hamilton, T.L. Lenz and M.S. Monaghan. An evidence-based review of fat modifying supplemental weight loss products. Journal of Obesity 2011.
155. Kerch, G. The potential of chitosan and its derivatives in prevention and treatment of age-related diseases. Marine Drugs 2015; 13(4): 2158-2182.
156. Amidi, M., E. Mastrobattista, W. Jiskoot and W.E. Hennink. Chitosan-based delivery systems for protein therapeutics and antigens. Advanced Drug Delivery Reviews 2010; 62: 59-82.
157. Scherlie, R., S. Buske, K. Young, B. Weber, T. Rades and S. Hook. In vivo evaluation of chitosan as an adjuvant in subcutaneous vaccine formulations. Vaccine 2013; 31: 4812-4819.
158. Wang, Y.-Q., J. Wu, Q.-Z. Fan, M. Zhou, Z.-G. Yue, G.-H. Ma and Z.G. Su. Novel vaccine delivery system induces robust humoral and cellular immune responses based onmultiple mechanisms. Advanced Healthcare Materials 2013; 3: 670-681.
159. Vicente, S., B. Diaz-Freitas, M. Peleteiro, A. Sanchez, D.W. Pascual, G.A. Fernandez and M.G. Alonso. A polymer/oil based nanovaccine as a single-doseimmunization approach. PLoS ONE 2013; 8: 62407.
160. Bento, D., H.F. Staats, T. Gonçalve and O. Borges. Development of a novel adjuvanted nasal vaccine: C48/80 associated with chitosan nanoparticles as a path to enhance mucosal immunity. European Journal of Pharmaceutics and Biopharmaceutics 2015; 93: 149-164.
161. Laffleur, F., A. Fischer, M. Schmutzler, F. Hintzen and A. Bernkop-Schnürch. Evaluation of functional characteristics of preactivatedthiolated chitosan as potential therapeutic agent for dry mouth syndrome. Acta Biomaterialia 2015; 21: 123-131.
162. Myriam, J., M. Alameh, D.De Jesus, M. Thibault, M. Lavertu, V. Darras, M. Nelea, M.D. Buschmann and A. Merzouki. Chitosan-based therapeutic nanoparticles for combination gene therapy and gene silencing of in vitro cell lines relevant to type 2 diabetes. European Journal of Pharmaceutical Sciences 2012; 45: 138-149.
163. Zhang, X., K. Mysore, E. Flannery, K. Michel, D.W. Severson, K.Y. Zhu and M. Duman-Scheel. Chitosan/Interfering RNA nanoparticle mediated gene silencing in disease vector mosquito larvae. Journal of Visualized Experiments 2015; 97: 52523.
164. Rosanne, R., J. Fergal O'Brien and C. Sally-ann. Chitosan for gene delivery and orthopedic tissue engineering applications. Molecules 2013; 18(5): 5611-5647.

Chapter 9

Carrageenan
A Novel and Future Biopolymer

F. Adam*, Md. A. Hamdan and S.H. Abu Bakar

Faculty of Chemical and Natural Resources Engineering
Universiti Malaysia Pahang, Malaysia

9.1 INTRODUCTION

There are a few local natural polysaccharides, such as carrageenan, gum, starch, and others, which have a variety of multiblock copolymers that can be used to develop composites and biopolymers. It possesses some unique properties, such as being an energy source compound and having medicinal function. It is also biodegradable. Carrageenan is a natural polymer that is easily available, nontoxic, cheap, renewable, biodegradable, and biocompatible raw material [1, 2].

9.1.1 Production of Carrageenan

Among the natural polymers, seaweed in the form of carrageenan is among the food polysaccharides with the existence of inter-sugar glycosidic linkages as their similarity [3]. Carrageenan is the generic name for a family of high molecular weight sulphated polysaccharides obtained from extraction of certain species of red seaweeds [4]. The word "carrageenan" is derived from the name of the country of origin of almost red seaweeds extracts for food and medicines that have been used for over 600 years – Carraghen, which is located at the south Irish coast [2].

In 2013, the global carrageenan industry was worth 762.35 million USD [5]. The largest producer for carrageenan is Philippines, which accounts for around 77% of the world's supply. The Asia Pacific market for carrageenan is extremely huge as compared to the North American and European markets. In fact, the Asia Pacific market for carrageenan was worth 575.2 million USD as of 2013, where China is the main exporter of carrageenan to both US and Europe [5].

World carrageenan production exceeded 56,000 tons as of 2013, and it has a very competitive market in many countries, such as Argentina, Canada, Chile, Denmark, France, Japan, Mexico,

**Corresponding author*: Email: fatmawati@ump.edu.my

Morocco, North Korea, Portugal, Russia, South Korea, Spain, and USA. The carrageenan market has witnessed a Compound Annual Growth Rate (CAGR) of 5.2% in 2015, and is expected to grow consistently. Carrageenan is also an important salad ingredient and additive in food, personal care, pharmaceuticals, and other industrial applications. Personal care products include cosmetics, hand and body lotions, shampoo, soap, toothpaste, and gel fresheners [5].

9.1.2 Present Application of Carrageenan in Food and Pharmaceutical Products

Seaweeds are the only source of phytochemicals, namely agar, carrageenan, and algin, which are widely used in various industries, such as food, confectionery, textiles, pharmaceutical, dairy, and paper industry [6]. It normally contains the sulphated polysaccharide which can be used to produce products with special physical properties relating to viscosity, texture, and gelling ability such as hydrogel [7, 8]. Carrageenan is extracted from red seaweed, *Chondrus crispus* and *Eucheuma* species. Meanwhile, kappa carrageenan is extracted from *Kappaphycus alvarezii*, as shown in Fig. 9.1. This type of species requires low capital, and needs relatively simple farming technology with short production cycle. The global production of carrageenan is about 6 million wet tones [9]. Furthermore, carrageenan has interesting pharmacology properties, such as inhibition of hepatitis A viruses [10]. Oligomers are a bioactive compound from carrageenan, which contain antiherpetic, anti-HIV (human immunodeficiency virus), antitumor, anti-inflammatory antioxidant, and anticoagulant properties [1, 11–13]. Besides, it has been used in agriculture as a fertilizer and the seaweed also has been investigated as a binder of heavy metal for pollution-control applications [8].

Figure 9.1 Dried *Kappaphycus alvarezii.*

Carrageenan is a regulated food additive and there is a concern over the maximum quantity (in mg) of carrageenan that can be added to food. The major players in this industry are Marcel Carrageenan, Seatech Carrageenan Company, and FMC BioPolymer [5]. A forecast expected that the Global Carrageenan will grow at a compound annual growth rate of about 7.8% during the time frame 2018–2022 [14].

Excellent functions as gelling, thickening, emulsifying, and stabilizing agents have assured the market demand of carrageenan. The demands increased in all areas of the foods, materials, and polymers application. The use of carrageenan for food has grown in the industrialized countries by at least 5–7% per annum [15].

9.2 RAW MATERIALS

9.2.1 Source of Carrageenan

Malaysia is sharing a strategic and geographical location of the ocean with some parts of Philippines, especially in Semporna, which is located on the East coast of Sabah. That ocean is suitable for seaweed farming. Seaweed is a microscopic algae found at the bottom of shallow coastal water and it grows in the intertidal, shallow, and deep sea areas up to 180 meters in depth [6]. There are several seaweed species found growing naturally on reefs in Semporna area at south of Sabah, in Banggi Island of the South China Sea and in Kudat area in north of Sabah [16]. Thus, seaweed cultivation has become an economically important natural resource for Malaysian commodity. Only in 2013, production of red seaweed as the raw material of carrageenan in Sabah accounted for 28% by volume (33,210 metric ton) and 3% by value (RM198.93 million) from the total marine aquaculture production [17]. In addition, the production of carrageenan from seaweed in 2013 was increased by volume to 110.0 metric tons as compared to 2012. Most seaweed products from Malaysia, especially carrageenan, are exported and the value is estimated to increase up to RM1.4 billion in 2020 [17].

9.2.2 Cultivation and Harvesting of Carrageenan

There are two methods to cultivate carrageenan – off-bottom line and floating raft method at water temperature between 25°C to 30°C in bright light condition. The sea water for the off-bottom line must be deep enough to ensure that the seaweed is not exposed at low tide. Meanwhile, for the floating raft method, it is suitable in protected areas where water current is weak or where the water is too deep for fixed bottom lines.

After harvesting, the seaweed is kept away from sand and dirt using drying mat or rack, and it is sun dried for two to three days. The best moisture content in seaweed is 35% and it is checked based on its firmness and bends [18]. Moisture content of more than 40% will rot the seaweed during storage, while if it is less than 35%, seaweed will become too firm, bouncy, and difficult to compress into bales. Upon drying, white salt-like crystals appear outside of the seaweed due to the evaporated seawater [18].

Upon processing, there are two main products of carrageenan, which are refined carrageenan and semi-refined carrageenan, as described by Food and Agriculture Organization of the United Nations (FAO) [19]. Upon production of refined carrageenan, seaweed is extracted using alkali, commonly sodium hydroxide, which cause chemical structure changes and increment of gel strength in the carrageenan [18]. Increasing the gel strength will lead to an increase in viscosity and polymerization of biopolymers. At this stage, sulphate groups is removed and formation of 3,6-anhydro-G-galactose increases. Besides that, there are a few methods to extract the carrageenan from the seaweed, which are alcohol precipitation, gel pressing, and freeze-thaw methods [1, 4]. Alcohol precipitation method can be used for any type of carrageenan, while gel pressing method is only suitable for kappa carrageenan.

Semi-refined carrageenan (SRC) refers to much shorter process and cheaper carrageenan. The first step of production of SRC is to wash everything out from seaweed that will dissolve in alkaline and water solution, leaving only carrageenan and insoluble matter behind [18]. The insoluble residue largely consisting of carrageenan and cellulose will then be dried and sold as SRC. In this process, seaweed does not undergo alkali extraction process. Kappa and iota carrageenan are commonly produced using this method, as it is the carrageenan that form gels with potassium ions. However, the market demand for iota is lower in comparison to kappa carrageenan. The final product from this SRC method is cheaper as compared to refined carrageenan because there is no special equipment required throughout the processing line, such as distillation equipment, gel-forming unit, refrigeration unit, and squeezer [18, 19].

Carrageenan in the form of refined and semi-refined carrageenan is a renewable polysaccharide for future biopolymers. It can be harvested in a very short cycle of three to six months for sustainable raw materials [15].

9.3 POLYMERIZATION PROCESS, CHEMICAL STRUCTURE, AND PRODUCTION

9.3.1 Polymerization and Structure of Carrageenan

Carrageenan contains the 3,6-anhydro group, which promotes helices formation, which connects two molecular chains to form a three-dimensional structure that is responsible for swelling properties of carrageenan and is suitable for development of biopolymers, biocomposite, hydrogel shell of microcapsules, beads, and wound dressing [20–24].

Crosslinking can be a part of polymerization of biopolymer [25]. It is a chemical modification where the formation of chemical links between molecular chains forms a three-dimensional network of connected molecules. There are two types of crosslinking, which are chemical crosslinking and physical crosslinking. The chemical crosslink hydrogels are the network with chains bond. Chemical, enzymatic, and radiation reactions are among the method that produced hydrogels with chains bond with better mechanical stability [26]. Meanwhile, the physical crosslink hydrogels are three-dimensional networks where the polymer chains bond through non-covalent interactions. It produces a simple physical entanglement which is sufficient to develop a stable hydrogel [27]. One of the ways to form physical crosslinking is hydrophobic hydrophilic interaction. The hydrophobic blocks of carrageenan are coupled to hydrophilic blocks, creating a polymer amphiphile.

Carrageenan molecules behave as highly aggregated molecules or in the form of a tight cluster. The structure is held by inter- and intra-molecular forces, such as hydrogen bonds, hydrophobic, and Van der Waals forces [28]. Breakage of these bonds will modify the properties of carrageenan, such as thermal stability, disintegration, and mechanical properties.

Structure of different type of carrageenan contains sulphate and hydroxyl functional group, as in Fig. 9.2, which make its property a pH-sensitive hydrocolloid. A study conducted on carrageenan hydrogel shows that swelling degree of carrageenan hydrogel biopolymer is at the highest value at high pH system [29]. At high pH, negative charge in the system and electrostatic repulsion becomes dominant. Diffusion of water molecule into the network causes the hydrogel to swell. Furthermore, in natural environment, the same charged groups will repel each other [30]. As a result, the network becomes more permeable to large molecules, and more water can penetrate the network, that leads to higher swelling degree.

9.3.2 Carrageenan Chemical Structure

Kappa, iota, and lambda carrageenan are soluble in hot water. Kappa and iota carrageenan are also soluble in cold water in the presence of sodium salt [31]. Besides, both carrageenan do not dissolve but swell in the presence of potassium and calcium salt in water. Lambda carrageenan has an ability to be soluble in cold water and gives a viscous solution which shows pseudo plasticity or shear thinning properties when it is pumped or stirred [31, 32]. Kappa carrageenan has a higher mechanical strength in comparison with iota and lambda carrageenan [33].

Carrageenan is mainly composed of D-galactose residues linked alternately in 3-linked-β-D-galactopyranose and 4-linked-β-D-galactopyranose units, and are classified according to the degree of substitution that occurs on their free hydroxyl groups [4]. There are three main varieties of carrageenan, which differ in their degree of salvation. The ester sulphated groups and bridges present give interesting properties, such as rheological, hydration, gel strength, texture, syneresis, synergism, melting and setting temperatures of the materials [34].

Kappa carrageenan undergoes polymerization process in form of gelation when it is cooled to between 40–60°C after mixing in hot solution, depending on the cations present [31, 35]. The carrageenan gel is thermally reversible and stable at room temperature [21, 36]. However, it can be remelted when it is heated to 5–20°C above the gelling temperature. Gelation of carrageenan will occur when a single helix of carrageenan molecule comes near a second identical single helix of carrageenan, thus forming a double helix structure. The double helices of carrageenan must aggregate to form a three-dimensional network of carrageenan [31]. Kappa carrageenan will attach to potassium ion, as shown in Fig. 9.3 to maintain its stability of the junction zone within the firm and brittle gel.

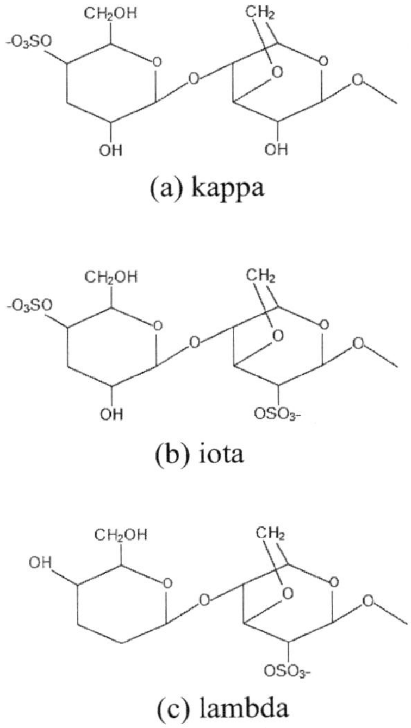

Figure 9.2 Chemical structure of (a) kappa, (b) iota, and (c) lambda carrageenan.

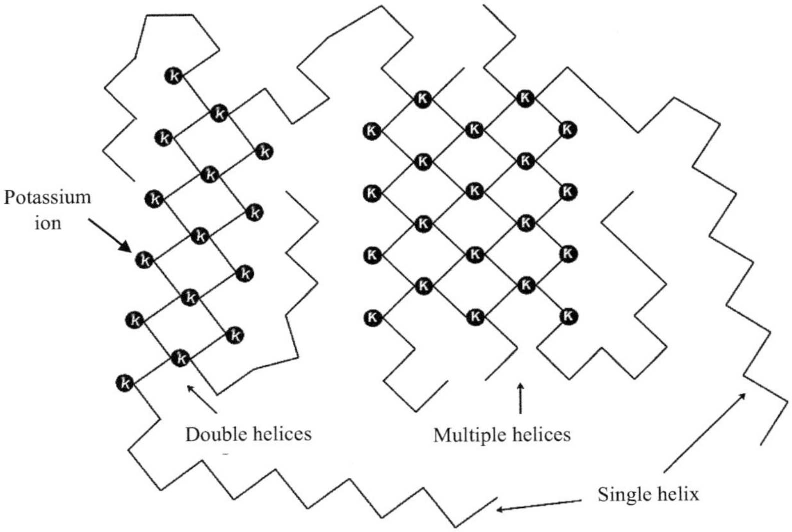

Figure 9.3 Gelation of kappa carrageenan with potassium ion.

A study found that upon heating of carrageenan, there was no significant swelling or hydration of particle until temperature exceeded 40–60°C [31]. Hydration of particles caused the rising of viscosity due to swollen particles, offering more resistance to flow. However, further heating up carrageenan solution to 75–80°C can produce a drop in viscosity value. Meanwhile, upon cooling, the viscosity of the solution increased, followed by gelation of carrageenan below temperatures between 40–50°C. Viscosity reflects the intermolecular interaction of the solution at molten solid state. High viscosity will create a stronger intermolecular interaction, such as hydrogen bonding in the bioplastic matrix, thus producing a strong solid material.

9.4 PROPERTIES

In a case study of carrageenan crosslinked biocomposite application for hard capsule, disintegration time was analysed to determine the ability of the biocomposite to degrade in water. The biocomposite hard capsule started to disintegrate after seven minutes in the disintegration test apparatus [37]. While the non-crosslink hard capsule swelled and showed no disintegration sign after 15 minutes. This difference of disintegration results is due to the establishment of physical crosslink or hydrogen bond interaction between kappa carrageenan and crosslinker in the sample, as per prediction from quantum mechanics calculation [38]. In the other case of chemical crosslinked in gelatine hard capsule that was stored at 40°C and 75% relative humidity (RH), was found to affect water solubility of the shell and thus retard the disintegration process [39].

Compared to chemical crosslink, the physical crosslink through H-bond may be easier to rupture and then may promote the hydration of the hard capsule shell during the disintegration process. H-bond is one of the solvation forces that is normally used in the ionic liquid application [40]. Thus, the physical crosslink through H-bond may be established between carrageenan and crosslinker, which affects the carrageenan matrix arrangement and promotes the disintegration of crosslinked carrageenan hard capsule.

Figure 9.4 represents two different grades of material for carrageenan powders, refined carrageenan (RC) and semi-refined carrageenan (SRC). The detail of both carrageenan types is summarized in Table 9.1. Viscosity and shear stress value are different in both types of carrageenan due to chemical structure changes and increment of gel strength [18].

Figure 9.4 Physical appearance of (a) refined carrageenan and (b) semi-refined carrageenan.

Table 9.1 Rheological properties of carrageenan.

	Range	*Unit*
Refined Carrageenan—Concentration: 1.5 w/v%		
Viscosity	200–260	mPas
Shear stress	70–100	Pa
Semi Refined Carrageenan—Concentration: 1.5 w/v%		
Viscosity	10–20	mPas
Shear stress	3–8	Pa

9.5 RELATIONSHIP BETWEEN STRUCTURE AND PROPERTIES

Carrageenan is a high molecular weight macromolecule from the polysaccharide group. Industrial grade of carrageenan extract has a molecular weight between 400–560 kDa [31, 41]. Meanwhile, semi-refined carrageenan (SRC) extract has molecular weight of approximately 615 kDa [42]. The 3,6-anhydro group in carrageenan promotes α helix formation, which is important for gelling [43]. It also increased its flexibility, which promotes swelling properties which are suitable for hydrogel or soft capsule production.

The chemical structure of kappa carrageenan can be determined by infrared spectroscopy characteristic. The characteristic infrared peak of kappa carrageenan is presented in Table 9.2 and Fig. 9.5. There are four basic functional group peaks for kappa carrageenan under IR spectroscopy, which are 3,6-anhydrogalactose-2-sulfate, D-galactose-4-sulfate, 3,6 anhydro-galactose, ester sulphate, and hydroxyl. Presence of these peaks will confirm the presence and purity of kappa carrageenan. As for an example, in iota carrageenan, a new peak appears at approximately 805 cm^{-1}, which indicates the presence of two sulphate ester groups on the anhydro-D-galactose residues, while for lambda carrageenan it shows a broad peak in wavelength of approximately 820–830 cm^{-1} [44].

Table 9.2 Characteristic IR peaks present in the kappa carrageenan [19, 30, 44, 45].

Functional group	*Molecular formula*	*FTIR peak* (cm^{-1})
3,6-anhydrogalactose-2-sulfate	$C_6H_{10}O_8S$	803–805
D-galactose-4-sulfate	$C_6H_{12}O_9S$	840–850
3,6 anhydro-galactose	$C_6H_{10}O_5$	925–935
Ester sulphate	S=O	1210–1260
Hydroxyl	O–H	3200–3600

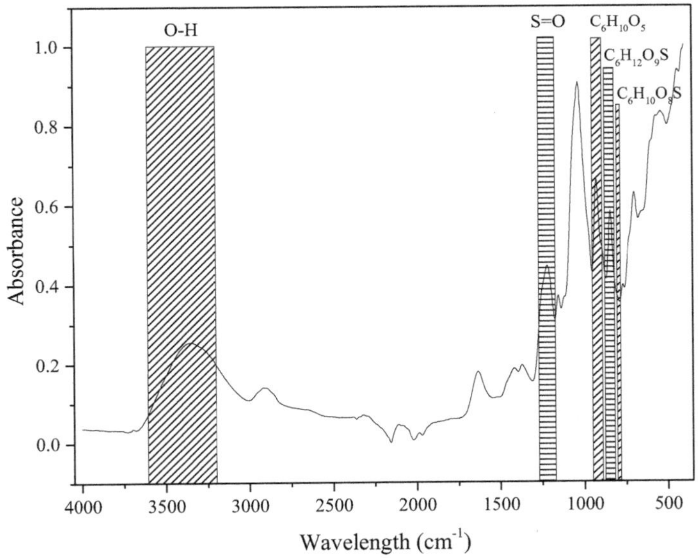

Figure 9.5 Infrared spectroscopy of raw carrageenan.

Thermal stability of carrageenan can be studied through thermogravimetric analysis (TGA). As shown in Fig. 9.6, the first stage of decomposition of carrageenan refers to the separation of water from the sample, starting from 40°C and reached the maximum at 150°C [46]. The second thermal event at 170°C represents the spontaneous decomposition of the carrageenan molecule by leaving of carbon dioxide and sulphur dioxide compounds [46]. The presence of sulphate group in the carrageenan structure causes the activation energy of carrageenan biocomposite to reduce [47]. At the last stage of decomposition, there is 17.74% of residue left in carrageenan in the form of ash [48].

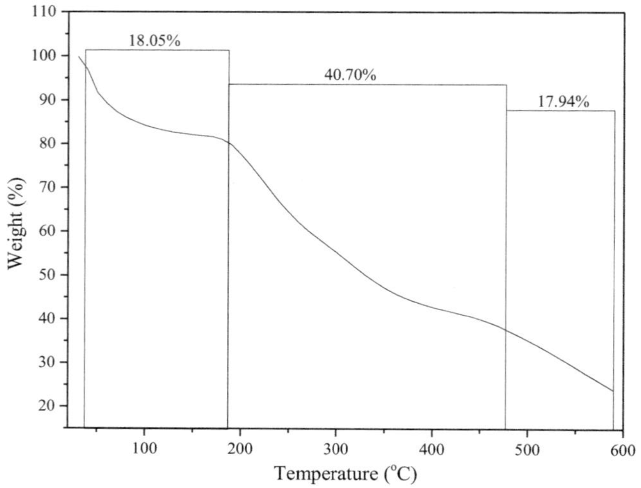

Figure 9.6 Thermogravimetric curve for raw carrageenan.

Carrageenan has an amorphous structure crystal, as shown in Fig. 9.7 [49, 50]. The X-ray diffractogram pattern of carrageenan has a weak and broad peak in the range of 2θ=20–22 [51]. The crystallinity structure shows that the carrageenan has a short chain structure and disorderly long chain structure [49].

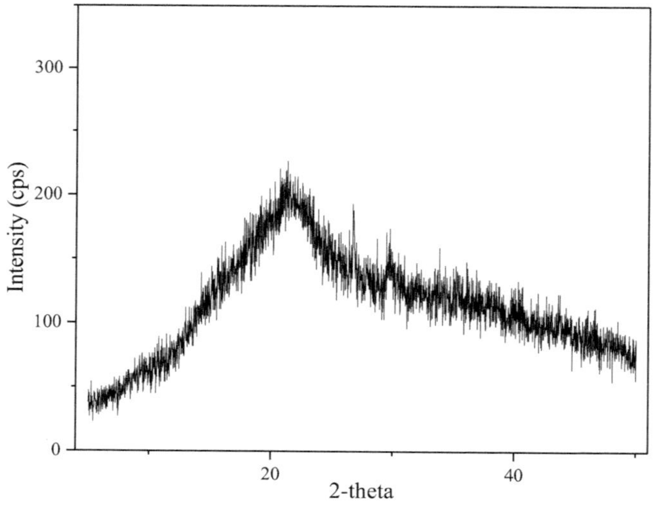

Figure 9.7 Crystallinity diffractogram of raw carrageenan.

9.6 MODIFICATIONS ACCORDING TO DRAWBACKS

Due to its inability to form a stable structure in the biocomposite application, the crosslinking method is applied to carrageenan solution. A new chemical bonding is created as a bridge between the carrageenan, crosslinker, and polymer to form a solid and stable biomaterial. This section will discuss the crosslinking of carrageenan to modify the properties, such as swelling, viscosity, and mechanical properties of the solution and solid form.

9.6.1 Chemical Crosslink Blending

Carrageenan has been incorporated with crosslinkers to produce a stable biopolymer. A few case studies have been conducted to understand the stability and mechanical effect from the incorporation of chemical and physical crosslinkers in the carrageenan. The presence of chemical crosslinkers in carrageenan, such as glutaraldehyde, genipin, and water-soluble carbodiimide (WSC) leads to an increase in the swelling ability of carrageenan, and causes the carrageenan to form hydrogel and does not disintegrate in water [30, 52–54]. The chemical crosslinked carrageenan is commonly used in the making of biofilm, biocomposite, and hydrogel for emulsion stabilization, syneresis control and bodying, binding, and dispersion purposes [55].

In glutaraldehyde-carrageenan hydrogel preparation, the carrageenan film was immersed in glutaraldehyde solution for chemical crosslinking reaction [29]. The study proved that carrageenan hydrogel crosslinked with chemical crosslinkers did not easily dissolve in water, as the film had lower swelling degree as compared to non-crosslinked film. The hydroxyl groups of carrageenan formed crosslinked structures with aldehydes of glutaraldehyde, thus creating bridges in the composite matrix [30]. The crosslinking structure was proven by the decrement of swelling degree.

Genipin, which is widely used in herbal medicine, was chemically crosslinked with kappa-carrageenan through blending process to produce a stable hydrogel for medical applications [22]. Incorporation of genipin in the formulation decreased the swelling degree of the hydrogel due to the increment of network stability of the modified hydrogel matrix. Addition of genipin increased the glass transition (T_g) of the composite, which represented the increment in the hydrogel stability [52].

Water-soluble carbodiimide contains ester linkages from carboxylic acid and hydroxyl groups which facilitated in chemical crosslinking reactions in alginate-carrageenan coating membrane film [54]. Alginate itself contains two carboxylic groups to crosslink alginate-alginate and alginate-carrageenan structures. Furthermore, carrageenan poses the carboxylic group that can participate in alginate-carrageenan crosslinking reaction.

9.6.2 Physical Crosslink Blending

Physical crosslinking can occur by formation of hydrogen bonding using dehydrothermal treatment, UV treatment, and gelatine-siloxane hybrid, but it is commonly applied in gelatine hydrogel [56]. In fact, physical crosslinker does not produce any chemical toxicity or reagent residuals, as compared to chemical crosslinker. However, there is still limited literature published regarding type and application of physical crosslinker in carrageenan, except using glyoxylic acid crosslinker [43, 57].

Carrageenan and glyoxylic acid as crosslinker was simulated at a molecular level to predict the possible location of the crosslinking in the conjugate complex for biofilm application, as shown in Fig. 9.8 [57]. The molecular electrostatic surface potential (MESP) analysis revealed that the red region, which represents the most negative electrostatic potential, was found around the oxygen and sulphur atoms in carrageenan molecule. Meanwhile, the blue region, which represents the most positive electrostatic potential, was found around the hydrogen atom, far from the double bond oxygen atoms in glyoxylic acid structure. Thus, these two regions might interact in the physical reaction process and form physical crosslink through hydrogen bond interaction. This was due to the carrageenan, which served as an electron donor, while the glyoxylic acid served as a proton donor to establish hydrogen bonding physical crosslinking [57]. Figure 9.9 represents the crosslinked section between carrageenan and crosslinker glyoxylic acid with bond length of 1.90 Å, using Gaussian calculation. The carrageenan and glyoxylic acid was crosslinked with hydrogen bonding [57].

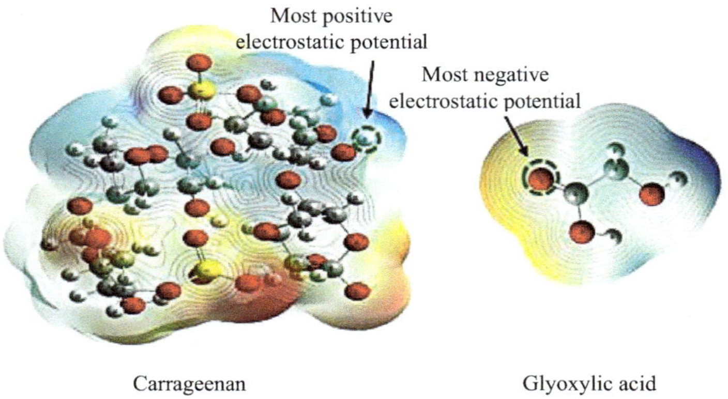

Figure 9.8 The molecular electrostatic potential (MESP) analysis for carrageenan and glyoxylic acid [57].

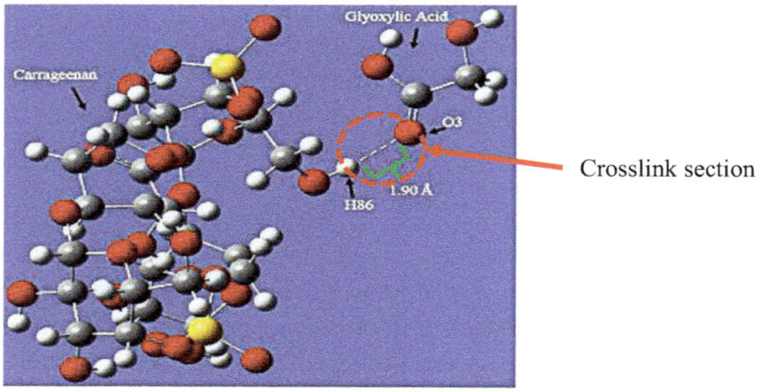

Figure 9.9 Crosslinking section between carrageenan and glyoxylic acid [57].

9.6.3 Additive in Carrageenan Biocomposite

Incorporation of fillers, such as cellulose, can increase the mechanical strength of the carrageenan biocomposite. Besides, the mechanical strength changes can be proven by the chemical analysis via infrared (IR) determination [19]. Addition of small amount carboxymethyl cellulose (CMC) and microcrystalline cellulose (MCC) in the carrageenan matrix increases the tensile strength, viscosity, intensity of hydroxyl group, and –CH stretching of the biocomposite film, as shown in Table 9.3. The intensity of the functional group can be determined by calculating the area under corresponding absorbance peaks [58].

Table 9.3 Properties of carrageenan film incorporated with cellulose filler for biofilm application in food packaging [19].

Formulation	***Tensile strength* (MPa)**	***Solution viscosity* (mPas)**	***IR intensity* (cm^{-1})**	
			OH	**CH**
Carrageenan	–	–	103.44	14.58
CMC	–	–	126.34	25.30
MCC	–	–	104.98	14.84
Carrageenan film	20.95	0.982	54.89	22.92
Carrageenan – CMC film	28.69	1.48	130.53	19.95
Carrageenan – MCC film	26.89	1.36	150.04	23.70

The result reflects that there is an occurrence of intermolecular interaction between the hydroxyl and CH stretching in CMC and MCC with the carrageenan matrix. The intermolecular interaction may be contributed by the presence of sulphate and hydroxyl group in carrageenan and other additives and toughening agents added in the biocomposite formulation. Tensile strength, viscosity, absorbance of hydroxyl group, and CH stretching has increased in the addition of cellulose filler in carrageenan matrix that lead to a strong interaction between the functional groups and the carrageenan matrix [19, 59, 60].

9.7 INDUSTRIAL APPLICATIONS

In food industry, carrageenan is commonly used as a gelling, stabilizing, and water carrying agents [61]. The function of carrageenan depends on its condition. Carrageenan is incorporated with protein to form gel or non-gel suspension and sediment. It can also form a transparent, turbid, elastic, brittle, high or low strength, thermal reversible, or irreversible type [61]. Besides, at different temperatures, carrageenan can form different properties of gel.

In addition, carrageenan is also used as water dessert gels and cake glazes. Incorporation of cellulose in making the gels and glazes reduce the rupture strength, thus forming more brittle and fragile strength with cloudy product [31]. Currently, carrageenan is widely used in the development of vegetarian products which have similar properties, appearance, and texture to commercial gelatine products, which act as stabilizer, filler, and binder [18, 31].

Carrageenan is also used in milk gels and flans as a stabilizer for evaporated milk and ice cream mixture [31, 33]. Carrageenan is proven to prevent protein and fat separation in milk and dairy products [18]. In the milk gel mixture, carrageenan forms a weak gel and interacts with positively charged amino acid in the protein. Addition of carrageenan in ice cream formulation is to prevent whey separation, while controlling the texture and ice crystal growth [18].

In meat products, carrageenan is used as a binder. Incorporation of carrageenan and locust bean gum in the canned pet food formulation produced an excellent and affordable product [18, 55].

In non-food applications, carrageenan is used as a thickener in toothpaste. The ability to bind water effectively and forming weak and stable water gel makes carrageenan a unique material for the toothpaste product [18, 33]. Besides, carrageenan is also used to retain the odor and structure of air freshener gels [18]. Carrageenan is mixed with potassium salt, water, and perfume to develop the semi-solid gel.

9.8 DEVELOPMENT AND FUTURE TRENDS

At present, carrageenan is widely used as a wrapping film in food products, such as sushi. The potential carrageenan can be further developed in food packaging applications. Incorporation of crosslinker and cellulose filler in carrageenan is suitable to produce a hard capsule, as for an example in Fig. 9.10.

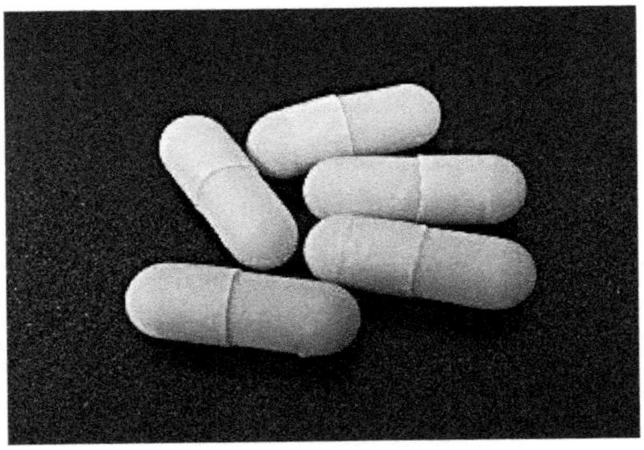

Figure 9.10 Carrageenan-based biocomposite for the hard capsule development.

The carrageenan from biocomposite formulation is also suitable and has potential to be a bioplastic product, as shown in Fig. 9.11. It can be an alternative material for bioplastic product due to its non-toxic properties and it can also degrade in soil and water. The bioplastic can solve the present marine plastic pollution and human-made debris [62]. Carrageenan bioplastic is a good solution for the massive plastic waste dumping in the landfill and ocean.

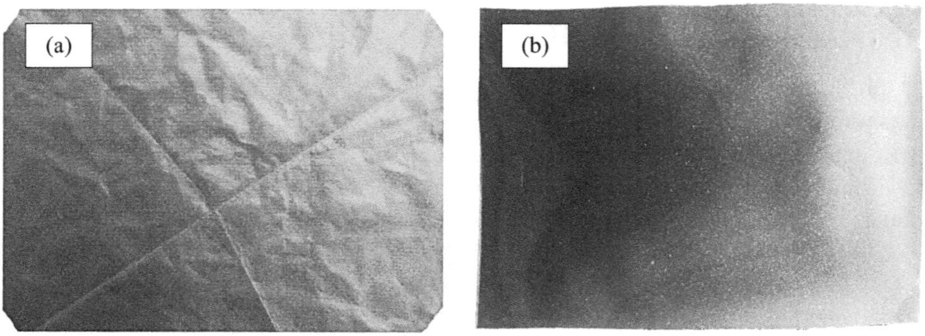

Figure 9.11 The modification of carrageenan biocomposite film in both a) and b) are suitable for bioplastic applications.

Carrageenan is also incorporated with polymers to develop hydrogel for biomedical and tissue engineering purpose [49]. Addition of carrageenan in the polyvinyl alcohol (PVA) polymer was found to overcome the issue of deformation resistance, thus preserving the shape of the hydrogel during lyophilization process without shrinkage [49].

Carrageenan plays an important role in the development of a wound dressing hydrogel using polyvinylpyrrolidone (PVP) and polyethylene glycol (PEG) [63], as shown in Fig. 9.12. The developed hydrogel has a soothing effect, thus it is not necessary to change the dressing too often [63].

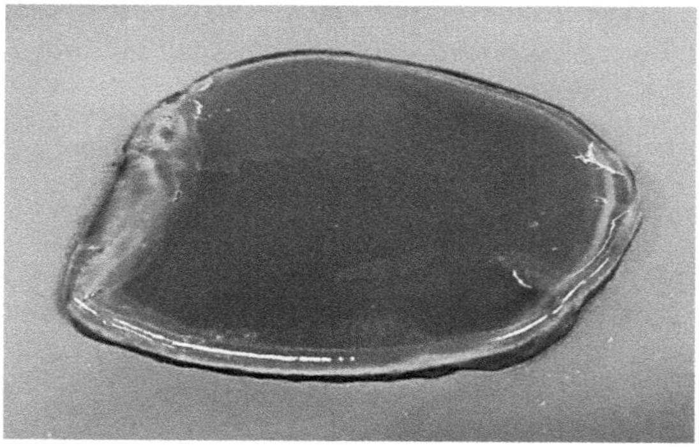

Figure 9.12 Hydrogel which has potential for wound dressing and biomedical purpose.

9.9 RECYCLING

At present, carrageenan is a non-recycling material, as it is easily biodegradable in water and sand. In water, carrageenan will disintegrate as the water will diffuse into the carrageenan matrix. Meanwhile in sand, due to changes in temperature, the organic content of carrageenan will degrade and decompose. Thus, this will be a good solution for plastic waste dumped in the landfill and marine pollution that causes the extinction of aquatic life.

At this age, usage of fossil polymers started to cause pollution to our environment. The source is non-renewable material, thus leading to an effort and cost for recycling the source. More carbon is released to the atmosphere and leads to many more consequences due to the plastic production and processing.

Using biopolymers, recycling is not a main concern because the properties of biopolymers can degrade and disintegrate. Application of carrageenan in bioplastics or biocomposites will reduce the recycling and pollution issue. Due to the mismanagement of fossil-based plastic waste, recycling is not an effective approach to be implemented in the developing countries around the globe. Therefore, future plastic industry demands for development of bioplastics from renewable and less carbon emission materials for future sustainability.

9.10 CONCLUSIONS

Carrageenan can be a future substitution of present fossil polymers, such as low density polyethylene in certain applications. Food wrapping plastics, such as low linear density polyethylene, can be replaced by the carrageenan-based bioplastic. The mechanical and thermal properties of carrageenan could be modified and designed to meet the product application of each biofilm application according to the consumer's specifications. Being excellently disintegrated and dissolved in water would be an advantage for carrageenan-based biopolymers to resolve the present plastic pollution in landfill and marine. Furthermore, it will be less toxic to the environment and aquatic life.

References

1. Prajapati, V.D., P.M. Maheriya, G.K. Jani and H.K. Solanki. Carrageenan: A natural seaweed polysaccharide and its applications. Carbohydrate Polymers 2014; 105(1): 97-112.
2. Necas, J. and L. Bartosikova. Carrageenan: A review. Veterinární Medicína (Praha) 2013; 58(4): 187-205.
3. Stephen, A.M. and S.C. Churms. Introduction. *In*: A.M. Stephen, G.O. Phillips and P.A. Williams (eds). Food Polysaccharides and Their Applications, 2nd Ed. CRC Press, Boca Raton, 2006, pp. 181-215.
4. Li, L., R. Ni, Y. Shao and S. Mao. Carrageenan and its applications in drug delivery. Carbohydrate Polymers 2014; 103(1): 1-11.
5. Research and Markets. Global Carrageenan Market-Trends and Forecasts: (2016-2021). 2017.
6. Gade, R., M. Siva Tulasi and V. Aruna Bhai. Seaweeds: A novel biomaterial. International Journal of Pharmacy and Pharmaceutical Sciences 2013; 5(Suppl. 2): 40-44.
7. Seedevi, P., S. Sudharsan, S.V. Kumar, A. Srinivasan, S. Vairamani and A. Shanmugam. Isolation and characterization of sulphated polysaccharides from Codium tomentosum (J. Stackhouse, 1797) collected from southeast coast of India. Advances in Applied Science Research 2013; 4(5): 72-77.
8. Cajipe, G.J.B. Utilization of seaweed resources. *In*: I.J. Dogma Jr., G.C. Trono Jr. and R.A. Tabbada (eds). Culture and Use of Algae in Southeast Asia: Proceedings of the Symposium on Culture and Utilization of Algae in Southeast Asia, 8-11 December 1981, Tigbauan, Iloilo, Philippines, 1990, pp. 77-79.
9. Valderrama, D., J. Cai, N. Hishamunda and N. Ridler. Social and economic dimensions of carrageenan seaweed farming. FAO Fisheries and Aquaculture Technical Paper no. 580, Food and Agriculture Organization of the United Nations, Rome, 2013.
10. Girond, S., J.M. Crance, H. Van Cuyck-Gandre, J. Renaudet and R. Deloince. Antiviral activity of carrageenan on hepatitis A virus replication in cell culture. Research in Virology 1991; 142(4): 261-270.
11. Carlucci, M.J., C.A. Pujol, M. Ciancia, M.D. Noseda, M.C. Matulewicz, E.B. Damonte and A.S. Cerez. Antiherpetic and anticoagulant properties of carrageenans from the red seaweed Gigartina skottsbergii and their cyclized derivatives: Correlation between structure and biological activity. International Journal of Biological Macromolecules 1997; 20(2): 97-105.
12. Yamada, T., A. Ogamo, T. Saito, H. Uchiyama and Y. Nakagawa. Preparation of O-acylated low-molecular-weight carrageenans with potent anti-HIV activity and low anticoagulant effect. Carbohydrate Polymers 2000; 41(2): 115-120.

13. Souza, B.W.S., M.A. Cerqueira, A.I. Bourbon, A.C. Pinheiro, J.T. Martins, J.A. Teixeira, M.A. Coimbra and A.A. Vicente. Chemical characterization and antioxidant activity of sulfated polysaccharide from the red seaweed *Gracilaria birdiae*. Food Hydrocoll. 2012; 27(2): 287-292.
14. Market Research. Carrageenan Sector: Worldwide Forecast until 2022. 2018.
15. Sade, A., I. Ali and Md. M.R. Ariff. The seaweed industry in sabah, East Malaysia. Jati, 2006; 11: 97-107.
16. Arman Shah, D.A. Seaweed farming in East Malaysia. Fisheries and Aquaculture Department, Food and Agriculture Organization of the United Nation 2010, pp. 2-3.
17. Safari, S. Prospects and Policy Review of seaweed as a High-Value Commodity in Malaysia. 2014.
18. McHugh, D.J. A Guide to the Seaweed Industry. 2003.
19. Hamdan, M.A. Investigation of Cellulose Incorporated Into Carrageenan Matrix For Hard Capsule Process Development. Universiti Malaysia Pahang, 2018.
20. Sherry Ku, M., W. Li, W. Dulin, F. Donahue, D. Cade, H. Benameur and K. Hutchison. Performance qualification of a new hypromellose capsule: Part I. Comparative evaluation of physical, mechanical and processability quality attributes of Vcaps Plus??, Quali-V?? and gelatin capsules. International Journal of Pharmaceutics 2010; 386(1-2): 30-41.
21. Hezaveh, H. and I.I. Muhamad. Controlled drug release via minimization of burst release in pH-response kappa-carrageenan/polyvinyl alcohol hydrogels. Chemical Engineering Research and Design 2013; 91(3): 508-519.
22. Meena, R., K. Prasad and A.K. Siddhanta. Development of a stable hydrogel network based on agar-kappa-carrageenan blend cross-linked with genipin. Food Hydrocoll. 2009; 23(2): 497-509.
23. Briones, A.V. and T. Sato. Encapsulation of glucose oxidase (GOD) in polyelectrolyte complexes of chitosan-carrageenan. Reactive & Functional Polymers Journal 2010; 70(1): 19-27.
24. Boateng, J.S., H.V. Pawar and J. Tetteh. Evaluation of in vitro wound adhesion characteristics of composite film and wafer based dressings using texture analysis and FTIR spectroscopy: A chemometrics factor analysis approach. RSC Advances 2015; 5(129): 107064-107075.
25. Simpson, W. and T. Holt. Gelation in addition polymerization. J. Polym. Sci. 1955; 18(89): 335-349.
26. Neradovic, D., C.F. Van Nostrum and W.E. Hennink. Thermoresponsive polymeric micelles with controlled instability based on hydrolytically sensitive N-isopropylacrylamide copolymers. Macromolecules 2001; 34(22): 7589-7591.
27. Gehrke, S.H. Synthesis and Properties of hydrogels used for drug delivery. *In*: G.L. Amidon, P.I. Lee and E.M. Topp (eds). Transport Processes in Pharmaceutical Systems, vol. 102. Marcel Dekker, Inc., New York, Basel, 2000, pp. 486-559.
28. Sletmoen, M. and B.T. Stokke. Review: Higher order structure of (1,3)-beta-D-glucans and its influence on their biological activities and complexation abilities. Biopolymers 2008; 89: 310-321.
29. Distantina, S., Rochmadi, M. Fahrurrozi and Wiratni. Hydrogels based on carrageenan extracted from *Kappaphycus alvarezii*. International Journal of Medical, Health, Biomedical, Bioengineering and Pharmaceutical Engineering 2013; 7(6): 244-247.
30. Distantina, S., R. Rochmadi, Md. Fahrurrozi and W. Wiratni. Synthesis of hydrogel film based on carrageenan extracted from Kappaphycus alvarezii. Modern Applied Science 2013; 7(8): 22-30.
31. Imeson, A.P. Carrageenan. *In*: G.O. Phillips and P.A. Williams (eds). Handbook of Hydrocolloids. Woodhead Publishing Ltd., UK, 2000, pp. 87-102.
32. Hernandez-Carmona, G., Y. Freile-pelegr, and E. Hernández-Garibay. Conventional and alternative technologies for the extraction of algal polysaccharides. *In*: H. Domínguez (ed.). Functional Ingredients from Algae for Foods and Nutraceuticals. Woodhead Publishing Series in Food Science, Technology and Nutrition, Woodhead Publishing, 2013, pp. 475-516.
33. Kelco, C.P. GENU Carrageenan Book. DK-4623 Lille Skensved, Denmark, 2001.
34. Santo, V.E., et al. Carrageenan-based hydrogels for the controlled delivery of PDGF-BB in bone tissue engineering applications. Biomacromolecules 2009; 10(6): 1392-1401.
35. Fakharian, M.-H., N. Tamimi, H. Abbaspour, A. Mohammadi Nafchi and A.A. Karim. Effects of k-carrageenan on rheological properties of dually modified sago starch: Towards finding gelatin alternative for hard capsules. Carbohydrate Polymers 2015; 132: 156-163.

36. Milani, J. and G. Maleki. Hydrocolloids in food industry. Food Ind. Process. - Methods Equip., 2012.
37. Hamdan, Md. A., F. Adam and K.N. Md. Amin. Investigation of mixing time on carrageenan-cellulose nanocrystals (CNC) hard capsule for drug delivery carrier. International Journal of Innovative Science and Research Technology 2018; 3(1): 457-461.
38. Abu Bakar, S.H. and F. Adam. Determination of physical crosslink between carrageenan and glyoxylic acid using density functional theory calculations. Malaysian Journal of Analytical Sciences 2017; 21(4): 979-985.
39. Brown, J., N. Madit, E.T. Cole, I.R. Wilding and D. Cade. Effect of cross-linking on the in vivo disintegration of hard gelatin capsule. Pharmaceutical Research 1998; 15(7): 1026-1030.
40. Mellein, B.R., S.N.V.K. Aki, R.L. Ladewski and J.F. Brennecke. Solvatochromic studies of ionic liquid/ organic mixtures. Journal of Physical Chemistry B 2007; 111: 131-138.
41. Cash, M.J. and S.J. Caputo. Cellulose derivatives. *In*: A. Imeson (ed.). Food Stabilizers, Thicheners and Gelling Agents. Wiley-Blackwell, United Kingdom, 2010, pp. 95-115.
42. Hoffmann, R., A. Russell and M. Gidley. Molecular weight distribution of carrageenans: Characterisation of commercial stabilisers and effect of cation depletion on depolymerisation. Gums Stabilisers Food Ind. 1996; 8: 137-150.
43. Ahmad, I.L. Synthesis and Characterization of Hydrogel for Hard Kappa Carrageenan Capsule through Glyoxal and Glyoxylic Acid Crosslinking. Universiti Malaysia Pahang, 2014.
44. Pereira, L., A. Sousa, H. Coelho, A. M. Amado and P.J.A. Ribeiro-Claro. Use of FTIR, FT-Raman and13C-NMR spectroscopy for identification of some seaweed phycocolloids. Biomolecular Engineering 2003; 20(4-6): 223-228.
45. Webber, V., S.M. De Carvalho, P.J. Ogliari, L. Hayashi, P. Luiz and P.L.M. Barreto. Optimization of the extraction of carrageenan from Kappaphycus alvarezii using response surface methodology. Ciência e Tecnologia de Alimentos 2012; 32(4): 1-7.
46. Thommes, M., W. Blaschek and P. Kleinebudde. Effect of drying on extruded pellets based on κ-carrageenan. European Journal of Pharmaceutics and Biopharmaceutics 2007; 31(2): 112-118.
47. Jamaludin, J., F. Adam, R.A. Rasid and Z. Hassan. Thermal studies on polysaccharides film arabic gum carrageenan. Chemical Engineering Research Bulletin 2017; 19: 80-86.
48. Food and Agriculture Organization. Carrageenan. 2001.
49. Zhang, Y, L. Ye, M. Cui, B. Yang, J. Li, H. Sun and F. Yao. Physically crosslinked poly(vinyl alcohol)-carrageenan composite hydrogels: Pore structure stability and cell adhesive ability. RSC Advances 2015; 5(95): 78180-78191.
50. Dai, W.G., L.C. Dong and Y.Q. Song. Nanosizing of a drug/carrageenan complex to increase solubility and dissolution rate. International Journal of Pharmaceutics 2007; 342(1-2): 201-207.
51. Liew, J.W.Y., K.S. Loh, A. Ahmad, K.L. Lim and W.R. Wan Daud. Synthesis and characterization of modified κ-carrageenan for enhanced proton conductivity as polymer electrolyte membrane. PLoS One 2017; 12(9): 1-15.
52. Hezaveh, H. and I.I. Muhamad. Effect of natural cross-linker on swelling and structural stability of kappa-carrageenan/hydroxyethyl cellulose pH-sensitive hydrogels. Korean Journal of Chemical Engineering 2012; 29(11): 1647-1655.
53. Liu, F., et al. Study of combined effects of glycerol and transglutaminase on properties of gelatin films. Food Hydrocoll. 2017; 65: 1-9.
54. Xu, J.B., J.P. Bartley and R.A. Johnson. Preparation and characterization of alginate-carrageenan hydrogel films crosslinked using a water-soluble carbodiimide (WSC). Journal of Membrane Science 2003; 218(1-2): 131-146.
55. Stanley, N. Production, properties and uses of carrageenan. *In*: D.J. McHugh (ed.). Production and Utilization of Products from Commercial Seaweeds. FAO Fisheries Technical Paper, Rome, 1987, 288: 116-146.
56. Eluvakkal, T., N. Shanthi, M. Murugan and K. Arunkumar. Extraction of antibacterial substances, galactofucoidan and alginate successively from the gulf of mannar brown seaweed Sargassum wightii greville ex J. Agardh. Indian Journal of Natural Products and Resources 2014; 5(3): 249-257.

57. Abu Bakar, S.H. and F. Adam. Determination of physical crossllink between carrageenan and glyoxylic acid using density functional theory calculations. Malaysian Journal of Analytical Sciences 2017; 21(4): 979-985.
58. Md. Amin, K.N. Cellulose Nanocrystals Reinforced Thermoplastic Polyurethane Nanocomposites. University of Queensland, 2016.
59. Haafiz, M.K.M., A. Hassan, Z. Zakaria, I.M. Inuwa, M.S. Islam and M. Jawaid. Properties of polylactic acid composites reinforced with oil palm biomass microcrystalline cellulose. Carbohydrate Polymers 2013; 98(1): 139-145.
60. Ummartyotin, S. and C. Pechyen. Microcrystalline-cellulose and polypropylene based composite: A simple, selective and effective material for microwavable packaging. Carbohydrate Polymers 2016; 142: 133-140.
61. Sidley Chemical, Co. Properties and Application of Kappa-Carrageenan. 2013. [Online]. Available: www.visitchem.com/properties-and-application-of-kappa-carrageenan/. [Accessed: 29-Nov-2018].
62. Reisser, J., J. Shaw, C. Wilcox, B.D. Hardesty, M. Proietti, M. Thums and C. Pattiaratchi. Marine plastic pollution in waters around Australia: Characteristics, concentrations, and pathways. PLoS ONE 2013; 8(11): e80466. https: //doi.org/10.1371/journal.pone.0080466.
63. De Silva, D.A., B.U. Hettiarachchi, L.D.C. Nayanajith, M.D.Y. Milani and J.T.S. Motha. Development of a PVP/kappa-carrageenan/PEG hydrogel dressing for wound healing applications in Sri Lanka. Journal of the National Science Foundation of Sri Lanka 2011; 39(1): 25-33.

Chapter 10

Natural Rubber and Bio-based Thermoplastic Elastomer

Wei Jiang

Department of Mechanical Engineering
Program of Materials Science and Engineering
The University of Texas at Austin
East Dean Keeton Street, ETC 3.128, Austin, TX 78712, USA

10.1 INTRODUCTION

Natural Rubber

Natural rubber (NR) is a biopolymer with unique elasticity produced by natural resources. It has been widely applied in many aspects of industrial and daily use, such as automotive, biomedical, and construction, etc. [1]. NR was recorded to be first used from 1300–300 BC in Yucatan Peninsula, Mexico [1]. There are several types of tropical plants, named rubber trees, producing NR. However, due to quality and yield reasons, about 99% of NR is harvested from the species *Hevea Brasiliensis*, which originated from Amazon River Valley and were distributed to south and South-east Asia in the 19th century [2]. The major content of NR is a polymer named cis-1,4-polyisoprene, composed of carbon and hydrogen atoms. NR is normally vulcanized to form crosslinked structure in actual applications. Due to the source, chemical content, molecular weight, and chemical structure of natural rubber, it features multiple outstanding advantages. NR is renewable due to its bio product nature. It has outstanding properties for engineering applications [2], such as low hysteresis loss (better than any synthetic rubber) under vibrating environment, good low temperature performance (T_g around –70°C), good bonding with metal surface, high anti-tearing and anti-abrasion performance, and easy processability. The major drawbacks of NR are its relative poor weather proof, acid resistance, and chemical resistance performances. To overcome or improve those drawbacks, multiple technical measures, such as chemical modification, blending with other polymers, compounding with functional additives, have been developed and will be introduced in later sections. With the described advantages and application history, it is possible that NR will continue contributing to the sustainability of the foreseeable future. The

For Correspondence: Email: weizyjiang@utexas.edu

investigation of bio-synthesis mechanism of NR, which endorses its unique properties, is likely to be a significant topic. The development of new applications of NR might draw more attentions from both academy and industry.

Bio-based Thermoplastic Elastomer

Thermoplastic elastomers (TPE) are a group of polymers featuring both good elasticity and easy processability. They have elastic performance like rubbers when being operated below their processing temperatures. TPE are not crosslinked and hence can be reprocessed and reshaped at their processing temperatures. TPE can be divided to two major categories: chemical-based TPE and bio-based TPE. This chapter only introduces bio-based TPE due to the chapter scope. Bio-based TPE can be achieved through blending of different polymers and synthesis of bio-based monomers. As a group of unique materials, bio-based TPE have some outstanding advantages, including good balance of elasticity and processability, recyclability, damping properties, good chemical resistance, excellent abrasion resistance, good low temperature performance, and environment compatibility [3]. The drawbacks of TPE are their relatively lower operation temperature compared to rubbers and higher price [3].

10.2 NATURAL RUBBER

NR is one of the most important natural resources. It is widely applied in more than 4,000 kinds of products in multiple fields including military, industry, medical, transportation, and consumer products [4]. Due to its unique low hysteresis performance, NR's role is not replaceable in many applications. One example is the tire industry, in which tire performance is improved by increasing the NR content (100% in aviation tires, >90% in heavy duty truck tires) [2]. This section will focus on the introduction of NR source, polymerization, modification, application, recycling, and future development.

10.2.1 Raw Materials

The plantation of rubber trees and production of NR mainly happened in South-east Asia, and provided more than 90% NR supply. The rest of the production is from South American and Africa. Currently, the rubber tree species, *Hevea Brasiliensis*, produces most of the commercial NR supply. However, due to the threat of disease and infection, it is risky to rely on only one species [5]. Out of the many NR production plants, two species named *Parthenium Argentatum* and *Taraxacum Kok-s*aghyz are under study as alternatives of *Hevea Brasilensis* because of their yield and anti-disease advantages [6, 7].

As strategic materials, the yield of NR increased by 100% (6 million tons to 12 million tons) from 1995 to 2015 [8]. However, the supply of NR is around 2 million tons in shortage because of the economy booming of multiple countries in the past decade. Development of future alternative rubber tree species and improvement of *Hevea Brasilensis* yield require the knowledge of NR production mechanism, which is still not very well established. Researches for understanding these mechanisms are needed to support the bio-genetic and plant breeding investigations [8].

10.2.2 Polymerization and Chemical Structure

Unlike other synthetic polymers, the polymerization of NR happens in plants, and is thus called bio-synthesis. Many potential pathways of the bio-synthesis mechanism have been investigated and proposed based on chemical analysis and structural analysis of NR molecules. To the date of art, it is commonly accepted that the monomer of NR is IPP, which is considered the product of isoprene (IP) and pyrophosphoric acid ($H_2P_2O_7$), as shown in Fig. 10.1 [9]. Isomerase enzyme is considered to function as the polymerization initiator, and metal ions (Mg^{2+}, Mn^{2+}, etc.) function as activators. The mechanism of NR bio-synthesis proven by Poulter and Riling in 1976 is shown in Fig. 10.2 [10]. The polymerization is initiated by isomerase enzyme to transfer some IPP to allylic carbocations with a pyrophosphate counter-anion. The isomerization IPPs will react with other IPPs through an addition reaction called prenyltransferase, and the addition will continue till termination. The prenyltransferase mechanism was proven by Hammet with fluorine substituted allylic pyrophosphates. The mechanism of termination in rubber trees is still not well understood till date.

The polymerization rate and molecular weight of NR are decided by IPP content, enzyme content, and activator content. A proper IPP to enzyme ratio combining proper activator content is essential to achieve high molecular weight. Proper activator content can maximize the rubber transferase rate, while extra activators can interact with rubber directly and prohibit transferase.

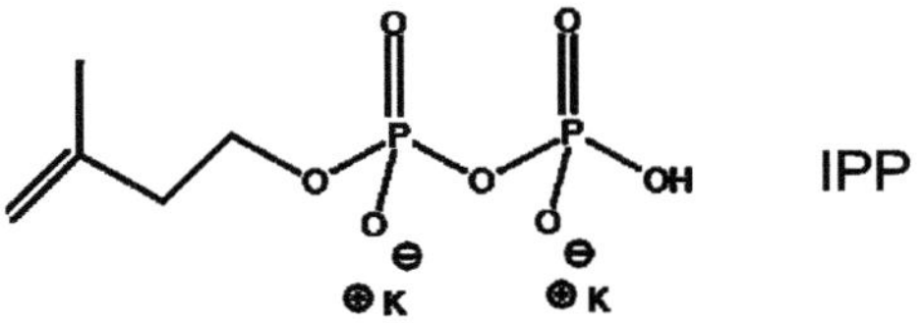

Figure 10.1 Chemical structure of IPP. K represents alkali metal ions.

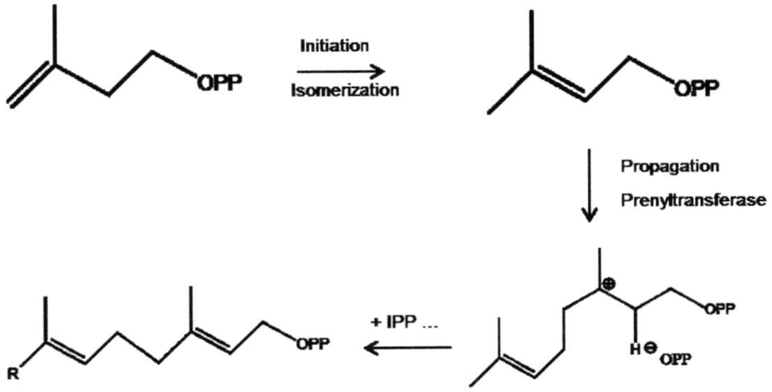

Figure 10.2 The schematic of terpenoid biosynthesis of NR proven by Poulter and Rilling [10]. OPP represents pyrophosphate counter-anions. R represents multiple IP units.

10.2.3 Modification of Natural Rubber

As a polymer with long chain and short branches (methyl groups), NR has excellent physical properties, including high elasticity, low heat generation (low loss factor), and high tensile strength. However, due to the high unsaturation and non-polar features, there are also some significant drawbacks on the properties of NR, including low oil and acid resistance, low heat and flame resistance, low abbreviation resistance, and low aging performance. There are typically three approaches to overcome the drawbacks of NR property: blending with synthetic polymers, compounding with inorganic additives, and chemical modification of NR molecules [2].

10.2.3.1 Chemical Modification

Among the NR modification approaches, chemical modification stands out to be the most efficient one because of its capability of modifying NR molecular structure, which is the root cause of NR properties. Chemical modifications of NR have two major goals, which are the improvement of NR properties, and changing of NR to other polymeric materials with specific applications. To achieve the goals, there are three major types of chemical modification.

- NR molecular structure or molecular weight modification without involving new atoms to NR
- Modification of carbon-carbon double bond by adding new atoms or chemical groups
- Grafting the allylic carbon with different chemical groups

Degradation

NR has a very long molecular chain, resulting in high viscosity, hence NR is hard to be processed by blending or compounding directly. Both physical and chemical approaches have been developed to reduce NR molecular weight, as is viscosity. Physical methods usually use shear force provided by a mechanical mixing device, such as open mill and internal mixer. Degradation degree can be controlled by mixing duration. NR can also be degraded using oxidative method. Oxidative agents include hydrogen peroxide, organic hydroperoxide, atmospheric oxygen, and ferric chloride-oxygen coupled with aromatic hydrazine or sulphanilic acid [11]. Effective degradation of NR using radical initators (potassium persulphate and propanal) was also reported [12]. NR can also be degraded with ozone or ultraviolet light in the presence of hydrogen peroxide [13].

Cyclisation

In NR cyclisation modification, unsaturated double bonds in polyisoprene react with acidic reagents or Lewis acids to add additional acid reagents and form cyclical groups [14]. The cyclical groups have less mobility, and hence can transform elastic NR to rigid materials. Proper degree of cyclisation can improve the hardness, strength, and abrasion resistance of NR.

Halogenation

One important category of NR chemical modification is halogenation, which is the introduction of halogen elements to the double bonds of isoprene units. Chlorination and bromination are two typical halogenation reactions. Halogenation can be conducted both with NR molecules or latex and on surfaces of vulcanized NR.

Chlorination of NR was first reported by Clark [15]. According to the works of many researchers, there are four major reactions of NR chlorination: addition, substitution, cyclisation, and crosslinking. The reaction process can be either radical or ionic. The typical approach of chlorination is to pass chloride gas through solution of NR or suspension of NR latex in water. The major advantages of chlorination of NR are the improvement of flame resistance and corrosion resistance performances. According to the thermogravimetric studies, chlorinated NR features a degradation temperature of 291°C in nitrogen atmosphere and thermo-oxidation temperatures of 294.5 and 487°C (two-step process) in air [16]. As a contrast, un-chlorinated NR has degradation temperatures of 200°C only. Chlorinated NR is corrosion-resistant to acid, alkali, wearing, aging, and sea water. Vulcanized NR sheets were also chlorinated, although the reaction can only occur on sheet surface due to the immiscibility of vulcanized NR. The advantage of surface chlorination is the improvement of surface flame resistance and corrosion resistance, while maintaining flexibility of the sheet. A typical halogenation schematic is illustrated in Fig. 10.3.

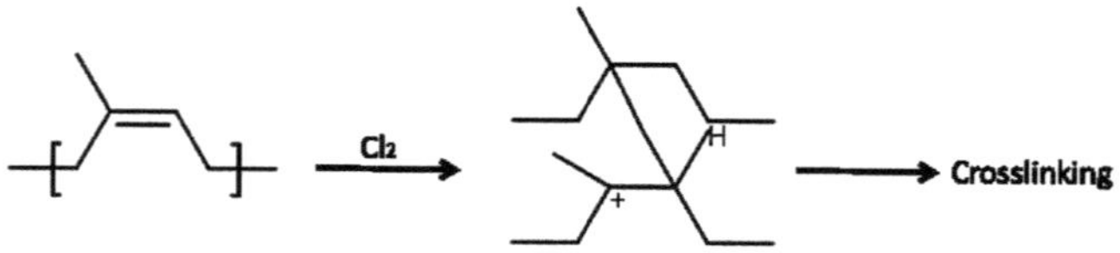

Figure 10.3 The schematic of natural rubber halogenations.

Multiple approaches to fabricate brominated NR have been reported. One approach applied N-bromosuccinimide to brominate NR through a radical process [17]; this approach generated substitutive bromination associated with cyclisation. Brominated NR was also obtained by treating NR latex with bromine gas. The drawback of bromination is its negative effects on NR thermal stability. It is found that when bromine content exceeds 47%, the brominated NR has a thermal decomposition temperature of 129°C, which is lower than that of the NR. Therefore, the application of brominated NR is currently not practical.

Hydrogenation

NR molecules contain isoprene groups, which are not chemically saturated and hence can react with hydrogen molecules (H_2) or hydrogen releasing chemicals. The existence of unsaturated double bonds in isoprene group results in lower thermal and oxidation resistance performances of NR, and therefore it is beneficial to overcome those drawbacks by saturating the double bonds via hydrogenation. There are two major approaches for NR hydrogenation: catalytic and non-catalytic [15]. Catalytic hydrogenation initiates the reaction between H_2 and double bonds in NR by catalyst. Typical catalysts include platinum, nickel, palladium, cobalt, molybdenum, cobalt oxide, calcium carbonate, and copper chromite, etc. [15]. NR with different hydrogenation degrees have been achieved by multiple research groups applying different catalysts/catalytic systems. Shahab and Basheer [18] reported the production of 23% hydrogenated NR using Rh(I) complex in benzene as a catalyst. Singha et al. [19] produced 100% hydrogenation of NR with $RhCl(PPh_3)_3$ catalyst, and investigated the effects of processing parameters. Hundred percent hydrogenated NR was also reported by Inoue et al. [20]. According to these researches, the increasing hydrogenation degree can result in higher thermal stability, hardness, and abrasion resistance of NR.

Hydrogenation of NR can also be achieved by non-catalytic techniques. A hydrogen releasing agent, di-imide, has attracted interest and been investigated [21]. Compared to the catalytic method, the non-catalytic method does not require extreme conditions, such as high temperature and pressure, and is relatively easy to conduct. Multiple resources can be applied to provide di-imide. Chemicals, such as hydrazine, arenesulphonylhydrazides, 1-thia-3,4-diazolidine 2,5-dione, and Toluenesulphonyl-hydrazide (TSH), can be used to generate di-imide through oxidization, thermal treatment, and photochemical irradiation [21]. Phinyocheep and Samran [22] utilized TSH as hydrogen sources to produce hydrogenated NR successfully. It is found that a longer reaction time would result in a higher hydrogenation degree. Ikeda et al. [23] investigated the hydrogenation of NR with di-imide from hydrazine oxidation. The mechanical performance of hydrogenated NR during vulcanization was also studied. It is reported that with a hydrogenation of 48%, the modified NR possessed better thermal stability and ozone resistance during vulcanization.

Epoxidation

Another chemical modification approach of NR is epoxidation, which converts the carbon-carbon double bonds to epoxides by oxidizing as shown in Fig. 10.4. Typical oxidation agents include hypochlorous acid, hydrogen peroxide, organic peracid, and air (or oxygen) [24]. The advantage of NR epoxidation is that it can improve NR properties, including air permeability, oil and solvent resistance, hardness, and damping (impact absorption and acoustic applications). Epoxidation of NR can be conducted in either organic solvent or aqueous environment, depending on applied oxidation agent [25]. The organic solvent approach is capable of epoxidizing most NR molecules because it can dissolve NR. As a contrast, NR latex are not dissolvable with water, hence epoxidation happens on latex surfaces only. It is found that the epoxidation degree is decided by processing conditions, including temperature, time, and oxidant concentration. Side reactions may be triggered during epoxidation and result in different side groups, such as hydroxyl, ester, and hydrofuran. An important application of epoxidated NR is as an intermediate reactant of other NR diversity.

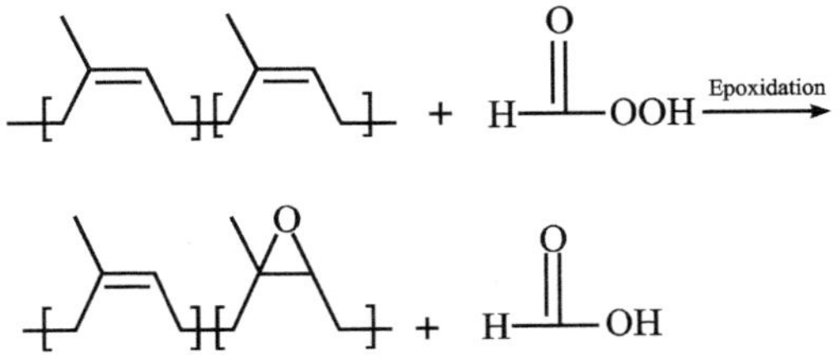

Figure 10.4 The schematic of natural rubber epoxidation.

The effect of epoxidation on NR properties depends on its degree. A number of measurement methods have been studied and applied on epoxidation degree determination. It is found that indirect degradation and elemental analysis show higher error level and consequent lower accuracy; these two techniques are mainly applied for indicative purposes only. Direct HBr titration has good accuracy at less than 15% (mole) epoxide content due to its reactions with epoxide when going higher. ^{1}H and ^{13}C NMR are mainly applied to analyze the epoxiation range of 20–75% (mole) [26]. It is found that there is a relationship between NR glass transition temperature (T_g) and epoxide content, therefore differential scanning calorimetry (DSC) technique can be used to determine epoxidation degree. An accurate method needs to be applied to obtain a calibrated engineering curve (between T_g and epoxide content) for accurate DSC epoxidation analysis.

The epoxidation process can modify the properties of NR significantly. NR property change depends on the level of expoxidation. It is found that higher epoxidation level will result in higher hardness of NR. The reason is that more epoxide contents will increase the difficulty of molecule movement. The NR oxidative stability can also be improved by epoxidation because it decreases the unsaturation level. The polarity of NR can be increased by epoxidation via introducing polar epoxide groups. Therefore, epoxidated NR has higher solubility in polar solvents and lower solubility in non-polar solvents, and hence can be applied as an oil-resistant material. The introduction of epoxide groups to NR molecular chains improves their T_g as well as their damping performance for potential acoustic applications.

Graft Copolymerization

Another category of important NR chemical modification is graft copolymerization as shown in Fig. 10.5. Among the graft copolymerization, addition of vinyl monomer to NR is one of the most interesting directions and has been widely investigated. Due to it containing both elastic molecular chains and plastic side groups, the graft modified NR can be applied as a thermoplastic elastomer or function as a compatibilizer for polymer blending applications.

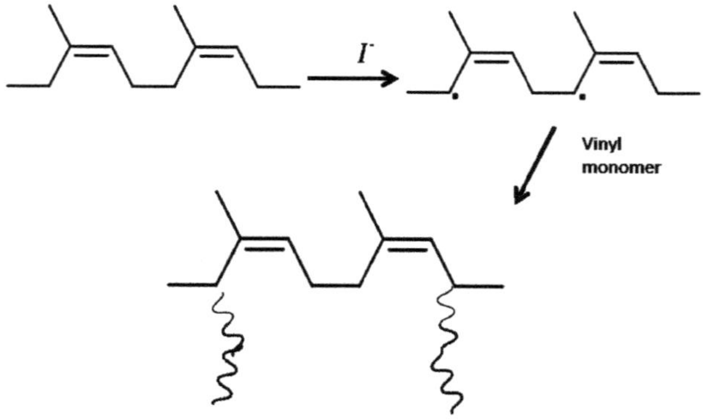

Figure 10.5 The schematic of natural rubber graft copolymerization.

Free radical process is the major methodology for NR graft copolymerization [27]. Normally, there are preservatives such as ammonia in NR, preventing the free radical process, and need to be removed. Redox catalyst systems were found to be able to overcome the prohibition effect. There are two opposite mechanisms to the graft copolymerization. One starts the initiation by adding radicals to the NR molecules; the other takes hydrogen atoms from NR molecules to form radical active sites. Many initiation and monomer systems for grafting NR have been investigated. Pukkate et al. [28] reported the grafting of polystyrene(PS) to NR. Tert-butyl hydroperoxide and tetraethylenepentamine were applied as a redox initiation system. The tensile strength of the PS modified NR was improved due to the formation of some nano-sized PS crystal regions, which function as solid fillers. Grafting of styrene and methyl methacrylate monomers onto NR were also studied by Man et al. [29]. As a contrast to PS modified NR, the monomer grafted

NR features lower tensile strength. The reason is that short side groups can constrain molecular chain rearrangement and reduce strain induced crystallization of NR.

Maleinisation

Another type of chemical modification of NR is maleinisation, which adds maleic anhydride (MA) to NR molecular chain and forms meleinised rubber as shown in Fig. 10.6. Two processes have been developed for maleinisation: radical reaction and electrocyclic reaction. Bacon and Farmer [27] first investigated the radical reaction using benzoyl peroxide as a radical initiator. It is found that the addition of MA happens on the allylic carbon instead of double bond carbon, since this addition can keep the unsaturation of NR molecules. Other chemicals, such as chlorobromodimethyl-hydantoin, azodiisobutyronitrile, and peroxides were also found to be catalysts of NR maleinisation [30]. Due to the existence of maleate group on NR molecule, vulcanized maleated NR has significantly improved solvent resistance and antiaging performance. However these properties are only close to those of a few synthetic polymers, hence maleated NR was not commercialized. Maleated NR is used more as a coupler of polymer blending.

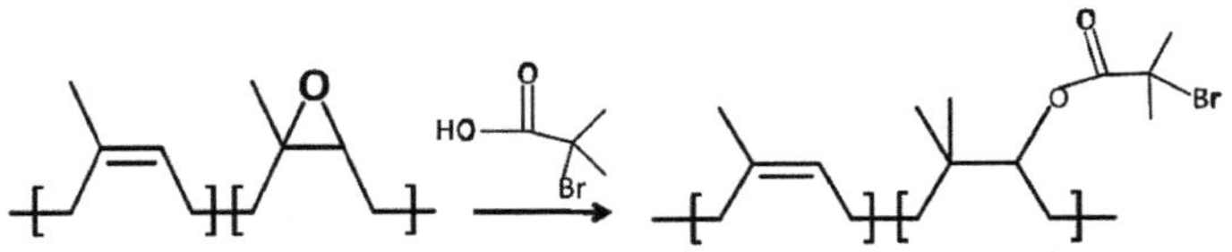

Figure 10.6 The schematic of natural rubber maleinisation.

10.2.4 Industrial Applications

10.2.4.1 Natural Rubber Application in Tire Industry

In the modern interconnected world, transportation is becoming more and more important. As an essential component, tire is facing an increasing role and demanding. Over 150 million tires were produced for vehicles and aviation applications in the past year. Due to its unique low hysteresis performance, NR's role is not replaceable in the tire industry. More than 11 million tons of NR was produced and used in tire industry in the same year [2].

To be used safely in transportation, tires must meet a number of functions. They need to support the vehicle or aircraft weight, accelerate the vehicle by providing rough surface, and turn the vehicle by providing the steering force while maintaining vehicle stability. Tires are also required to provide vibration insulation from the road and stop the vehicle in time by supplying enough braking force. The unique properties of NR match the tire industry requirements, and are widely applied.

10.2.4.2 Pressure Sensitive Adhesives Application

Pressure sensitive adhesive is an important category of materials due to its easy use and wide applications [31]. NR-based adhesives count for almost 30% of all adhesives, and are widely applied in both industry and construction applications. NR-based pressure sensitive adhesives are processed mainly in two ways, which are mill method and solution method [31]. The mill method uses mechanical force to reduce NR molecular weight, and the latter one breaks down molecular weight chemically. One major type of NR-based adhesive is epoxidized NR, which can be prepared by reaction between peracid and NR. Two grades of epoxidized NR have been commercialized, which contain 25 mol% and 50 mol% epoxidation. The increased epoxidation content would result in the increasing of density, polarity, T_g, and reduction of strain inducted crystallization [31].

Three key factors of adhesives are tack, peel, and shear performance, which are typically characterized by tack test, peel test, and shear test, respectively. One important parameter affecting those performances is the thickness of the adhesive. Poh and Kwo [32] found that all the performances increase with adhesive thickness. Proper amount of tackifier and filler content in adhesives can also increase tack and peel strength, however too much tackifier and filler will increase the hardness and decrease adhesion. The introduction of tackifier and filler to adhesive will decrease rubber content, hence result in declined shear strength. Commonly applied tackifiers include Cl resin and phrpetro resin. Typical fillers include silica, magnesium oxide, zinc oxide, calcium carbonate, barium chloride, and kaolin [32]. NR molecular weight also affects adhesion performances significantly. Proper molecular weight optimizes tack, peel, and shear performances by providing proper viscosity. A high molecular weight will cause increased entanglement among chains and decreased adhesion performance.

10.2.4.3 Fiber Reinforced Natural Rubber Composites

Polymer-fiber composites are attracting more and more attention recently because of their outstanding mechanical properties. Renewable and environmentally friendly materials are becoming critical in recent decades. NR is a potential matrix material for composites because of its unique damping performance and bio-material nature. As most elastic materials, NR has outstanding flexibility, but not enough mechanical strength for industry and construction applications. It is necessary and meaningful to introduce reinforcement into NR to form composites featuring both high flexibility and strength. Both long fibers and relatively short fibers can be applied as fillers [33]. The content of short fibers in NR composites is normally higher than that of long fibers, because enough short fibers are needed to establish fiber to fiber contact for good load transference from elastomer matrix. NR composites properties mainly depend on rubber content, fiber content, fiber types, and fiber orientation.

Fiber reinforced NR composites feature a number of advantages, including high strength-density ratio, high corrosion resistance, and high modulus-density ratio. On the other hand, those composites also have disadvantages, such as highly orientation determined strength and relatively high manufacturing cost. Both organic and inorganic fibers can be applied as fillers, among which natural fibers are attracting more and more attention due to their wide source and renewability. The drawback of NR is their hydrophilic surface, which results in poor adhesion with hydrophobic elastomer matrix. Natural fibers normally require surface modification to enhance the fiber-matrix adhesion. Natural fibers include animal fibers and plant fibers [34]. Plant fibers have a much wider source and can withstand higher operation temperatures, hence it is the major focus.

10.2.5 Development and Future Trends

As a strategic material, NR is mostly harvested from Hevea from limited regions in the world, which possesses a high risk of supply shortage. To overcome this issue, diverse technical approaches are under investigation. Methodologies of synthesizing isoprene with NR-like properties are under development. The mass production of NR with other crops is also a hot topic and a lot of efforts have been conducted.

NR has been widely applied in a number of fields; however it has some drawbacks, such as poor abrasion resistance, low oil resistance, and low heat resistance, etc. Those drawbacks significantly limit the extension of NR applications. Although chemical modifications have been investigated to overcome the drawbacks, further research is still needed to develop more efficient approaches and environmentally friendly chemical agents. One potential modification is the addition of SO_2 to NR- the product can be used as a fire retardant agent or raw materials for carbon foam fabrication. The challenge is the stability of SO_2-NR. Another interesting direction of NR chemical modification is the fabrication of low molecular weight molecules with specific functional groups. Those functional small molecules have potential applications, such as coatings materials and adhesives. The most recent development is the application of N_2O as a NR breaking down agent. One more important application of NR is as a polymer electrolyte membrane host. PMMA-grafted NR, sulphonated DPNR-graft-PS, and sulphonated hydrogenated DPNR-graft-PS have been recently developed for this application.

The application of bio-based fibers and NR to green composites fabrication is critical based on both mechanical performance and environment considerations. A lot of natural fibers, including both plant fibers and animal fibers, have been used to fabricate polymer composites. However, the weak adhesion between fiber and NR is a challenge. The surface modification of natural fibers with more efficient and green approaches needs to be further investigated.

10.2.6 Recycling

Raw NR is a gel-like material, which has very low peel resistance, shear resistance, abrasion resistance, and solvent resistance. Raw NR needs to be vulcanized to form crosslinked three-dimension (3-D) network for some applications, such as tire, acoustic materials, and vibration insulation materials. The 3-D structure endorses NR outstanding mechanical performances, however it also prevents vulcanized NR from being melt reprocessed or solvent reprocessed. The recycling of vulcanized NR has been a challenge since it was first applied.

One old method of treating used NR materials is landfill. The landfill consumes a large area of land and can introduce multiple environmental problems. Another traditional way of treating used NR is grinding with a mixer or extruder [35]. The rubber particles from grinding can be used as polymer composite fillers or materials to improve concrete elasticity. Vulcanized NR can also be transformed to fuel with vacuum pyrolysis because of its high calorie content. Approaches to break crosslinking in NR using microwave and ultrasonic have also been investigated. The product can be reprocessed due to the elimination of crosslinking bonds. Another important and green method of recycling NR is biodegradation, which will cause bacteria to break down crosslinking bonds [35]. However the drawback of this approach is its low efficiency and potential organism pollution problem. In general, the recycling of vulcanized NR is still a challenge; efficient and green approaches need to be further developed.

10.3 BIO-BASED THERMOPLASTIC ELASTOMER

10.3.1 Raw Materials

Thermoplastic elastomers are produced mainly through two approaches: blending of plastic and uncrosslinked elastic polymers and direct synthesis from monomers. A number of bio-based plastics and elastomers can be applied in thermoplastic elastomer fabrication. Example composites include poly(lactic acid)/NR [36], polycaprolactone/epoxidized NR [37], poly(lactic acid)/polyurethane [4], polypropylene/NR [38], polystyrene/NR [39], and poly(decalactone) diols/polyurethane [40], etc. Some thermoplastic elastomers have also been obtained by synthesizing bio-based monomers, such as cyclic diol/succinate oligomer [41], cyclic diol/alkylthiosuccinate oligomer [41], macrolide [42], methyl 12HS [42], and soybean oil [43], etc.

10.3.2 Production Process and Property

TPE are processed either through compounding of several polymers or polymerization of monomers. The structure of compounded TPE is shown in the left picture of Fig. 10.7. Two or more polymers were heated up to their T_g, and then mixed by mechanical shearing force provided by an extruder or a mill. Elastomers are introduced to plastics to form the shown island-sea structure, in which elastic phase is normally the islands. The size of elastic phase is typically in the range of several to hundreds of micrometers, depending on processing conditions and application requirements. The elastic phase introduces the high flexibility and hysteresis into the blend, while the plastic phase keeps the advantages of reprocessability, high hardness, and high strength. The structure of polymerized TPE is shown in the right picture of Fig. 10.7, which is consisted of plastic segment and elastic segment. The physical dimensions of the plastic segments are normally in nanometers, which is much smaller than those of elastic phases in compounded TPE.

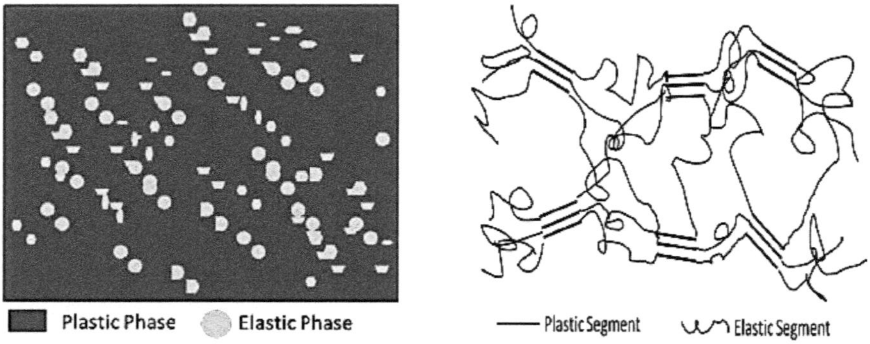

Figure 10.7 Thermoplastic elastomer structure through polymer compounding (left) and monomer polymerization (right).

The properties of TPE are decided by their structures, which include the ratio between plastic component and elastic component, the size of elastic phase or segment, the molecular weight of each component, and the molecular structure of each component. In general, an increasing elastic component content can improve the flexibility, softness, tensile enlongation of TPE. However, a high elastomer load will result in the decline of tensile strength and reprocessability. The size of elastic phase or segment also plays an important role on the mechanical properties. When composition is fixed, a smaller elastic phase/segment size will result in the simultaneous increasing of strength and elongation of thermoplastic elastomer. Proper molecular weights of both plastic component and elasti component will optimize the mechanical properties of formed TPE. Too high or low molecular weight are not favored, because they may decrease the flexibility, hardness, strength, and operation temperature of TPE. A number of thermoplastic elastomer fabrication techniques will be introduced in the following sections.

10.3.2.1 Bio-based Thermoplastic Elastomers through Compounding

Bio-based polymers have attracted a huge interest in recent years. Those polymers are produced from natural resources, and hence are renewable and environmentally friendly. Within those polymers, there are both plastics and elastomers, the combination of which will keep the advantages of both components. Many plastics/elastomers combinations have been compounded and investigated. Tanrattanakul and Bunkaew investigated the fabrication of poly(lactic acid) (PLA) and NR thermoplastic elastomer [36]. The NR content was controlled at 60 percent. A melt blending in an internal mixer was followed by injection molding to process sheet samples. To enhance the plasticity of PLA, several plasticizers were applied and their effects on mechanical properties of the blend were investigated. It was found that plasticizers generally improved both strain and breaking strength.

Mishra et al. developed thermoplastic elastomer using polycaprolactone (PCL) and epoxideized NR, and investigated its application as a heat shrinkable material [37]. PCL is a biodegradable thermoplastic material with good processibility, while epoxideized NR is also a biodegradable material with good oil resistance and gas sealing. The themoplastic elastomer blend was also mixed in an internal mixer at 160°C and injection molded at 180°C [37]. Dicumyl peroxide was applied as a crosslinking agent to induce the crosslinking of epoxidized NR and enhance the heat shrinkability. It was found that polymer chains could be oriented by stretching at its processing temperature, and the orientation could be frozen by cooling. When being heated, the polymer chains tend to shrink to their initial high entropy position. Crosslinked points were found to serve as memory points to enhance shrinkage. The tensile strength and elongation at break were also improved by the crosslinking.

Feng and Ye investigated the morphologies and mechanial properties of PLA and polyurethane thermoplastic elastomer [4]. Thermoplastic polyurethane elastomer is a material with outstanding properties, featuring high strength, flexibility, abbrassion resistance, biocompatibility, and toughness. It has been widely using in multiple ares including electrical, electronic, midication, and automobile. It was found that the introduction of polyurethane into PLA can improve its toughness. Obvious yield was observed from the PLA/polyurethane blend.

Bio-based thermoplastic elastomer was also developed by Ismail and Suryadiansyah using polypropylene (PP), NR, and recycled rubber [38]. The blending was achieved by dispersing rubber particles to PP in a mixer at proper processing temperature. It was found that the recycled rubber/PP blend has higher tensile strength but lower elongation than NR. Asaletha et al. developed polystyrene (PS) and NR thermoplastic elastomer [39]. It was found that the increased rubber content could result in decreased tensile and tear strength, but increased impact strength.

10.3.2.2 Bio-based Thermoplastic Elastomers through Monomer Polymerization

Besides mechanical blending of polymers, polymerization is the other approach of fabricating bio-based TPE. Since this approach starts from monomers, it can generate much smaller elastic domains than the blending method. Smaller elastic domains are beneficial for improving uniformity, flexibility, and elongation at break of the thermoplastic elastomer. A number of fossil-based monomers have been used to produce thermoplastic elastomer, which are not green processes from the perspectives of materials, process, and catalysts. The synthesis of bio-based and biodegradable TPE applying enzymatic catalyst was investigated. Kobayashi and Matsumura [42] reported the synthesis of poly(dodecanolide-12-hydroxystearate (12HS)) and poly(pentadecanolide-12HS) using Lipase as enzymatic catalyst. In the products, macrolide functions as the rigid segment and 12HS is the elastic segment. Thermoplastic elastomer with molecular weight of 290,000 gram per mole was achieved. It is found that there is a relationship between monomer units randomness and polymerization time. A long polymerization can increase the randomness. The content of 12HS is the determining parameter of product crystallinity and mechanical properties. The increasing 12HS content will decrease crystallinity, melting temperature (T_m), Young's modulus, and tensile strength. The elongation at break will be increased. The product was proved to be biodegradable, and its degradability increases with increasing 12HS content. It was illustrated that the bio-based poly(macrolide-12HS) can be potentially applied as biodegradable thermoplastic elastomer.

10.3.3 Industrial Applications

A number of bio-based TPE have been applied in multiple industrial fields. A polyether-block-amide block copolymer produced by Arkema from castor oil has a hardness of 72 Shore D. It has been widely used in electronics, automotive, and sportswear. DuPont developed a renewable Hytrel RS thermoplastic elastomer using a corn sugar derivative as raw material. Hytrel RS contains up to 60% bio-based material, and has a Shore D hardness of 83. It has been widely applied in different applications, such as airbag door, energy damper, tubing boot, and hose. GLS developed two classes of bio-based TPE, which found applications in sports equipment, such as ski binding and caster wheels. OnFlex developed the BIO 5300 series compounds with good injection molding and extrusion capability. These compounds also feature high abrasion resistance, impact resistance, tensile, and tear performances. BIO 5300 can be applied in a wide range of applications, such as sports equipment, door panel skin, and interior trim.

10.3.4 Future Trend

The use of renewable raw materials in development of thermoplastic elastomer has attracted a lot of interest and is becoming more important. The most applicable raw materials can be categorized to carbohydrate, vegetable oil, lignin, sugar cane, solid plant waste, and waste oil. With the developed bio-based TPE as introduced in this chapter, there are still some factors and challenges waiting to be considered and addressed. One major challenge is the production of bio-based TPE in large scale economically. It will be important to develop low cost logistics for bio-based raw materials, new high throughput manufacturing approaches, new bio-based high efficiency catalysts (microbial/enzyme), and efficient methodologies to process and recover downstream products. Another major focus of future research will be the improvement of performances of bio-based TPE. For instance, some chemical additives can be developed to improve the performance, such as adhesion, abrasion resistance, oil resistance, flame retardant, and processability, etc. Addition of nanoparticles to TPE is another technical option. Similar to the functions of enhancing fossil polymers, nanoparticles, such as graphene, carbon nanotube, nanoclay, silica, and silver can improve a lot of performances, including flame retardant, microbial resistance, thermal stability, solvent absorption, and biodegradation resistance. One additional concern is the potential supply competition among bio-based thermoplastic elastomer raw materials, bio-based fuel, and food.

10.4 CONCLUSIONS

NR features the advantages of unique elasticity and renewability. NR and its derivatives have been widely applied in the industrial field, such as tires, gaskets, acoustic insulation, vibration insulation, and adhesives. However, NR also features some drawbacks, including relatively poor weather resistance, acid resistance, oil resistance, and chemical resistance performances. A number of technical measures, such as chemical modification, blending with other polymers, compounding with functional additives, have been developed to overcome those drawbacks and extend NR applications. NR will potentially continue contributing to the sustainability of foreseeable future. To achieve this goal, the investigation of bio-synthesis mechanism of NR, the culturing of new NR producing crops, the development of efficient and green chemical modification approaches, and the exploration of rapid and environmentally friendly NR recycling techniques will be major future focuses.

Bio-based TPE combine the advantages of easy reprocessing of conventional TPE and renewability of bio sources. Bio-based TPE find wide applications in industries such as electronics, automotive, and sports equipment. However, further efforts are needed to improve bio-based TPE performances chemically or physically. In order to scale up the production of bio-based TPE, issues such as logistic cost, manufacturing efficiency and cost, downstream processing, and raw materials supply need to be addressed.

References

1. Hofmann, W. Rubber Technology Handbook. Hanser Publishers, Distributed in the USA by Oxford University Press, 1989.
2. Kohjiya, S. and Y. Ikeda (eds). Chemistry, Manufacture and Applications of Natural Rubber. Woodhead Publishing, Cambridge, UK, 2014.
3. Legge, N.R., G. Holden and H. Schroeder (eds). Thermoplastic Elastomers: A Comprehensive Review. Carl Hanser Verlag, Munich, 1987, pp. vii + 574.
4. Feng, F. and L. Ye. Morphologies and mechanical properties of polylactide/thermoplastic polyurethane elastomer blends. Journal of Applied Polymer Science 2011; 119(5): 2778-2783.

5. Firmino, A., H. Tozze Jr and E. Furtado. First report of *Ceratocystis fimbriata* causing wilt in *Tectona grandis* in Brazil. New Disease Reports 2012; 25: 24. http: //dx.doi.org/10.5197/j.2044-0588.2012.025.024.
6. Cornish, K. The separate roles of plant cis and trans prenyl transferases in cis-1, 4-polyisoprene biosynthesis. European Journal of Biochemistry 1993; 218(1): 267-271.
7. Cornish, K. Biochemistry of natural rubber, a vital raw material, emphasizing biosynthetic rate, molecular weight and compartmentalization, in evolutionarily divergent plant species. Natural Product Reports 2001; 18(2): 182-189.
8. Mooibroek, H. and K. Cornish. Alternative sources of natural rubber. Applied Microbiology and Biotechnology 2000; 53(4): 355-365.
9. Lodish, H., D. Baltimore, A. Berk, S.L. Zipursky, P. Matsudaira and J. Darnell. Molecular Cell Biology, 3rd Ed. Scientific American Books, New York, 1995.
10. Poulter, C.D. and H.C. Rilling. Prenyltransferase: The mechanism of the reaction. Biochemistry 1976; 15(5): 1079-1083.
11. Nor, H.M. and J. Ebdon. Ozonolysis of natural rubber in chloroform solution Part 1. A study by GPC and FTIR spectroscopy. Polymer 2000; 41(7): 2359-2365.
12. Phetphaisit, C. and P. Phinyocheep. Kinetics and parameters affecting degradation of purified natural rubber. Journal of Applied Polymer Science 2003; 90(13): 3546-3555.
13. Radhakrishnan Nair, N., S. Thomas and N. Mathew. Liquid natural rubber as a viscosity modifier in nitrile rubber processing. Polymer International 1997; 42(3): 289-300.
14. Riyajan, S.A., D.J. Liaw, Y. Tanaka and J.T. Sakdapipanich. Cationic cyclization of purified natural rubber in latex form with a trimethylsilyl triflate as a novel catalyst. Journal of Applied Polymer Science 2007; 105(2): 664-672.
15. Brydson, J.A. Rubber Chemistry. Applied Science Publishers, London, 1978.
16. Cai, Y., S.D. Li, C.P. Li, P.W. Li, C. Wang and M.Z. Lv and K. Xu. Thermal degradations of chlorinated natural rubber from latex and chlorinated natural rubber from solution. Journal of Applied Polymer Science 2007; 106(2): 743-748.
17. Xue, X., Y. Wu, F. Wang, J. Ling and X. Fu. Preparation and thermal stability of brominated natural rubber from latex. Journal of Applied Polymer Science 2010; 118(1): 25-29.
18. Shahab, Y. and R. Basheer. Nuclear magnetic resonance spectroscopy of partially saturated diene polymers. I. 1H-NMR spectra of partially hydrogenated and partially deuterated natural rubber, gutta percha, and cis-1, 4-polybutadiene. Journal of Polymer Science: Polymer Chemistry Edition 1978; 16(10): 2667-2670.
19. Singha, N.K., P. De and S. Sivaram. Homogeneous catalytic hydrogenation of natural rubber using $RhCl(PPh_3)_3$. Journal of Applied Polymer Science 1997; 66(9): 1647-1652.
20. Inoue, S.i. and T. Nishio. Synthesis and properties of hydrogenated natural rubber. Journal of Applied Polymer Science 2007; 103(6): 3957-3963.
21. Samran, J., P. Phinyocheep, P. Daniel, D. Derouet and J.Y. Buzare. Spectroscopic Study of Di-Imide Hydrogenation of Natural Rubber. Macromolecular Symposia 2004; 216(1): 131-144. Wiley Online Library.
22. Samran, J., P. Phinyocheep, P. Daniel and S. Kittipoom. Hydrogenation of unsaturated rubbers using diimide as a reducing agent. Journal of Applied Polymer Science 2005; 95(1): 16-27.
23. Ikeda, Y., P. Phinyocheep, S. Kittipoom, J. Ruancharoen, Y. Kokubo and Y. Morita, K. Hijikata and S. Kohjiya. Mechanical characteristics of hydrogenated natural rubber vulcanizates. Polymers for Advanced Technologies 2008; 19(11): 1608-1615.
24. Roberts, A.D., (ed.). Natural Rubber Science and Technology. Oxford University Press, Oxford and New York 1988.
25. Ng, S.-C. and L.-H. Gan. Reaction of natural rubber latex with performic acid. European Polymer Journal 1981; 17(10): 1073-1077.
26. Burfield, D.R., K.-L. Lim, K.-S. Law and S. Ng. Analysis of epoxidized natural rubber. A comparative study of dsc, nmr, elemental analysis and direct titration methods. Polymer 1984; 25(7): 995-998.
27. Bacon, R. and E. Farmer. The interaction of maleic anhydride with rubber. Rubber Chemistry and Technology 1939; 12(2): 200-209.

28. Pukkate, N., Y. Yamamoto and S. Kawahara. Mechanism of graft copolymerization of styrene onto deproteinized natural rubber. Colloid and Polymer Science 2008; 286(4): 411-416.
29. Man, S.H.C., A.S. Hashim and H.M. Akil. Properties of styrene-methyl methacrylate grafted DPNR latex at different monomer concentrations. Journal of Applied Polymer Science 2008; 109(1): 9-15.
30. Mitov, Z. and R. Velichkova. Modification of styrene-isoprene block copolymers – 3. Addition of maleic anhydride – mechanism. European Polymer Journal 1993; 29(4): 597-601.
31. Khan, I. and B. Poh. Natural rubber-based pressure-sensitive adhesives: A review. Journal of Polymers and the Environment 2011; 19(3): 793-811.
32. Poh, B. and H. Kwo. Peel and shear strength of pressure-sensitive adhesives prepared from epoxidized natural rubber. Journal of Applied Polymer Science 2007; 105(2): 680-684.
33. Abdelmouleh, M., S. Boufi, M.N. Belgacem and A. Dufresne. Short natural-fibre reinforced polyethylene and natural rubber composites: Effect of silane coupling agents and fibres loading. Composites Science and Technology 2007; 67(7): 1627-1639.
34. Jacob, M., S. Thomas and K. Varughese. Biodegradability and aging studies of hybrid biofiber reinforced natural rubber biocomposites. Journal of Biobased Materials and Bioenergy 2007; 1(1): 118-126.
35. Myhre, M. and D.A. MacKillop. Rubber recycling. Rubber Chemistry and Technology 2002; 75(3): 429-474.
36. Tanrattanakul, V. and P. Bunkaew. Effect of different plasticizers on the properties of bio-based thermoplastic elastomer containing poly (lactic acid) and natural rubber. Express Polymer Letters 2014; 8(6): 387-396.
37. Mishra, J.K., Y.-W. Chang and D.-K. Kim. Green thermoplastic elastomer based on polycaprolactone/ epoxidized natural rubber blend as a heat shrinkable material. Materials Letters 2007; 61(17): 3551-3554.
38. Ismail, H. Thermoplastic elastomers based on polypropylene/natural rubber and polypropylene/recycle rubber blends. Polymer Testing 2002; 21(4): 389-395.
39. Asaletha, R., M. Kumaran and S. Thomas. Thermoplastic elastomers from blends of polystyrene and natural rubber: morphology and mechanical properties. European Polymer Journal 1999; 35(2): 253-271.
40. Tang, D., C.W. Macosko and M.A. Hillmyer. Thermoplastic polyurethane elastomers from bio-based poly (δ-decalactone) diols. Polymer Chemistry 2014; 5(9): 3231-3237.
41. Yagihara, T. and S. Matsumura. Enzymatic synthesis and chemical recycling of novel polyester-type thermoplastic elastomers. Polymers 2012; 4(2): 1259-1277.
42. Kobayashi, T. and S. Matsumura. Enzymatic synthesis and properties of novel biodegradable and biobased thermoplastic elastomers. Polymer Degradation and Stability 2011; 96(12): 2071-2079.
43. Zhu, L. and R.P. Wool. Nanoclay reinforced bio-based elastomers: Synthesis and characterization. Polymer 2006; 47(24): 8106-8115.

Chapter 11

A Life Cycle Assessment of Protein-based Bioplastics for Food Packaging Applications

A. Jones[1], S. Sharma*[1] and S. Mani[2]

[1]Department of Textiles, Merchandising & Interiors
University of Georgia, Athens, GA, USA
[2]College of Engineering, University of Georgia, Athens, GA, USA

11.1 INTRODUCTION

A relatively new area of materials science and polymer chemistry is searching for proteins from biomass as a potential source of raw material in the production of plastics. When proteins are utilized in the production of a given plastic, this plastic is then classified as a bioplastic [1]. With the use of proteins in the production of plastics comes a change of properties of the resulting plastic, making it more desirable for certain applications. One area is food packaging, in which bioplastics have been examined. When it comes to determining what type of material to be used to produce food packaging, there are certain properties that are examined. In terms of cost and ease of manufacture, polyethylene is the most common choice when it comes to making traditional food packaging plastics, as it is the plastic that is most easily made with addition polymerization. Food packaging plastics made from polyethylene tend to fall into two different categories: high density polyethylene (HDPE) for more durable applications, such as the packaging for milk and juices; and low density polyethylene (LDPE), which is more suitable for flexible lids and squeezable bottles for food storage[2]. However, there are major downsides to the use of polyethylene in food packaging, one of which is the lack of recycling after the packaging has been used. In 2009, it was estimated that only 9.7% of food package wastes generated from HDPE out of 620,000 tons was recycled. On the other hand, negligible or limited amount of food package wastes from LDPE was recycled out of 800,000 tons of total wastes generated in 2009 [3].

This lack of biodegradability of traditional food packaging plastics also results in the gradual leaching of plastics from land to the ocean. It has been estimated that in 2010, between 4.8 and 12.7 million metric tons of plastic waste entered into the ocean from coastal countries [4]. Due to the movement of ocean currents, large gyres have formed on the surface of the ocean in

**Corresponding Author*: Email: ssharma@uga.edu

certain locations. The ocean's currents cause the formation of gyres by generating an anticyclonic rotation in an isolated area, where the surface water will be pulled into the middle of the gyre, then submerged underneath the plastic at the core of the gyre [5]. The formation of the gyres will result in the gradual shifting of concentration of plastics on the surface of the ocean, where certain areas will have a much higher concentration of plastic, while others will have very little. For instance, in the North Pacific gyre, it was determined that in certain areas the concentration of plastics on the surface reached 10^6 per square kilometer, while outside these areas, there was very little plastic found [6]. When examined further, it was found that the major convergence of plastics on the surface of the ocean will occur in subtropical latitudes, as this area is also witness to a convergence of surface currents [7]. In order to limit the amount of ocean surface waste that is generated, it is necessary to examine other types of plastics that can be utilized.

Another major downside that is posed by the use of traditional plastics in food packaging is the lack of inherent antibacterial properties within the plastic to prevent spoilage. While the true ost of food spoilage hasn't been calculated accurately, it has been estimated to run into billions of dollars a year, and that is only taking into account the lost value of the food that was to be consumed[8]. Food spoilage is due to the microbial spoilage that will take place in food when it encounters microflora during storage. Organism types, such as yeast, bacteria, and fungi, are able to consume the nutrients provided by the food, and when temperature and pH conditions are favorable, the food will end up becoming spoiled [9]. In order to prevent bacterial growth, the addition of food preservatives into the plastic itself has been examined as a potential answer on how to limit food spoilage when packaged. When food preservatives, such as sodium benzoate or sodium nitrite, are incorporated into traditional food packaging plastics at a concentration of 15%, inhibition of Gram + bacteria and fungi was determined [10]. However, with the use of additives in plastic production comes the drawback of potential leeching of said food preservatives once the plastics are disposed of in a landfill.

With these limitations in the current plastics that are utilized in food packaging applications, alternative materials such as proteins are being examined for their potential use. One protein of interest is the albumin protein that is derived from the egg of a hen, which contains the lysozyme enzyme. When the lysozyme is isolated from the albumin protein, it is determined that the enzyme will exhibit antibacterial activity against bacteria that is common in spoiled food such as *Listeria monocytogenes* [11]. Another protein that has been examined for potential food packaging application is zein, one of the protein constituents derived from maize. When additives such as nisin or lauric acid are incorporated with zein in producing a plastic film, it has been found that antimicrobial effects are exhibited by the protein plastic films [12]. Further studies of protein-based plastics also point to antibacterial potential, as it has been found that bacteria that is common in causing food spoilage are unable to grow on the surface of albumin-based plastics [13].

These protein-based plastics may also pose an advantage in comparison with petroleum-based plastics, due to the fact that they possess the potential of biodegradation over time. This capability of biodegradability can decrease the environmental burden that will occur due to plastic use, as the CO_2 emission levels generated during biodegradable plastic composting are lower than the incineration of traditional plastics [14]. However, the precursor steps that are necessary to produce the biodegradable plastics have been found to have a much greater environmental impact than that of traditional plastics, as it will require more energy and water to produce the raw materials of biodegradable plastics [15], as well as generate higher levels of photochemical ozone formation during landfill disposal [16]. Further analysis must be conducted and improvements in the production of raw materials must be made in order to produce a viable alternative to petroleum-based plastics.

To justify the use of proteins in the production of food packaging plastics in place of traditional plastics, it will be necessary to conduct life cycle assessments of the technologies to be utilized. Based on previous life cycle assessment, it has been found that plastic food packaging has a lower environmental impact in comparison with other traditional packaging technologies such as glass [17], but had a greater environmental impact in comparison with materials such as recycled paper [18]. However, a gap in the life cycle assessment of food packaging is present due to the fact that materials that are produced by alternative, sustainable feedstock have not been taken into account, such as plastics that are derived from proteinaceous sources. Therefore, the objectives of this study were to (1) conduct a cradle-to-grave life cycle assessment of protein-based bioplastics manufactured from polyethylene, albumin protein, and zein protein for food packaging applications, and to (2) assess the environmental impacts of protein-based bioplastics over conventional food plastics.

11.2 METHODS

11.2.1 Bioplastics Manufacturing Process

To produce a bioplastic, it is necessary to first denature a protein in order to change its raw globular form to a more linear structure. In order to perform denaturation, it will be necessary to apply both heat and pressure in order to ensure that the protein will denature in a way in which it will form a plastic at the end of the process [27]. The heat at which the protein will denature varies based on the type of protein used [13], but after heating it is crucial to maintain pressure while the material is cooled to ensure that the material will maintain its solid plastic state [19]. After cooling, a completed plastic is generated and can be modified in further processing.

11.2.2 Goals and Scope Definition

We use life cycle assessments (LCA) to compare the various inputs and outputs that are associated with the production of food packaging products in the United States. For this LCA, the analysis will be limited to the production of raw materials and the production of the food packaging, in a cradle-to-grave approach. This limitation is made in order to allow for a model that can be developed that will not depend on any steps necessary outside the scope of gathering and utilizing raw materials.

For this LCA, we compared the production of food packaging plastic from albumin from hen egg white (Alb), zein from corn protein, and low density polyethylene (LDPE). Life cycle data for the production, harvesting, transport, storage, and use of the materials utilized in bioplastic production was included. The design models of the conversion of raw materials into plastics are illustrated in Figs 11.1–11.3, with a full cradle-to-grave analysis conducted. The entire production process took into account all the energy and emissions associated with the various products whose use is necessary in the production of the raw materials, such as fertilizers, chicken feed, and petroleum. In this LCA, the effective functional unit was one kilogram (kg) of plastic produced, as this can serve as a good fit when compared to the standards that are used in other LCAs conducted on food packaging applications [20, 21]. Detailed mass and energy balances were conducted to estimate all necessary inputs and outputs to manufacture both protein-based bioplastics and its counterpart from fossil sources.

11.2.3 The Production Use, and Disposal of Food Packaging Plastics

To begin the process of converting biostocks into bioplastics, the production, harvesting, and transport of raw materials used for the production of albumin [22a, 22b], zein from corn [23, 24], LDPE, and glycerol [25] were estimated through the gathering of previously generated data.

Then, the produced biostocks were transported to the location where the bioplastics can be produced. For the protein-based plastics, it was necessary to denature the proteins. Protein denaturing occurs when the base structure of the protein is modified to the extent where tertiary

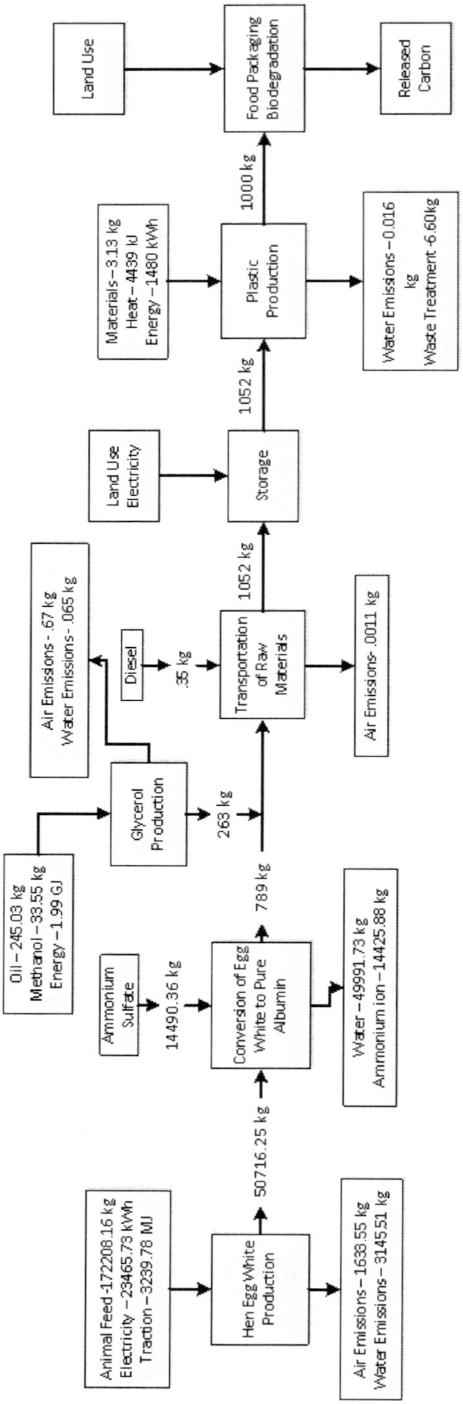

Figure 11.1 Process design of conversion of egg white to food packaging plastic and disposal.

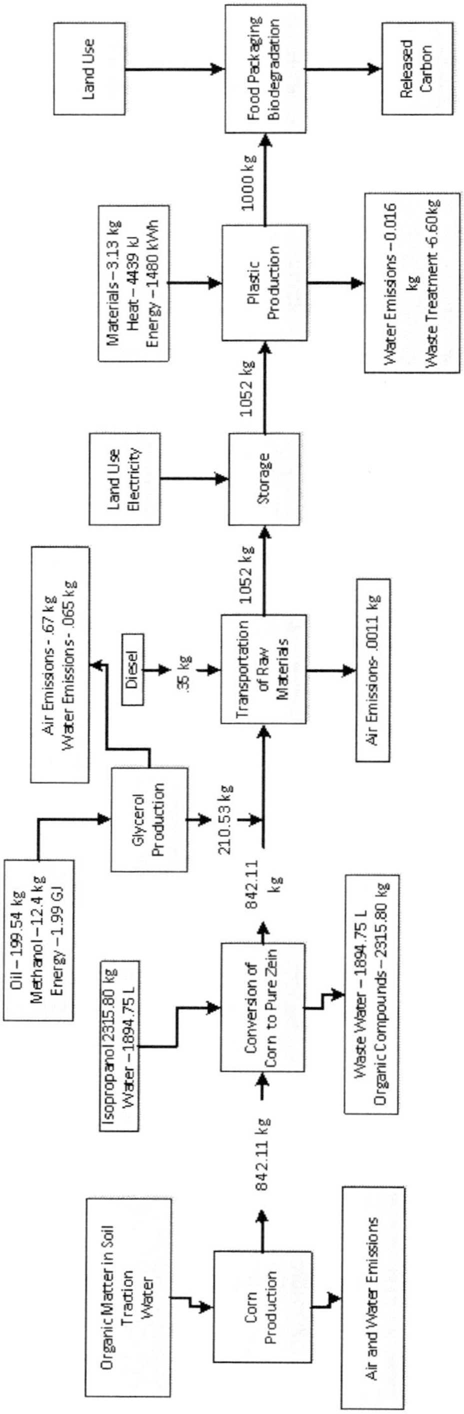

Figure 11.2 Process design of converting zein protein to food packaging plastic and disposal.

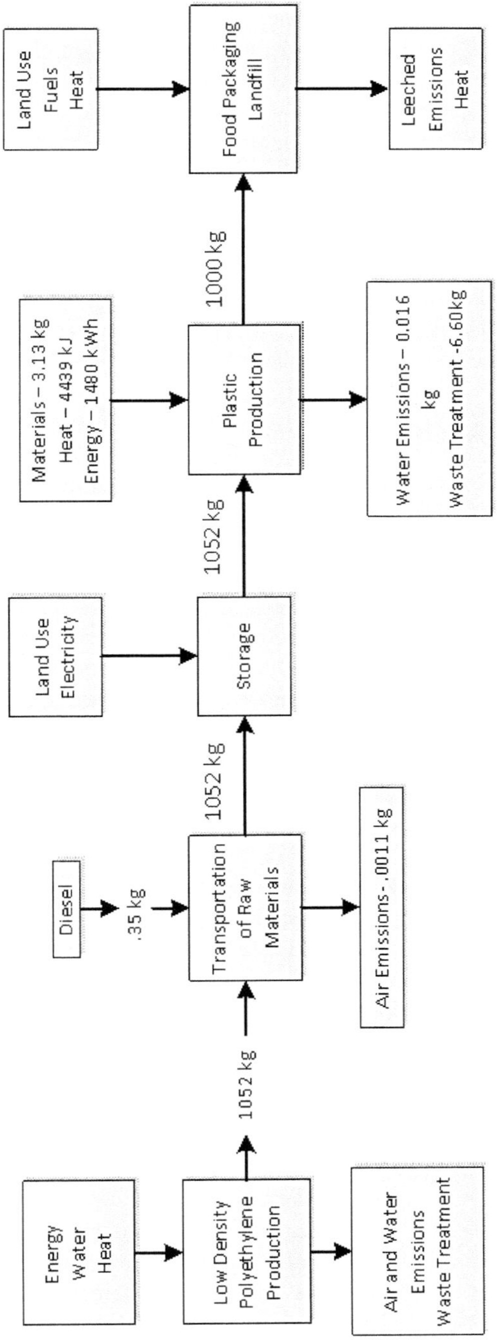

Figure 11.3 Process design of conversion of low density polyethylene to food packaging plastic and disposal.

and secondary structure is lost due to the elimination of weak stabilizing bonds throughout the protein [26]. This modification is achieved with the use of agents, such as radiation, heat, organic solvents, and strong acids and bases. For protein-based plastics, the use of both heat and pressure in the formation of plastics is a widely used method [27, 19, 28]. It was also necessary to mix a plasticizer with the protein to form the plastic, and glycerol was utilized as the plasticizer based on the results of past studies. For LCA purposes, the albumin-based plastic compositions consisted of a 75–25 mixture of albumin and glycerol, while the zein bioplastic consisted of an 80–20 mixture of zein and glycerol. As for the LDPE, it was assumed to be converted using the same process (without plasticizer), with a processing temperature in the range of 105 to 115°C.

After the raw materials had been converted into plastic and become utilized by the consumer, it was necessary to dispose of the materials. As the amount of food packaging plastic made from LDPE that was recycled was negligible [3], for this model the LDPE plastic were disposed of by the use of landfill. When it comes to the protein-based plastics, they pose the great advantage of being naturally biodegradable, as pure protein-based plastic films in nature could degrade completely in less than three weeks [29]. In this study, we assumed that the protein-based bioplastics would be completely degraded, whereas the LDPE plastics would be landfilled [20].

11.2.4 Inventory Analysis

A series of cradle-to-grave scenarios were developed and utilized to determine the environmental impact of the use of each raw material in food packaging applications. Data for many of the production steps for the production and disposal of the food packaging plastics was generated through the SimaPro software database, as it has gathered pertinent data for the examination for these processes in other applications [30]. For more specific settings in the processes, additional information was gathered based on the purification of zein from corn [23] and albumin from egg white [22a], as well as the production of glycerol to be utilized for the plasticization of protein-based plastics [25] and the environmental impact of biodegrading the materials [20].

For albumin protein production from egg, we included all the inputs of feed to allow the chickens to produce eggs, as well as the electricity and natural gas needed to incubate and transport the eggs. While the eggs themselves were the main output of interest in this process, we also included the air and water emission outputs as well, due to the farming of the chickens. After harvesting the eggs, all the inputs and outputs for the egg were shifted to the albumin protein production process. In this process, albumin was precipitated from egg white through the use of ammonium sulphate, a process that could be scaled up for large industrial operations [22b]. It must be noted that there was a 6.6% conversion rate of albumin from the eggs, as the rest of the egg matter consisted of the shell, the yolk, and water.

Zein production process included the cultivation of corn, harvesting and transporting, and further processing into Zein protein. All the inputs necessary for the entire production process were accounted for in this LCA study. Zein protein was extracted from dry corn mixed with water and 2-propanol[24]. The overall mass yield of Zein protein was 3.9% of the total mass of harvested dry corn [23].

All necessary inputs for the manufacturing of LDPE and glycerol and all air, water, and land emissions were obtained from US LCI database [25].

Once the raw materials had been produced, the raw materials were transported to the intended production facilities where the plastics could be made. To transport the materials by rail, it was necessary to utilize certain inputs, such as the energy generated by oil and the metals necessary to construct the transportation infrastructure. In terms of the outputs for this process, there were air, water, and soil emissions, as well as the final waste flows (all in kg), which were included with the main output of sending a certain amount of material a certain number of kilometers. For albumin protein production, the egg producer Foodonics International Inc. was utilized as the supplier of raw albumin material. Foodonics is based in Hoboken, Georgia, which is 385 km away from Athens, Georgia. In terms of the corn production utilized for the zein protein, corn farms from Fulton County, Illinois, were used as the producer, as farms from this area have been analyzed in the past for LCA of corn production [32]. Farms in this county were approximately 1060 km away from Athens, GA, which may have an impact on the amount of fuel needed to produce the plastic. As for the LDPE, Trison International Inc. will be utilized as the producer of the LDPE pellets. Since LDPE plastic pellet production has often been performed in areas where petroleum production occurs[31], the LDPE pellets came from an area where the petroleum industry was well established. Due to this requirement Trison International, based out of Houston, Texas, was chosen as the supplier for LDPE. Their distance from the production plant (1400 km) may have been a greater environmental cost in terms of transport, in comparison with the albumin and zein plastics being made.

Before the plastics were produced from the raw materials, it was necessary to store the materials in a warehouse to ensure that constant production was possible. This step in the process had very similar inflows due to the fact that it was necessary to build the warehouse on land close to the production facilities, as well as provide energy to the warehouse for climate control. The outputs for this step in the process besides the ability to store material included the air, water, and soil emissions (in kg) that occurred due to the power provided to the warehouse.

With the production of plastic now possible, the analysis of other inputs that were necessary to allow for the production to occur was conducted. The added inputs required for plastic production included the resources that were avoided when producing the plastics, as well as the heat and energy required for processing the raw material and forming plastics with high heat and pressure. It was because of this energy use that besides the plastic produced, there were also emissions to the air, water, and ground (in kg). The process diagrams of albumin and zein-based bioplastics are shown in Figs 11.1 and 11.2.

Once the plastic material was used by the consumer, it was then necessary to take into account the environmental impacts that will occur when the plastic is disposed of. For the LDPE plastics, landfill disposal was utilized, as this is one of the common ways in which LDPE material is taken care of. In this final step of the process, land use is an important input, as well as the amount of materials that are avoided when the land is being used to store waste (in kg, MJ, or m^3). With the use of landfills there are large amounts of emissions generated (in kg), as well as trace amounts of other waste flows having an effect on environmental impact. The process diagram of the life cycle of LDPE-based plastic is illustrated in Fig. 11.3. In contrast, the albumin and zein-based plastics will gradually biodegrade, which limit the environmental impact of plastic disposal when reflected in the assessment.

11.2.5 Impact Assessment Methods

After data collection, it was then necessary to analyze the inputs and outputs of plastic production through the use of the SimaPro 7.3 Life Cycle Assessment software. The data was entered into the software based on the type of plastic being produced, as certain types of plastic required much different raw materials, and were separated as such. Once entered and separated, a cradle-to-grave assessment was possible through the use of both the Tool for the Reduction and Assessment of Chemical and other environmental Impacts (TRACI 2 V3.03), as developed by the EPA, as well as the Building for Environmental and Economic Sustainability (BEES) software that has been developed by the National Institute of Standards and Technology. TRACI analyzed the data and determined the various environmental impacts that each process of plastic use would have into nine separate categories: global warming (kg CO_2 eq.); acidification (H+ moles eq.); carcinogens (kg benzene eq.); noncarcinogens (kg toluene eq.); respiratory effects (kg PM 2.5 eq.); eutrophication (kg N eq.); ozone depletion (kg CFC-11 eq.); ecotoxicity (kg 2,4-Dichlorophenoxyacetic acid); and smog (g NO_x eq.). This initial analysis is crucial, as it serves as a good indicator on which steps in the plastic production will have the greatest environmental impact. After this initial analysis, the three types of plastics will be compared based on the key impacts of global warming potential, acidification, eutrophication, and the release of noncarcinogens into the environment, as well as solid waste generated and the amount of water needed to produce one kg of packaging plastic material. These four impacts have been chosen due to the fact that they indicate the levels of CO_2 emissions, environmental changes due to plastic production, as well as the unintentional leeching of materials into the environment. An example of the environmental impact that plastic production has (in this case, the conversion of plastic from raw materials and global warming gas emissions) is illustrated in Fig. 11.4.

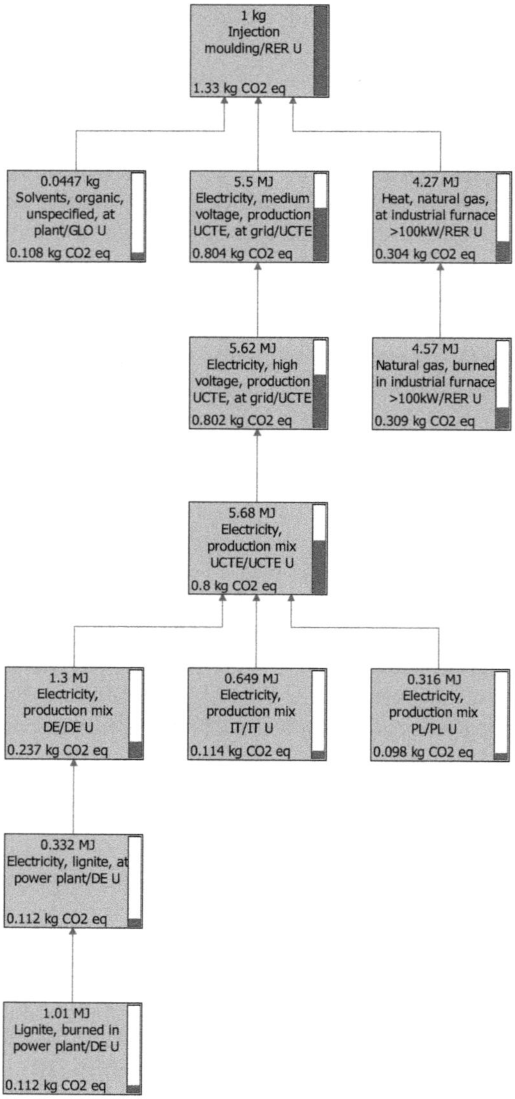

Figure 11.4 Flow chart of CO_2 emissions generated in the operation of a injection molding process[30].

11.2.6 Sensitivity Analysis of Impact Assessment

In order to determine how sensitive the results are to varying conditions in the production of protein-based plastics, alternate scenarios have been utilized. Alternate scenarios have been run to model how the harvesting of raw materials to be converted to pure albumin or zein protein for plastic production would affect environmental impacts. In this scenario, the raw material yield for albumin or zein protein feedstocks were elevated by 20% in one TRACI analysis, and decreased by 20% in another analysis to determine how sensitive the data generated will be to agricultural yield changes. These results will be compared to both the initial protein conversion rate and the LDPE plastic in terms of environmental impact, as this will indicate if said rates would be able to diminish or enhance the impacts of use of the plastics.

Another area that is crucial to examine is the rate at which raw material can be converted into the proteins that can be utilized in plastic production, as this is a resource-intensive step for both albumin and zein plastic production. In a similar fashion as is examined in raw material yield alternate scenarios, the conversion rates of raw material feedstock into pure proteins were elevated by 20% in one TRACI analysis and decreased by 20% in another analysis, with the results compared to the initial rate scenarios (as well as the environmental impact of LDPE production).

11.3 RESULTS AND DISCUSSION

11.3.1 Initial Cradle to Grave Analysis of Albumin, Zein, and LDPE Plastics

The results of the cradle-to-grave analysis of the production of each plastic type are shown in Figs 11.5 through 11.7. For the albumin-based bioplastic production, a key finding is that the three major processes that determine environmental impact are the harvesting of eggs for raw material, the conversion of the egg whites into albumin, and the logistical operations of transport and storage of the raw materials prior to production. The egg production and albumin purification processes dominate the environmental categories of acidification, eutrophication, and ozone depletion, as these environmental impact take into account the costs of feeding the chickens to produce the eggs, as well as the chemicals needed to convert egg whites into pure albumin. As for the logistics process, it dominates the other categories, as the transportation of raw material will result in the use of fossil fuels, and the required construction of storage facilities will require land transformation and additional emissions.

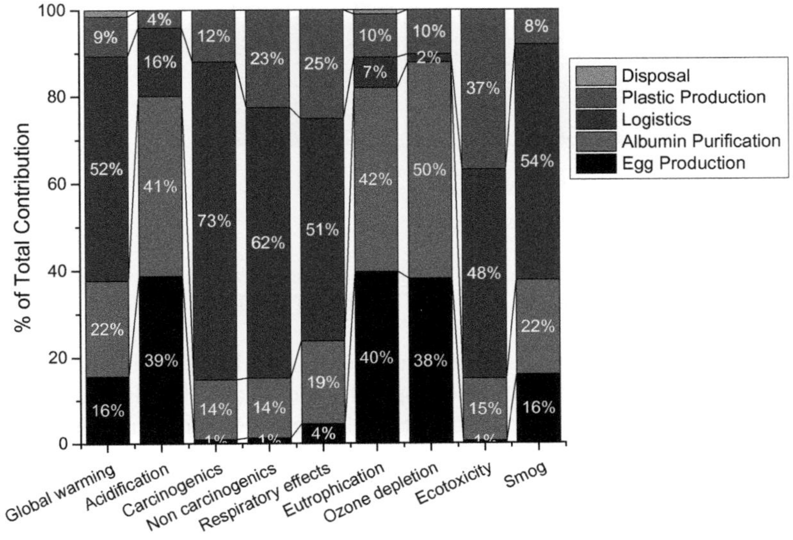

Figure 11.5 Comparison of albumin plastic use processes on the environment through the use of TRACI 2 (V 3.03).

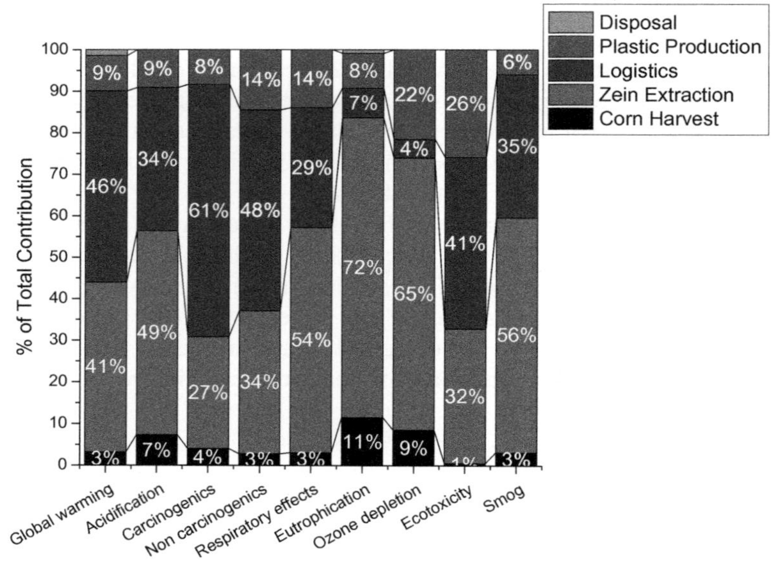

Figure 11.6 Comparison of zein plastic use processes on the environment through the use of TRACI 2 (V 3.03).

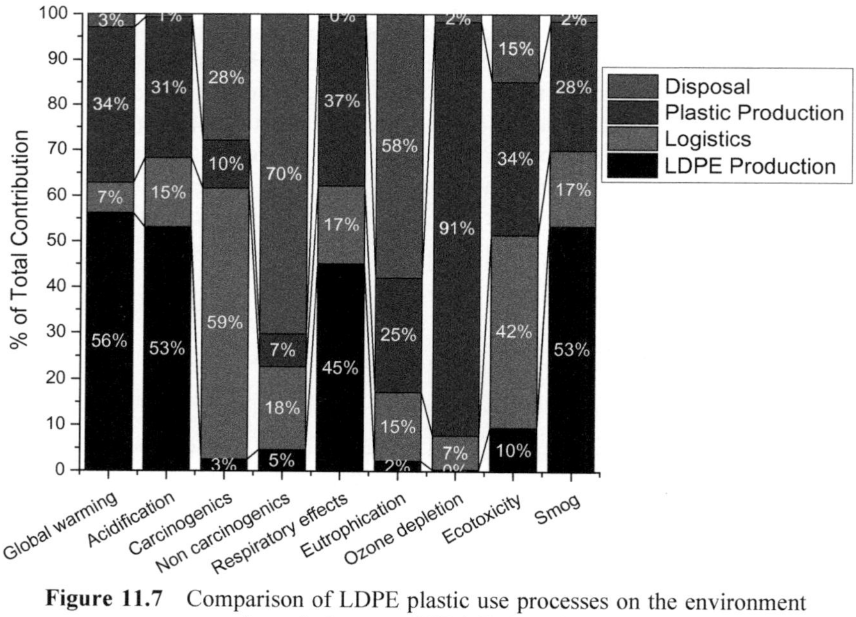

Figure 11.7 Comparison of LDPE plastic use processes on the environment through the use of TRACI 2 (V 3.03).

The processes of zein extraction and logistics are the two key contributors in environmental impacts when it comes to the production and disposal of zein-based bioplastics. Zein extraction is especially impactful when it comes to the ozone depletion, smog, and eutrophication impacts, as the drying of the harvested corn and the chemical treatment of the resulting dry corn will require additional fossil fuel use and chemical treatments that will result in harmful waste emissions. The logistics impact for zein is similar to that of the logistics of albumin, with greater impact in global warming, carcinogens, and noncarcinogens due to the transport of zein protein and resulting storage facilities.

With LDPE plastic use analysis, it is found that there is no overall dominant process that will dictate the environmental impact of LDPE plastic usage. For instance, while the production of LDPE will result in higher emissions of global warming gasses, acidification potential, and smog, the eventual disposal of the LDPE plastics has a substantial impact on the amount of noncarcinogens released and eutrophication potential that is possible. These findings for LDPE disposal can be explained through the gradual leeching of the plastic over time in a landfill, as additional materials utilized in LDPE production will contribute to a high level of emissions over time.

11.3.2 Comparison of Albumin, Zein, and LDPE Production in Initial Scenario

An expected finding is the lower water usage that is a result of LDPE plastic use and disposal, as shown in Table 11.1. LDPE plastics do not require large amounts of water in order to harvest its raw material, so there will be less water usage overall. However, an unexpected finding made is that LDPE plastic usage will also result in less solid waste generated, with LDPE solid waste generation half of the solid waste generated in zein plastic usage, and much lower in comparison with albumin plastic usage. The lower levels of solid waste generation can also be drawn to a lack of an intensive biomass generation step, as well as a lack of process needed to convert a raw material into a usable input.

Table 11.1 Analysis of water usage and solid waste generation in plastic production and disposal through BEES 2 (V 3.03).

Raw material type	*Water usage* (l)	*Solid waste generated* (Mg)
Albumin	9570	214.89
Zein	12800	2.60
LDPE	8740	1.77

When notable air emissions from plastic production are examined, the use of LDPE plastics is responsible for lower levels of all of the gases, including fossil-based carbon, as shown in Table 11.2. The decreased levels of petroleum use can be pointed to the fact that both the albumin and the zein required the use of diesel-powered tractors in order to harvest and transport raw materials from the field to a location where they can be utilized. The usage of zein plastics will end up with the highest levels of fossil fuel emissions, as natural gas required to dry the corn before zein extraction will contribute even more to greenhouse gas emissions.

Table 11.2 Analysis of notable air emissions from plastic production through the use of TRACI 2 (V 3.03).

Notable air emission (in kg CO_2 eq.)	*Albumin*	*Zein*	*LDPE*
Carbon dioxide	2.988185	0.949504	3.0955
Carbon dioxide, fossil	3.64116	5.139164	0.033975
Carbon dioxide, land transformation	0.194442	0.210398	0.007234
Dinitrogen monoxide	5.908183	0.199046	0.014261
Methane, biogenic	7.3004	3.695287	0.003736
Methane, fossil	1.725141	0.550021	0.509105

As for the water emissions from plastic production are examined, the use of albumin plastics will be responsible for the lowest levels of BOD5 and COD, while LDPE will have the lowest levels of nitrate and phosphate emissions, as illustrated in Table 11.3. Albumin bioplastics are able to limit BOD5 and COD emissions due to the fact there is a lack of major wastewater generation processes, such as fertilizer use or plastic leeching in a landfill. In terms of the LDPE plastics, they are able to limit the release of nitrate and phosphate into water due to the fact that fertilizer use is not necessary in LDPE raw material production, severely limiting the generation of the two emissions.

Table 11.3 Analysis of notable air emissions from plastic production through the use of TRACI 2 (V 3.03).

Notable water emission **(in kg N eq.)**	***Albumin***	***Zein***	***LDPE***
Ammonium, ion	0.001690278	0.000122	0.00042
BOD5, Biological Oxygen Demand	8.31177E-05	0.015919	0.002804
COD, Chemical Oxygen Demand	0.000138499	0.016027	0.011618
Nitrate	0.015681662	0.005415	0.000239
Phosphate	0.011800321	0.014438	0.009731

After comparing the three types of plastic use for greenhouse gas warming potential (GWP) and noncarcinogen (NC) emissions, it is the use of LDPE plastics that will cause the lowest levels of GWP emissions, as shown in Fig. 11.8. This finding is due to the fact that there is no conversion process necessary for LDPE raw material use, as well as the lack of glycerol usage in the logistics process of LDPE plastic production. When comparing albumin and zein plastics, it is the process of converting corn into zein that will cause zein bioplastic to emit the highest levels of GWP gasses. When the three plastic types are compared for NC emissions, the LDPE plastics will emit almost 250 kg toluene eq. more of emissions when compared to the other plastics. This is due to the usage of landfilling for the LDPE plastic, as this will lead to long term leeching of NC wastes from the plastic into the areas surrounding the landfill.

When comparing the plastic production processes for their potential in causing acidification and eutrophication in Fig. 11.9, it is found that LDPE plastic use will result in lower potential risks in both of the emission categories. This lowered risk is due to the fact that the protein-based plastics require glycerol as a plasticizer to form a plastic, as well as a relative lack of acidification and eutrophication impacts on the use of raw materials in LDPE production. For the two protein-based plastics, it is found that albumin will have a higher acidification potential in comparison with the zein-based plastics. This increased rate of acidification potential is due to the environmental impact of raising chickens that will produce eggs, as well as the chemical process that is required to convert the egg white into pure albumin protein. When comparing the eutrophication potentials, it is the albumin bioplastic use that will have lower levels of eutrophication. While the raw material gathering process of albumin will lead to higher levels of eutrophication, it is the conversion process of corn to zein that will lead to higher levels of eutrophication-causing emissions.

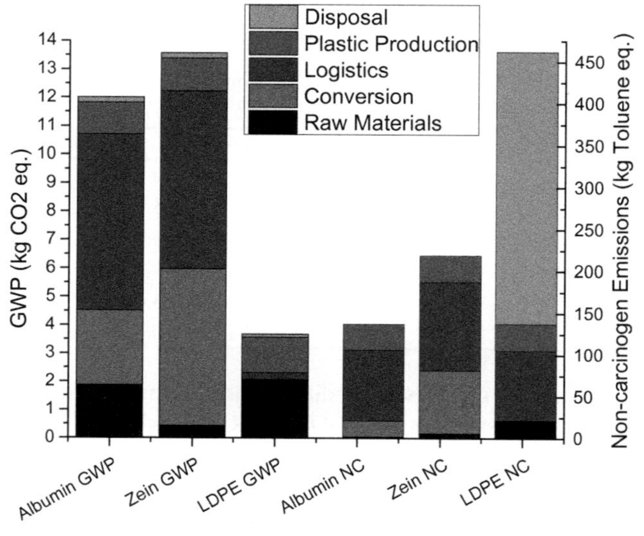

Figure 11.8 Comparison of greenhouse warming potentials and non-carcinogens between processes in plastic production.

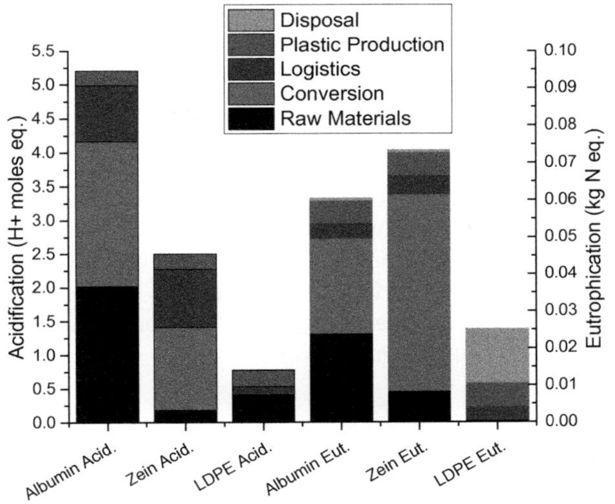

Figure 11.9 Comparison of acidification and eutrophication potentials between processes in plastic production.

11.3.3 Sensitivity Analysis of Protein-based Bioplastics

11.3.3.1 Harvest Level Sensitivity

After conducting the LCA under normal conditions, we were then able to modify certain aspects of the production of protein-based plastics to determine how sensitive the LCA is to certain data modifications. For protein-based plastics, it can be possible to change the amount of raw material that could be harvested through differing the yields from the raw material. To determine the sensitivity of the LCA data, we conducted an analysis of the production and disposal of protein-based plastics based on both an increase and decrease of possible raw material biomass yield by 20 percent.

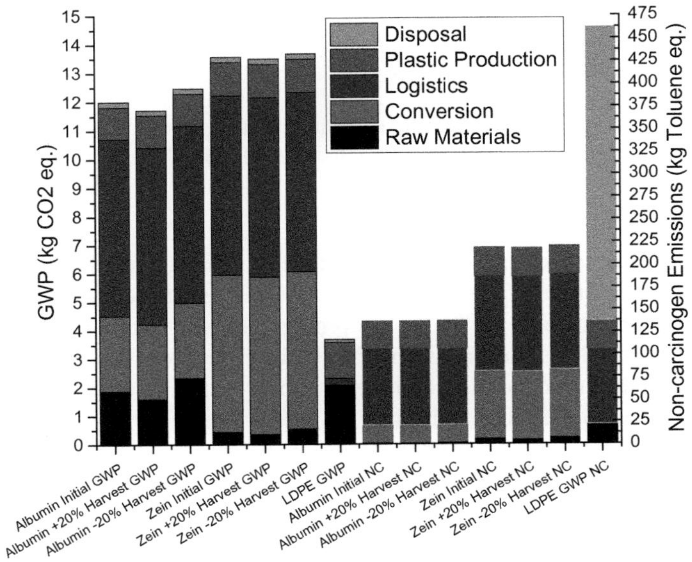

Figure 11.10 Comparison of global warming potential and non-carcinogen emissions of plastic production with bioplastics of varying harvesting rates of raw material through the use of TRACI 2 (V 3.03).

With modifying the harvest levels of raw material biomass in Figs 11.10 and 11.11, there is a marginal change in all four of the emission categories examined for zein protein. This is to be expected, as the harvesting of corn for zein production was not a major contributor to any of the emission categories when they were examined in the initial scenario. The same is not true for the modification of egg harvest levels and emissions, as the egg harvest process was a major contributor to all of the emission categories to be examined. As a result, when the egg harvest levels are modified, there is a noticeable change in the emissions levels, with a rise in emissions when the harvest is lower, as well as a decrease in emissions when harvest levels are increased. It must be noted that the changes in emissions levels are not substantial enough to improve its standing when compared to LDPE plastic use, as the modification of harvest levels will not decrease the emissions to an extent where bioplastic production will generate less GWP, acidification, or eutrophication causing emissions.

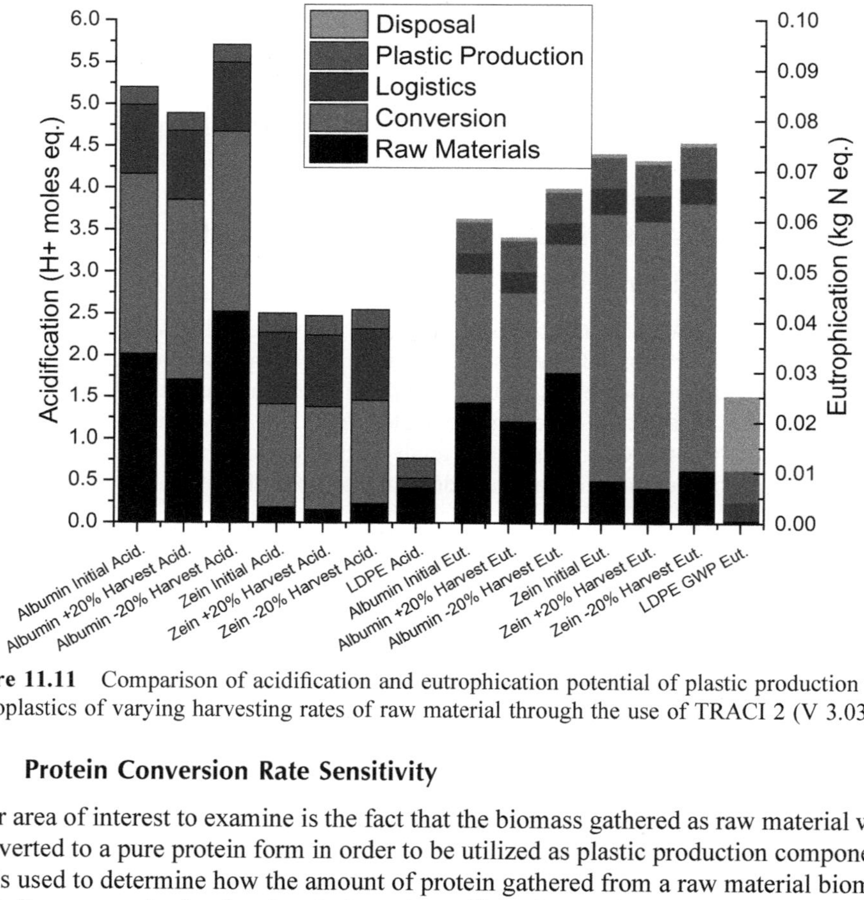

Figure 11.11 Comparison of acidification and eutrophication potential of plastic production with bioplastics of varying harvesting rates of raw material through the use of TRACI 2 (V 3.03).

11.3.3.2 Protein Conversion Rate Sensitivity

One other area of interest to examine is the fact that the biomass gathered as raw material will need to be converted to a pure protein form in order to be utilized as plastic production component. This analysis is used to determine how the amount of protein gathered from a raw material biomass will have an influence on the levels of emissions that will be incurred. For this analysis, the protein conversion rates have been increased in one scenario by 20%, and decreased in the other scenario by 20%, then compared to the initial values.

In this analysis of modifying protein conversion yields, we find that there is no significant change in emission levels when the protein yields are increased or decreased, as illustrated in Figs 11.12 and 11.13. This finding is most likely due to the fact that it will take a much higher change in protein conversion rates in order to witness significant changes, and that it would require substantial modification of the biomass itself before conversion in order for the yield amounts to change to that extent.

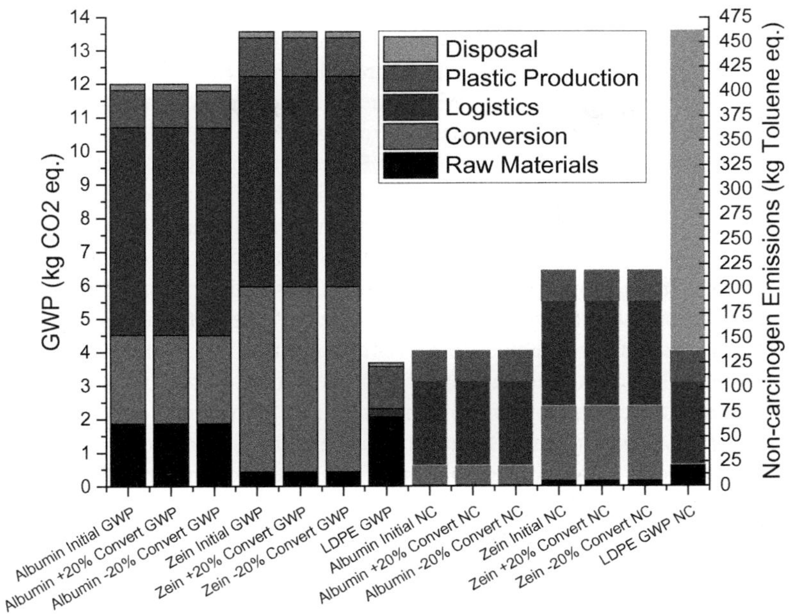

Figure 11.12 Comparison of the impact of varying levels of conversion rates on GWP and NC emissions through the use of TRACI 2 (V 3.03).

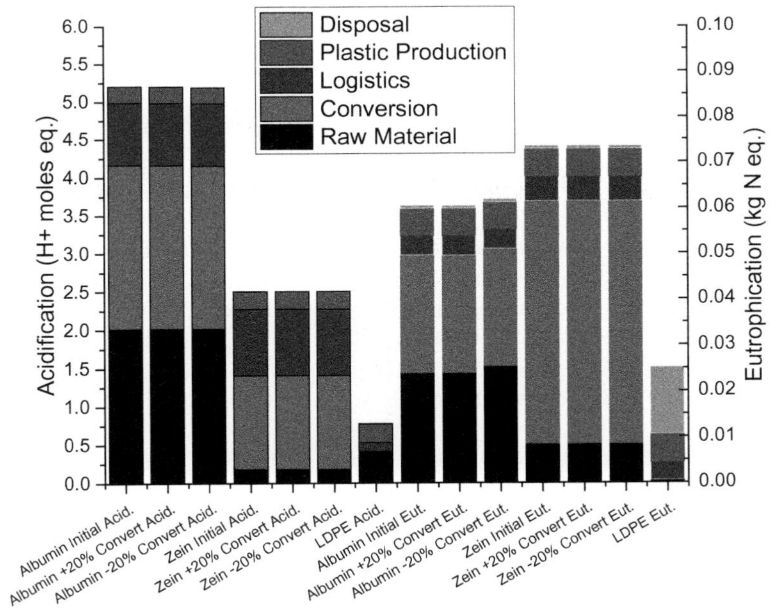

Figure 11.13 Comparison of the impact of varying levels of conversion rates on acidification and eutrophication emission potentials through the use of TRACI 2 (V 3.03).

11.4 CONCLUSIONS AND SUGGESTIONS FOR FURTHER RESEARCH

This study examines the LCA of albumin, zein, and LDPE-based food packaging plastic containers produced in Athens, GA. For this study, environmental impacts are investigated with a life cycle that entails cradle-to-grave scenarios based on the plastic produced. In the general scenario the protein-based food packaging plastics will emit more emissions that involve global warming gases, acidification, and eutrophication, but do provide an advantage when compared to LDPE in the area of non-carcinogen emissions. When each plastic production process is examined, it is found that the harvesting of eggs, the conversion of egg white to pure albumin, and the logistics processes have the greatest environmental impact for albumin plastics, while for zein production, the conversion of corn into pure zein and the logistics that are involved with zein bioplastic production have the highest environmental impacts. When alternate scenarios, such as harvest yield modification, protein conversion, and production location modifications are examined, it is found that modifying these properties does not yield a drastically different end result.

In terms of additional research that can be conducted based on the results of this study, one crucial area of potential research could be the continued analysis and optimization of corn and feedstock harvesting. The environmental impact of harvesting is crucial in the environmental impact of the protein-based plastics, so any improvements in this area would be crucial in developing a more environmentally friendly plastic from alternative sources. One other major area of interest would be the analysis of how to convert biomass into pure protein form in a way that will limit environmental impacts, as this step in the process of converting biomass into plastic needs to be improved in order to make protein plastic use ecologically feasible.

References

1. Vert, M., Y. Doi, K. Hellwich, M. Hess, P. Hodge, P. Kubisa, M. Rinaudo and F. Schué. Terminology for biorelated polymers and applications (IUPAC Recommendations 2012). Pure Applied Chemistry 2012; 84(2): 377-410.
2. Marsh, K. and B. Bugusu. Food Packaging—Roles, Materials,and Environmental Issues. Journal of Food Science 2007; 72(3): 39-55.
3. EPA. Municipal Solid Waste in the United States: 2009 Facts and Figures. Agency, U.S.E.P., Ed. Washington, DC, 2010.
4. Jambeck, J.R., R. Geyer, C. Wilcox, T.R. Siegler, M. Perryman, R. Andraday, R. Narayan and K.L. Law. Plastic waste inputs from land into the ocean. Science 2015; 347(6223): 768-771.
5. Gross, M. Plastic waste is all at sea. Current Biology 2013; 23(4): R135-137.
6. Law, K.L., S.E. Morét-Ferguson, D.S. Goodwin, E.R. Zettler, E. DeForce, T. Kukulka and G. Proskurowski. Distribution of surface plastic debris in the Eastern Pacific Ocean from an 11-year data set. ACS Environmental Science and Technology 2014; 48(9): 4732-4738.
7. Law, K.L. and S.E. Morét-Ferguson, N.A. Maximenko, G. Proskurowski, E.E. Peacock, J. Hafner and C.M. Reddy. Plastic Accumulation in the North Atlantic Subtropical Gyre. Science 2010; 329(5996): 1185-1188.
8. Stratford, M. Food and beverage spolage yeasts. *In*: G.H.F. Amparo Querol (ed.). The Yeast Handbook. Springer-Verlag, Berlin, 2006; pp. 335-379.
9. Gram, L., L. Ravn, M. Rasch, J.B. Bruhn, A.B. Christensen and M. Givskov. Food spoilage–interactions between food spoilage bacteria. International Journal of Food Microbiology 2002; 78(1-2): 79-97.
10. Vartiainen, J., E. Skytta, J. Enqvist and R. Ahvenainen. Properties of antimicrobial plastics containing traditional food preservatives. Packaging Technology and Science 2003; 16(6): 223-229.
11. Hughey, V.L., P.A. Wilger and E.A. Johnson. Antibacterial activity of hen egg white lysozyme against listeria monocytogenes Scott A in foods. Applied and Environmental Microbiology 1989; 55(3): 631-638.
12. Hoffman, K.L., I.Y. Han and P.L. Dawson. Antimicrobial effects of corn zein films impregnated with nisin, lauric acid, and EDTA. Journal of Food Protection 2001; 64(6): 885-889.
13. Jones, A., A. Mandal and S. Sharma. Protein-based bioplastics and their antibacterial potential. Journal of Applied Polymer Science 2015; 132(18): 41931-41940.

14. Yuki, S. Life cycle assessment of biodegradable plastics. Journal of Shanghai Jianotong University (Science) 2012; 17(3): 327-329.
15. Chaffee, C. and B.R. Yaros. Final Report; Progressive Bag Alliance, 2007.
16. Khoo, H.H. and R.B.H. Tan. Environmental impacts of conventional plastic and bio-based carrier bags Part 2: end-of-life options. International Journal of Life Cycle Assessment 2010; 15(4): 338-345.
17. Humbert, S., V. Rossi, M. Margni, O. Jolliet and Y. Loerincik. Life cycle assessment of two baby food packaging alternatives: glass jars vs. plastic pots. International Journal of Life Cycle Assessment 2009; 14(2): 95-106.
18. Zabaniotou, A. and E. Kassidi. Life cycle assessment applied to egg packaging made from polystyrene and recycled paper. Journal of Cleaner Production 2003; 11(5): 549-559.
19. Jones, A., M.A. Zeller and S. Sharma. Thermal, mechanical, and moisture absorption properties of egg white protein bioplastics with natural rubber and glycerol. Progress in Biomaterials 2013; 2: 12. https: //doi.org/10.1186/2194-0517-2-12.
20. Bohlmann, G.M. Biodegradable packaging life-cycle assessment. Environmental Progress 2004; 23(4): 342-346.
21. Perugini, F., M.L. Mastellone and U. Arena. A life cycle assessment of mechanical and feedstock recycling options for management of plastic packaging wastes. Environmental Progress 2005; 24(2): 137-154.
22. (a) Stevens, L. Egg white proteins. Comparative Biochemistry and Physiology Part B: Comparative Biochemistry 1991; 100(1): 1-9.
(b) Chick, H., and C.J. Martin. The precipitation of egg-albumin by ammonium sulphate. A contribution to the theory of the "Salting-Out" of proteins. Biochemical Journal 1913; 7(4): 380-398.
23. Shukla, R. and M. Cheryan. Zein: The industrial protein from corn. Industrial Crops and Products 2001; 13: 171-192.
24. Wilson, C.M. Isoelectric focusing of zein in agarose. Cereal Chemistry 1984; 61(2): 198-200.
25. Franklin, A. Cradle-to-gate life cycle inventory of nine plastic resins and four polyurethane precursors. Eastern Reseach Group, Inc., Prarie Village, KS, 2011.
26. IUPAC. Compendium of Chemical Terminology, 2nd Ed. Blackwell Scientific Publications, Oxford, 1997.
27. Sue, H.J., S. Wang and J.L. Lane. Morphology and mechanical behaviour of engineering soy plastics. Polymer 1997; 38(20): 5035-5040.
28. Sothornvit, R., C.W. Olsen, T.H. McHugh and J.M. Krochta. Formation conditions, water-vapor permeability, and solubility of compression-molded whey protein films. Journal of Food Science 2003; 68(6): 1985-1999.
29. González, A., M.C. Strumia and C.I.A. Igarzabal. Cross-linked soy protein as material for biodegradable films: Synthesis, characterization and biodegradation. Journal of Food Engineering 2011; 106: 331-338.
30. SimaPro SimaPro 7.3, 2015.
31. Sailors, H.R. and J.P. Hogan. History of Polyolefins. Journal of Macromolecular Science: Part A—Chemistry 1981; 15(7): 1377-1402.
32. Kim, S., B.E. Dale and R. Jenkins. Life cycle assessment of corn grain and corn stover in the United States. The International Journal of Life Cycle Assessment 2009; 14(2): 160-174.

Chapter 12

Bio-polyurethane and Others

M.A. Sawpan

Composite Materials Research, Pultron Composites Limited
Gisborne 4010, New Zealand

12.1 INTRODUCTION

The origin of polyurethane dates back to the beginning of World War II, when it was first developed as a replacement for rubber. Professor Dr. Otto Bayer (1902–1982) and his co-workers at IG Farben in Leverkusen, Germany, first made polyurethanes in 1937 [1, 2]. During the war, polyurethane coatings were used for the impregnation of paper and the manufacture of mustard gas resistant garments, high-gloss airplane finishes, and chemical and corrosion resistant coatings to protect metal, wood and masonry. By the end of the war, polyurethane coatings were being manufactured and used on an industrial scale, and could be custom formulated for specific applications. By the mid-50s, polyurethanes could be found in coatings and adhesives, elastomers and rigid foams. It was not until the late-50s that comfortable cushioning flexible foams were commercially available. Subsequent decades saw many further developments, and today we are surrounded by polyurethane applications in every aspect of our everyday lives. While polyurethane is a product that most people are not overly familiar with, as it is generally 'hidden' behind covers or surfaces made of other materials, it would be hard to imagine life without polyurethanes.

Polyurethanes represent 5% of the worldwide polymer consumption, but the dynamics of their growth is constantly high, around 4–6%. World consumption of polyurethanes was in the order of 8 million tons in 2000 [3], and is expected to reach 18 million tons by 2016 [4]. Polyurethanes can be found in liquid coatings and paints, tough elastomers, such as roller blade wheels, rigid insulation, soft flexible foam, elastic fiber, or as an integral skin. The main field of polyurethane application is the furniture industry; around 30% of the total polyurethanes produced worldwide are used for the production of mattresses from flexible slab-stock foams. Automotive manufacture is the second important application for flexible and semi-flexible polyurethanes (seat cushioning, bumpers, sound insulation, and so forth). Rigid polyurethane foams are used in thermal insulation of buildings and refrigerators, cold stores, pipe insulation, refrigerated transport, thermal in chemical and food industries. The polyurethane elastomers are used for shoe soles, footwear, athletics shoes, pump and pipe linings, industrial tyres, microcellular elastomers, etc. Polyurethane adhesives, sealants, coatings, and fibers represent another group of polyurethanes with specific applications. Figure 12.1 presents the breakdown of polyurethane usages.

For Correspondence: Email: moyeenbd@gmail.com; Tel: +64 6 867 8582; Fax: +64 6 867 8542

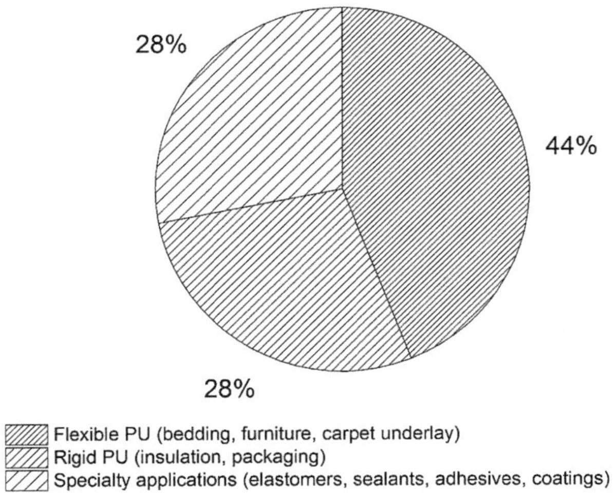

Figure 12.1 The main applications of polyurethane.

12.2 POLYURETHANE APPLICATIONS

Polyurethane can be prepared to show a wide range of properties, from those of flexible elastomers to those rigid crosslinked polymers, which can be used to fit in different applications [5]. Polyurethane packaging foam [6] can provide more cost-effective, form-fitting cushioning that uniquely and securely protects items that need to stay safely in place during transit. Polyurethane packaging is widely used to safely protect and transport many items. A versatile on-site solution for many packaging challenges, polyurethane packaging can save time and be more cost-effective by providing a custom-fit container with each shipment [45].

Polyurethanes are an important component in major appliances that consumers use every day. The most common use for polyurethanes in major appliances is rigid foams for refrigerator and freezer thermal insulation systems. Rigid polyurethane foam is an essential and cost-effective material that can be used for meeting required energy ratings in consumer refrigerators and freezers. The good thermal insulating properties of rigid polyurethane foams result from the combination of a fine, closed-cell foam structure and cell gases that resist heat transfer.

Polyurethanes are widely used in the automobile industry. In addition to the foam that makes car seats comfortable, bumpers, interior ceiling sections, spoilers, doors, and windows all use polyurethanes. Polyurethane reduces weight and increases fuel economy, corrosion resistance, insulation, and sound absorption. It has been estimated that the 1995 model cars used an average of 13.6 kg of polyurethane material per car [7].

Polyurethanes are commonly used in a number of medical applications [8–11], including catheter and general purpose tubing, hospital bedding, surgical drapes, wound dressings, and a variety of injection-moulded devices. Their most common use is in short-term implants. Polyurethane use in medical applications can be more cost-effective, and provide for more longevity and toughness.

Polyurethanes are frequently used in building and construction applications because of their excellent strength-to-weight ratio, insulation properties, durability, and versatility. It helps to conserve natural resources and preserve the environment by reducing energy usage. Polyurethane is widely used in floors, flexible foam padding cushions of carpet. In the roof, reflective plastic coverings over polyurethane foam can bounce sunlight and heat away, helping the house stay cool, while helping reduce energy consumption. Polyurethane building materials add design flexibility to new homes and remodelling projects.

Polyurethanes play an important role in advanced materials, such as composite wood. Polyurethane-based binders are used in composite wood products to permanently glue organic materials into oriented strand board, medium-density fiberboard, long-strand lumber, laminated-veneer lumber, and even strawboard and particleboard [12–14].

The use of flexible foam as a bonding material for fabric primarily started in 1961, when the apparel industry began to employ it [7]. Fine threads of polyurethanes were used to combine with nylon to make lightweight, stretchable garments. Over the years, polyurethanes have been improved and developed into spandex fibers, polyurethane coatings, and thermoplastic elastomers. Due to advances in polyurethane technology, a broad range of polyurethane apparel from man-made skins and leathers are being used for garments, sports clothes, and a variety of accessories.

Polyurethanes are frequently used in the electrical and electronics industries to encapsulate, seal, and insulate fragile, pressure-sensitive, microelectronic components, underwater cables, and printed circuit boards [15–17]. Polyurethane potting compounds are specially formulated to meet a diverse range of physical, thermal, and electrical properties. They can protect electronics by providing excellent dielectric and adhesive properties, as well as exceptional solvent, water, and extreme temperature resistance.

Polyurethane, mostly in the form of flexible foam, is one of the most popular materials used in home furnishings, such as furniture, bedding, and carpet underlay. As a cushioning material for upholstered furniture, flexible polyurethane foam works to make furniture more durable, comfortable, and supportive.

About 14.1% polyurethane is used for specialty applications, such as protective and decorative coatings, adhesives, caulks, sealants, and so on. Protective and decorative polyurethane coatings are used for different substrates, such as wood, plastic, metal, leather, and textiles [7].

12.3 POLYURETHANE CHEMISTRY

Polyurethanes are obtained by the poly-addition reaction of a diisocyanate (or polyisocyanate) and an oligomeric polyol (low molecular weight polymer with terminal hydroxyl groups), resulting in the formation of linear, branched, or crosslinked polymers [18]. Schematic reaction between a diisocyanate and diol is shown in Fig. 12.2. The high reactivity of the isocyanate group can be contributed to the positive charge of the carbon atom in the cumulated double bond system of its N=C=O group. Resonance structures of an aromatic isocyanate makes it clear to understand (Fig. 12.3). The electron deficiency on the carbon explains the reactivity of isocyanates towards nucleophilic attack, therefore most reactions take place across the C=N bond, as shown in Fig. 12.3(a). The structure in Fig. 12.3(c) becomes important when R is aromatic, in which case the negative charge on the nitrogen will be distributed throughout the benzene ring, reducing further the electron change on the central carbon of the isocyanate. This is also the reason why aromatic isocyanates, such as diphenylmethane diisocyanate (MDI) and toluene diisocyanate (TDI), are more reactive than aliphatic isocyanates, such as hexamethylene diisocyanate (HDI)

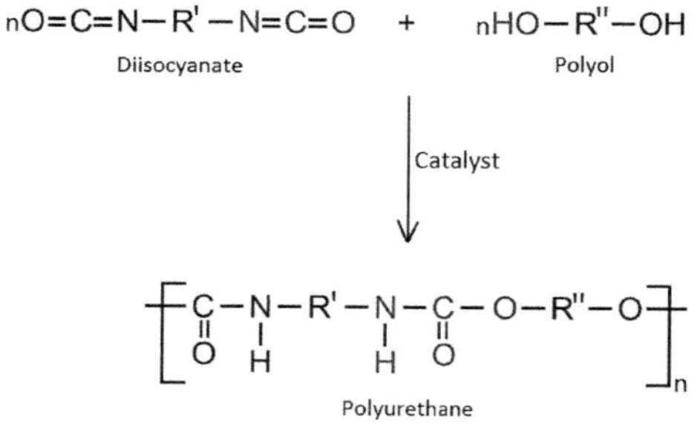

Figure 12.2 Schematic representation of the reaction between diisocyanate and diol.

and isophorone diisocyanate (IPDI). As a general outline, any electron-withdrawing group linked with R will increase the positive charge on carbon, thereby increasing reactivity of the isocyanate group towards nucleophilic attack. On the other hand, electron donating groups will decrease the reactive of isocyanate groups.

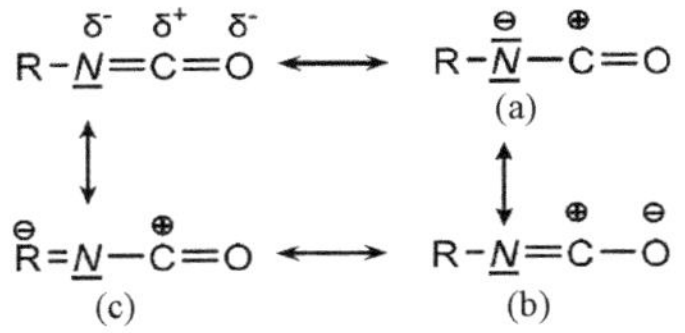

Figure 12.3 Resonance structures of isocyanate group.

In polyurethanes, the isocyanate component can be aromatic or aliphatic. The most important isocyanates, covering the majority of polyurethane applications, are aromatic isocyanates: TDI and MDI. TDI is produced from toluene via nitration, hydrogenation, and phosgenation, as shown in Fig. 12.4. TDI is commercialised using a mixture of 2,4 and 2,6 isomers or 2,4 TDI as pure isomers. The isomers differ markedly with respect to their structure as well as reactivity. 2,6-TDI isomer is symmetric compared to 2,4-TDI isomer, and is therefore expected to form hard segments which have better packing characteristics. The reactivity of the ortho position in the 2,4-TDI isomer is known to be approximately 12% of the reactivity of the isocyanate group in the para position due to the steric hindrance caused by the methyl group. However, when the reaction temperature approaches 100°C, steric hindrance effects are overcome and both the positions react at nearly the same rate. In comparison, the isocyanate groups on the 2,6 isomer have equal reactivities when both groups are unreacted. However, after one of the isocyanate groups reacts, the reactivity of the second group drops by a factor of around 3.

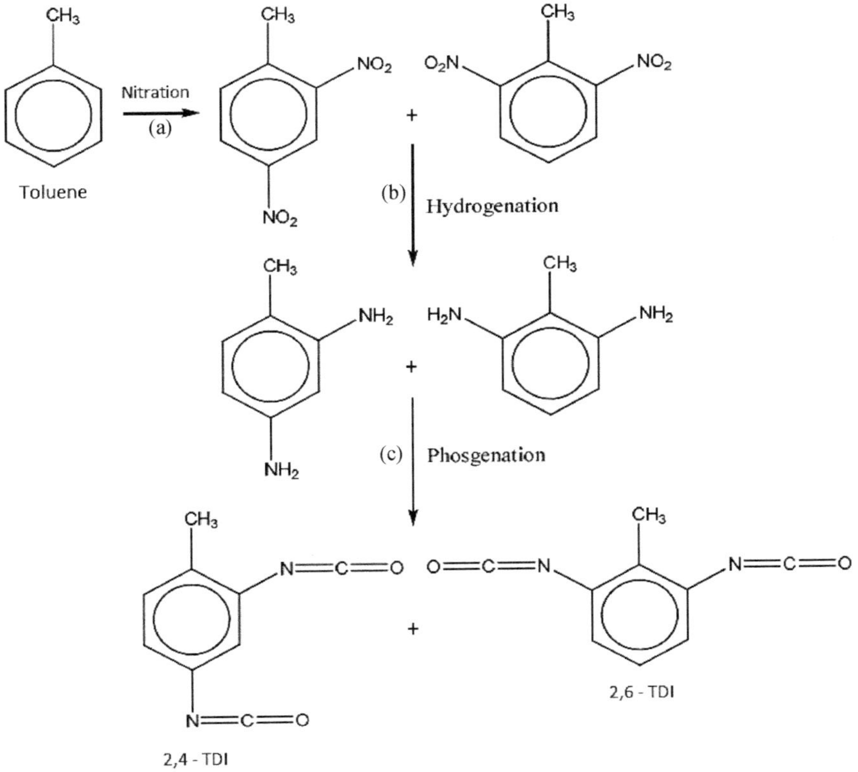

Figure 12.4 Schematic representation of synthesis route of toluene diisocyanate (TDI).

The other important isocyanate MDI is commercialised in various forms and functionalities, the most important of which are: pure MDI and polymeric MDI (Fig. 12.5). Pure MDI is commercialised mainly as 4,4′ isomers, but it is possible to use 2,4′ and 2,2′ isomers. On the other hand, polymeric MDI is a complex mixture of isomers and oligomers (i.e., trimer, hexamer, pentamer, and higher adducts) with isocyanate equivalent weight of 133 and a number-average isocyanate functionality of 2.7.

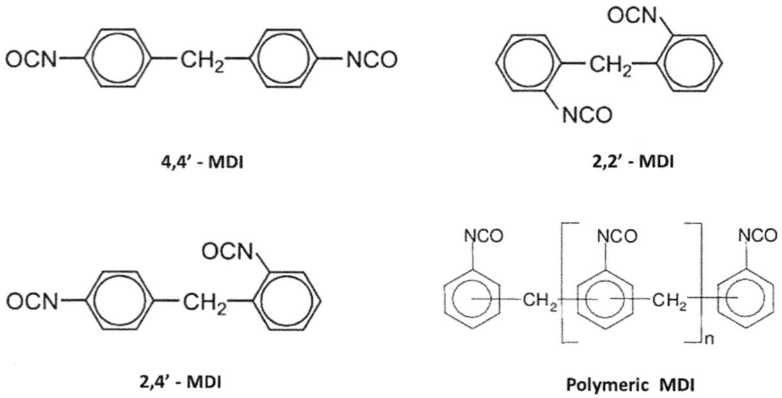

Figure 12.5 Chemical structure isomers of diphenylmethane diisocyanate (MDI).

Aliphatic isocyanates, such as HDI, IPDI, and 4,4′-methylene dicyclohexyl diisocyanate (HMDI) are used to a much lesser extent, and only for special applications. The aliphatic isocyanate group is less reactive than the aromatic. Aliphatic isocyanates are used only if special properties, such as light stable coatings and elastomeric properties, are required for the final product.

12.4 STRUCTURE OF POLYURETHANE

Polyurethanes are an important class of polymers and exhibit an exceptionally versatile range of properties and applications. In order to meet specific requirements, their structures can be tailored by selecting appropriate polyols and polyisocyanates. In industry, only a few polyisocyanates are commonly used, while a variety of polyols are available. Therefore, the selection of polyol typically determines the properties of the created polyurethane. The most important properties are: molecular weight, intra/inter-molecular force, stiffness of chain, crystallinity, and crosslinking [7, 19].

The molecular weight of the oligomeric polyol used for polyurethane manufacture has a profound effect on the properties of the resulting polymer. Polyols used for polyurethane manufacture are divided from the molecular structural point of view into two groups. In the first group there are the low molecular weight polyols having unitary and concrete molecular weight, such as ethylene glycol, propylene glycol, neopentyl glycol, and glycerol. The second group of polyols for polyurethane contains low molecular weight polymers (oligomers with a maximum molecular of 10,000 dalton) with terminal hydroxyl groups characterised by an average molecular weight [7, 19]. A hard polyurethane is formed when the molecular weight of polyol is low, elastic and flexible polyurethane forms from high molecular weight polyols, and intermediate molecular weight leads to semi-rigid or semi-flexible properties [7]. Intra/inter-molecular secondary forces, such as Van der Waals force, hydrogen bond, London dispersion forces, dipole interaction, and ionic force, create cumulative cohesive energy, which causes an intense effect on the physical and mechanical properties of the resulting polyurethane, because the repeating units have various cohesive energies of functional groups from polyol [19]. Molecular flexibility depends on the freedom of rotation around single bonds in the main chain of the polymer molecule. Restrictions in this free rotation reduce the flexibility. Aromatic groups in the polyol can strongly reduce the molecular flexibility, which causes high softening temperature, hardness, and a decrease in elasticity of polyurethane. In general, a majority of the oligomeric polyols are amorphous. Some oligomeric polyols, such as polytetramethylene glycol, polyethylene glycol, and polyethylene adipate show crystallinity. Crystallinity decreases molecular flexibility and mobility of the polymer chains; consequently softening temperature, hardness, and tensile strength increase; and elongation, solubility, and flexibility decrease. The degree of crosslinking depends on functionality and molecular weight of oligomeric polyols. Similar to crystallinity, crosslinking decreases flexibility, solubility, and elongation; and increases hardness, tensile strength, softening temperature, and modulus of elasticity [7, 19].

Apart from polyols, there are indigenous reactions of polyisocyanate that can contribute to significant crosslinking during polymerization of polyurethane (Fig. 12.6). When polyurethane is formed in the presence of excess polyisocyanate, active hydrogen of urethane reacts with the isocyanate to form allophanate crosslink. The formation of allophanate is a high temperature (>110°C), reversible reaction. When polyurea is formed in the presence of excess polyisocyanates, similar to the allophanate formation, the active hydrogen of polyurea reacts with polyisocynates to generate a biuret. Also similar to the allophanate formation, the reaction between urea and isocyanate is a reversible reaction and needs high temperature too (>110°C). Heterocyclic isocyanurate can be formed from trimerisation of isocyanates that takes place in the presence of a special catalyst (e.g., potassium acetate). The isocyanurate rings act as extremely stable crosslinks in the polyurethane network. Isocyanurates are frequently used to produce polyurethanes of high temperature stability and flame retardance [19, 20].

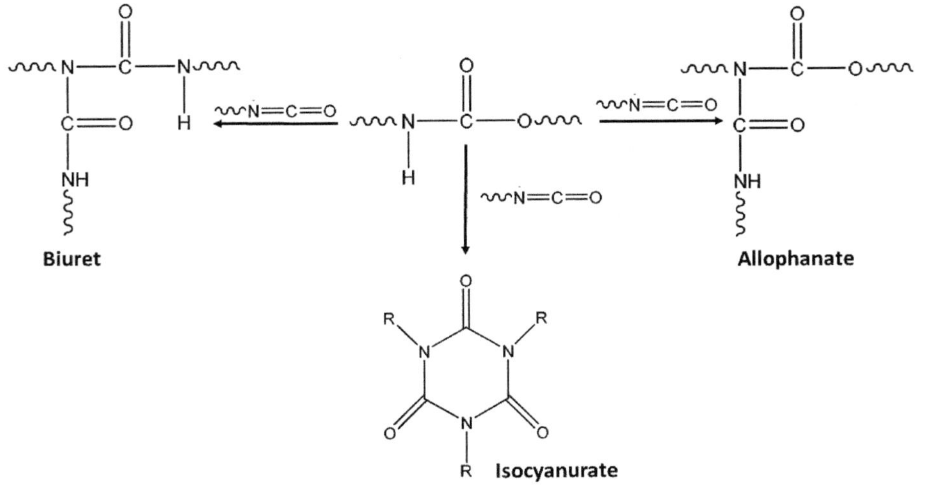

Figure 12.6 Crosslinking reactions of polyisocyanate.

12.5 RAW MATERIALS OF POLYURETHANE

The principal raw materials currently used for the production of diisocyantes are derived mainly from petroleum and its derivatives [21]. It is possible to produce aliphatic diisocyanates from dimerised fatty acid which is bio-based, and such commercial products can be found in the market. However, these polyurethanes can only be used as coatings and are not suitable for foam applications [22, 23]. On the other hand, the polyol component has the potential to be bio-based in some applications. These polyols diversify the industry's supply options and help mitigate the effects of uncertainty and volatility of petroleum supply and pricing. Since polyurethane chemistry is wide-ranging in terms of both feedstock possibilities and applications, the following sections will endeavor mainly to present the preparation of polyols from vegetable oils and their derived polyurethane.

12.5.1 Bio-based Polyol for Polyurethane

The development of polyols based on renewable resources played an important role in the polyurethane industry. Bio-based polyols can be drop-in replacements for existing polyols, as they have similar structures and properties, but are derived from renewable raw materials rather than petrochemicals, as well as substitutes for existing polyols that have slightly different structures and properties, or new compounds with entirely new functionality and performance characteristics. Depending on the preferred building blocks and their origin, the renewable content of commercially available bio-based polyols varies between 30–100%. Consequently, the renewable content of polyurethane based on different formulations also varies substantially among different products and applications, with a range of 8–70% [4, 23].

Preparation of polymeric materials from renewable resources is of great significance economically and ecologically. Vegetable oils are becoming extremely important as renewable resources for the preparation of polyols required for the polyurethane industry. The vegetable oils, such as soybean oil, castor oil, sunflower oil, palm oil, rapeseed oil, olive oil, linseed oil, canola oil, and so on, with a worldwide production of around 156 million tons per year (in 2012) are used mainly in human food applications (76%), in technical applications (19.5%) such as soap and oleochemical industry, and in other applications (1.5%) [24, 25]. The properties of polyurethanes are closely tied to the properties of the polyols from which they are produced, so it is critical that bio-based polyols offer the performance needed for a given polyurethane application. In many cases vegetable oil-based polyols do not have the necessary functionality (hydroxyl groups) in their native form to be useful for polyurethane manufacture, so this needs to first be introduced by chemical manipulation [26].

Glycerol is an important starter for the synthesis of bio-based polyol, which is produced by the hydrolysis of triglycerides (esters of glycerol with fatty acid), as available abundantly in vegetable and animal sources [19, 27]. The chemical structure of a triglyceride is shown in Fig. 12.7. Bio-based polyurethanes have been generally prepared by first converting the triglycerides of fatty acid into polyols, followed by reaction with diisocyanates [24].

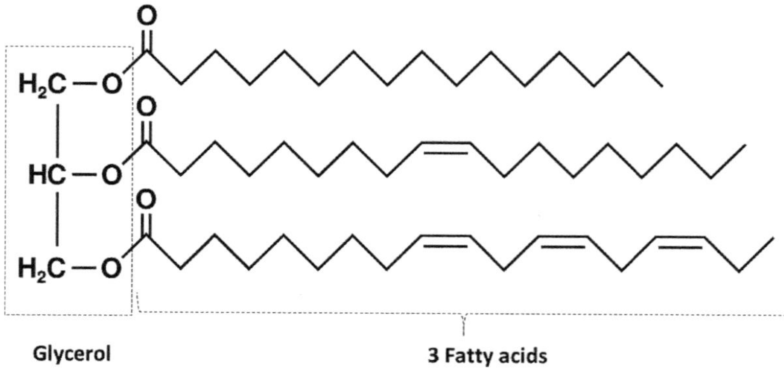

Figure 12.7 Chemical structure of a triglyceride.

Depending on the origin and type of fatty acids in vegetable oils, such as soybean oil [28, 29], castor oil [30], palm oil [21, 31], canola oil [32], sunflower oil [33], corn oil, and linseed oil [34], the fatty acid side chains contain carbon numbers ranging from 8 to 24 and carbon-carbon double bond numbers from 0 to 5 [35, 36]. Some fatty acids have specific functional groups, for example hydroxyl-containing ricinoleic acid from castor oil. In all vegetable oils the general reactive sites are esters and carbon–carbon double bonds, which can be modified to give a variety of monomers. Some common fatty acids are shown in Fig. 12.8.

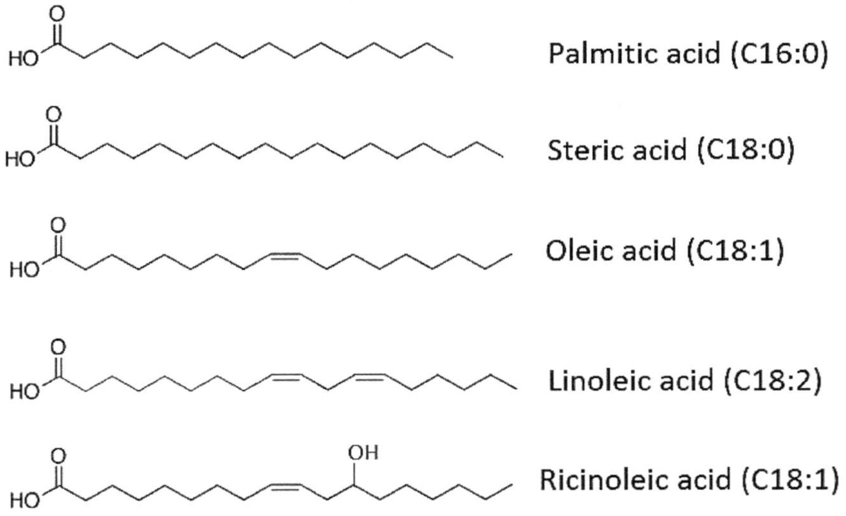

Figure 12.8 Chemical structures of different fatty acids of vegetable oils.

Castor oil (Fig. 12.9) is a pale yellow and viscous liquid, derived from the bean of the castor plant (*ricinus communis*) that grows in tropical and subtropical regions. This oil is one of the best industrial choices due to its unique fatty acid composition and ready availability [5, 37, 38]. It is unique among all fats and oils, in that:

i) it contains an 18-carbon hydroxylated fatty acid with one double bond
ii) ricinoleic acid comprises approximately 92–95% of the fatty acid composition
iii) high degree in product uniformity and consistency
iv) it is a nontoxic, biodegradable, renewable resource

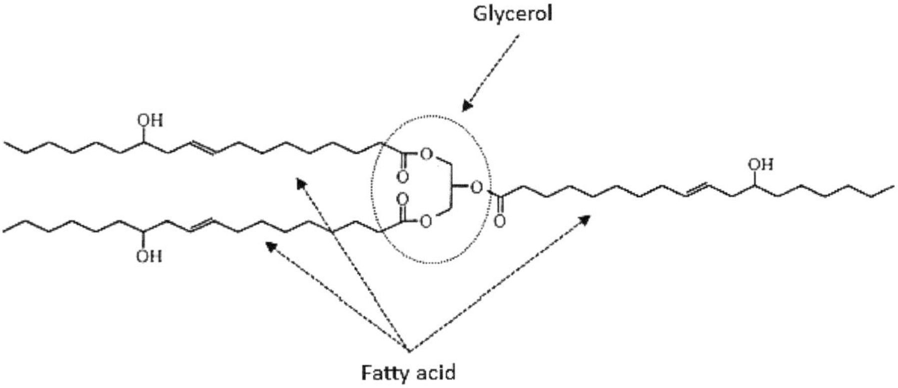

Figure 12.9 Chemical structure of castor oil.

The hydroxyl groups in castor oil account for a unique combination of physical properties, for example (i) relatively high viscosity and specific gravity, (ii) solubility in alcohols in any proportion, and (iii) limited solubility in aliphatic petroleum solvents. High purity castor oil may be used as a polyol to produce polyurethane coating, adhesives, and casting compounds. The monoglyceride of castor oil is found to be a better choice as a triol over the oil itself in the synthesis of hyperbranched polyurethane. Hyperbranched polymers are a unique class of polymers with many unusual and desirable properties, such as low viscosity, high solubility, good reactivity, easy processing, etc.

As mentioned earlier, most vegetable oils do not have the necessary functionality (hydroxyl groups) in their native form to be useful directly for polyurethane manufacture. The preparation of polyols from vegetable oils can be categorised into two groups: (i) reaction through esteric groups, and (ii) reaction through double bonds [39, 40].

12.5.1.1 Transesterification and Transamidation of Vegetable Oil

Transesterification and transamidation reactions use the ester moieties in the structures of vegetable oils to produce polyols. Polyols produced from vegetable oils by transesterification with glycerol are a mixture of monoglycerides, diclycerides, and triglycerides of unsaturated fatty acids. Among these components, monoglycerides, which contain two hydroxyl groups per molecule, play an important role in polyurethane production. Pre-polymer is formed by the reaction between diols (i.e., monoglycerides) and diisocyanates. The resulting unsaturated pre-polymers are crosslinked by radical polymerisation of the double bonds in a crosslinked network. This process is used for the manufacture of urethane alkyd coatings. As the fatty acid portions make up the larger proportion (about 90% fatty acids to 10% glycerol) of the triglycerides, most of the physical and chemical properties result from the effects of the various fatty acids esterified with glycerol [41, 42].

Like the transesterification process, transamidation with amines, usually with diethanolamine, can also transfer vegetable oils, such as linseed, soybean, rapeseed, and sunflower oils, into diethanol fatty acid amides for producing polyurethane products. The fatty acid diethnolamides are bifunctional compounds and improve the compatibility of various polyols with reasonable physical and mechanical properties of polyurethane.

Different reactions of double bonds are used for the conversion of unsaturated triglyceride of vegetable oils to produce polyols. The most commonly used reactions are (i) epoxidation, (ii) ozonolysis, (iii) hydroformylation, and (iv) metathesis.

12.5.1.2 Epoxidation of Vegetable Oil

Epoxidation of carbon–carbon double bonds, followed by oxirane ring opening with amines, carboxylic acids/halogenated acids, or alcohols is used for the preparation of commercial vegetable oil based polyols for polyurethane [36, 43, 44]. A schematic representation of epoxidation of vegetable oil followed by oxirane ring-opening is presented in Fig. 12.10. Prilezhaev reaction, a chemical reaction of an alkene with a peroxy acid, is one of the most common methods of epoxidation [45]. The most widely used peroxy acid for the oxidation of the alkene is meta-Chloroperoxybenzoic acid, due to its stability and good solubility in most organic solvents. In industrial process, epoxidation is carried out using peracetic acid, which is prepared from acetic acid and H_2O_2. There are also many catalytic methods of epoxidation which utilise H_2O_2 as an oxidant, with transition metal compounds as catalysts. In the report [46], epoxidation of methyl linoleate was studied using catalytic amounts of methyltrioxorhenium (4 mol%) along with pyridine and H_2O_2 over 4 hours. The catalyst loading could be lowered to as little as 1 mol% at the expense of the time taken to reach full conversion. In another work [47], amorphous Ti/SiO_2 was used as a catalyst in conjunction with H_2O_2 in the presence of tert-butanol for the epoxidation of soybean oil and its methyl esters. Other methods of epoxidation include the use of tungsten-based catalysts38 and the use of enzymes [48]. The properties of polyols produced by epoxidation depend on several production variables, including feedstock characteristics and the types of ring-opening agents. Vegetable oils with a higher degree of unsaturation produce polyols with higher hydroxyl functionalities, resulting in polyurethanes with higher crosslinking density, higher glass transition temperature, and higher tensile strength [25].

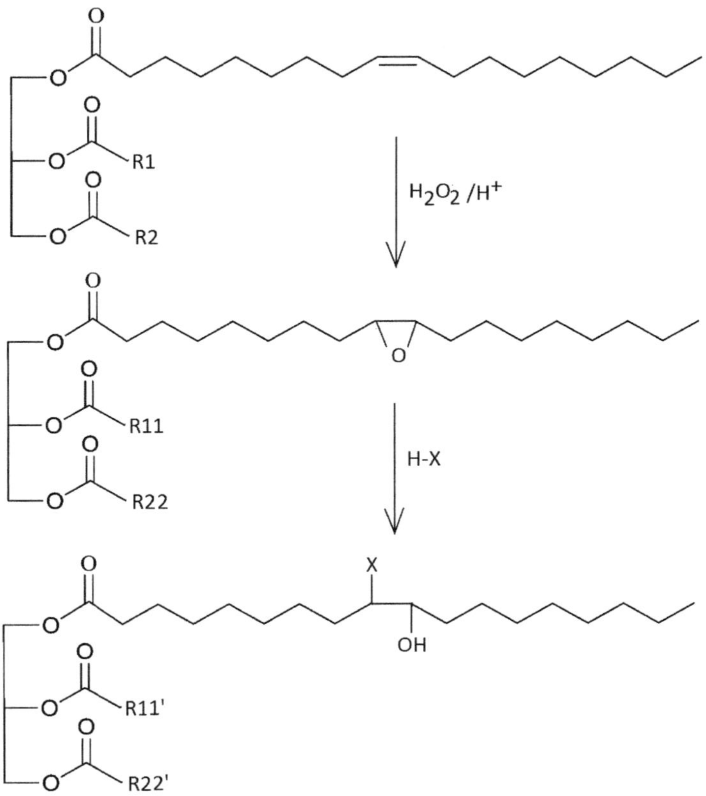

Figure 12.10 Epoxidation of vegetable oil for the production of bio-polyols (R1, R2, R11, R22, R11′, R22′ are fatty acid side chains of vegetable oils).

12.5.1.3 Ozonolysis of Vegetable Oil

Ozonolysis of vegetable oil into ozonides, followed by reduction to aldehyde, and finally to alcohol creates polyols that contain primary hydroxyl groups that are more reactive towards isocyanates than secondary hydroxyl groups [49, 50]. In addition, the hydroxyl groups of polyols prepared using this method are located at the end of the fatty acid chains, which make all the aliphatic chains in the macromolecular network, resulting in rigid polyurethane. Plant oils, including trilinolein (or triolein), low-saturation canola oil, and soybean oil, could be ozonolyzed in methylene chloride or methanol, and then hydrogenated sequentially to form polyols [49]. In the report [50], canola oil was ozonolyzed in water to produce ozonide, that was then reduced to polyol in tetrahydrofuran. The ozonolysis with $CaCO_3$ was also reported on soybean oil in the presence of ethylene glycol [51]. An example of the reaction of vegetable oil with ozone and ethylene glycol is presented in Fig. 12.11.

12.5.1.4 Hydroformylation of Vegetable Oil

Hydroformylation is one of the most important chemical transformations of vegetable oils into polyols [52]. The hydroformylation route, adding a carbon atom on the chain, leads to primary hydroxyl functions [53], compared to epoxide ring opening route that leads to a polyol with a mixture of primary and secondary hydroxyl functions. In the process, double bonds in vegetable oil structures are first converted to aldehydes at 70–130°C via rhodium or cobalt catalysed hydroformylation by 1:1 mixture of CO and H_2, and then to hydroxyl groups via the hydrogenation of aldehyde [36, 54, 55]. A schematic of hydroformylation reaction of vegetable oil is presented in

Fig. 12.12. Hydroformylated polyols (polyols derived via the hydroformylation and hydrogenation pathway) show shorter gel time and better curing efficiency during the reaction with isocyanates for polyurethanes when compared to epoxidized polyols [56]. Polyurethane foams with greater rigidity can be produced from hydroformylated polyols when mixed with a crosslinker such as glycerol [36].

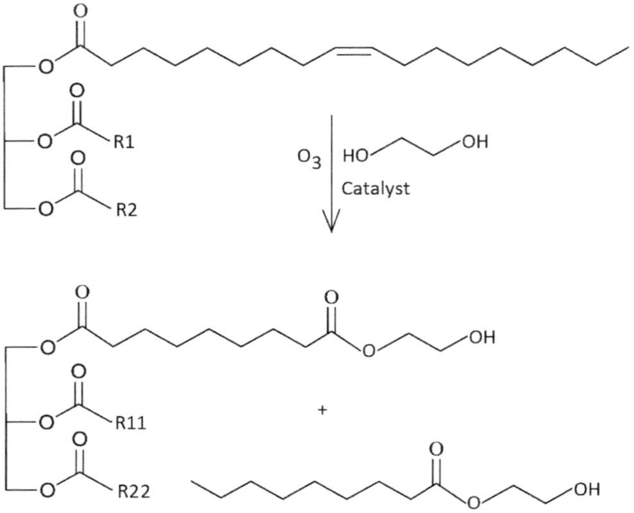

Figure 12.11 Ozonolysis of vegetable oil for the production of bio-polyols (R1, R2, R11, and R22 are fatty acid side chains of vegetable oils).

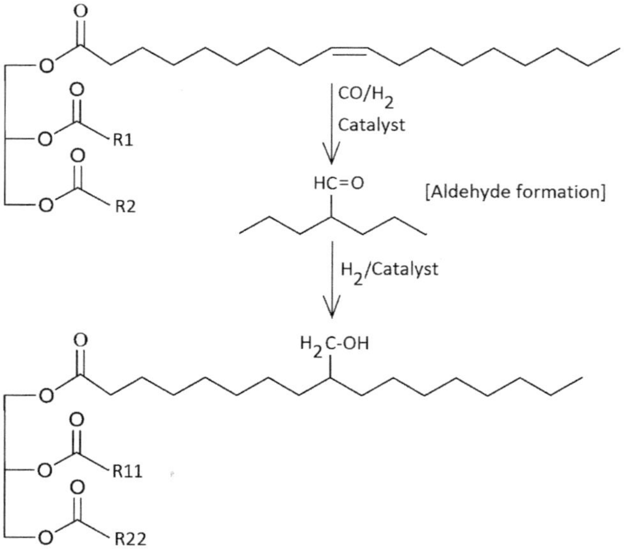

Figure 12.12 Hydroformylation of vegetable oil for the production of bio-polyols (R1, R2, R11, and R22 are fatty acid side chains of vegetable oils).

12.5.1.5 Metathesis of Vegetable Oil

Metathesis is an interesting reaction for the conversion of vegetable oils into polyols for polyurethanes [57]. Long-fatty acid chain diols can be obtained from vegetable oils in two steps – by a metathesis reaction followed by reduction, with ruthenium as the catalyst [57, 58]. The unsaturated part can be transformed into polyol through epoxidation-alcoholysis route. For example, metathesis reaction of trioleine (triester of glycerol with oleic acid) with ethylene being catalysed by ruthenium catalyst resulted in triglyceride with terminal double bonds, which then transformed into polyol by epoxidation, followed by alcoholysis with methanol. Metathesis of vegetable oils, such as olive oil, soybean oil, and linseed oil results in viscous high-molecular-weight oils, so-called stand oils, with outstanding drying properties due to their pronounced unsaturated character [59].

12.5.2 Commercial Bio-polyol for Polyurethane

Commercialisation of a wider range of higher-performing, cost-competitive bio-based polyols from both established polyol manufacturers and newer companies focused on the production of renewable materials will drive further growth of the sustainable polyurethane segment. The bio-polyol market is worth US$3,077 million for a consumption base of 1,104 kilotons by 2018, as per MarketsandMarkets. North America is currently the largest consumer of bio-polyols. Corn and soybean are primarily used as feedstock to produce bio-based polyols in North America. Market demand is driven from end-user industries, such as automotive, packaging, and furniture. The automotive industry in North America is the largest consumer of bio-polyols. Almost 70% vehicles manufactured by Ford Motor Company in North America contained seat components that use bio-polyols. Cargill Incorporated, DOW Chemical Company, Johnson Controls Incorporated, etc. are among the largest manufacturers in North America. Polyurethane rigid foams are currently the biggest application, with consumption expected to increase from 456.8 kilotons by 2018. Asia-Pacific is currently the fastest growing consumer of bio-polyol based polyurethane rigid foams.

Industrial groups, such as BASF, Cargill, Oleon, and Huntsman, have developed their own bio-based polyester polyols from vegetable oils, more than 20 different types of which are currently available commercially. The use of these molecules makes it possible to obtain polyurethanes with a bio-based content of 60–70%, by weight [4, 60].

Cargill is the first company to commercialize bio-based polyols on a large scale in the flexible polyurethane foam market. Cargill makes BiOH® polyols by converting the carbon-carbon double bonds in unsaturated vegetable oils to epoxide derivatives, and then further converting these derivatives to polyols using mild temperature and ambient pressure. These polyols provide excellent reactivity and high levels of incorporation, leading to high-performing polyurethane foams. These foams set a new standard for consistent quality with low odor and color. Foams containing these polyols retain their white color longer without ultraviolet stabilizers. They are also superior to foams containing only petroleum-based polyols in standard tests.

Croda's has developed a range of completely bio-based polyester polyols known as Priplast, to meet the growing market demand for high performance, renewable building blocks for coatings and adhesives. Priplast 3238 is a fully amorphous, di-functional polyol created especially for reactive polyurethane applications. Priplast 3293 is a semi-crystalline type for higher modulus and strength reactive polyurethane adhesives. Priplast 3286 is amorphous with a higher functionality for greater crosslink. These products can help to meet the positive environmental image of sportswear and consumer electronics, along with the renewability and lightweight targets in the transportation industry. They offer durability, UV and hydrolysis resistance, moisture protection barrier, and flexibility at low temperatures.

Dow Polyurethanes has developed bio-based polyol called RENUVA™ for adhesive and sealant applications including: construction, automotive, flexible packaging, footwear, and wood furniture. RENUVA™ dispels unwanted odors and addresses the needs of the industry by offering excellent performance and high levels of renewable content. According to life-cycle analysis, the manufacturing of bio-based polyols with RENUVA™ is greenhouse gas neutral and uses 60% fewer fossil fuel resources than the manufacturing of conventional polyols.

Vertellus Performance Materials produces castor oil derived polyols designed for polyurethane coating systems with the trade name Polycin® D, T, M and GR series (D for diols, T for triols, and

M for multi-functional polyols), with a molecular weight ranging from 370 to 3,500 g/mol and a hydroxyl content from 400 down to 35. The GR series is derived only from the building blocks of castor oil and results in a wide range of physical properties. The Polycin DTM series polyols are based on ricinoleic acid combined with glycerine.

Myriant's bio-succinic acid can be used to produce polyester polyols with renewable content up to 100%, which is comparable in performance to adipic acid polyols. Using succinic acid polyols offers formulators renewable content at a competitive, stable long term price. Myriant has produced a line of developmental polyester polyols.

Soy based polyol marketed as Agrol® has been developed by BioBased Technologies in 2005. Agrol® is available in a range of functionalities that can be used in a variety of polyurethane applications. The typical hydroxyl values are from 340 down to 70 mg KOH/kg and the typical molecular weight is between 560 and 2,070 g/mol. Soy polyols perform like their petrochemical counterparts and enable manufacturers to increase the sustainability of end products without sacrificing performance. In some cases, products are even enhanced with lighter weight, more strength, and better durability. It has multiple functionalities that enable manufacturers to formulate applications ranging from flexible to case to rigid foams.

Bayer Material Science has developed a soybean polyol system for polyurethane products, Baydur 730S. This soy-based polyurethane is used for flexible foam systems. In addition, Bayer announced its bio-based polyols for rigid foams, with 40–70% bio-content, high functionality (2–5), and molecular weights of 140–280 g/mol; these rigid foams may be used for building and refrigeration insulations [61].

Mitsui Chemicals has launched cost-competitive bio-polyols based on castor oil. Eighty percent of the world's castor oil is produced from non-edible plants, thus ensuring a stable feedstock supply. The new manufacturing plant, with an 8,000 tons per year capacity, started commercial production of bio-polyols in January 2016.

Huntsman has developed a new product Glycerin carbonate (GC) called JEFFADDTM having 73% bio-content, and is a unique hydroxyl-functional carbonate having carbonate and hydroxyl reactive sites. Each reactive site opens pathways to utilize GC in polyurethanes. One component polyurethane systems blocked with GC are fully polymerized when reacted with amines. This is an advantage to conventional blocked systems which release blocking group when polymerized. In addition to bio-content, the hydroxyl group in the hard segment enhances hydrogen bonding and improves strength of the cured material.

Jointly DuPont Tate & Lyle Bio Products and BioAmber have developed multiple bio-based polyols using bio-succinic acid and bio-based 1,3-propanediol via the fermentation of corn. These polyols are the building blocks that deliver high performance in a variety of polyurethane applications, from footwear and waterproof films to artificial leather and coating, adhesives, and elastomer applications.

Polyurethane company Merquinsa has commercialized renewable sourced thermoplastics urethane (TPU)–Pearlthane® and Pearlbond Eco, which are made from 40–95% vegetable oil and fatty acids. The TPUs can be used as adhesives, coatings, and injection specialties, especially for electronics and footwear. The performance of these bio-based TPUs is claimed to be equal to or in some cases even better than petrochemical TPUs [61].

12.6 NON-ISOCYANATE POLYURETHANE

The synthesis of non-isocyanate polyurethane (NIPU) has recently gained an increasing attention in chemical industry. Health and safety concerns associated with isocyanate chemistry motivate the search for non-phosgene routes for diisocyanate preparation [62, 63]. NIPU is based on the reaction of polycyclocarbonates and polyamines [64]. NIPU crosslinks are obtained by the reaction between polycyclocarbonate oligomers and aliphatic or cycloaliphatic polyamines with primary amino groups. The resulting materials are called polyhydroxyurethane (PHU), since they also contain hydroxyl groups throughout the macromolecular chain (Fig. 12.13). These hydroxyl groups thus create inter and intramolecular hydrogen bonds with urethane group, which confers higher chemical and physical strength to these PHU [65, 63].

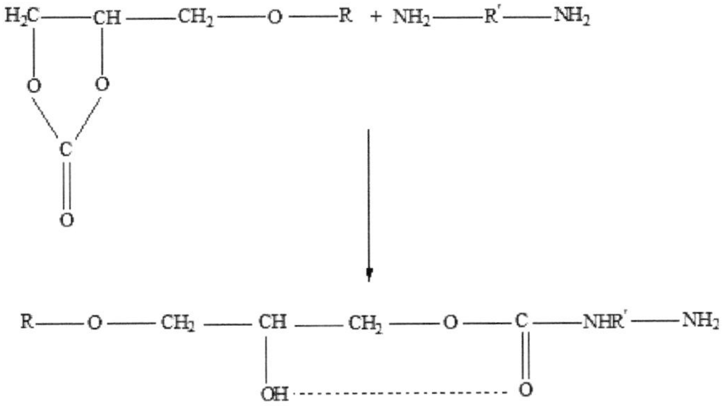

Figure 12.13 Polyhydroxy urethane (PHU) formation through reaction between cyclocarbonate and amine.

Cyclocarbonate oligomers can be obtained through different methodologies [63, 66], such as polymerization of unsaturated cyclocarbonate monomers, copolymerization of unsaturated cyclocarbonate monomers with vinyl ester monomers, reaction of oligomeric chlorohydrin ether with carbonate of alkaline metals, reaction of oligomeric polyols with an acid chloride of carbonic acid, and addition of carbon dioxide into cyclic ethers (epoxides or oxetanes) in the presence of catalyst. However, the most common and effective method is carbon dioxide/epoxides reaction system because of the availability of materials and simplicity of the process [62, 67]. A wide range of catalyst systems have been developed for CO_2 incorporation reactions for the synthesis of cyclocarbonates, such as alkali metal salts [68, 69], onium salt [70], metal complexes [71], ionic salts [72], organic amine [73], polymer based amino acids [74], and so on, under high pressure or mild conditions.

The development of bio-based isocyanates is also receiving a lot of attention because if achieved, then entirely bio-based polyurethanes could be obtained. The main approaches for bio-based synthesis of polyurethanes are: (i) the ring-opening of bifunctional five-membered cyclic carbonates, commonly synthesized by reaction of inexpensive and abundant CO_2 with oxiranes, such as epoxidized soybean oil or limonene, with diamines (Fig. 12.14), and (ii) a self-condensation approach of AB-type monomers, bearing hydroxyl groups with *in situ* formed isocyanates via Curtius rearrangement [75].

Bio-based non-isocyanate urethane was obtained by the reaction of a cyclocarbonate synthesized from a modified linseed oil and an alkylated phenolic polyamine from cashew nut shell liquid. The incorporation of functional cyclic carbonate groups to the triglyceride units of the oil was done by reacting epoxidized linseed oil with carbon dioxide in the presence of a catalyst [76]. Polycondensation of castor oil derived dimethyl dicarbamates and diols using 1,5,7-Triazabicyclo[4.4.0]dec-5-ene (TBD) as catalyst has been reported to be an efficient strategy to obtain non isocyanate-based and fully renewable polyurethanes [75]. In another work, bio-based polyurethane was prepared by reacting carbonated soybean oil with diamines [77]. Carbonated soybean oil was reacted with different diamines, such as 1,2-ethylenediamine, 1,4-butylenediamin, and 1,6-hexylenediamine at 70°C for 10 hours, and then for 3 hours at 100°C, and polyurethane was formed. In different processing conditions, bio-based polyurethanes were synthesized by reaction of carbonated soybean oil (CSBO), either with bio-based short diamines or amino-telechelic oligoamides derived from fatty acids to achieve, respectively, thermoset or thermoplastic NIPUs.

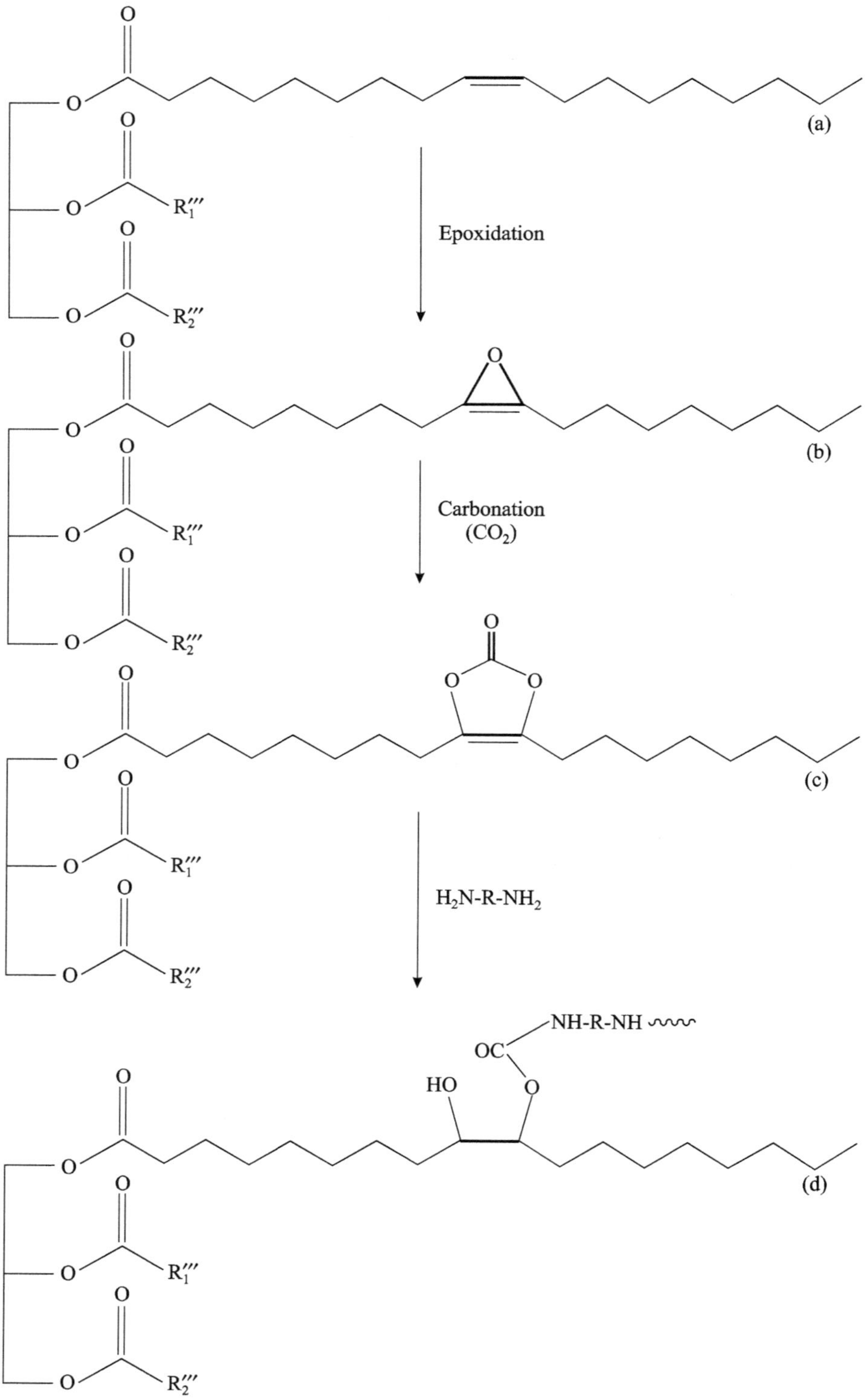

Figure 12.14 Schematic of reaction steps of formation of bio-based polyurethane from (a) vegetable oil, (b) epoxidized vegetable oil, (c) cyclocarbonated vegetable oil, and (d) non-isocyanate bio-polyurethane. R_1, R_2, R_1', R_2', R_1'', R_2'', R_1''', and R_2''' are fatty acid side chains of vegetable oils.

Carbonated vegetable oils were obtained by metal-free coupling reactions of CO_2 with epoxidized soybean oils under supercritical conditions (120°C, 100 bar). Soybean oil triglycerides were also brominated at the allylic positions with N-bromosuccinimide, followed by reacting with AgNCO to synthesize soybean triglyceride multi-isocyanates [78]. Reaction of epoxidized soybean oil with CO_2 at optimized conditions could result in CSBO in high conversion and low level of residual epoxy [79]. Catalyzed by tetrabutylammoniumbromide (5 mol% with respect to epoxy groups), 94% epoxidized soybean oil could be transformed to cyclocarbonated soybean oil [80]. In another case, CSBO with a conversion of 98.6% can be synthesized using $SnCl4.5H_2O$/ tetrabutylammonium bromide catalysts [81].

12.7 CONCLUSIONS

Polyurethanes, one of the most versatile and intensively used industrial materials, have been successfully employed in different applications, such as foams (flexible, semi-rigid, and rigid), elastomers, adhesives, coatings, fibres, thermosets, and thermoplastics. Preparation of polymeric materials, such as polyurethanes from renewable resources, is of great significance economically and ecologically. Polyurethane from renewable resources are gaining acceptance because of some attractive properties related to the specific structures of the vegetable oils and benefits for the environment protection and production sustainability. Vegetable oils, such as soybean oil, canola oil, palm oil, and castor oil are becoming extremely important as renewable resources for the preparation of polyols required for polyurethane manufacturing. Bio-based polyurethanes have been generally prepared by first converting the vegetable oils into polyols, followed by reaction with diisocyanates. The preparation of polyols from vegetable oils can be realized through several approaches.

Vegetable oils are abundant and inexpensive raw materials, offering some socio-economic advantages. Currently, vegetable oil-based polyurethanes are in high demand due to the depletion of petroleum reserves, the escalating price of petroleum products, and environmental concerns. Efforts to achieve a complete bio-based content include research into the development of bio-based isocyanates and vegetable oil-based polyols, for the production of polyurethanes with specific properties. However, bio-based polyurethanes may still have higher processing costs associated with the biomass feedstock. There is good potential for this situation to change over the next few years with advances in bio-processing technology.

References

1. Sharma, V. and P.P. Kundu. Condensation polymers from natural oils. Progress in Polymer Science 2008; 33: 1199-215.
2. Sharmin, E. and F. Zafar. Polyurethane: An introduction. *In*: F. Zafar and E. Sharmin (eds). Polyurethanes, InTech, Croatia, 2012, pp. 3-16.
3. Vilar, W.D. Chemistry and Technology of Polyurethanes, 3rd Ed., Vilar Consultoria Técnica Ltda, Lagoa, Rio de Janeiro, RJ, Brazil, 2002.
4. Nohra, B., L. Candy, J.-F. Blanco, C. Guerin, Y. Raoul and Z. Mouloungui. From petrochemical polyurethanes to biobased polyhydroxyurethanes. Macromolecules 2013; 46: 3771-3792.
5. Mosiewicki, M.A., G.A. Dell'Arciprete, M.I. Aranguren and N.E. Marcovich. Polyurethane foams obtained from castor oil-based polyol and filled with wood flour. Journal of Composite Materials 2009; 43: 3057-72.
6. Mispreuve, H.L.S.A., U. Tribelhorn and S. Waddington. Polyurethane foam for packaging applications. *In*: US5484820 A, 1996.
7. Szycher, M. Szycher's Handbook of Polyurethanes, CRC Press, 2012.

8. Guan, J., K.L. Fujimoto, M.S. Sacks and W.R. Wagner. Preparation and characterization of highly porous, biodegradable polyurethane scaffolds for soft tissue applications. Biomaterials 2005; 26: 3961-3971.
9. Fujimoto, K., M. Minato, S. Miyamoto, T. Kaneko, H. Kikuchi, K. Sakai, M. Okada and Y. Ikada. Porous polyurethane tubes as vascular graft. Journal of Applied Biomaterials 1993; 4: 347-354.
10. Santerre, J.P., K. Woodhouse, G. Laroche and R.S. Labow. Understanding the biodegradation of polyurethanes: From classical implants to tissue engineering materials. Biomaterials 2005; 26: 7457-70.
11. Chiu, H.-T., C.-Y. Chang, H.-W. Pan, T.-Y. Chiang, M.-T. Kuo and Y.-H. Wang. Characterization of polyurethane foam as heat seal coating in medical pouch packaging application. Journal of Polymer Research 2012; 19: 1-12.
12. Fornasieri, M., J.W. Alves, E.C. Muniz, A. Ruvolo-Filho, H. Otaguro, A.F. Rubira and G.M. de Carvalho. Synthesis and characterization of polyurethane composites of wood waste and polyols from chemically recycled pet. Composites Part A: Applied Science and Manufacturing 2011; 42: 189-195.
13. Ashori, A. and A. Nourbakhsh. Characteristics of wood–fiber plastic composites made of recycled materials. Waste Management 2009; 29: 1291-1295.
14. Gao, Z., D. Wu, W.Su and X. Ding. Polyurethane-solid wood composites prepared with various catalysts. I. Mechanical properties and dimensional stabilities. Journal of Applied Polymer Science 2009; 111: 1293-1299.
15. Carter, D.G., D.J. Green and M.C. Collins. Use of reactive hot melt adhesive for packaging applications. *In*: US5018337 A, 1991.
16. Schuler, N.L., J.D. Jenks, M.P. Lasecki and H. Donley. Polyurethane/polyurea elastomer coated steel battery box for hybrid electric vehicle applications. *In*: CA 2264412; 2001.
17. Gurunathan, T., C.R.K. Rao, R. Narayan and K.V.S.N. Raju. Polyurethane conductive blends and composites: synthesis and applications perspective. Journal of Materials Science, 2013; 48: 67-80.
18. Dieterich, D. and K. Uhlig. Polyurethanes. *In*: Ullmann's Encyclopedia of Industrial Chemistry. Wiley-VCH Verlag GmbH & Co., KGaA, 2000.
19. Ionescu, M. Chemistry and Technology of Polyols for Polyurethanes. Rapra Technology, Shrewsbury, UK, 2005.
20. Szycher, M. Szycher's Handbook of Polyurethanes, 2nd Ed. CRC Press, Boca Raton, U.K., 2013.
21. Chian, K.S. and L.H. Gan. Development of a rigid polyurethane foam from palm oil. Journal of Applied Polymer Science, 1998; 68: 509-515.
22. Petrović, Z.S. Polyurethanes from vegetable oils. Polymer Reviews 2008; 48: 109-155.
23. Shen, Li, J. Haufe and M.K. Patel. Product overview and market projection of emerging bio-based plastics. *In*: Universiteit Utrecht., 2009.
24. Alagi, P. and S.C. Hong. Vegetable oil-based polyols for sustainable polyurethanes. Macromolecular Research 2015; 23: 1079-1086.
25. Zlatanić, A., C. Lava, W. Zhang and Z.S. Petrović. Effect of structure on properties of polyols and polyurethanes based on different vegetable oils. Journal of Polymer Science Part B: Polymer Physics 2004; 42: 809-819.
26. Clark, A. Low-cost synthesis and evaluation of polymers prepared from oilseed rape and Euphorbia Lagas-cae oils. *In*: University of Warwick, 2001.
27. Schäfer, H.J. 1997. 'Fatty Acid and Lipid Chemistry. Edited by F. Gunstone. 252 pages, Blackie Academic & Professional, 1996. Lipid/Fett, 99: 449-449.
28. Ni, B., L. Yang, C. Wang, L. Wang and D.E. Finlow. Synthesis and thermal properties of soybean oil-based waterborne polyurethane coatings. Journal of Thermal Analysis and Calorimetry 2010; 100: 239-246.
29. Campanella, A., L.M. Bonnaillie and R.P. Wool. Polyurethane foams from soyoil-based polyols. Journal of Applied Polymer Science 2009; 112: 2567-2578.
30. Zhang, C., Y. Xia, R. Chen, S. Huh, P.A. Johnston and M.R. Kessler. Soy-castor oil based polyols prepared using a solvent-free and catalyst-free method and polyurethanes therefrom. Green Chemistry 2013; 15: 1477-1484.
31. Chuayjuljit, S., A. Maungchareon and O. Saravari. Preparation and properties of palm oil-based rigid polyurethane nanocomposite foams. Journal of Reinforced Plastics and Composites 2010; 29: 218-225.

32. Kong, X. and S.S. Narine. Physical properties of polyurethane plastic sheets produced from polyols from canola oil. Biomacromolecules 2007; 8: 2203-2209.
33. Palaskar, D.V., A. Boyer, E. Cloutet, J.-F. Le Meins, B. Gadenne, C. Alfos, C. Farcet and H. Cramail. Original diols from sunflower and ricin oils: Synthesis, characterization, and use as polyurethane building blocks. Journal of Polymer Science Part A: Polymer Chemistry 2012; 50: 1766-1782.
34. Sharma, V., J.S. Banait and P.P. Kundu. Swelling kinetics of linseed oil-based nanocomposites. Journal of Applied Polymer Science 2009; 114: 446-456.
35. Pfister, D.P., Y. Xia and R.C. Larock. Recent advances in vegetable oil-based polyurethanes. ChemSusChem 2011; 4: 703-717.
36. Li, Y., X. Luo and S. Hu. Polyols and polyurethanes from vegetable oils and their derivatives. *In*: Bio-based Polyols and Polyurethanes, Chapter 2. Springer International Publishing, 2015.
37. Mutlu, H. and M.A.R. Meier. Castor oil as a renewable resource for the chemical industry. European Journal of Lipid Science and Technology 2010; 112: 10-30.
38. Thakur, S. and N. Karak. Bio-based tough hyperbranched polyurethane-graphene oxide nanocomposites as advanced shape memory materials. RSC Advances 2013; 3: 9476-9482.
39. Gunstone, F.D. Chemical reactions of fatty acids with special reference to the carboxyl group. European Journal of Lipid Science and Technology 2001; 103: 307-314.
40. Corma, A., S. Iborra and A. Velty. Chemical routes for the transformation of biomass into chemicals', Chemical Reviews 2007; 107: 2411-2502.
41. Abdullah, B.M. and J. Salimon. Epoxidation of vegetable oils and fatty acids: catalysts, methods and advantages. Journal of Applied Sciences 2010; 10: 1545-1553.
42. Eychenne, V., Z. Mouloungui and A. Gaset. Thermal behavior of neopentylpolyol esters: Comparison between determination by TGA-DTA and flash point. Thermochimica Acta 1998; 320: 201-208.
43. Guo, A., Y. Cho and Z.S. Petrović. Structure and properties of halogenated and nonhalogenated soy-based polyols. Journal of Polymer Science Part A: Polymer Chemistry 2000; 38: 3900-3910.
44. Caillol, S., M. Desroches, G. Boutevin, C. Loubat, R. Auvergne and B. Boutevin. Synthesis of new polyester polyols from epoxidized vegetable oils and biobased acids. European Journal of Lipid Science and Technology 2012; 114: 1447-1459.
45. Hilker, I., D. Bothe, J. Prüss and H.J. Warnecke. Chemo-enzymatic epoxidation of unsaturated plant oils. Chemical Engineering Science 2001; 56: 427-432. https://polyurethane.americanchemistry.com/Applications/.
46. Du, G., A. Tekin, E.G. Hammond and L.K. Wood. Catalytic epoxidation of methyl linoleate. Journal of the American Oil Chemists' Society 2004; 81(5): 477-480.
47. Campanella, A., M.A. Baltanas, M.C. Capel-Sanchez, J.M. Campos-Martin and J.L.G. Fierro. Soybean oil epoxidation with hydrogen peroxide using an amorphous Ti/SiO_2 catalyst. Green Chemistry 2004; 6: 330-334.
48. Orellana-Coca, C., J.M. Billakanti, B. Mattiasson and R. Hatti-Kaul. Lipase mediated simultaneous esterification and epoxidation of oleic acid for the production of alkylepoxystearates. Journal of Molecular Catalysis B: Enzymatic 2007; 44: 133-137.
49. Petrović, Z.S., W. Zhang and I. Javni. Structure and properties of polyurethanes prepared from triglyceride polyols by ozonolysis. Biomacromolecules 2005; 6: 713-719.
50. Narine, S.S., X. Kong, L. Bouzidi and P. Sporns. Physical properties of polyurethanes produced from polyols from seed oils: I. Elastomers. Journal of the American Oil Chemists' Society 2007; 84: 55-63.
51. Narayan, R., D. Graiver, K.W. Farminer, P.T. Tran and T. Tran. Novel modified fatty acid esters and method of preparation thereof. *In*: US20100084603 A1, 2010.
52. Raquez, J.M., M. Deléglise, M.F. Lacrampe and P. Krawczak. Thermosetting (bio)materials derived from renewable resources: A critical review. Progress in Polymer Science 2010; 35: 487-509.
53. Dahlke, B., S. Hellbardt, M. Paetow and W.H. Zech. Polyhydroxy fatty acids and their derivatives from plant oils. Journal of the American Oil Chemists' Society 1995; 72: 349-353.
54. Guo, A., D. Demydov, W. Zhang and Z.S. Petrovic. Polyols and polyurethanes from hydroformylation of soybean oil. Journal of Polymers and the Environment 2002; 10: 49-52.

55. Petrović, Z.S., A. Guo, I. Javni, I. Cvetković and D.P. Hong. Polyurethane networks from polyols obtained by hydroformylation of soybean oil. Polymer International 2008; 57: 275-281.
56. Guo, A., W. Zhang and Z.S. Petrovic. Structure–property relationships in polyurethanes derived from soybean oil. Journal of Materials Science 2006; 41: 4914-4920.
57. Mol, J.C. and R. Buffon. Metathesis in oleochemistry. Journal of the Brazilian Chemical Society 1998; 9: 1-11.
58. Mol, J.C. Metathesis of unsaturated fatty acid esters and fatty oils. Journal of Molecular Catalysis 1994; 90: 185-199.
59. Boelhouwer, C. and J.C. Mol. Metathesis reactions of fatty acid esters. Progress in Lipid Research 1985; 24: 243-267.
60. Desroches, M., M. Escouvois, R. Auvergne, S. Caillol and B. Boutevin. from vegetable oils to polyurethanes: synthetic routes to polyols and main industrial products. Polymer Reviews 2012; 52: 38-79.
61. Sherman, L.M. Polyurethanes: Bio-based Materials Capture Attention. *In*: Plastics Technology, 2007.
62. Camara, F., S. Benyahya, V. Besse, G. Boutevin, R. Auvergne, B. Boutevin and S. Caillol. Reactivity of secondary amines for the synthesis of non-isocyanate polyurethanes. European Polymer Journal 2014; 55: 17-26.
63. Guan, J., Y. Song, Y. Lin, X. Yin, M. Zuo, Y. Zhao, X. Tao and Q. Zheng. Progress in study of non-isocyanate polyurethane. Industrial & Engineering Chemistry Research 2011; 50: 6517-6527.
64. Figovsky, O., L. Shapovalov, A. Leykin and R. Potashnikova. Recent advances in the development of non-isocyanate polyurethanes based on cyclic carbonates. *In*: PU Magazine, 2013.
65. Cornille, A., J. Serres, G. Michaud, F. Simon, S. Fouquay, B. Boutevin and S. Caillol. Syntheses of epoxyurethane polymers from isocyanate free oligo-polyhydroxyurethane. European Polymer Journal 2016; 75: 175-189.
66. Webster, D.C. and A.L. Crain. Synthesis and applications of cyclic carbonate functional polymers in thermosetting coatings. Progress in Organic Coatings 2000; 40: 275-282.
67. Besse, V., F. Camara, C. Voirin, R. Auvergne, S. Caillol and B. Boutevin. Synthesis and applications of unsaturated cyclocarbonates. Polymer Chemistry 2013; 4: 4545-4561.
68. Liang, S., H. Liu, T. Jiang, J. Song, G. Yang and B. Han. Highly efficient synthesis of cyclic carbonates from CO_2 and epoxides over cellulose/KI. Chemical Communications 2011; 47: 2131-2133.
69. Wang, J.-Q., J. Sun, C.-Y. Shi, W.-G. Cheng, X.-P. Zhang and S.-J. Zhang. Synthesis of dimethyl carbonate from CO_2 and ethylene oxide catalyzed by K_2CO_3-based binary salts in the presence of H_2O. Green Chemistry 2011; 13: 3213-3217.
70. Wang, J.-Q., D.-L. Kong, J.-Y. Chen, F. Cai and L.-N. He. Synthesis of cyclic carbonates from epoxides and carbon dioxide over silica-supported quaternary ammonium salts under supercritical conditions. Journal of Molecular Catalysis A: Chemical 2006; 249: 143-148.
71. Buchard, A., M.R. Kember, K.G. Sandeman and C.K. Williams. A bimetallic iron(iii) catalyst for CO_2/epoxide coupling. Chemical Communications 2011; 47: 212-214.
72. Foltran, S., J. Alsarraf, F. Robert, Y. Landais, E. Cloutet, H. Cramail and T. Tassaing. On the chemical fixation of supercritical carbon dioxide with epoxides catalyzed by ionic salts: An *in situ* FTIR and Raman study. Catalysis Science & Technology 2013; 3: 1046-1055.
73. Kerry Yu, K.M., I. Curcic, J. Gabriel, H. Morganstewart and S.C. Tsang. Catalytic coupling of CO_2 with epoxide over supported and unsupported amines. The Journal of Physical Chemistry A 2010; 114(11): 3863-3872.
74. Qi, C., J. Ye, W. Zeng and H. Jiang. Polystyrene-supported amino acids as efficient catalyst for chemical fixation of carbon dioxide. Advanced Synthesis & Catalysis 2010; 352: 1925-1933.
75. Unverferth, M., O. Kreye, A. Prohammer and M.A.R. Meier. Renewable non-isocyanate based thermoplastic polyurethanes via polycondensation of dimethyl carbamate monomers with diols. Macromolecular Rapid Communications 2013; 34: 1569-1574.
76. Mahendran, A.R., N. Aust, G. Wuzella, U. Müller and A. Kandelbauer. Bio-based non-isocyanate urethane derived from plant oil. Journal of Polymers and the Environment 2012; 20: 926-931.
77. Javni, I., D.P. Hong and Z.S. Petrović. Soy-based polyurethanes by nonisocyanate route. Journal of Applied Polymer Science 2008; 108: 3867-3875.

78. Çaylı, G. and S. Küsefoğlu. Biobased polyisocyanates from plant oil triglycerides: Synthesis, polymerization, and characterization. Journal of Applied Polymer Science 2008; 109: 2948-2955.

79. Tamami, B., S. Sohn and G.L. Wilkes. Incorporation of carbon dioxide into soybean oil and subsequent preparation and studies of nonisocyanate polyurethane networks. Journal of Applied Polymer Science 2004; 92: 883-891.

80. Wilkes, G.L., S. Sohn and B. Tamami. Nonisocyanate polyurethane materials, and their preparation from epoxidized soybean oils and related epoxidized vegetable oils, incorporation of carbon dioxide into soybean oil, and carbonation of vegetable oils. *In*: US7045577 B2, 2004.

81. Li, Z., Y. Zhao, S. Yan, X. Wang, M. Kang, J. Wang and H. Xiang. Catalytic synthesis of carbonated soybean oil. Catalysis Letters 2008; 123(3-4): 246-251.

Chapter 13

Keratin-based Bioplastic from Chicken Feathers
Synthesis, Properties, and Applications

A. Gupta*, B.Y. Alashwal, Md. S. Bala and N. Ramakrishnan

Faculty of Chemical and Natural Resources Engineering, Universiti Malaysia Pahang, Gambang 26300, Kuantan, Malaysia

13.1 INTRODUCTION

Since their introduction in the late 1970s, plastic carrier bags have become a common element in today's life [1]. Bioplastics, made from renewable resources, are bio-based and biodegradable. They are harmless to the environment [2]. Chicken feathers are available abundantly as waste, and keratin protein extracted from it can be regenerated into various forms for biotechnological applications, such as films, sponges, fibers, alone or blended with other natural or synthetic polymers [3].

Protein-based biomaterials have become the interest among researchers for many biomedical and biotechnological applications due to their ability to function as a synthetic extracellular matrix that facilitates interface interactions. Several proteins have been investigated in the formulation of naturally-derived biomaterials, including collagen, albumin, gelatin, fibroin, and keratin. Keratin-based materials showed potential applications in the biomaterial field due to their intrinsic biocompatibility, biodegradability, mechanical durability, and natural abundance. In addition, the chicken feathers consist of about 90% of keratin protein. Chicken feathers are a by-product which are available in huge amounts, but only used in a small portion. Furthermore, chicken feathers are a very inconvenient and troublesome waste product of the poultry industry.

According to Barone and Schmidt [4], the main chemical structure of chicken feather is keratin protein. Since the chicken feather fiber is mainly made of the structural protein keratin, its chemical durability is primarily determined by keratin. Based on research, chicken feathers could potentially be used for protein fiber production [4]. Keratin protein is contained in chicken feathers and has some advantages in comparison with other proteins. First, the fibrous feather keratin can stretch approximately 6% at breaking, unlike hair keratin that can stretch twice its length. Feather

**For Corresponding*: Email: arun@ump.edu.my

keratin is also a special protein. It has a high content of cysteine (7%) in the amino acid sequence and cysteine has –SH groups that cause sulfur-sulfur (disulfide) bonding. This high content of cysteine makes keratin a stable protein by forming a network structure through joining adjacent polypeptides by disulfide cross-links. The feather keratin fiber is semi-crystalline and made up of crystalline fiber phase and amorphous protein matrix phase linked to each other. Moreover, the feather fiber shows good durability and resistance to degradation. This is because keratin has extensive cross-linking and strong covalent bonding within its structure [5]. Apart from that, chicken feathers are available abundantly as a poultry waste. Thus, the depletion of petroleum resources and increasing environmental awareness has triggered development of next generation materials that are environment-friendly and/or available resourcefully to meet the increasing demand for plastics. It is important for the global community to have an alternative for the product derived from petroleum oil, such as bioplastics. The possible applications of keratin bioplastic in a variety of technologies and products can result in their sustainable conversion into high-value cost-effective materials for converting the chicken feather waste into useful products using less quantity, humanity safe, and eco-friendly chemicals.

13.2 PRE-TREATMENT METHOD OF THE CHICKEN FEATHERS

There are some studies which used different methods for cleaning chicken feathers.

Schrooyen and other researchers [6, 7] used petroleum ether for the removal of grease from chicken feathers. The cleaning process was continued in the Soxhlet apparatus for 12 hours in a temperature range of 40–60°C [6, 8]. Afterwards, feathers were washed many times with water mixed with the laundry detergent and sodium chlorite for the removal of manure, blood, extra mass, or body parts. Successively, clean feathers were spread on galvanized iron sheets and dried under the sun for three days. Al-Musallam used chicken feathers in his study and cleaned them initially with tap water to remove debris, which was followed by water having 0.1% Tween 80 (v/v). They were then washed, and afterwards, feathers were cut and defatted by soaking them in diethyl ether (99.7%) for 24 hours. Excess diethyl ether was washed out with distilled water [9]. Rad et al. reported a similar method, but for the removal of impurities, chicken feathers were washed with hot water mixed with detergent for 60 minutes, with the last step repeated twice [7]. Some other researchers washed with water, filtered, dried, and used in pieces or particle form [10, 11]. In the same way, Flores-Hernández washed the chicken feathers with water and used ethanol for cleaning to make them white, odor free, and sanitized [12, 13].

Chicken feathers are collected from chicken processing plants. The chicken feathers are soaked in ether for 24 hours to remove any oils, grease, or stains. Soap water will then be used to wash the feathers before proceeding to dry them under sunlight. After the feathers are dry, they are cut up into smaller fragments by blending before keeping them carefully in well-closed containers or plastics at room temperature.

13.3 EXTRACTION METHOD OF KERATIN

For the past several years, a number of methods have been developed to extract keratin using oxidative and reductive chemical processes. Some researchers have attempted to hydrolyze keratin in aprotic ionic liquids, but they are expensive [14, 15]. Wang and Cao [15] extracted keratin from the feathers by using hydrophobic ionic liquids. Ji et al. [46] used the ionic liquid to dissolve feathers to obtain the keratin. Chicken feather-based particles were prepared by dissolving and regenerating waste feathers in an ionic liquid of 1-butyl-3-methylimidazolium chloride [(BMIM)Cl]. There are some other methods of dissolution of feathers, such as steam explosion [16, 17], and the Shindai method [18]. They use detergents for washing, and protease enzyme for the removal of proteins and bacteria medium. Among them, some researches involve the use of reducing agents (dithiothreitol and 2-mercaptoethanol) [19, 20] which can cleave the disulfide bonds of the keratin. However, the most frequent and familiar method is the use of alkaline solution as the reducing agent. The present work will use this method.

Fifty gram of the chicken feathers is taken in a beaker and mixed with 1000 ml of 1 N sodium hydroxide solution. The solution is heated and maintained at 60°C for 6 hours with continuous stirring. The solution will then be filtered and centrifuged at 10,000 rpm for a period of 10 minutes. To make it particle free, the supernatant liquid will be carefully collected and filtered using filter paper. Furthermore, 700 g of ammonium sulfate is used to dissolve in deionized water. Then the solution is kept stirring until all the ammonium sulfate particles dissolve. To make it particle free, the solution is then filtered [21].

13.4 PROTEIN PRECIPITATION AND PURIFICATION

The filtrate solution that had been previously collected will be placed in a beaker and stirred. 2 N hydrochloric acid is added drop-wise to the solution to make the pH neutral. The solution is again centrifuged at 10,000 rpm for 5 minutes and filtered using cellulose filter paper (Grade 1, Whatman) to ensure it is free of particles. The solid particles collected are washed by adding them to 100 ml deionized water and stirring. The solution is then centrifuged at 10,000 rpm for 5 minutes and the solids are collected carefully. The collected solid particles are then dissolved in 100 ml of 2 M sodium hydroxide solution. The solution is then centrifuged again at 10,000 rpm for 5 minutes, and all the liquids are collected carefully and stored, while the solids are discarded. The precipitating, washing, and dissolving steps are repeated 3 times [7].

13.5 POLYMERIZATION PROCESS AND PRODUCTION

The preparation of protein films from keratin extracted from wool and human hair has been used for several years. A variety of studies has been devoted to the improvement of the properties of keratin films by blending keratin with natural [8, 22] and synthetic polymers [23–25]. Yao Dou and his group have used polyvinyl alcohol (PVA) as the coalescent agent for the preparation of feather keratin-based films (Keratin/Polyvinyl Alcohol blend films cross-linked by Dialdehyde Starch) [25].

Yamauchi et al. [26] were among the first to begin to investigate the properties of products made from extracted wool keratins, and in doing so, described the physiochemical and biodegradational properties of solvent-cast keratin films. Tonin et al. [16] discovered the relationship between poly (ethylene oxide) (PEO) and keratin blended films in order to develop a keratin-based material with improved structural properties. Although pure keratin films are too fragile for practical use, the addition of glycerol resulted in a transparent, relatively strong, flexible, and biodegradable film [26].

In recent work, a 60 ml of keratin solution was poured into a beaker and three various concentrations of glycerol (2 w%, 5 w%, and 10 w%) were added. The two components were mixed continuously under constant magnetic stirring by heating at 60°C for 5 hours. The mixture was poured into an aluminum weighing boat and dried in the oven at 60°C for 24 hours.

13.6 PHYSICAL AND CHEMICAL PROPERTIES

13.6.1 Fourier Transform Infrared Spectroscopy (FTIR)

FTIR investigation can be used as an effective tool to assess the structural changes in proteins [27]. Particularly, there is amide I band in the range between 1600 and 1700 cm^{-1} and amide II band in the region of 1510 and 1580 cm^{-1} which provide useful information. Amide I, which is the most intense absorption band in proteins, is useful for the analysis of the protein secondary structure and arises mainly from C=O stretching, with a minor contribution from C−N stretching. The amide II band originates from the N−H bending and C−H stretching vibrations [28]. FTIR analysis of bioplastic synthesized from keratin is shown in Fig. 13.1. FTIR analysis of keratin is shown in Fig. 13.2. It is clearly seen that similar peaks found for keratin are also found for all three bioplastics. Therefore, bioplastic formed is almost totally from keratin and some weight percent of glycerol.

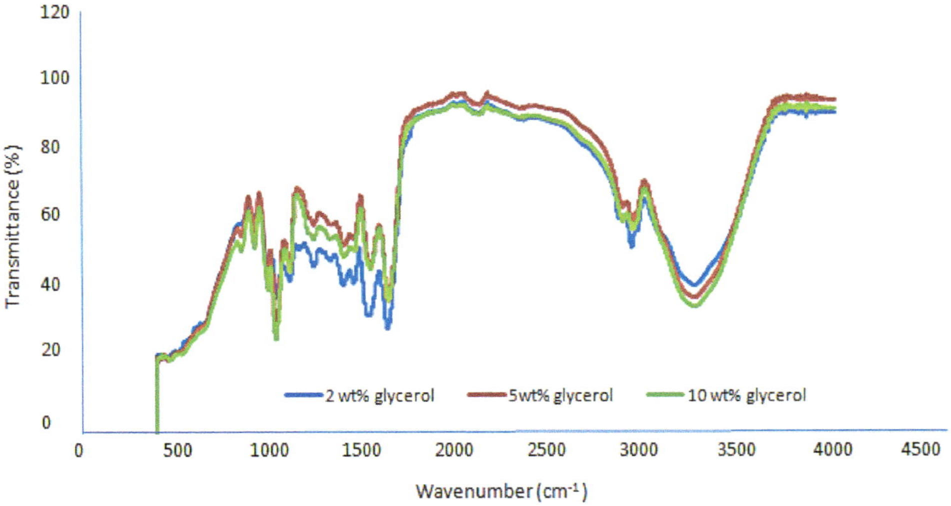

Figure 13.1 FT-IR Graph of bioplastics at different concentrations of glycerol.

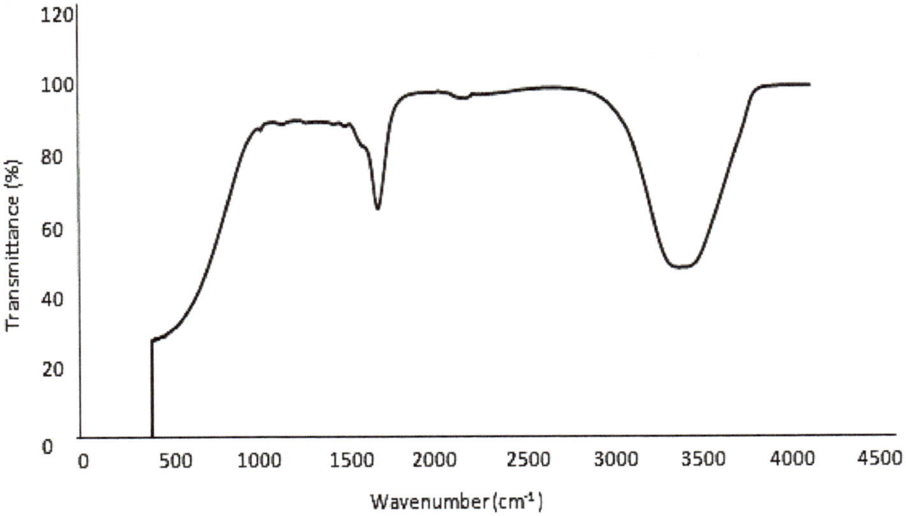

Figure 13.2 FT-IR graph of keratin, which is used as base material.

In the presence of glycerol, a shift peak towards higher wavenumber values was observed, which indicates a decreased *β*-sheet interaction with glycerol and the promotion of disordered structures. Similar results have been reported in previous investigations when glycerol was used to plasticize in pea, cereal sorghum, maize, and wheat proteins [29]. The amide II band is related with N–H bending and C–H stretching vibrations. Although it is much less conformationally sensitive than amide I, it is much more sensitive to the environment of the N–H group [30]. Therefore, the amide II band can be used to deduce changes to the environment of the N–H groups and respond to differences in hydrogen-bonding environments [31]. In general, stronger hydrogen-bonded N–H groups absorb at higher frequencies. As compared to the neat quill, decrease in absorption intensity centered at 1515 cm^{-1} can be seen in the presence of glycerol plasticizer. However, the relative intensity at 1540 cm^{-1} increases, and this increase is more prominent in the presence of glycerol. Glycerol, containing more hydroxyl groups, can form a higher number of hydrogen bonds, thus increasing the amount of hydrogen bonded peptide groups. Therefore, it was observed at a higher wavenumber.

The spectrum of cluster peaks below 3000 cm^{-1} down to around 2850 cm^{-1} are the peaks from the stretching of sp^3-hybridised C–H bonds. The alcohol O–H group peak is around or about 3300 cm^{-1}. A sharp peak right around 1700 cm^{-1} could possibly mean that there is a carboxylic acids group in the mixture. Nevertheless, the criteria of the peaks meet the criteria of the existence of an ester group, which are carbonyl peak around 1740 cm^{-1}, alkyl peaks below 3000 cm^{-1}, and a very strong peak around 1200 cm^{-1} for esters, which represent the C–O single bond vibration [32].

13.6.2 X-ray Diffraction (XRD)

Figure 13.3 shows comparison of XRD analysis with different concentrations of glycerol. The graph shows peaks which indicate the strong crystalline characteristics. The shift to higher angles indicates a decrease in the corresponding interlayer spacing, which means that the blend component has an ordered structure. The increase of d-spacing shows that this blend has a less ordered structure, and thus crystallization becomes more difficult [33]. It is shown in Fig. 13.3 that bioplastic film synthesized with 2 wt% of glycerol has the highest 2θ values and the lowest d-spacing, which means it has the most ordered structure among the bioplastic film. It may be due to the lower glycerol content in the mixtures and thus dispersed better [34]. Besides, bioplastic film synthesized with 5 wt% and 10 wt% glycerol showed almost similar results. The bioplastic film with 10 wt% glycerol showed the lowest d-spacing compared with other glycerol compositions. It can be seen from the figure that strong crystalline characteristics are shown by the bioplastic film, which is made up of the lowest glycerol amount.

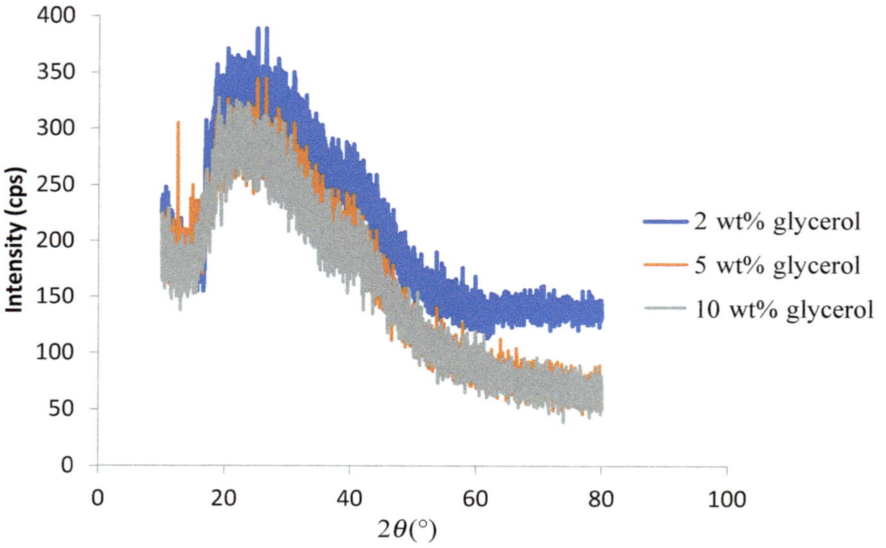

Figure 13.3 Comparison of XRD analysis with different concentrations of glycerol.

13.6.3 Scanning Electron Microscope (SEM)

Figure 13.4 shows the scanning of SEM analysis of bioplastics film, (a) 2 wt% of glycerol, (b) 5 wt% of glycerol, and (c) 10 wt% of glycerol. Figure 13.4(a) shows very good compatible morphologies without edge, cavity, and holes. This phenomenon shows a good adhesion between the components with a diffused polymer-plasticizer interface, which is attributed to the occurrence of chemical interactions between keratin and glycerol [27]. Thus, the glycerol was well-dispersed to form a homogeneous matrix with evident signs of plasticization in the keratin matrix, without separation at the interface producing single phase morphology [35].

SEM analysis of bioplastic film with 2 wt% glycerol [Fig. 13.4(b)] and with 10 wt% glycerol [Fig. 13.4(c)] shows the presence of undissolved particles. As compared to 5 wt% glycerol bioplastic film [Fig. 13.4(b)], 10 wt% glycerol bioplastic film [Fig. 13.4(c)] shows more voids and undissolved particles. The presence of empty microvoids may be due to the fact that the degree of dispersion of the plasticizer in the polymer matrix is better at lower plasticizer content, and the tendency to form empty voids and phase separation increases when the glycerol content increases [8].

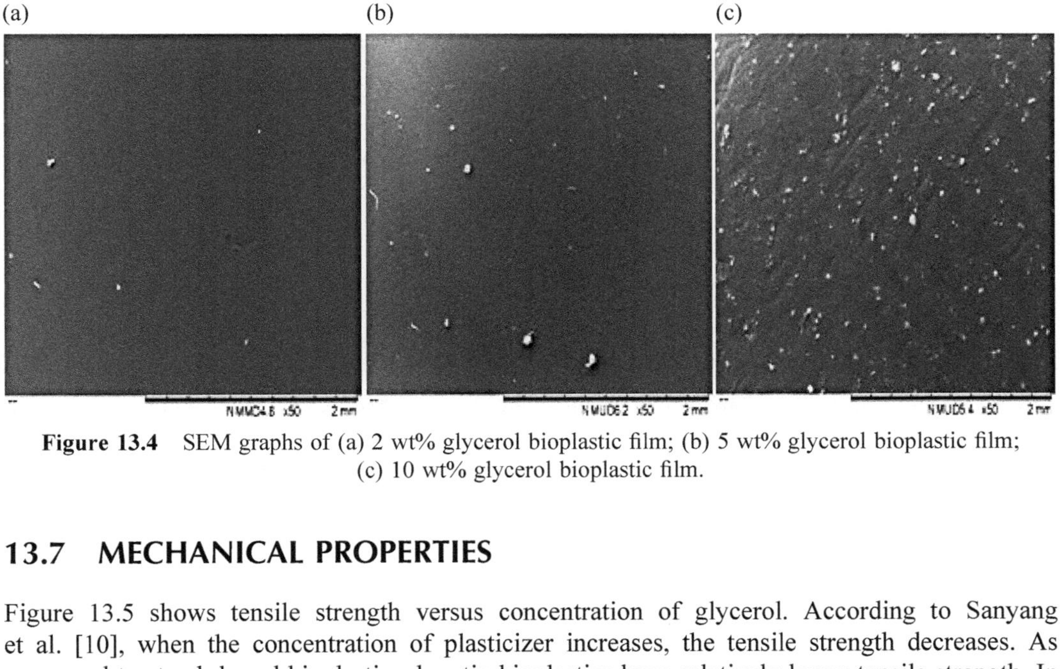

Figure 13.4 SEM graphs of (a) 2 wt% glycerol bioplastic film; (b) 5 wt% glycerol bioplastic film; (c) 10 wt% glycerol bioplastic film.

13.7 MECHANICAL PROPERTIES

Figure 13.5 shows tensile strength versus concentration of glycerol. According to Sanyang et al. [10], when the concentration of plasticizer increases, the tensile strength decreases. As compared to starch-based bioplastics, keratin bioplastics have relatively lower tensile strength. In contrast, the tensile strength value for 15 wt% of glycerol-plasticized sugar palm starch film shows 9.59 MPa, while for 10 wt% glycerol concentration keratin bioplastic film is just 0.0409 MPa [36]. When compared to citric acid crosslinked films, citric acid crosslinked feather keratin films have a much higher tensile strength of 230 MPa [22]. Tensile strengths of keratin based plastics films in this work generally have a lower tensile strength value as compared to other biodegradable plastics, such as starch-based and citric acid crosslinked plastics film.

Figure 13.6 shows the graph of Young's Modulus versus concentration of glycerol. It is shown that with the increase of glycerol concentration, the Young's Modulus decreases. As compared to starch-based bioplastics, keratin bioplastics have lower tensile modulus, while for keratin based film with 10 wt% glycerol concentration, 2 wt% is 7.5593 MPa [36].

Young's Modulus is the function of stress and strain. Consequently, strain is the function of elongation at break, therefore it is validated that the Young's Modulus has an inverse behavior. Young's Modulus of keratin-based plastics produced in this work films generally has lower values as compared to other biodegradable plastics, such as starch-based and citric acid crosslinked plastics film [35].

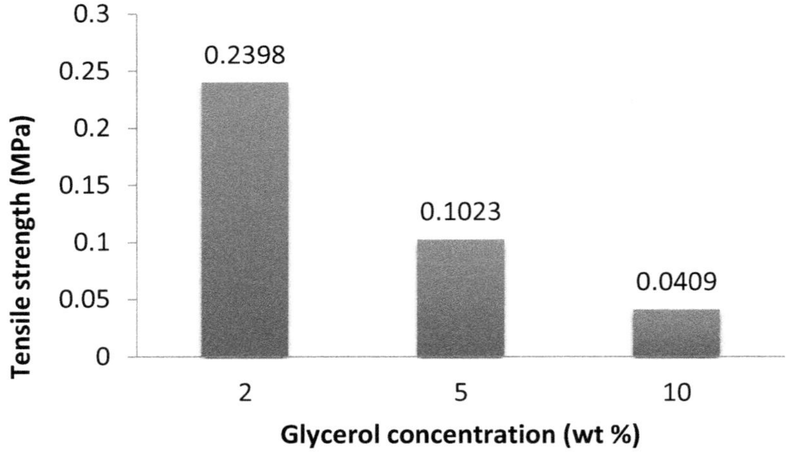

Figure 13.5 Graph of tensile strength at different concentrations of glycerol.

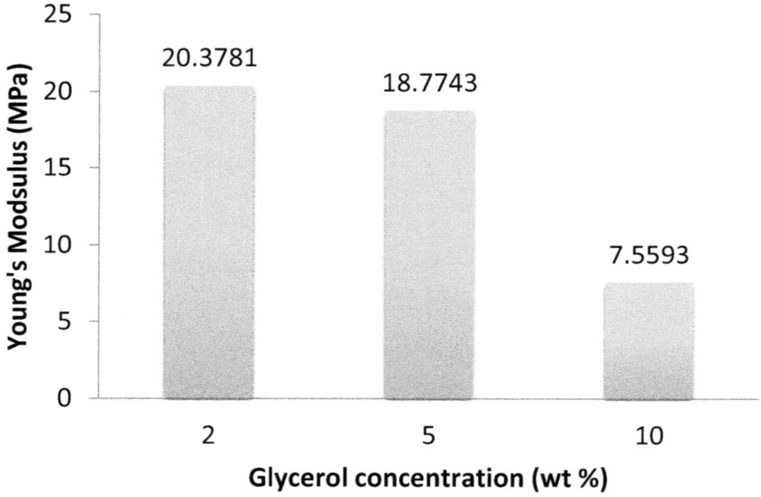

Figure 13.6 Graph of Young's Modulus at different concentrations of glycerol.

As the concentration of glycerol increases, tensile strength and Young's Modulus of keratin bioplastic films decreases. These behaviors agree with other studies on the effects of glycerol concentration on protein-based films: wheat gluten and soy proteins [23], peanut proteins [37], and fish muscle proteins [24], among others.

13.8 THERMAL PROPERTIES

13.8.1 Thermogravimetric Analysis (TGA)

It is shown that the bioplastics synthesized from keratin with different glycerol concentrations have the same trend of weight loss. These degradation trends are almost identical to the research done where there are two step weight losses for plasticized film [38].

Two step weight loss steps can be seen in case of quill material only. The weight loss in the first stage (near 100°C) for the quill was due to the evaporation of residual water, whereas the second step (between 250 and 600°C) was mainly due to the degradation of the quill keratin. The degradation of each plasticized resin consisted of three weight loss steps. The first gradual weight loss (below 150°C) is due to the evaporation of moisture, the second (between 150 and 250°C) is attributed to the plasticizer evaporation, and the final weight loss (by higher than 250°C) is due to the decomposition of quill material. Plasticizers act as reducing hydrogen bonding, van der Waals, or ionic interactions that hold polymer chains together, by forming plasticizer-polymer interactions. Plasticizers allow additional free volume to the system, and/or by causing a physical separation of adjacent chains and by acting as lubricants between chains [38]. There is a research that is almost similar to current study which shows that starch-based films have three thermal degradation events, where the initial stage started at a temperature less than 100°C, the second stage happened ~125°C to 290°C, and the third staged happened at heating beyond 290°C, with the Ton set occurring around 300°C. Therefore, bioplastics, synthesized from keratin with glycerol as plasticizer, have thermal stabilities similar to starch-based films.

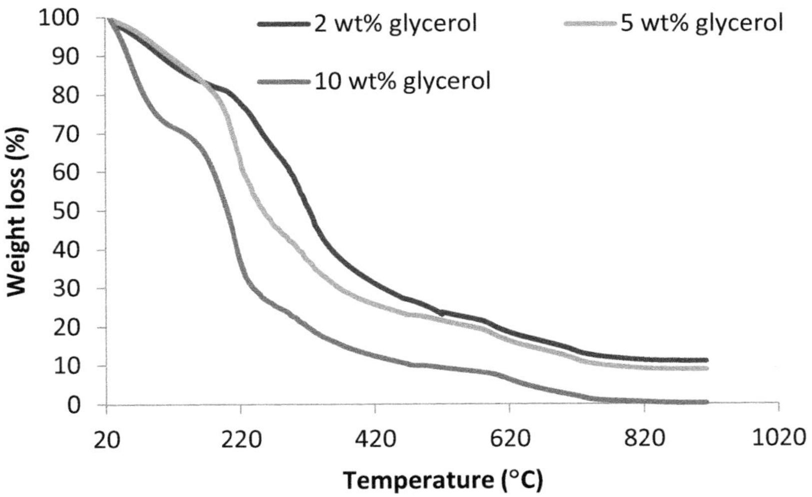

Figure 13.7 TGA Curves at different concentrations of glycerol.

The presence of glycerol, which are small molecules that are dispersed homogeneously in the PLA polymer, acts as a barrier sheet to prevent oxidation, prevent the permeability of volatile degradation out from the blend materials, and helps delay the thermal degradation process, and this proves that glycerol plays an important role in degradation temperature [39]. The increase of glycerol concentration significantly increases the thermal degradation rate at constant temperature. However, in the present study, increase in glycerol concentration has a significant effect on the thermal stability thermal degradation rate, as well as thermal stability, which is similar to previous research [10].

13.8.2 Differential Scanning Calorimetry (DSC)

Generally, glass transition temperature (T_g) of protein-based films increases with chain stiffness and with an increase in inter and intra-molecular attractive forces. Molecular interaction decreases with plasticizer decreases, as well as decreasing T_g values [24]. The increase in T_g provoked an increase in the resistance and rigidity of films, as previously reported by Honary and Orafai [40]. A linear increase in puncture forces is observed when T_g values varied from 40 to 80°C, in gelatin-based films plasticized with glycerol [41].

The thermal properties of keratin bioplastics with different concentrations of glycerol were studied by DSC in the range from room temperature to 250°C, as shown in Fig. 13.8. The DSC curves showed that the melting temperature is around 27°C. According to Reddy et al. [22], unmodified chicken feathers, which is almost similar to that used in this work, did not show any melting peak, whereas cyanoethylated chicken feathers had an endothermic melting peak at around 167°C, that should be due to the introduction of cyan group onto chicken feathers.

By definition, T_g is the temperature at which the forces binding the molecules in the matrix are relaxed to allow large-scale molecular movement [42]. The addition of plasticizers into the plastic film reduce the T_g, but according to Sanyang et al. [10], the decrease of T_g values are insignificant when the plasticizer concentrations increase, which is like the current study.

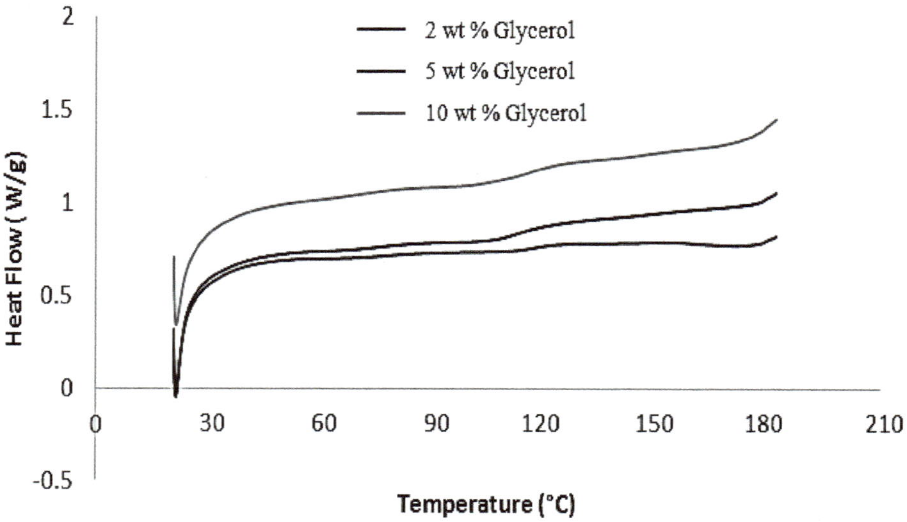

Figure 13.8 DSC curves at different concentration of glycerol.

13.9 BIODEGRADABILITY STUDY

Biodegradability of the bioplastic film by incubating in protease enzyme solution was investigated, and is shown in Figs 13.9–13.11.

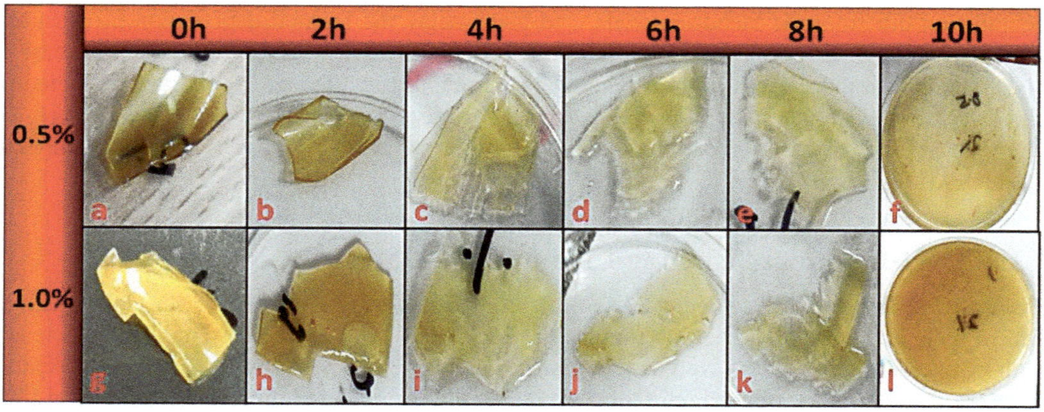

Figure 13.9 Appearance for 2 wt% glycerol bioplastic film degradation in 0.5 wt% and 1 wt% stock solution.

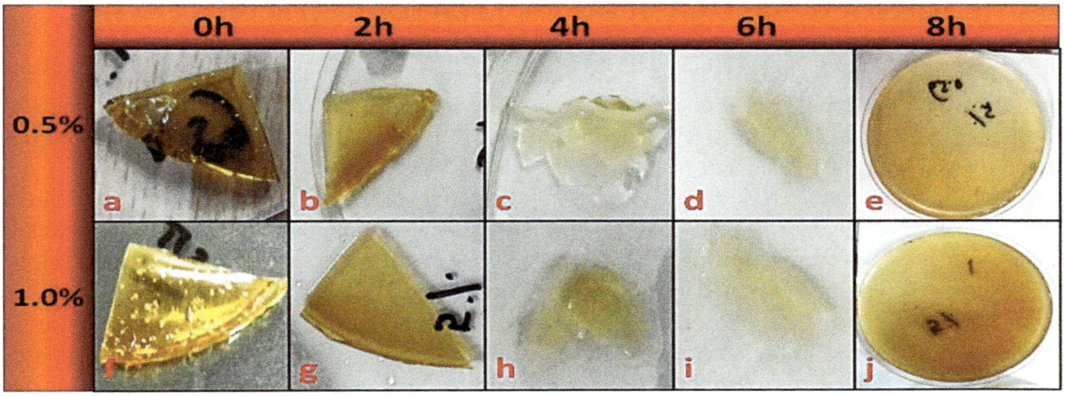

Figure 13.10 Appearance for 5 wt% glycerol bioplastic film degradation in 0.5 wt% and 1 wt% stock solution.

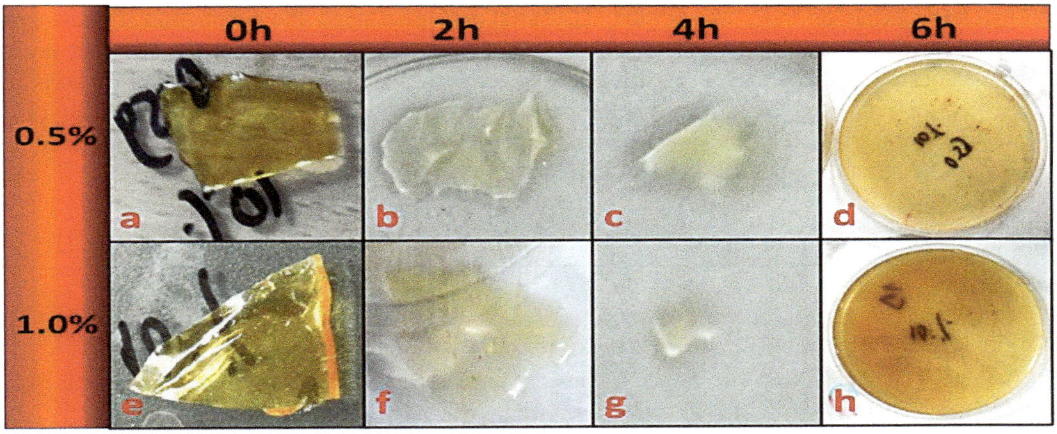

Figure 13.11 Appearance for 10 wt% glycerol bioplastic film degradation in 0.5 wt% and 1 wt% stock solution.

For 1 wt% stock solution, the plastic film is seen to be degraded faster compared to 0.5 wt% stock solution. The amount of plastic film degrading by time for 1 wt% stock solution is higher compared to 0.5 wt% stock solution. It can be concluded that the amount of enzyme present affects the rate of biodegradability.

Besides, compositions of plastic film also affect the biodegradability rate. When the concentration of glycerol increases, lesser time is taken for the plastic film to degrade. It can be seen in Fig. 13.9 that 2 wt% glycerol-based film takes about 10 hours to be degraded fully in stock solution. This may be due to high strength and strong bonding between plasticizer and keratin within the film. In contrast, for the highest composition of glycerol, which is for 10 wt% glycerol (Fig. 13.11), it just took about 6 hours for the plastic film to completely degrade. This may be due to poor strength and bonding between keratin and glycerol within the film. The total time taken for the plastic film to fully degrade is the same for all concentrations in 1 wt% and 0.5 wt% stock solution. The amount of enzyme does not have a significant effect on biodegradability rate. After observation for about 1 day, the bioplastic film is found to be fully degraded within 10 hours. It proves that the product of this research is totally biodegradable and harmless to the environment and living things.

13.10 GENERAL APPLICATIONS OF KERATIN-BASED BIOPLASTIC

The possible applications of keratin bioplastic in a variety of technologies and products can result in their sustainable conversion into high-value materials and products in the presence of cost-effective technologies for converting the chicken feather waste into useful products using less quantity, humanity safe, and eco-friendly chemicals.

In addition, extracted keratins can form self-assembled structures that regulate cellular recognition and behavior. These qualities have led to the development of keratin biomaterials with applications in wound healing, drug delivery, tissue engineering, trauma and medical devices.

During the course of Ichiro Izawa and Masaki Inagaki's research, wool and hair keratin were used in the biological function in tissue engineering, and they devised a method and procedures to develop the site- and phosphorylation state-specific antibodies. These antibodies have facilitated understanding of the cytoskeletal organization, signal transduction, and transcriptional mechanisms, as well as clinical diseases, and will be further utilized for developing new drugs [43].

Xiao-Chun Yin and his group studied the feather keratin films and found load and release drugs. The resultant feather keratin biopolymer films were pH-responsive and showed controllable drug-release behavior [44]. In addition, C. Tonin had developed the production of new bio-compatible materials of PEO/keratin suitable for a number of applications in different areas, ranging from medical applications (scaffolds, drug delivery membranes), to filtration and absorption equipment, by film casting, electro-spinning, and conventional wet-spinning techniques [16]. Moreover bioplastic based on 1,8-octanediol-plasticized feather keratin is used as bio material for food packaging and biomedical applications [45].

References

1. Yu, R. and Y.D. Hang. Kinetics of direct fermentation of agricultural commodities to L (+) lactic acid by Rhizopus oryzae. Biotechnology Letters 1989; 11(8): 597-600.
2. Andrews, R. and M.C. Weisenberger. Carbon nanotube polymer composites. Current Opinion in Solid State and Materials Science 2004; 8(1): 31-37.
3. Khot, S.N., J.J. Lascala, E. Can, S.S. Morye, G.I. Williams, G.R. Palmese, S.H. Kusefoglu, R.P. Wool. Development and application of triglyceride-based polymers and composites. Journal of Applied Polymer Science 2001; 82(3): 703-723.
4. Barone, J.R. and W.F. Schmidt. Polyethylene reinforced with keratin fibers obtained from chicken feathers. Composites Science and Technology 2005; 65(2): 173-181.
5. Blanchard, C.R., S.F. Timmons and R.A. Smith. Keratin-based hydrogel for biomedical applications and method of production. Google Patents, Aug-1999.
6. Schrooyen, P.M.M., P.J. Dijkstra, R.C. Oberthür, A. Bantjes and J. Feijen. Partially carboxymethylated feather keratins. 2. Thermal and mechanical properties of films. Journal of Agricultural and Food Chemistry 2001; 49(1): 221-230.
7. Gupta, A., N.B. Kamarudin, C.Y.G. Kee and R.B.M. Yunus. Extraction of keratin protein from chicken feather. Journal of Chemistry and Chemical Engineering. 2012; 6(8): 732-737.
8. Silverajah, V.S.G., N.A. Ibrahim, W.M.Z.W. Yunus, H.A. Hassan and C.B. Woei. A comparative study on the mechanical, thermal and morphological characterization of poly (lactic acid)/epoxidized palm oil blend. International Journal of Molecular Sciences 2012; 13(5): 5878-5898.
9. Al-Musallam, A.A., D.H. Al-Gharabally and N. Vadakkancheril. Biodegradation of keratin in mineral-based feather medium by thermophilic strains of a new *Coprinopsis* sp. International Biodeterioration & Biodegradation 2013; 79: 42-48.
10. Sanyang, M.L., S.M. Sapuan, M. Jawaid, M.R. Ishak and J. Sahari. Effect of plasticizer type and concentration on tensile, thermal and barrier properties of biodegradable films based on sugar palm (*Arenga pinnata*) starch. Polymers (Basel). 2015; 7(6): 1106-1124.

11. Fan, J., T.-Da Lei, J. Li, P.-Yu Zhai, Y.-H. Wang, Fu-Y. Cao and Y. Liu. High protein content keratin/ poly (ethylene oxide) nanofibers crosslinked in oxygen atmosphere and its cell culture. Materials & Design 2016; 104: 60-67.
12. Wrześniewska-Tosik, K. and J. Adamiec. Biocomposites with a content of keratin from chicken feathers. Fibres and Textiles in Eastern Europe 2007; 15(1): 106-112.
13. Flores-Hernández, C.G., A. Colín-Cruz, C. Velasco-Santos, V.M. Castaño, J.L. Rivera-Armenta, A. Almendarez-Camarillo, P.E. García-Casillas and A.L. Martínez-Hernández. All green composites from fully renewable biopolymers: Chitosan-starch reinforced with keratin from feathers. Polymers (Basel). 2014; 6(3): 686-705.
14. Idris, A., R. Vijayaraghavan, A.F. Patti and D.R. MacFarlane. Distillable protic ionic liquids for keratin dissolution and recovery. ACS Sustain. Chemical Engineering 2014; 2(7): 1888-1894.
15. Wang, Y.-X. and X.-J. Cao. Extracting keratin from chicken feathers by using a hydrophobic ionic liquid. Process Biochemistry 2012; 47(5): 896-899.
16. Tonin, C., A. Aluigi, C. Vineis, A. Varesano, A. Montarsolo and F. Ferrero. Thermal and structural characterization of poly (ethylene-oxide)/keratin blend films. Journal of Thermal Analysis and Calorimetry 2007; 89(2): 601-608.
17. Zhao, W., R. Yang, Y. Zhang and L. Wu. Sustainable and practical utilization of feather keratin by an innovative physicochemical pretreatment: High density steam flash-explosion. Green Chemistry 2012; 14(2): 3352-3360.
18. Nigam, P. Microbial enzymes with special characteristics for biotechnological applications. Biomolecules 2013; 3(3): 597-611.
19. Schrooyen, P.M.M., P.J. Dijkstra, R.C. Oberthür, A. Bantjes and J. Feijen. Partially carboxymethylated feather keratins. 1. Properties in aqueous systems. Journal of Agricultural and Food Chemistry 2000; 48(9): 4326-4334.
20. Shavandi, A., T.H. Silva, A.A. Bekhit and A.E.-D.A. Bekhit. Keratin: dissolution, extraction and biomedical application. Biomaterials Science 2017; 5(9): 1699-1735.
21. Husain, M.S.B., A. Gupta, B.Y. Alashwal and S. Sharma. Synthesis of Hydrogel Using Keratin Protein from Chicken Feather. *In*: Conference ICCEIB2018, Kuala-Lumpur Malaysia, 2018, 193, no. Peer Reviewed, p. 16.
22. Reddy, N., L. Chen and Y. Yang. Biothermoplastics from hydrolyzed and citric acid crosslinked chicken feathers. Mater. Sci. Eng. C 2013; 33(3): 1203-1208.
23. Gennadios, A., A.H. Brandenburg, J.W. Park, C.L. Weller and R.F. Testin. Water vapor permeability of wheat gluten and soy protein isolate films. Industrial Crops and Products 1994; 2(3): 189-195.
24. Sobral, P.J.A., F.C. Menegalli, M.D. Hubinger and M.A. Roques. Mechanical, water vapor barrier and thermal properties of gelatin based edible films. Food Hydrocolloids 2001; 15(4-6): 423-432.
25. Husain, M., A. Gupta, B. Alashwal and S. Sharma. Synthesis of PVA/PVP based hydrogel for biomedical applications: A review, Energy Sources, Part A: Recovery, Utilization, and Environmental Effects 2018; 40(20): 2388-2393.
26. Yamauchi, K., A. Yamauchi, T. Kusunoki, A. Kohda and Y. Konishi. Preparation of stable aqueous solution of keratins, and physiochemical and biodegradational properties of films. Journal of Biomedical Materials Research 1996; 31(4): 439-444.
27. Sharma, S. and A. Gupta. Sustainable management of keratin waste biomass: applications and future perspectives. Brazilian Archives of Biology and Technology 2016; 59: e16150684... http://dx.doi.org/10.1590/1678-4324-2016150684.
28. Jackson, M. and H.H. Mantsch. The use and misuse of FTIR spectroscopy in the determination of protein structure. Critical Reviews in Biochemistry and Molecular Biology 1995; 30(2): 95-120.
29. Duodu, K.G., J.R.N. Taylor, P.S. Belton and B.R. Hamaker. Factors affecting sorghum protein digestibility. Journal of Cereal Science 2003; 38(2): 117-131.
30. Barth, A. Infrared spectroscopy of proteins. Biochim. Biophys. Acta (BBA)-Bioenergetics 2007; 1767(9): 1073-1101.
31. Almutawah, A., S.A. Barker and P.S. Belton. Hydration of gluten: A dielectric, calorimetric, and Fourier transform infrared study. Biomacromolecules 2007; 8(5): 1601-1606.

32. Yua, H.M., A. Gupta, R. Gupta and M.S.B. Husain. Investigation of bioplastic properties developed from acrylate epoxidized soybean oil through ring opening polymerization process. Journal of Chemical Engineering and Industrial Biotechnology 2017; 1: 29-41.
33. Wang, J., S. Hao, T. Luo, Z. Cheng, W. Li, F. Gao, T. Guo, Y. Gong and B. Wang. Feather keratin hydrogel for wound repair: Preparation, healing effect and biocompatibility evaluation. Colloids Surfaces B Biointerfaces 2017; 149: 341-350.
34. Alashwal, B.Y., M.S. Bala, A. Gupta, S. Sharma and P. Mishra. Improved properties of keratin-based bioplastic film blended with microcrystalline cellulose: A comparative analysis. Journal of King Saud University - Science 2019; (InPress). https://doi.org/10.1016/j.jksus.2019.03.006.
35. Ramakrishnan, N., S. Sharma, A. Gupta and B.Y. Alashwal. Keratin based bioplastic film from chicken feathers and its characterization. International Journal of Biological Macromolecules 2018; 111: 352-358.
36. Suyatma, N.E., L. Tighzert, A. Copinet and V. Coma. Effects of hydrophilic plasticizers on mechanical, thermal, and surface properties of chitosan films. Journal of Agricultural and Food Chemistry 2005; 53(10): 3950-3957.
37. Jangchud, A. and M.S. Chinnan. Peanut protein film as affected by drying temperature and pH of film forming solution. Journal of Food Science 1999; 64(1): 153-157.
38. Ullah, A., T. Vasanthan, D. Bressler, A.L. Elias and J. Wu. Bioplastics from Feather Quill. Biomacromolecules 2011; 12(10): 3826-3832.
39. Sharma, S., A. Gupta, A. Kumar, C.G. Kee, H. Kamyab and S.M. Saufi. An efficient conversion of waste feather keratin into ecofriendly bioplastic film. Clean Technologies and Environmental Policy 2018; 1-11.
40. Honary, S. and H. Orafai. The effect of different plasticizer molecular weights and concentrations on mechanical and thermomechanical properties of free films. Drug Development and Industrial Pharmacy 2002; 28(6): 711-715.
41. Thomazine, M., R.A. Carvalho and P.J.A. Sobral. Physical properties of gelatin films plasticized by blends of glycerol and sorbitol. Journal of Food Science 2005; 70(3): E172–E176.
42. Zhang, Y., Q. Liu and C. Rempel. Processing and characteristics of canola protein-based biodegradable packaging: A review. Critical Reviews in Food Science and Nutrition 2018; 58(3): 475-485. doi: 10.1080/10408398.2016.1193463.
43. Izawa, I. and M. Inagaki. Regulatory mechanisms and functions of intermediate filaments: A study using site-and phosphorylation state-specific antibodies. Cancer Science 2006; 97(3): 167-174.
44. Yin, X.-C., F.-Y. Li, Y.-F. He, Y. Wang and R.-M. Wang. Study on effective extraction of chicken feather keratins and their films for controlling drug release. Biomaterials Science 2013; 1(5): 528-536.
45. Liu, S., K. Huang, H. Yu and F. Wu. Bioplastic based on 1, 8-octanediol-plasticized feather keratin: A material for food packaging and biomedical applications. Journal of Applied Polymer Science 2018; 135(30): 46516.
46. Ji, Y., J. Chen, J. Lv, Z. Li, L. Xing and S. Ding. Extraction of keratin with ionic liquids from poultry feather. Separation and Purification Technology 2014; 132: 577-583.

Index